FACING WEST:

The Metaphysics of Indian-Hating and Empire-Building

by

Richard Drinnon

University of Oklahoma Press
Norman and London

Library of Congress Cataloging-in-Publication Data

Drinnon, Richard.
Facing west : the metaphysics of Indian-hating and empire-building / by Richard Drinnon.
p. cm.
Originally published: Minneapolis : University of Minnesota Press, c 1980.
Includes bibliographical references and index.
ISBN 0-8061-2928-X
1. Indians of North America—Public opinion. 2. Indians of North America—Civil rights. 3. Public opinion—United States. 4. Race discrimination—United States. 5. United States—Territorial expansion. 6. United States—Race relations. I. Title.
E98.P66D74 1997
305.8′00973—dc20 96-38834
CIP

Permissions acknowledgments can be found on p. 572.

The paper in this book meets the guidelines for permanence and durability of the Committee on Production Guidelines for Book Longevity of the Council on Library Resources, Inc. ∞

 Published by the University of Oklahoma Press, Norman, Publishing Division of the University. Manufactured in the U.S.A. First edition, 1980. First printing of the University of Oklahoma Press edition, 1997.

1 2 3 4 5 6 7 8 9 10

For our grandson Saul,
who has lovely dark-brown skin
and some understanding already
of what this book is all about

Acknowledgments

AS always, I am hopelessly indebted to Anna Maria Drinnon, grandmother of the young man to whom this book is dedicated, critic, coeditor, and companion in all that matters, including the vision quest these pages represent. In this our fifth undertaking together, John F. Thornton provided encouragement from the outset, nudged me toward the final title, curbed my excesses and wanted to delete more, and again acted as an incisive and exemplary editor. Except for the uncurbed vagaries that remain, despite his best efforts, this work is the result of a truly cooperative effort. Judy Gilbert, another veteran of joint projects, patiently typed and retyped the manuscript and again proved herself a friend who cared enough to take pains.

In 1963–64 I held a Faculty Research Fellowship to study patterns of American violence. War, assassinations, and other studies, including a book of my own, intervened, compelled revision of my plans, and delayed fulfilling my obligations for the grant. However belatedly, I wish to express now my gratitude to the Social Science Research Council for financing the beginnings of this book. I am also grateful to the Trustees of Bucknell University for summer grants that helped in the research and writing.

Fellow historians, colleagues and students alike, have read or heard sections of the manuscript. I have already thanked most of them for their criticisms; in specific instances I also acknowledge their help elsewhere. It has been invaluable, especially for a work that presumes to carry its theme across the span of Anglo-American history. No doubt I have been guilty not of minor poaching but of major trespass. Specialists in the various fields and eras will not and should not excuse my inevitable blunders—the most I can hope is that they will point them out with the legendary forbearance of the community of scholars. Whether they forbear or not, I am much obliged to them, for they frequently guided me to the primary sources on which this study is based. I am indebted to ethnohistorians for what is by now an impressive body of writing on red–white relations, to those who have studied the complexities of racism, to biographers of figures discussed below and to editors of their memoirs and letters—and in fact to all those cited in the notes and bibliographical essay. Even those authors with whom I am in profound disagreement have helped me identify problems and formulate what I thought about them and about specific individuals, events, and themes.

Welcome harbingers of an overdue reinterpretation of our past have been four books that came out while this book was at various stages of preparation. Richard Slotkin's *Regeneration through Violence: The Mythology of the American Frontier, 1600–1860* (1973) and Francis Jennings's *Invasion of America: Indians, Colonialism, and the Cant of Conquest* (1975) came to my desk after I had substantially completed my chapters on the seventeenth century, but elsewhere I have taken advantage of their insights and findings. (For an appraisal of Jennings's contributions, see my "Ravished Land" in *The Indian Historian*, IX [Fall 1976], 24–26.) In *Fathers and Children: Andrew Jackson and the Subjugation of the American Indian* (1975), Michael Paul Rogin imaginatively harnessed together depth psychology and political theory to relate the man to his times and vice versa. I drew directly on Rogin for my understanding of Jackson and the First Seminole War and have been instructed by his example elsewhere. (For more extended analysis of Rogin's work, see my evaluation in the *New York Times Book Review*, June 15, 1975.) Finally, Robert F. Berkhofer, Jr., *The White Man's Indian: Images of the American Indian from Columbus to the Present* (1978) relates very directly to my themes and, in its own way, fills in the gap created below when I leave the mainland in the 1890s to go island-hopping out in the Pacific. (Again, for more extended analysis, see my "Red Man's Burden," *Inquiry*, I [June 26, 1978], 20–22.) I have been reassured by the reflection that all our hands have been, in their very different ways, feeling the same elephant.

FACING WEST FROM CALIFORNIA'S SHORES

Facing west from California's shores,
Inquiring, tireless, seeking what is yet unfound,
I, a child, very old, over waves, towards the house of
 maternity, the land of migrations, look afar,
Look off the shores of my Western sea, the circle almost
 circled;
For starting westward from Hindustan, from the vales of
 Kashmere,
From Asia, from the north, from the God, the sage, and the
 hero;
From the south, from the flowery peninsulas and the spice
 islands,
Long having wander'd since, round the earth having wander'd,
Now I face home again, very pleas'd and joyous,
(But where is what I started for so long ago?
And why is it yet unfound?)

—Walt Whitman,
Leaves of Grass, 1860–61

Contents

Preface to the 1997 Edition

GOING back to the beginning always highlights continuities. Just the other day the living-color cover of the *New York Times Book Review* carried a likeness of one of the Great Captains looking out over the fountain of waters at the Continental Divide. "UNDAUNTED COURAGE," read the legend. "Meriwether Lewis is still a hero in Stephen E. Ambrose's story of the opening of the American West" (March 10, 1996). "Opening" land that had been from time immemorial empty, vacant, "closed," is a very old story indeed, older than William Bradford's *Of Plymouth Plantation* (1642) and his positing therein an unpeopled wilderness, the "vast and unpeopled countries of America." With an undaunted tenacity that this book traces through the centuries, Americans still cling to the assumptions behind this preposterous story. Prophetically relevant today is John Quincy Adams's observation after the Louisiana Purchase: "Westward the star of empire takes its way, in the whiteness of innocence."

Former Secretary of Defense Robert S. McNamara, head of the team who pushed the empire westward all the way to Indochina, very recently has cried out with mea culpas that seemingly announce his defection from the closed ranks of the guiltless. "We were wrong," he admits in his memoir, *In Retrospect* (1995), "terribly wrong." Acting according "to what we thought were the principles and traditions of this nation," McNamara and his empire-builders made mistakes—"mostly honest," he claims—the foremost of which was their total failure to identify the nationalist core of the Vietnamese drive to unify their country. "I had never visited Indochina," he acknowledges, "nor did I understand or appreciate its history, language, culture, or values." Moreover, thanks to the purging of top State Department Asia hands in the McCarthy fifties, he and other officials in the Kennedy and Johnson administrations "lacked experts for us to consult to compensate for our ignorance about Southeast Asia." But this supposed dearth of "experts" is itself suggestive. McNamara still does not grasp that his imperial ignorance of other cultures and peoples, especially colored, is as American as the Pledge of Allegiance. It is precisely because he was acting according to "the principles and traditions of this nation" (which I analyze throughout this study) that the Vietnamese were as unknown to him as the Seminoles were to Andrew Jackson and the Filipinos to William McKinley.

McNamara tells us to focus on one question. "The right question is, did you rely on the wrong strategy—conventional military tactics instead of winning the hearts and minds of the people—and the answer to that is yes. It was totally wrong." But here he is simply recycling the counterinsurgency thesis of his erstwhile special assistant, Edward G. Lansdale, the master sceneshifter I try to keep up with in my concluding chapters, and in my 1990 preface. What went terribly wrong was not the empire's westward destination but the flawed strategy that kept it from getting there and to all those hearts and minds.

Here shines still the whiteness of innocence.

In my text I used the word *genocidal* as descriptive of John Endicott's orders before he sailed for Block Island in 1636 (p. 34) and to characterize Puritan intentions toward New England tribes (p. 44). Later, in a note, I made explicit that I was drawing directly on the definition of genocide formulated by the United Nations General Assembly in 1946 (p. 324n). In a 1991 essay, "The Pequot War Reconsidered" (*New England Quarterly*, LXIV, 206–24), Steven T. Katz remonstrates against my usage and that of other errant historians. A professor of Jewish studies at Cornell University, Katz starts by discarding the UN definition in favor of his own, "more stringent," use of the term to mean "an intentional action aimed at the complete physical eradication of a people." He uses that as a springboard for giving Puritans the benefit of more doubtful factors than we can possibly consider here, and finishes by brandishing his conclusion: "Such factors suggest that while the British could certainly have been less thorough, less severe, less deadly in prosecuting their campaign against the Pequots, the campaign they actually did carry out, for all its vehemence, was not, either in intent or execution, genocidal." Why not? Well, because his stringent definition had ruled it out from the beginning. Some Pequots having escaped the vehemence, their tribe was not completely eradicated. "As recently as the 1960s, Pequots were still listed as a separate group residing in Connecticut"—and Katz might have added, some survivors were still there playing bingo as recently as the 1990s. In *American Holocaust* (1992; p. 318n), David E. Stannard wonders what Professor Katz would say to a professor of Native American studies who went out of his way to write an essay claiming the Holocaust was not genocidal, "because, after all, some Jews survived—a number of them even live in Connecticut today."

What defensive professors say about genocide can be unintentionally revealing. What protective politicians do about it, or rather did not do for forty years, is perhaps still more significant.

On December 9, 1948, the General Assembly of the United Nations unanimously passed a "Genocide Convention" to prevent (and punish) the mass killing of racial, ethnic, religious, and national groups. It became international law in 1951 upon ratification by the necessary twenty member nations, and scores of others soon signed on. But the United States left it bottled up in the Senate Foreign Relations Committee for decades and withheld ratification until 1988. Leo Kuper, who is after Raphael Lemkin the leading student of genocide, cites this extraordinary delay and the obvious reluctance of the United States to ratify this basic human rights convention as framework for just the right question: "Did it fear that it might be held responsible, retrospectively, for the annihilation of Indians in the United States, or its role in the slave trade, or its contemporary support for tyrannical governments engaging in mass murder?"* Who knows? But the senators made this an even better question by making U.S. ratification not only late and grudging but also conditional: they excluded the jurisdiction of the International Court and thereby made certain that neither they nor their constituents would ever be hauled before an international penal tribunal to answer for alleged genocidal acts against the Pequots or any other targets of what I call Indian-hating. The survivors are left to draw what comfort they can from the likelihood that the overbearing whiteskins have at least betrayed consciousness of guilt.

In March 1968 an American task force swept down the Batangan peninsula by the South China Sea, burning hamlets and killing fleeing Vietnamese. "We were out there having a good time," said a

* Leo Kuper, "The United States Ratifies the Genocide Convention," in *The History and Sociology of Genocide*, ed. Frank Chalk and Kurt Jonassohn (New Haven: Yale University Press, 1990), pp. 422–25.

participant I quoted (p. 454). "It was sort of like being in a shooting gallery." In February 1991 Americans on a carrier in the Persian Gulf launched interminable air strikes against Iraqis fleeing down the roads out of Kuwait. A dispatch from the deck of the USS *Ranger* had one pilot saying it was a "turkey shoot" and another likening the targets to "fish in a barrel." So once more, a year after the 1990 edition of this book, American cowboys were out there in "Indian country" having a good time.

Yet the westward star of empire had gone hopelessly off course. Those following in Lewis and Clark's footsteps could lay claim to the Pacific Ocean as an extension of the American West—as "my Western sea," in Walt Whitman's words—and those who carried this possessiveness all the way to Asia and the South China Sea could dreamily contend that they had been going in the right direction. But on around to the Persian Gulf in the Middle East? The last traces of the West as an area or place went up in smoke over the electronic battlefields of Desert Storm. True West had become a traveling shooting gallery. By engulfing the other cardinal points it had become global. It was everywhere and bound nowhere. It was ungrounded.

With landless Wests of the next millennium incoming at the speed of missiles and rockets, we have good reason to hunker down here and listen to a voice from the Northern Cheyennes, a people for whom *place* has always mattered. "To us, to be Cheyenne means being one tribe—living on our own land," says John Wooden Legs. "Our land is everything to us. . . . It is the only place where Cheyennes remember same things together." Without places that matter that way, this voice would have us reflect, we need no missiles and rockets to make us homeless.

So then let us once again go back to the beginning.

Richard Drinnon
Port Orford, Oregon
July 1996

Preface to the 1990 Edition

ORIGINALLY I had closed this study with an insight from *Lame Deer: Seeker of Visions*: "I have seen pictures of Song My, My Lai, and I have seen pictures of Wounded Knee—the dead mothers with their babies," said the Sioux shaman. "And I remember my grandfather, Good Fox, telling me about the dead mother with a baby nursing at her cold breast, drinking that cold milk. My Lai was hot and Wounded Knee was icy cold, and that's the only difference" (1972; p. 69). Lame Deer's words underscored, from a Native American point of view, my theme of "Indian-hating," the white hostility that for four centuries had exterminated "savages" who stood in the path of Anglo-American expansion. But my Job-like editor rightly pointed out that the extract hardly provided an upbeat note on which to conclude and, besides, if readers had not been persuaded of the inner identities of the two massacres by that point, nearly five hundred pages along, they never would be. I agreed, cut the paragraph from the galleys, and concluded instead by voicing what I thought was a modest, sharply qualified hope: with Indians taking the lead, "Americans of all colors might just conceivably dance into being a really new period in their history."

Enter Ronald Reagan, the Hollywood cowboy, former lieutenant in the 14th Cavalry Regiment, and campaigner for cutbacks in federal programs.

First reconsider the doom of those proto-victims called Pequots: in part one I accompany the punitive Puritan expeditions of 1637 that made some four hundred Pequots "as a fiery Oven" in their village near the Mystic River and later finished off three hundred more in the mud of Fairfield Swamp. For these proceedings, said

Captain John Underhill, "we had sufficient light from the word of God." At the head of parts two through five I place epigraphs that suggest how every generation of Anglo-Americans down to the present has followed the pattern set then and has repeated, with minor variations, such justifications for burning natives in their villages and rooting them out of their swamps. Understandably the Puritans had not intended to have that pattern of oppression associated with the tribe they destroyed. Rather they sought, as Captain John Mason reported, "to cut off the Remembrance of them from the Earth." The name Pequot had become extinct, declared the Connecticut General Assembly, and no survivors should carry it. Yet scattered individuals did, and both the name and the remembrance lived on in them, in a recent lawsuit they joined to reclaim lands taken illegally, and finally in a bill, unanimously approved in both houses of the Congress, that would have given them $900,000 to buy eight hundred acres of their own reservation. In April 1983 President Reagan vetoed this settlement of the Pequot claim on the grounds that $900,000 was "too much to pay" these improbable survivors.

Reagan handed down his Pequot veto just a few months after his Secretary of the Interior had pronounced all such reservations socialist abominations. "If you want an example of the failures of socialism, don't go to Russia," urged James Watt. "Come to America, and see the American Indian reservations. . . . If Indians were allowed to be liberated they'd go out and get a job and that guy [the tribal leader] wouldn't have his handout as a paid government official" (*Washington Post,* January 20, 1983). Watt's reasoning echoed that behind the 1950s "termination" bills, the series of acts that "liberated" nearly a hundred tribes and bands from their inherited treaty rights, trust status, and reservations. Cut off from their land sanctuaries, they lost those places where Indians "remember same things together," as John Wooden Legs once said of his tribe's Cheyenne Reservation. All too aware of these precedents, Indians mounted a sharp protest that moved Watt to say that he did not say what he had said, as it were, and his boss tried to outline a policy that would pursue "self-government for Indian tribes without threatening termination" (*New York Times,* January 25, 1983). But many tribal peoples remained unbelievers: "The Reagan Policy Is Termination," editorialized *Wassaja,* the national Indian newspaper (July/August 1983).

Reagan and his officials did seek to abrogate the treaties that supposedly guaranteed the Indian land base and cut the Indian budget by forty percent (domestic budget cuts for all other Americans amounted to about twelve percent). In this mean-spirited milieu racial attitudes hardened, and proliferating anti-Indian organizations lobbied for termination and circulated hate materials, including posters of an arrow-pierced fish saying "I'LL BE GLAD WHEN THOSE DAMN INDIANS QUIT FISHING AND GO BACK ON WELFARE" and bumper stickers urging "SAVE A FISH: SPEAR AN INDIAN."

Finally on June 30, 1988, after eight years in office, Reagan came as close as he would ever come to issuing an Indian policy statement. Maybe "we" had made a mistake in "giving" Indians reservations, the president told a Russian audience in Moscow:

> Maybe we made a mistake in trying to maintain Indian cultures. Maybe we should not have humored them in that, wanting to stay in that primitive life style. Maybe we should have said: No, come join us. Be citizens along with the rest of us.

Maybe "we" should have dealt with the other tribes as the Connecticut colonists had tried to deal with the Pequots—to wit, "to cut off the Remembrance of them from the Earth." With marvelous compression, the president had revealed his impenetrable ignorance of the status and circumstances of his red constituents—Indians had been citizens since 1924 and their reservations had not been government gifts—while voicing the historic certitude that "we" white Americans have a natural right to say yes or no to these aboriginal neighbors. The exact measure of Anglo-American racial arrogance has been "our" arrogation of the right to decide whether they should be "humored" in their "primitive life style" or have it wiped out.

Woe are "we," or should be. In his Indian statement, the great communicator surely spoke for most of "the rest of us" white Americans and thereby made this study of Indian-hating as timely as yesterday's headlines. And had we collectively stepped back or even inched back from the primordial tradition of disrespect he so effortlessly articulated, then the modest hope of my 1980 conclusion would seem less embarrassingly immodest today.

Four centuries after Columbus, Frederick Jackson Turner had pronounced the continental frontier "gone, and with its going has closed the first period of American history." In his seminal 1893 paper, Turner found that thus far, to the Pacific and the present, the dominant fact in American life had been the expansion of its land frontier, "the meeting point between savagery and civilization." Now with the passing of "free" land, the problem he pursued in another essay was how to find a substitute somewhere else—in a "remoter West," in education, in science. The historian of the frontier drew hope from the current call for a more assertive foreign policy, for an interoceanic canal, for enhanced sea power, "and for the extension of American influence to outlying islands and adjoining countries." What was to stop the Winning of the West from taking to the high seas?

In part five I track Turner's outriders of empire to the outlying islands and adjoining countries he had waved them toward. There they waged wars of subjugation shaped by attitudes toward native peoples ("goo-goos" and "gooks") that made the massacres in Vietnam's "Indian Country" in the 1960s consistent with those of Filipinos in Batangas at the turn of the century and with those of Indians on the continent earlier—the My Lai–back–to–Wounded Knee connection drawn by Lame Deer so aptly across space and time.

Yet Vietnam represented a novel break in this historical sequence. For the first time Anglo-Americans had hurled themselves westward against a human barrier that proved insurmountable. The roll back of empire from the far Pacific rim marked a decisive reversal and the close of Turner's second period in American history. "I saw that we had overreached ourselves," said *Look* foreign editor J. Robert Moskin on his return from a trip to Vietnam. "America's historic westward-driving wave has crested" (November 18, 1969). In chapter XXX I quoted Moskin approvingly but hedged by adding quickly that what Turner said of the second period applies with a vengeance to the third period of American history: "He would be a rash prophet who should assert that the expansive character of American life has now entirely ceased." Indeed.

For all my self-protective hedging, I still was unprepared for the swiftness and the effectiveness of the ensuing campaign to restore confidence in "the expansive character of American life," Turner's circumlocution for our imperial destiny. Forestalling any possibility of a pause for root-and-branch rethinking and reimagining, weavers of mystification and blurrers of memory rushed in to patch over the break in narrative sequence and then rewrote the history of the immediate past as though nothing novel had occurred. Scholarly fabulists even contrived to have American "innocence" emerge, unblemished as ever. In *America in Vietnam,* the most elaborate apology, Guenter Lewy reached "the reasoned conclusion" that "the sense of guilt created by the Vietnam war in the minds of many Americans is not warranted" (1978; p. vi). *Now you see the million bodies, now you don't . . .*

In *A Bright Shining Lie*, a best-selling and award-winning requiem for our lost colony, Neal Sheehan recently observed that "South Vietnam, it can truly be said, was the creation of Edward Lansdale" (1988; p. 138). The claim that the legendary Lansdale "created" South Vietnam is far less improbable than it might seem, as I try to show in my own chapters on the senior CIA operative who became our best-known secret agent and, in the eyes of many, our preeminent authority on counterinsurgency in the Third World. With what Sheehan calls his "brilliantly led counterrevolution" against the Filipino peasants just behind him, Lansdale moved on to Vietnam in 1954, made Ngo Dinh Diem president, and helped repulse the first wave of his new Friday's enemies. Without Lansdale, Sheehan argued persuasively, "the American venture in Vietnam would have foundered at the outset" (p. 137). That it kept afloat allowed Sheehan's flawed subject, "the good colonialist" John Paul Vann, to come aboard years later, determined to "emulate his hero, Lansdale" by learning how to manipulate his native advisees (pp. 75, 76).

Lansdale liked to say that "counterinsurgency" was another word for "brotherly love." In practice, it was another word for manipulation, and the hoary objective at its core was manipulation of expendable natives into furthering American interests while convincing them they were furthering their own. Had we won over the Vietnamese people, argued Lansdale and other covert warriors such as John Vann and Robert Komer, then we might have adopted the guerrillas' own tactics to crush them and win the war. In *America in Vietnam* Guenter Lewy made their counterinsurgency thesis his own as he deferentially drew on Lansdale and the others to blame the loss of the war primarily on the high command's big-unit attrition strategy.

Held the creator of South Vietnam and its possible savior, had American officialdom only listened, Lansdale added these singular tributes to his other distinctions and became still more legendary in the postwar years. In an adulatory biography *Edward Lansdale: The Unquiet American*, the military historian Cecil B. Currey joined Sheehan and Lewy in claiming that Lansdale had virtually invented Diem and years later might have put his tottering regime on a stable political footing, had he not been ignored and undercut by the American mission at every turn: "This gray, unassuming man with the Mount Rushmore head may have held the keys to American success in his hands and yet no one listened" (1988; p. 317). If we had only listened, that is to say, we might still be out there building Lansdale's "Third Force" on the farthest edge of our cravings for empire.

In May 1984 the Pentagon invited Lansdale to address a two-day conference on new tactics to use against the Sandinista government in Nicaragua and the insurgents in El Salvador.* According to Currey, he condemned the current reliance on heavy weapons and conventional forces, and proposed his counterinsurgent alternative, "a broad campaign of psychological operations" (p. 346). A few months after the conference, the House Select Committee on Intelligence made public two documents proposing "psychological operations" that I believe may have come directly from the emeritus proconsul's bag of dirty tricks: a CIA covert-warfare manual encouraged assassinations in Nicaragua similar in every essential to the targeted killings in Cuba Lansdale had worked out for the Kennedys in the early 1960s; and a CIA comic book for Nicaraguan citizens came up with freewheeling suggestions for sabotage that were suggestive of his psychological-warfare operations in the Philippines and Indochina—down to and including stopping up toilets with sponges and putting dirt into gas tanks. Even Lansdale's hallmark ignorance of target populations was matched by the comic book's droll advice that Nicaraguans steal mail from mailboxes; but as one anti-Sandinista pointed out, "in Nicaragua, we don't have any mailboxes" (*New York Times*, October 19, 1984).

So the empire-building is ongoing and as timely as yesterday's headlines. In the 1970s I could still suppose that the American invaders would eventually learn from Vietnam the great folly of trying to make the world over in their own white image. But it proved no such lesson to Lansdale, who died in 1987 convinced that he knew what it would have taken to win in Vietnam, and no such lesson to Lewey and Sheehan and Currey and no doubt most of the rest of us. "The United States now faces the possibility of increased involvement in Central America," Currey cautions. "Perhaps it is time, finally, to apply the lessons of Edward G. Lansdale" (p. 282). Currey has been comforted to find "an identified cadre of 'Lansdalians' " who rushed to pick up the torch, including John

*In "Victims of Our Visions," a thoughtful review of this study, Piero Gleijeses had a "minor criticism"—he wondered why my account of white American experience with "inferior races" did not include "Hispanics, especially Mexicans. . . . Drinnon does not explain his failure to include them in the book" (*Washington Post Book World*, September 28, 1980). The failure stemmed from my decision, based on considerations of focus and space, to exclude Hispanics and not follow some of the same empire-builders into their countries south of the border. Had I done so by adhering less rigidly to my East-West axis, readers might have more readily anticipated this export of Lansdale's lessons to Latin America. In fine, I agree with the criticism and fear Gleijeses too generous in calling it "minor."

Singlaub and "a young Marine lieutenant colonel named Oliver North [who] came also to view himself as a Lansdalian, thinking of himself as a Lansdale of the 1980s" (p. 347).

Whether or not Graham Greene modeled the protagonist in *The Quiet American* (1955) on the then Colonel Lansdale, Alden Pyle was an astonishing likeness, and the secret agent rightly saw himself carrying on Pyle's great dream of Americanizing the world.* Pyle's deadly "innocence" also anticipated the next generation of Lansdalians, for to this day they too go abroad "like a dumb leper who has lost his bell, wandering the world, meaning no harm" (p. 36).

From my closing sentence hangs, for now and forever, my impenitent hope of a new day: with Indians taking the lead, "Americans of all colors might just conceivably dance into being a really new period in their history." This was "an inappropriate ending," reproved a friendly reviewer: "Drinnon's own work is testimony to the importance of less dancing and more analysis" (*Inquiry*, September 22, 1980, p. 26). But was it?

Equally unmoved by the prospect of discerning the land's rhythms, a moderately hostile reviewer disliked my "dark, dancing tribesmen" and undertook to teach me the realities of tribal life: "One cannot look to Native American societies, past or present, for models of life without repression, projection, sadism, ethnocentrism (God is Red?), and violence. The Noble Savage is also a racist stereotype" (*New Mexico Historical Review*, October 1981, p. 413). But "savages," noble or ignoble, had entered my pages only as inner demons of the white invaders, so had I really earned this admonition?

Finally, a thoroughly disgusted reviewer lamented my want of propriety:

> When [Thomas L.] McKenney reveals the sexual limitations of his Quaker upbringing after observing an Indian dance, Drinnon makes some revelations of his own. If McKenney had let go, surmises

*In June 1987 Greene told a correspondent that he had not used Lansdale "at all as my character and I really have no information about him"—quoted in a letter from Christopher Robbins to the *New York Times Book Review*, June 18, 1989. For Lansdale's reasons for believing that Pyle was based on him, see Currey, pp. 197–98. Not altogether persuaded that the novelist's memory has served him well in this instance, I have limited my response to a few rewordings that flag this contested ascription.

Apart from these rewordings in chapter XXVIII, "The Quiet American," and a handful of typographical corrections throughout, the text of this edition remains unchanged.

> Drinnon, "a dancing counterpart might have leaped out of him, joined the circle, chanted, copulated, and run off into the free and boundless forest." McKenney may well have entertained such infantile fantasies, but at least he had the good sense to repress them. [*Western Historical Quarterly*, October 1981, pp. 433–34]

But what if the first head of our Indian service had been a great voluptuary who welcomed this embrace of the shaggy New World? What if he had acted out the fantasy—including that hair-raising copulation—would that have been so awful? Worse than awful, answer this vexed historian and the long line of his predecessors with the good sense to share his loathing of the body's rhythms, though they used other words to condemn them. In the beginning, Increase Mather plainly had "such infantile fantasies" or something like them in mind when he warned in his *Danger of Apostacy* (1679) that "people are ready to run wild into the woods again and to be as Heathenish as ever, if you do not prevent it."

Yes, on this Turtle Continent God may indeed be Red, for all we non-Indians know. Unwilling to let go, grasp the outstretched hands, and join the circle, we have seen in those feet thumping and caressing the earth only what we must be prevented from becoming, wild running bodies, dusky dancing reminders of our own inescapable mortality. Outside the circle always, we have walled ourselves off from experiencing these ritual reaffirmations of the relatedness of everything. Repressing our yearnings, we have failed even to *see* the existence of a tribal metaphysics that calls into question the universality and benificence of our cherished binary oppositions, including subjective/objective, imagination/understanding, reason/passion, spirit/flesh, and not least, civilization/savagery. As opposed to the white metaphysics I pursue in this book, the shut-out red metaphysics affirms rather than negates, and asks us to lift ourselves to the Sioux truth: "We are all related."

To lift ourselves so high would be, no doubt, to fall down into the circle again, into Mather's heathenism, my disgusted critic's infantilism, and assorted aberrations we have all been warned against, including animism, romanticism, and those "primitive life styles" that worry presidents. It would elicit courteous calls for "less dancing and more analysis," and blunt reminders that "the Noble Savage is also a racist sterotype." So be it, for somehow we must transcend a recorded history that has sung hallelujahs for the conquerors of our red relatives. We must learn to speak a new language, the language of the body, as Martha Graham called the dance. Or rather, we shall have to relearn an old language, as the shaman I have quoted knew we would: Human beings "have forgotten the secret knowledge of their bodies, their senses . . . their dreams" (*Lame Deer*, p. 157).

I have called the rediscovery of that secret knowledge my "modest" and my "impenitent" hope. In 1925, the year of my birth, the poet William Carlos Williams called it, more bleakly, our only hope: "However hopeless it may seem, we have no other choice: we must go back to the beginning; it must all be done over . . ." (*In the American Grain*, p. 215).

Let us then go back to the beginning.

Richard Drinnon
November 1989

Introduction

"HERE," said Cotton Mather, "hath arisen light in darkness." Here, said Jonathan Swift, hath arisen sure signs of blight: "Remark your commonest Pretender to a Light *within*, how dark, and dirty, and gloomy he is *without*."

In *A Tale of a Tub* and its magnificent appendix, "The Mechanical Operation of the Spirit" (1696–98), Swift showed that the Puritans' heavenly light had been deflected within by their cardinal principle, *"the Corruption of the Senses is the Generation of the Spirit"*:

> Because the Senses in Men are so many Avenues to the Fort of Reason, which in this Operation is wholly block'd up. All Endeavours must be therefore used, either to divert, bind up, stupify, fluster, and amuse the *Senses*, or else to justle them out of their Stations; and while they are either absent, or otherwise employ'd or engaged in a Civil War against each other, the Spirit enters and performs its Part.

Exalting the spirit while ingeniously suppressing the senses, such Protestant enthusiasts fled from their animal bodies just as the New England Puritans, or "God's afflicted Saints" as they called themselves, had fled from the complex historical process across the Atlantic. Yet their "higher" and "lower" natures faced each other across a common frontier, so that the higher they soared above their flesh the further they plunged "into the lowest Bottom of Things; like one who travels the *East* into the *West;* or like a strait Line drawn by its own Length into a Circle."

My own tale is merely part of the larger story of asceticism marching out into the modern world—building economies, founding nation states, and conquering empires—all the while corrupting the senses and fashioning the "iron cage" for humankind Max Weber lamented in *The Protestant Ethic and the Spirit of Capitalism* (1904–5): "For the last stage of this cultural development," he concluded, "it might be truly said: 'Specialists without spirit, sensualists without heart; this nullity imagines that it has attained a level of civilization never before achieved.' " In the several parts of this volume, I trace this cultural development mainly among the immigrants who became Anglo-Americans but ask the reader here at the outset to bear in mind their common background and attitudes and assumptions with those Europeans who stayed at home or spread out to colonize other corners of the globe. This awareness of the worldwide context of comparable phenomena becomes all the more necessary in the light of our national habit of seeing U.S. history as being so exceptional as to be unique.

Distinctive patterns of response to unfamiliar surroundings did emerge, to be sure, and elucidating these is the task of this book. Yet for it not to miscarry, that elucidation must be grounded in the vigilance I invoke, grounded in watchful recognition that the record of history is nearly barren of authentically novel responses to novel circumstances. In an imaginative piece in the April 1965 *Speculum* titled "The Legacy of the Middle Ages in the American Wild West," Lynn White, Jr., easily demonstrated "our detailed and massive continuity with the European Middle Ages," including prototypes of the revolver, barbed wire, the windmill, and other paraphernalia for Winning the West, down to the rope thrown by lynching parties over a limb of the nearest lone pine: "In most societies," White pointed out, "there are clear rubrics for execution, a tradition of propriety as to the forms of killings committed by the group, which the group feels deeply impelled to follow, perhaps, because to follow them makes the past share the guilt of the execution. To know the subliminal mind of a society, one must study the sources of its liturgies of inflicting death" (XL, 199). Societies are known by their victims. On the more inclusive level of wholesale killings and hurtings, "the metaphysics of Indian-hating," as Herman Melville called that national animosity, provided just such a liturgy of inflicting death and just such sources for studying the European-derived subliminal mind.

A critical constant of that subliminal mind manifested itself as the will to power over "fallen" nature. Long before their first landfall, European immigrants were alienated from the "howling Wilderness" that had to be mastered in themselves and in their new surroundings. Yet Native Americans were bodies *in* that wildness, indwellers of the very animal world the newcomers so arrogantly sought to rise above. The collision between indigenous nature and imported asceticism had the disastrous consequences that the ethnobotanist Melvin Gilmore expressed succinctly decades ago: "The people of the European race in coming into the New World have not really sought to make friends of the native population," he observed, "or to make adequate use of the plants, or the animals indigenous to this continent, but rather to exterminate everything they found here and to supplant it with plants and animals to which they were accustomed at home" ("Uses of Plants by Indians of the Missouri River Region," *Thirty-third Annual Report of the Bureau of American Ethnology, 1911–12*). In the several parts of this book I undertake an analysis in depth of a series of encounters with "the West," from the Massachusetts Bay Colony and Block Island across the Alleghenies to the Mississippi; on across the Rockies to the Pacific slope; out from there onto island stepping-stones; and touching down most recently in Asia—a straight line of march drawn by its length almost into a circle.

2

Place has always mattered to plants and animals. Mainstream historians pay it their due when they stress the importance of both place and *time,* yet how often the sensual surfaces of organisms in particular settings fail to grace their pages. In effect trying to leave the growth-and-decay cycle of their own bodies behind—not unlike the religious enthusiasts Swift indicted—these Eurocentric historians are the guardians of a linear, continuous, irreversible Time of perpetual progress, in which place is largely irrelevant. They are the secular heirs of Judeo-Christian teleology with its reified Time, which had and has little or nothing to do with the cycles of organisms. "Christian religion and the Western idea of history are inseparable and mutually self-supporting," noted Vine Deloria, Jr., in *God Is Red* (1973):

> To retrench the traditional concept of Western history at this point would mean to invalidate the justifications for conquering the Western Hemisphere. Americans in some manner will cling to the traditional idea that they suddenly came upon a vacant land on which they created the world's most affluent society. Not only is such an idea false, it is absurd. Yet without it both Western man and his religion stand naked before the world.

To avoid similar absurdity, I have adopted a view of time that is more akin to the one advanced by Duwamish Chief Seattle in a great and sad speech of 1854: "Tribe follows tribe, nation follows nation, like the waves of the sea."

My study begins with Robert Lowell in the 1960s, swings rapidly back past Nathaniel Hawthorne in the 1830s to Thomas Morton and John Endicott in the 1620s, then turns about and returns more slowly past Thomas Jefferson and the Founding Fathers, past Herman Melville and others to the volunteers in the Philippines at the turn of the century, and concludes with the nation-builders in Indochina in the 1960s and a little beyond. But like a tree trunk's concentric rings, my completed cycle contains smaller cycles—five parts like so many Indian round dances of life and death—within which time internal to individuals, groups, and complex events receives the scrupulous attention it merits—after all, as a Western historian I have a certain vested interest in Time. But since I perceive the same sort of massive continuity that Lynn White perceived between the Middle Ages and the Wild West, I occasionally swing or shuffle from one episode to another, to make or renew acquaintances and point to consistencies in preceding and succeeding generations. To reflect that only six generations of their family separated the immigrant Henry Adams in the 1630s from the historian Henry Adams in the 1890s, for instance, puts the past into a less linear framework; and to learn that successive waves of Adamses had a relatively fixed view of natives says much about their society's subliminal mind. If the reader does not become too disoriented by this process, my heretical method will, I hope, make the present help explain the past and the past help write the history of the present.

On a spatial level, the strategy of this book is embarrassingly simple-minded. I propose we tag along with these European immigrants as they become citizens of the republic and push the American empire into the setting sun, see what they do and listen attentively to what they have to say, and then try to understand all that, as well as their unspoken convictions. Since the psychic history of generations is embedded both in what they say and in what they leave unsaid, language is both the subject and the medium of this inquiry. As faithful amanuensis and interpreter of the historical actors featured in this work, I introduce as many quotations of their prose and verse as space will allow, so they can in a sense be heard speaking their own lines. But why Endicott and Hawthorne and the others in my cast? Because individuals are the living substance of history, the carriers and creators, sometimes, of the attitudes and ideas of their generations. Just as a thoughtful writer on the history of architecture must, I focus on selected edifices as microcosms of larger trends and styles in that generational cycle. Individuals appear because in my judgment they embody mainstream attitudes and assumptions, as do John Adams and Jefferson, or because they swim against the current, as do Morton and Melville. They were not chosen because they are easy targets—such killers as John Underhill, George Armstrong Custer, and William Laws Calley have in fact only walk-on or minor parts. On the other hand, such overt philanthropists as Thomas Loraine McKenney and such covert philanthropists as Edward Geary Lansdale have leading roles, since they are more elusive, more interesting, and I venture, more instructively revealing. If my choices have been reasonably judicious, then they will enable us to identify significant national patterns of deracination and extermination that other students of our past can build upon, modify, correct, or refute—as they can, of course, my understanding of three concepts that merit special mention here.

1. *Repression.* For the Saints and their descendants "going native" has always been tantamount to "going nature." "People are ready to run wild into the woods again and to be as Heathenish as ever," warned Increase Mather in *The Danger of Apostacy* (1679), "if you do not prevent it." Prevent it the Saints did, by converting bodies from instruments of rhythmic pleasure into instruments of domination and aggression. They withdrew libidinal energies from individuals and from the family level—witness their practice of placing their own children with other families for fear of "spoiling" them through natural affection. Simultaneously they harnessed these withdrawn energies to their economy and church–state, which in time became the American empire. The consequences were not entirely denied by the Saints themselves, as my discussion of William Bradford shows. Their system of repressions resulted in the corruption of the senses Jonathan Swift saw at the time for what it was. The implied correlation between repression and violence, a central theme in this book, is in accord with evidence accumulated by students working with the familiar "frustration–aggression hypothesis" and is more vividly illustrated by the bottled-up sexual and other longings periodically rising up out of Attica and other prisons across the country—and across the world, for that matter. In our modern "carceral" society, as Michel Foucault has demonstrated in *Discipline and Punish* (1977), "Time penetrates the body and with it all the meticulous controls of power." For the relatively free body in a less repressive society, on the other hand, we can turn to Melville's novel *Typee* (1846) for a definition of "joy" in the South Pacific: "interchange-

able days." Or consider Henry David Thoreau's prescription in *A Week on the Concord and Merrimack Rivers* (1849): "We need pray for no higher heaven than the pure senses can furnish, a purely sensuous life."

Plainly, sexual repression was not a purely imaginary phenomenon Sigmund Freud foisted upon an unsuspecting world. To a significant degree he merely fleshed out in psychological terms insights already expressed by imaginative writers, poets, and artists, some of whom are important for my own understanding of progress and impoverishment. Nonetheless, my deep indebtedness to Freud for illuminating the discontents of "civilization" will be clear to any reader of these pages. I remain indebted even though Freud himself accepted linear time, stages of society, and other Eurocentric notions I reject. Following his own courageous example, I go beyond the master to treat his findings as a set of working hypotheses—some to be accepted if they further our inquiries, others to be rejected if they do not. In my mind the empirical evidence at hand remains always primary and always finally resistant to any attempt to force it into the scholastic system built by Freud's disciples on his ground-breaking inquiries. *Repression,* in the nontechnical sense in which I use the term, refers not only to restraint and denial of genital gratification, though that is surely an important component, but also to social controls and constraints imposed upon the desires and needs of the *whole* body. Here a good illustration would be the fifty long years the U.S. Bureau of Indian Affairs refused to let the Plains Indians experience their Sun Dance. On this level, Freud's ideas have equal importance and dovetail with Johan Huizinga's discussion of dancing as "sacred play" and with Max Weber's analysis of the antithetical Calvinist drive "to bring the things of the flesh under His will."

2. *Racism.* One of the varieties of Western racism has been what I call, following Melville's lead, the metaphysics of Indian-hating, those deadly subtleties of white hostility that reduced native peoples to the level of the rest of the fauna and flora to be "rooted out." It reduced all the diverse Native American peoples to a single despised nonwhite group and, where they did survive, into a hereditary caste. In its more inclusive form, Western racism is another name for native-hating—in North America, of "niggers," "Chinks," "Japs," "greasers," "dagoes," etc.; in the Philippines, of "goo-goos"; and in Indochina, of "gooks."

Winthrop Jordan has justly remarked that the term *racism* is terribly hard to define. I hazard a provisional definition, in chapter 5, that stresses social prejudices and discriminatory *actions* or learned emotions and learned *behavior*. There, I adduce my reasons for rejecting the accepted wisdom that racist practice waited upon systematic racist theory, the nineteenth-century "scientific" ideology rationalizing Western imperialism and colonialism. (Accepting such wisdom is roughly equivalent to saying—though the parallel is more benign—that the practice of birth control waited upon Margaret Sanger to coin the term.) But my definition and argument are better left to their evidential context. Here, I should distinguish racism from *ethnocentrism*, the inner-centeredness of a group united by social and cultural ties—political and economic institutions, language, myths, and the like. Unlike a racial group, an ethnic group may have mixed ancestry and relatively few hereditary ties. It may evince *religious bigotry* and other forms of intolerance in its dealings with other groups—as the English with the Irish—but without identifiable hereditary characteristics separating the groups, bona fide racism remains stillborn. The valiant efforts of the English to turn their ethnocentric feelings of superiority over the "black" Irish into racism ultimately failed, despite their best efforts, since the Celts remained at most "*white* niggers" in their eyes. In fact the Irish proved useful to the English in keeping real nonwhites ("wogs") in line—Rudyard Kipling's Kim in the work of that title (1901) was "a Sahib and the son of a Sahib" who had served in an Irish regiment and, precisely because of that *white* ancestry, was qualified and destined to "command natives."

The *origins* of racism remain relatively obscure. This study undertakes to make them less so with findings that buttress those of other scholars who have traced their deep roots in the Western psyche. The evidence suggests that the roots lie there intertwined with more general repressive attitudes toward nature and the body, and with concomitant associations of dark skin *color* with filth, death, and radical evil generally. Out of this psychosexual complex arose the generic *native*, that despised, earthy, animalic, suppressed "shadow self" projected by the Western mind.

The *consequences* of racism are more tangible. Anchored however obscurely in the unconscious, it became a key component of the national theology, from the Bay Colony's New Israel to the republic's Manifest Destiny and white man's burden and New Frontier. In the national experience race has always been of greater importance than *class*, the cornerstone of European property-based politics. Racism defined natives as nonpersons within the settlement culture and was in a real sense the enabling experience of the rising American empire: Indian-hating identified the dark **others** that white settlers were not and must not under any circumstances become, and it helped them wrest a continent and more from the hands of these native caretakers of the lands. This book is about that racism.

3. *"Civilization."* Western writers, including Freud, have not used this term interchangeably with *culture*, so that other peoples might have other "civilizations." Instead, they have used it to distinguish Western superculture, or the one true "civilization," from so-called primitive cultures. This ethnocentricity with its unmistakable racist overtones led to what Weber properly scoffed at, the "nullity" imagining that it had "attained a level of civilization never before achieved." To guard against such nonsense, I have fenced in this expansionist term with ironic quotation marks.

3

On a clear day you can see Block Island from the place on the Rhode Island coast where I wrote most of this book. It is a visible reminder that Captain John Endicott's 1636 expedition to kill all the adult males on the island was an expedition to the Saints' first West after the Bay Colony. Other Wests followed hard after, in the Connecticut Valley and across the continent to the "New England of the Pacific," as historian John Fiske called Oregon. I grew up on the Pacific slope there, the son of belated pioneers.

Norwegian immigrants, my mother's people, the Tweeds, settled first in Iowa and then, after a cyclone had leveled their farm buildings, continued their westering to the Pacific, relocating in the lush Willamette Valley. That was in the 1890s, just as Frederick Jackson Turner had announced the passing of the frontier and "free" land. Of Irish and German stock, my father, John Henry, was born and raised near the Cumberland Gap in the Boone and Crockett country of Kentucky and Tennessee. A good man with an ax and a rifle, Jack, as we called him, valued vigilance and always slept with a loaded revolver under his pillow or close at hand. He worked his way west at the turn of the century and, Turner notwithstanding, took up "free" land in the Columbia River basin.

I can still vividly recall one of Jack's stories about homesteading on those dry lands. It pivoted around the popular local pastime that made all "squaws" in the field, except the very old, fair sexual game for mounted ranchers. A twist in the usual upshot of "squaw chasing" came one day, it so happened, when one quick-thinking quarry squatted down and threw sand up into her "privates," as Jack always called them, before her ardent pursuer could haul her in with his lariat. The image of the thwarted rider of the purple sage and perhaps memories of the old days always raised belly laughs from Sherm Wilcox and Jack's other homesteading cronies, as though they were hearing the tale for the first time. I myself liked it hugely, which is to admit that I also grew up an unthinking and unfeeling heir to the contempt for the land and its peoples that I have portrayed in this book. As a boy I shot two of the few remaining native pheasants in the Willamette Valley, just as perhaps a few decades earlier I might have delightedly shot two of the remaining Native Americans. In college, Turner became one of my favorite historians, and from him I assuredly learned nothing that would have led me to question the triumphant saga of my people's march across a land that was deemed vacant save for the few wretches whose "savagery" had to be ended.

I mention these autobiographical fragments not to establish my bona fides as a sourdough but to present my own early unconscious presuppositions. In a way every inquiry into Western history, into reified repressed Time, implicitly raises the question: How have we become so alienated from ourselves and from the land? This study differs slightly in raising that question explicitly. But putting it to myself, I have tried to rethink and refeel our past, tried to extricate myself from the strictly *internal* perspective of Western "civilization."

"The white man does not understand the Indian for the reason he does not understand America," said Sioux Chief Standing Bear. "The roots of the tree of his life have not yet grasped the rock and soil. The white man is still troubled with primitive fears." In the chapters that follow, the reader may be certain, however, that as a white man I locate myself—all but a painfully extracted sliver of myself—*within* the process under scrutiny. My goal is not to represent, indict, and condemn this common heritage—though judgments are both inevitable and desirable—but to unearth, recall, and respond. My aim has been to be guided by sensuous reason and an ethic of honesty rather than to practice a specious "objectivity."

On every frontier the "perennial rebirth" of national fable has yet to happen. Standing Bear's contention that rootless whites were still landless sojourners on this continent was echoed over a quarter of a century ago by Felix S. Cohen in the *American Scholar* (1951–52): "The real epic of America is the yet unfinished story of the Americanization of the White Man, the transformation of the hungry, fear-ridden, intolerant men that came to these shores with Columbus and John Smith." And he went on: "The American way of life has stood for 400 years and more as a deadly challenge to European ideals of authority and submissive obedience in family life, in love, in school, in work, and in government." Still standing with that challenge is the invitation of Native Americans: "The lands wait for those who can discern their rhythms." If it is too much to expect that this book will contribute to that perennially desired regeneration, I still cherish the hope that it may be another step toward the lands and—once again in Thoreau's words—toward giving our history "some copper tints and reflections at least."

Weekapaug, Rhode Island
July 1978

PART ONE

Maypoles and Pequots

Moreover, when saints call themselves sinners, they are not so wrong, considering the temptations to instinctual satisfaction to which they are exposed in a specially high degree—since, as is well known, temptations are merely increased by constant frustration, whereas an occasional satisfaction of them causes them to diminish, at least for the time being.

—SIGMUND FREUD,
Civilization and Its Discontents, 1930

CHAPTER I

The Maypole of Merry Mount

Jollity and gloom
were contending for an empire.

—NATHANIEL HAWTHORNE,
Twice-Told Tales, 1837

IN MAY 1968 Robert Lowell's play *Endecott and the Red Cross* opened at the American Place Theater in New York. One of a trilogy called *The Old Glory*, it appeared at just the right time. That was the spring of the Columbia University sit-ins and, across the Atlantic, of the insurrectionary Paris May days. That Easter, a half-block from Lowell's apartment in the West Sixties, Central Park danced with Indian-clad hippies and yippies making merry. On television the Vietnam Show dragged on, while half a world away U.S. soldiers and marines complained bitterly about an enemy "who would not stand up and fight" but, elusive as the play of shadows, glided back into the jungles and villages of what American officers liked to call "Indian Country." Ties between these current happenings and the poet–playwright's historical drama were nearly palpable.

Set in the Massachusetts Bay Colony of the 1630s, *Endecott and the Red Cross* was an imaginative reconstruction of the actual suppression of Thomas Morton and his followers and of their dispersal from Merry Mount near Wollaston, the site of present-day Quincy, Massachusetts. Proving ancestral distastes perennial, Lowell's nostrils took offense at the pleasure-loving Morton, as though he sniffed a yippie Abbie Hoffman who had unexpectedly ambled on stage from the streets of Greenwich Village. As Lowell presented him, Morton was a sloppy, fat, two-faced rogue who sold guns and liquor to the Indians while pretending to believe that "the blue-assed Puritans" hated him merely because of his love for the Book of Common Prayer. Even the Anglican priest Blackstone, his ally, choked on Morton's salacious doggerel about having free maids, white or red, in the forest or in bed, declaring in revulsion that there had to be some decency: "There must be some boundary between the Indian and English subjects." And though the priest and the invading Puritan captain agreed on little else, that was exactly why Endicott had marched his expedition from Salem—to establish such a boundary, to stop the infernal Maypole dancing and enforce moral decorum, and to put an end to the horror of drunken whites mating with red women. To their conventional English minds, it was as if Morton had gone native in a big way; proof existed in the number of his red friends among the revelers seized by the Puritans.

The expedition's chaplain, Elder Palfrey, warned Endicott not to consider these Indians harmless nor to underestimate the threat they posed:

> There are three thousand miles of wilderness behind these Indians, enough solid land to drown the sea from here to England. We must free our land of strangers, even if each mile is a marsh of blood!

To this man of God the Indians were a plague that "must be smothered if we want our children to live in freedom." He demanded they be taken out of sight and shot. Endicott later actually so ordered, with the promise that after his men had finished killing the Indians, they might then start burning down the houses of Merry Mount. Tomorrow would be soon enough to burn the native village itself.

It always was. As the actors spoke their lines in the American Place Theater, traditional burnings and killings had just culminated at My Lai, though the massacre there would not become public knowledge until 1969. But to look forward so prophetically, which is to say realistically, Lowell had first had to look backward for basic historical truths about those immigrants who became Anglo-Americans. As he saw them, Morton and Endicott were archetypes in the development of the national character.* Repelled by the former, the poet had him appear as a one-dimensional expression of the pleasure principle, an eruption of pure id. Attracted to Endicott almost in spite of himself, Lowell had that Puritan share his own predilection for antinomies: Endicott surprisingly turned out to be an irresolute and mild man, forced to suppress imprudent sympathies for his victims and unsettling memories of his former life as a courtier of King James: "Why did I come to this waste of animals, Indians, and nine-month win-

* See "Notes and Bibliographical Essay" at the conclusion of the book for a running commentary on sources mentioned in the text and on others used as general background in each chapter.

ters?" Yet, when circumstances seemed to demand action, act he did. At the play's end Endicott stood victorious atop Merry Mount, in control of the present and heir to the future.

2

In March 1837, thirteen decades closer to this world of the Puritan fathers, Nathaniel Hawthorne's short story "The Maypole of Merry Mount" appeared in *Twice-Told Tales*. In great measure the inspiration and foundation of Lowell's play, it, too, had ties to current events. The Second Seminole War was dragging on, with U.S. soldiers and marines trying to "root the Indians out of their swamps" and damning them for not standing up and fighting. The year before, President Andrew Jackson had urged that Seminole women and children be tracked down and "captured or destroyed," but even those tactics had not brought the Indians to heel. From Baltimore the *Niles Weekly Register* continued to hope, nevertheless, that "the miserable creatures will be speedily swept from the face of the earth." Elsewhere preparations were under way to send the Cherokees on their Trail of Tears to the West. Worthy forerunners of this Jacksonian America, then, were Endicott and his band as they appeared in Hawthorne's short story. "Their weapons were always at hand to shoot down the straggling savage," he wrote of the Saints. "When they met in conclave, it was never to keep up the old English mirth, but to hear sermons three hours long or to proclaim bounties on the heads of wolves and the scalps of Indians."

In Hawthorne's hands that curious incident at Merry Mount two hundred years earlier came to resemble a sort of primal Woodstock nation of epochal significance. Nothing less than the future of the national character was at stake. "Jollity and gloom were contending for an empire."

On one side were the children of Pan, though Hawthorne sensibly questioned whether they had come directly from classical Greece: "It could not be that the fauns and nymphs, when driven from their classic groves and homes of ancient fable, had sought refuge, as all the persecuted did, in the fresh woods of the west. These were Gothic monsters, though perhaps of Grecian ancestry." That the great god was their ancestor was more certain than probable, however, for one comely youth "showed the beard and horns of a venerable he-goat," while another appeared scarcely less monstrous with the head and antlers of a stag on his shoulders. Green boys and glee-maidens all, they loved twenty different colors "but no sad ones"; they wore "foolscaps and had little bells appended to their garments, tinkling with a silvery sound responsive to the inaudible music of their gleesome spirits."

Flower children who dressed their Maypole in blossoms "so fresh and dewy that they must have grown by magic on that happy pine tree," they acted out a wild philosophy of pleasure that made games of their lives. "Once, it is said, they were seen following a flower-decked corpse with merriment and festive music to his grave." On the joyous occasion Hawthorne described, the jovial priest Blackstone crowned a "lightsome couple" Lord and Lady of the May and prepared to join them in pagan wedlock, for they were "really and truly to be partners for the dance of life." In other, quieter times, the Merry Mounters were said to have sung ballads, performed juggling tricks, played jokes on each other, and when bored by their nonsense, "made game of their own stupidity and began a yawning match." Yet their gay veneration of the Maypole was quite serious: they danced around it "once, at least, in every month: sometimes they called it their religion or their altar, but always it was the banner-staff of Merry Mount." Theirs was no narrow creed of the elect, however, for sometimes they could be seen playing around the great shaft and exerting themselves to entice a live bear into their circle or to make a grave Indian share their mirth.

Onto this sun-bright field rushed the Saints, "black shadows" sworn to eternal enmity against such mirth, and at the same time "waking thoughts" bent on scattering the fantasies of such dreamers. They displayed a taste for the funereal, the colors of death; their songs were psalms; their festivals, fast days. Among them, "woe to the youth or maiden who did but dream of a dance." Mingling with the Merry Mounters that nuptial night, they suddenly turned on their unsuspecting hosts and captured the dancers one and all. Their leader's command followed swiftly:

> Wherefore bind the heathen crew and bestow on them a small matter of stripes apiece as earnest of our future justice. Set some of the rogues in the stocks to rest themselves so soon as Providence shall bring us to one of our own well-ordered settlements where such accommodations may be found. Further penalties, such as branding and cropping of ears, shall be thought of hereafter.

The lovelock and long glossy curls of the bridegroom challenged Puritan proprieties: "And shall not the youth's hair be cut?" asked ancient Palfrey. "Crop it forthwith, and that in the true pumpkin-shell fashion," answered his commander. No unmanly sentiments tempted the latter to heed the youth's pleas to spare his bride, do what they would with him: " 'Not so,' replied the immitigable zealot. 'We are not wont to show an idle curiosity to that sex which requireth the stricter discipline.' " As for the dancing bear, " 'shoot him through the head!' said the energetic Puritan. 'I suspect witchcraft in the beast.' "

The zealot at the head of the invaders was, of course, John Endicott, "the severest Puritan of all who laid the rock-foundation of New England." As indecisive as an avalanche, Hawthorne's Endicott was seemingly much less complex than Lowell's and much more grim and forceful, so that "the whole man, visage, frame and soul, seemed wrought of iron gifted with life and thought, yet all of one substance with his head-piece and breastplate." To be sure, the early love of the bridal pair softened him a little and led him to command that after the boy's hair was cropped, they be brought along "more gently than their fellows." But this order betrayed no second thoughts or inner misgivings. To Endicott's mind the boy was a likely recruit for his armed band, since he promised to become "valiant to fight and sober to toil and pious to pray"; the girl was a welcome addition, since she had the makings of a fit "mother in our Israel." They merited special treatment, for one day they would join Israel in its holy war for the American wilderness: "But now shall it be seen that the Lord hath sanctified this wilderness for his peculiar people," as Endicott had already declared. "Woe unto them that would defile it!"

Yet Hawthorne saw in this man of iron, with his unswerving singleness of purpose, an essential side that Lowell passed by. The latter, in his preoccupation with inner hesitancies and divisions, created a figure not unlike himself and his age, a figure torn by doubts and frustrations, by ambiguous feelings of historical guilt, and by treacherous sympathies for the victims of New Israel. Hawthorne did not lack ambivalences of his own, of course, and was very much a man of *his* generation; still he burrowed within himself and back through the centuries to come up with an insight of critical importance: Endicott had *enjoyed* being a key part of the system of repressions. From all their deposits of self-denial he and other Puritan fathers had made furtive withdrawals of gratification in the suppression of others—in cropping ears, placing reprobates in stocks, and branding and whipping them. That clearly was what Hawthorne meant when he said that the whipping post "might be termed the Puritan Maypole." Around it the Saints got their kicks, albeit of a different sort from those of the Gay Sinners.

3

In all of American literature there is perhaps no greater metaphoric epiphany than when Hawthorne's Endicott, immediately after capturing the dancers, assaults the Maypole with his keen sword:

> Nor long did it resist his arm. It groaned with a dismal sound, it showered leaves and rosebuds upon the remorseless enthusiast, and finally, with all its green boughs and ribbons and flowers, symbolic of departed pleasures, down fell the banner-staff of Merry Mount. As it sank, tradition says, the evening sky grew darker and the woods threw forth a more somber shadow.

There it all was, a tableau as memorable as "The Castration of Uranus," except this time the victim was Pan and he was emasculated in an American setting: The fall of the Maypole shadowed forth, as Endicott proclaimed triumphantly, "the fate of light and idle mirthmakers among us and our posterity." It was Hawthorne's genius to see an orgiastic dimension in Endicott's wanton act, to cast the act as a sexual assault, and to have him carry it out with all the frenzied sadism of a "remorseless enthusiast."

Any doubts that Hawthorne was consciously clothing his insight in this splendid symbolism are put to rest by the one regret he allowed Endicott to voice:

> I thought not to repent me of cutting down a Maypole . . . yet now I could find in my heart to plant it again and give each of these bestial pagans one other dance around their idol. It would have served rarely for a whipping post.

Endicott's suppressed sexuality rose up here, in his vision of whipping pagans around the upright symbol of their bestiality. And just as his pleasure was perverse, Hawthorne made plain, so were many other Puritan violent delights.

CHAPTER II

Thomas Morton

> After this they fell to great licentiousness and led a dissolute life, pouring out themselves into all profaneness. And Morton became Lord of Misrule.
>
> —WILLIAM BRADFORD,
> *Of Plymouth Plantation*, 1642

IN "The Maypole," Hawthorne said, "The facts recorded on the grave pages of our New England annalists have wrought themselves almost spontaneously into a sort of allegory." Of course facts rarely work themselves into any kind of symmetry, and it was only while digging for the "deep and aged roots" of his family that the writer had found those facts in the first place. Nevertheless, in his obsessive search for a usable personal past Hawthorne had unearthed the collective repressions that surfaced on his pages with such near spontaneity. His art fused these emergent repressions with facts into an allegory that has profound historical meaning. It was as though the long-silenced Thomas Morton, the very antithesis of his grave New England annalists, had finally been allowed, in the gay and colorful part of the story, to say a few words on his own behalf.

For almost two decades, from 1627 through 1645, the colonial authorities had sought to silence Morton forever. Their straining efforts to retch him out of their systems by sending him off in chains might well be thought of as a prototypical case. It pretty much set the pattern for an unending series of attempts to purge the country, always once and for all, of rebels and heretics, savages and barbarians, familists and antinomians, loose livers and free lovers, and all other undesirable species. It came first, and it was no less pregnant with issues than the cases against Roger Williams, Anne Hutchinson, the Quakers, and those that followed. That it remains so largely unknown today is the measure of the triumph of Morton's enemies: they discovered that if they could not spit him out of the country to be forgotten, they could swallow him, as it were, by having him rot away in prison to be forgotten. Almost.

The campaign to silence Morton had three distinct stages. The first dated almost from his arrival in 1625. He traded extensively with the Indians, prospered, and raised his Maypole in 1627. The following year Captain Miles Standish and eight men from nearby Plymouth invaded Merry Mount, captured its host, and hauled him off to their settlement. From there, in the late summer of 1628, he was shipped to England (in the charge of John Oldham) to stand trial for selling guns and spirits to the Indians. The allegations and evidence against him were so insubstantial, however, that the case collapsed before it reached the courtroom. Within a year from his deportation Morton returned mockingly to Plymouth and soon reestablished himself in his old home at Merry Mount. About Christmas of 1629 John Endicott, then in charge of Massachusetts Bay, tried to have Morton arrested for not submitting to his own "good order and government," but the unrepentant scoffer "did but deride Captain Littleworth," as he called Endicott, and easily eluded his pursuers.

The second stage dated from the fall of 1630: Shortly after the arrival of Governor John Winthrop and the first wave of the Great Migration, the enlarged body of magistrates met for the first time and issued a warrant for Morton's arrest. He was brought before them on September 17, for which session Endicott came down from Salem. Never was a kangaroo court more summary. Morton's words of self-defense were cut off short so he could hear the verdict against him. He was to be set in the stocks; his goods were to be seized to pay for his transportation back to England, to meet his debts, and most curiously, "to give satisfaction to the Indians for a canoe hee unjustly tooke away from them; and . . . his howse, after the goods are taken out, shal-be burnt downe to the ground in the sight of the Indians, for their satisfaction, for many wrongs hee hath done them from tyme to tyme." These were the men, as Morton ironically observed in his own account, who had "come prepared to ridd the Land of all pollution." In imposing sentence, he added, Winthrop had explained that Merry Mount was to be burned to the ground "because the habitation of the wicked should no more appeare in Israell." Four months later Morton was finally hoisted over the side of the *Handmaid* by tackle, since he refused to go aboard voluntarily. Only then, as he was shipped off into exile a second time, was his house burned down —in his sight and not simply in sight of the Indians—leaving behind, as Morton recorded, "bare ashes as an emblem of their cruelty."

Again set free in England, for this set of spindly charges had even less chance of surviving an Atlantic passage than the first, Morton promptly mounted a counterattack. A solicitor and member of Clifford's Inn before his emigration in the early 1620s, he was competent, angry, and possessed of an energy that made him a more formidable adversary than Winthrop and Endicott could have anticipated. He handled legal work for Sir Ferdinando Gorges, the dominant figure in the Council for New England and active foe of the Bay Puritans, and spearheaded an assault on the charter of the Massachusetts Company. By 1634 he and his associates had been so successful that the colony was in a state of panic that their patent and powers would fall to an expedition headed by Gorges, the newly appointed governor-general, and seconded by his aide-de-camp, the despised Morton. And it was during this heady period, with his high-handed enemies squirming, that Morton wrote *New English Canaan,* to which we shall turn presently. At all events, apparently only lack of funds kept the expedition from being launched, with Gorges, Morton, and their associates bidding fair to win all the stakes in both Old and New England. But their fortunes were tied to those of Archbishop Laud and Charles I; the rush of events toward civil war put an end to their hopes. In the summer of 1643 Morton reappeared at Plymouth as nothing more, in the unkind words of his editor, "than a poor, broken-down, disreputable, old impostor, with some empty envelopes and manufactured credentials in his pocket."

The third stage was short and bitter. Morton was closely watched in his temporary refuge at Plymouth. When he made preparations to go to Maine in the spring of 1644, Endicott, by then governor, had a warrant issued for his arrest but was unable to serve it. A few months later, however, Endicott's officers grabbed Morton, perhaps as he tried to slip through the province, and brought him before the court of assistants in September 1644. This time his alleged offense was having appealed to the King's Privy Council against the actions of the colonial government. When even this singular "crime" could not be proved, he was thrown into jail until more evidence could be accumulated. He was still locked away in May 1645 when he addressed an appeal to the court to "behould what your poor petitioner hath suffered in these parts"; the list of sufferings concluded with his most recent, "the petitioner coming into these parts, which he loveth, on godly gentlemen's imployments, and your worshipps having a former jelosy of him, and a late untrue intelligence of him, your petitioner hath been imprisoned manie Moneths and laid in Irons to the decaying of his limbs." Lying chained in an unheated cell through a New England winter might have broken the health of a much younger man. Now ill and enfeebled—or "old and crazy," as Winthrop would have it—Morton was fined and turned loose to die somewhere, as he finally obliged the authorities by doing a couple of years later.

Out of this medley of unproved accusations and summary judgments emerged one awkward fact: the colonial authorities never once had compelling evidence that Morton had committed *any* punishable offense under English law. It was not a crime, at least on any formal level, to fraternize with natives. It was not a crime to have a Maypole and especially not since 1618 when King James had issued a decree encouraging Maypole dancing. And least of all was it a crime to petition the Crown through proper legal channels for redress of grievances. To move against Morton the Puritans were thus forced into indirection and extralegality, into veiling the true source of their fear and hatred of him—his true offense against "ye Massachusetts Magistrats," as Samuel Maverick observed, "was he had touched them too neare."

The least trumped up and most serious of all the counts was that he had sold guns to the Indians. This was never established in open court, and in fact there was no English law against it—King James's proclamation (1622) against the practice did not have the status of law and was in any event of doubtful application to Morton. But say that he had traded firearms to the Indians for furs, as he almost certainly had, and then, as William Bradford angrily recorded, had given the Indians instructions in their use. So what? Why prohibit one set of human beings something permitted another? The Saints affirmed they wished to live in harmony with the Indians and bring them Christian light: What then was more logical than for them also to share their technology with red friends who could thereby more efficiently share their wilderness? Alas, the logic was not that of sharing. The planters were *colonizers*. They were the cutting edge of a colonial empire that was currently subjugating Ireland and moving to apply that experience to North America. To arm those about to be conquered struck them as illogical to the point of madness. They knew that were they despoiled of their lands and subjected to foreign discipline, they would use the guns in their hands. It took little imagination for them to sense that others would do likewise, especially if the others were Indians.

Of course the planters and their kinsmen preferred to put the matter the other way round: not their expansion but their very existence was threatened. They warned Morton, according to William Bradford, that "the country could not bear the injury he did" by trading pieces to the Indians; "it was against their common safety and against the King's proclamation." And this became the fixed view of the matter. Four generations later John Adams forcibly restated its essentials. Morton's fun, he wrote in 1802,

> his songs and his revels were provoking enough, no doubt. But his commerce with the Indians in arms and ammunition, and his instructions to those savages in the use of them, were serious and dangerous offenses, which struck at the lives of the new-comers, and threatened the utter extirpation of all the plantations.

Three generations still further along, Charles Francis Adams, Jr., approvingly quoted his great-grandfather's dictum in the course of editing the 1883 edition of *New English Canaan,* conceding as well that Morton's suppression was not a question of law but one of self-preservation:

> Yet it is by no means clear that, under similar circumstances, he would not have been far more severely and summarily dealt with at a later period, when the dangers of a frontier life had brought into use an unwritten code, which evinced even less regard for life than, in Morton's case, the Puritans evinced for property.

But the dangers of frontier life had seemed real enough to Bradford and the colonists of his day as they acted on the leading assumption of the vigilante code the Adamses later defended. Throughout, I venture, from 1628 to 1883 and after, the unwritten code assumed Indians not to be persons, who might be responsive to kindness and fair dealing, but "savages," who would inevitably use any available weapon to strike at the lives of newcomers, those bearers of "civilization."

2

Were it not for *New English Canaan,* there we would have to leave Morton, a curious footnote in the accounts of the first English settlements, known only through the hostile pages of Bradford and Winthrop as a man who, for gain and out of spite, transgressed the American Way. And indeed the book did narrowly escape the oblivion to which its author had been consigned. It excited little or no public interest when it was published in Amsterdam in 1637. It may not have reached the colonies in Morton's lifetime, though Bradford seems to have seen a copy, perhaps one that passed from hand to hand in Plymouth before it disappeared or was destroyed. John Quincy Adams finally ran it down in Europe and brought it home, but in 1825 this copy was still the only one known to exist in North America. Toward the end of the century, when Charles Francis Adams, Jr., worked with the family copy to produce an annotated edition, only a dozen or so others were known to have survived in the various public and private collections on both sides of the Atlantic. But survive somehow the book did, and that was a great good thing: It was like the man—sprawling and disorderly, abrasive and sometimes obtuse, yet usually intelligent, erudite and observant, overflowing with wit and high spirits, richly suggestive.

Morton's ironic title took his "courteous reader" back beyond the Judeo-Christian tradition, with its conquistador hostility toward nature, to the first Promised Land and to the heathen Canaanites who lived happily therein till driven off or exterminated by the Israelites. Spiritual descendants of the latter, the Saints had stepped off their ships into what they could only see as a menacing waste. As Bradford mused, "What could they see but a hideous and desolate wilderness, full of wild beasts and wild men." *What could they see indeed?!*—only another land of milk and honey, retorted *New English Canaan,* only an enchanting green land full of game and friendly red men.

Morton shared the old English passion for field sports in full measure; he was a good shot and avid fisherman, and had scarcely arrived before he called his dogs and took to the woods and streams and bays. Everywhere he discovered astonishing fecundity: Often he had a thousand wild geese "before the mouth of my gunne." The black ducks were so plentiful it was the custom of his house "to have every mans Duck upon a trencher; and then you will thinke a man was not hardly used." Wild turkeys sallied by his door in great flocks, there to be saluted by his gun in preparation for "a turne in the Cooke roome." Never was he without venison, winter or summer, flesh which was "farre sweeter then the venison of England." As for dangerous wild beasts, there were no lions; bears, though numerous, were not to be feared, since they "will runne away from a man as fast as a little dogge." The New England coast so abounded with local cod, the inhabitants "doe dunge their grounds" with them. Oysters were at the entrance of all the rivers, fat and good: Morton had "seene an Oyster banke a mile at length." Of mussels and clams there were an infinite store. Striped bass teemed in the rivers and bays, so that "I my self, at the turning of the tyde, have seene such multi-

NEW ENGLISH CANAAN
OR
NEW CANAAN.

Containing an Abſtract of New England,

Compoſed in three Bookes.

The firſt Booke ſetting forth the originall of the Natives, their Manners and Cuſtomes, together with their tractable Nature and Love towards the Engliſh.

The ſecond Booke ſetting forth the naturall Indowments of the Country, and what ſtaple Commodities it yealdeth.

The third Booke ſetting forth, what people are planted there, their proſperity, what remarkable accidents have happened ſince the firſt planting of it, together with their Tenents and practiſe of their Church.

Written by Thomas Morton of Cliffords Inne gent, *upon tenne yeares knowledge and experiment of the Country.*

Printed at AMSTERDAM,
By JACOB FREDERICK STAM.
In the Yeare 1637.

New English Canaan, Facsimile of the Title Page, 1637.

Bishop Fish, from Konrad Gesner's *Icones animalium*, 1560. In *Shakespeare's Life and Times* (1967), Roland Mushat Frye pointed out that the monstrous creature shown here was published several times during the playwright's lifetime and must have been the model for the costuming of the half-man, half-beast Caliban in *The Tempest* (1611). The figure also gives some hint, I believe, as to how the Puritan patriarchs saw the ithyphallic Morton —unfortunately we have no portrait of the "Lord of Misrule." (By permission of the Folger Shakespeare Library.)

tudes passe out of a pound, that it seemed to mee that one might goe over their backs drishod."

For this unspoiled moment, continent and man fused, the power of the one to arouse awe embraced by the nearly commensurate imagination of the other. Morton's lyrical summary of his findings should be known by every student of America's past:

> The more I looked, the more I liked it. And when I had more seriously considered of the bewty of the place, with all her faire indowments, I did not thinke that in all the knowne world it could be paralel'd, for so many goodly groues of trees, dainty fine round rising hillucks, delicate faire large plaines, sweete cristall fountaines, and cleare running streams that twine in fine meanders through the meads, making so sweete a murmuring noise to heare as would even lull the senses with delight a sleepe, so pleasantly doe they glide upon the pebble stones, jetting most jocundly where they doe meete and hand in hand runne downe to Neptunes Court, to pay the yearely tribute which they owe to him as soveraigne Lord of all the springs. Contained within the volume of the Land, [are] Fowles in abundance, Fish in multitude; and [I] discovered,

> besides, Millions of Turtledoves one the greene boughes, which sate pecking of the full ripe pleasant grapes that were supported by the lusty trees, whose fruitfull loade did cause the armes to bend: [among] which here and there dispersed, you might see Lillies and of the Daphnean-tree: which made the Land to mee seeme paradice: for in mine eie t'was Natures Masterpeece; Her chiefest Magazine of all where lives her store: if this Land be not rich, then is the whole world poore.

Partly promotional tract for the schemes of Morton and his patron Gorges and partly anti-Puritan polemic, *New English Canaan* was also much more. The work represented an authentic and almost singular effort of the European imagination to extract a sense of place from these new surroundings or, better, to meet the spirit of the land halfway. There was thus absolutely no reason to question the sincerity of his 1645 petition, wherein he stated he had returned to these parts "which he loveth." Like the Indians, he loved the wilderness the Saints hated.

A prime reason the country was so beautiful and commodious, more like an English park than the delusory waste of the Saints, was the Indian practice of firing the underbrush every spring and fall: without that, Morton pointed out, "it would be all a coppice wood, and the people would not be able in any wise to passe through the Country out of a beaten path." In this and the other sections that made up the first of the three parts of *New English Canaan*, he proved himself a shrewd and sympathetic field observer of Indian manners and customs. To be sure, he did not always rise above colonial bigotry, as in his flat statement that the natives had no religion, a view he accepted on the authority of Sir William Alexander, and in his ethnocentric, Bradford-like celebration of the plague of 1616–17 that swept away so many natives the place became "so much the more fitt for the English Nation to inhabit in, and erect in it Temples to the glory of God." Nevertheless, he placed the ethnographic chapters at the head of his work and in enthusiastic detail paid the Indians the uncommon tribute of taking their culture seriously.

The natives of Morton's pages had much to teach the European immigrants. "These people are not, as some have thought," he wrote, "a dull, or slender witted people; but very ingenious, and very subtile." They made their wampum or money from clam shells, built wigwams that looked like the houses of the "wild Irish," tanned the skins of deer and other animals into "very good lether," had excellent midwives who helped their women "have a faire delivery, and a quick," and even had their own physicians and surgeons, who "doe make a trade of it, and boast of their skill where they come."

Morton related an anecdote of one powah, or medicine man: He "did undertake to cure an Englishman of a swelling of his hand for a parcell of biskett, which being delivered him hee tooke the party greived into the woods aside from company, and with the helpe of the devill, (as may be conjectured,) quickly recovered him of that swelling, and sent him about his worke againe." The playful conjecture may have reflected Morton's sober belief, for he said elsewhere of the medicine men, "some correspondency they have with the Devil out of all doubt." But he was himself an ingenious and subtle man, quite capable of parodying the Puritan conviction that the Indians were in league with the devil. I see no other way of reading his conclusion that the Indians lived rich and contented lives, wanting nothing needful, "the younger being ruled by the Elder, and the Elder by the Powahs, and the Powahs . . . by the Devill; and then you may imagine what good rule is like to be amongst them."

Sir William Alexander's views notwithstanding, Morton discovered that in fact the Indians were not altogether lacking in religion and that they had traditions that spoke of the creation and of the immortality of the soul. His account of their mournings for the dead had an ethnographic precision not to be found in John Bradbury, Henry Marie Brackenridge, or other explorers who centuries later encountered the same phenomena among the trans-Mississippi tribes: The natives of New England held it impious to deface graves and their monuments, Morton wrote, and "have a custome amongst them to keepe their annals and come at certaine times to lament and bewaile the losse of their freind; and use to black their faces, which they so weare, instead of a mourning ornament, for a longer or a shorter time according to the dignity of the person."

The grave of the mother of one chief, Sachem Chickatawbut, had two bearskins sewed together and propped up over it as a monument, he recorded, "which the Plimmouth planters defaced because they accounted it an act of superstition." To all appearances he joined the Indians in accounting this act a wanton desecration of a sacred place. Subsequently Chickatawbut had a vision that Morton tried to translate for his readers. He pictured the chief standing before his angry tribespeople speaking eloquently of his holy experience:

> When last the glorious light of all the skey was underneath this globe, and Birds grew silent, I began to settle, (as my custome is,) to take repose; before mine eies were fast closed, mee thought I saw

> a vision, (at which my spirit was much troubled,) and, trembling at that doleful sight, a spirit cried aloude behold, my sonne, whom I have cherisht, see the papps that gave thee suck, the hands that lappd thee warme and fed thee oft, canst thou forget to take revenge of those uild people that hath my monument defaced in despitefull manner, disdaining our ancient antiquities and honourable Customes? See now the Sachems grave lies like unto the common people of ignoble race, defaced; thy mother doth complaine, implores thy aide against this theevish people new come hether; if this be suffered I shall not rest in quiet within my everlasting habitation. This said, the spirit vanished; and I, all in a sweat, not able scarce to speake, began to gett some strength, and recollect my spirits that were fled.

If not the first, this has to be one of the very earliest attempts by Europeans to catch the metaphors and rhythms of Indian oratory. More remarkably still, as Morton rendered the speech, notwithstanding its overlay of Anglicisms, it proved him capable of entering into the lives of the victims of colonial aggression and of sympathizing with how they felt as persons.

3

So did *New English Canaan* pose the ultimately subversive question: Who were the real "uild people"? The Indians? They were at home in the land, treated Morton and other planters hospitably, shared what they had (as in "Platoes Commonwealth"), danced as a form of communal art, and derived other innocent delights from living in their bodies. Or the Saints? They hated the land, had already massacred some of its inhabitants, defaced their graves and otherwise abused their hospitality, clutched avariciously at property and things, forbade dancing, and generally denied the pleasures of their bodies. Even the careless reader could not miss Morton's answer: "I have found the Massachusetts Indian[s] more full of humanity then the Christians; and haue had much better quarter with them; yet I observed not their humors, but they mine." He perceived at its inception the stereotype of the treacherous savage and rejected it out of hand. The Saints demanded of every newcomer, he wrote, full acceptance of "the new creede" that "the Salvages are a dangerous people, subtill, secreat and mischeivous; and that it is dangerous to live seperated, but rather together: and so be under their Lee, that none might trade for Beaver, but at their pleasure, as none doe or shall doe there."

Beside the radicalism of Morton's challenge, Roger Williams's questioning of Puritan title to Indian land seems innocuously liberal. Morton asserted the superior humanity of the Indians and then went dangerously far toward establishing that claim by living among them in amity. As a living example he undermined, just as they were establishing it, the colonizers' notion of the treacherous savage and their need to see themselves as a tightly knit armed band of Christians perched on the edge of hostile territory.

Now we return to the verdict of Winthrop, Endicott, and the other magistrates in 1631. Morton's house was to be "burnt downe to the ground in the sight of the Indians, for their satisfaction, for many wrongs hee hath done them from tyme to tyme." The judgment showed commendable Christian concern for the natives, one the magistrates doubtlessly hoped would reach its mark in England, but it really had to be tipped over on its head to be understood rightly. His house was to be burned for the many *rights* he had done the Indians from time to time. He had willfully violated the racist core of the magistrates' nascent code by hunting with the Indians, trading them guns, enjoying their culture with them, dancing with them, sleeping with them—all just as if they were persons.

As it happened, the Indians responded humanely and were not at all satisfied by the sight of Morton's house burning to the ground. While he was being shipped off in irons, he later learned, "the harmeles Salvages, (his neighboures,)" at Merry Mount, came up to the Saints, grieved to see these arsonists at work, "and did reproove these Eliphants of witt for their inhumane deede. . . ."

CHAPTER III

John Endicott

> . . . a fit instrument to begin this
> Wildernesse-worke, of courage bold
> undaunted, yet sociable, and of a chearfull
> spirit, loving and austere, applying himselfe
> to either as occasion served.
>
> —EDWARD JOHNSON,
> *Wonder-Working Providence*. 1653

FOR GOOD REASON John Endicott figured centrally in "The Maypole of Merry Mount" while Morton was nowhere to be seen. More scrupulous than Lowell in his use of sources, Hawthorne curbed any temptation to achieve dramatic unity by staging a showdown that did not and could not have occurred. In fact Morton was on his way into exile for the first time, in the late summer of 1628, when Endicott landed with the Puritan forerunners at Salem, then Naumkeag. Shortly afterward Endicott did lead an expedition to Merry Mount and no doubt regretted his inability to get his hands on "mine Hoste" who had thus unavoidably been detained elsewhere. According to Bradford, Endicott merely rebuked the remnant of Morton's band "for their profaneness and admonished them to look there should be better walking"—that is, they should walk the straight and narrow path of Puritan virtue. Beyond that we know next to nothing, save for the one detail Hawthorne found so critically important. Seeing the magnificent banner-staff still standing, Endicott fell upon it and did in historical truth cause "that Maypole to be cut down."

Endicott was in charge of the colony from 1628 until the summer of 1630, when he turned his authority over to Governor John Winthrop and became an assistant, in which post he served from 1630 to 1634, 1636 to 1640, and 1645 to 1648. He was deputy governor from 1641 to 1643 and in 1650 and 1654. He was governor in 1644, in 1649, from 1651 to 1653, and from 1655 until his death in 1665. In 1637 he was elevated to the Council for Life (later declared unconstitutional), on which the only other members were Winthrop and Thomas Dudley. Throughout his life, clearly, he en-

joyed the high esteem of his fellow Saints. As Lawrence Shaw Mayo pointed out in his filial biography, it may be significant that Endicott was born in 1588, the year of the defeat of the Spanish Armada: "It was also the natal year of John Winthrop. Who will say that fundamentally 1588 is not the most important date in the history of Massachusetts? If it had not been for the defeat of the Armada and the birth of John Endecott and John Winthrop, where should we be today?"

Where indeed? And there are other good questions. Was Endicott the relatively irresolute leader depicted by Lowell? Or was he the iron-willed zealot who appeared in Hawthorne's pages and in the old portrait of the American Antiquarian Society? Or was he but the "fit instrument" another admirer, Captain Edward Johnson, saw in him, "loving or austere . . . as occasion served"?

It appears that almost invariably the occasions of his public life served austerity and not love, presumably reserved for family and close associates. Before Winthrop's arrival Endicott had not only cut down Morton's Maypole but had also deported two Puritan brothers whom he pronounced "schismatical" for opposing his ideas of church government. As an assistant he physically assaulted Thomas Dexter of Lynn, for which act he apologized to Winthrop, acknowledging that he had been "too rash in strikeing him, understanding since that it is not lawful for a justice of the peace to strike. But if you had seen the manner of his carriadge, with such daring of mee with his armes on kembow &c. It would have provoked a very patient man." Never a patient man, he verbally assaulted three Anabaptists from Rhode Island who came before the General Court in 1651. After sentence had been imposed, they had had the temerity to ask what law they had broken. In his *Ill Newes from New-England* (1652), John Clark, one of the defendants, recalled Governor Endicott's response: that they had denied infant baptism; then "being somewhat transported [he] broke forth and told the prisoners they really deserved death." He held them to be trash and declared the magistrates of Massachusetts "would not have such trash brought into their jurisdiction."

Endicott earned his reputation as "a greater persecutor." As head of the commonwealth he shouldered primary responsibility

◀ *Governor John Endicott*, n.d. (1665?). By an unknown hand, this was an earlier portrait than that of the American Antiquarian Society mentioned in the text—defenders complained that the latter made their hero appear "a cold and narrow bigot." Here Endicott still appears Hawthorne's Puritan of Puritans. (Courtesy of the Commonwealth of Massachusetts.)

for whipping, branding, and banishing, and ultimately for executing the "open and capitall blasphemers" called Quakers. His General Court hanged four members of the Society of Friends. With the Restoration, the unrepentant governor explained why in a letter to Charles II. The Quakers were "open enemies to government itself" in any other than the hands of their friends; he and his assistants had, therefore, "to passe a sentence of banishment against them upon pain of death." Such was their turbulency, however, he had at last, "in conscience both to God and man . . . to keep the passage with the point of the sword held toward them." And since the Quakers had wittingly rushed on the point of the sword by "their owne act, we with all humility conceive a crime bringing their blood upon their owne head." Unresponsive to this humble appeal for sympathy, Charles commanded Endicott to cease any current actions against Quakers and to "forbear to proceed any farther therein."

2

Every shred of believable evidence supports Hawthorne's view of the man. Oddly enough the novelist approached Endicott as if he had *New English Canaan* at hand, a most unlikely possibility. Still, had it been available, Hawthorne would have found it most engaging. Of course Morton could hardly have given him or us a "balanced" view of the man he called "Captaine Littleworth," for Endicott truly "had an akeing tooth at mine Host of Ma-re Mount" and proved it time and again. Yet just as obviously Morton had a direct personal interest in understanding his inveterate adversary. His prescient chapter, "Of the manner how the Seperatists doe pay debts to them that are without," anticipated Hawthorne's key insights.

Morton knew well the case of "an honest man, one Mr. Innocence Fairecloath," since he had helped present it to the Privy Council in the course of his attack on the Massachusetts charter. "Fairecloath" or Philip Ratcliff, as he was known in the flesh, had come over as an agent of Matthew Cradock. Members of the Salem congregation found his beliefs "without" their church, however, so "disdained to be imployed by a carnall man, (as they termed him,) and sought occasion against him, to doe him a mischeife." They worked their way into his debt and, when he sought to collect, sent him "an Epistle full of zealous exhortations to provide for the soule; and not to minde these transitory things that perished with the body." The counsel moved Ratcliff to exclaim in a moment of unguarded anger: "Are these youre members? if they be all like these, I beleeve the Divell was the setter of their Church."

Endicott promptly charged blasphemy and determined Ratcliff would be "made an example for all carnall men to presume to speake the least word that might tend to the dishonor of the Church of Salem; yea, the mother Church of all that holy Land." In court the charges against him multiplied, "seeing hee was a carnall man, of them that are without." In June 1631, according to Winthrop's *Journal*, Ratcliff was sentenced to be whipped, have his ears cut off, pay a fine of forty pounds, and be perpetually banished—all this for "most foul, scandalous invectives against our churches and government."

Morton dwelt on the pleasure the punishment gave the Saints. "Shackles," or the deacon of the church at Charlestown, sobbed and wept with Ratcliff, "and his handkercher walkes as a signe of his sorrow for Master Fairecloaths sinne, that hee should beare no better affection to the Church and the Saints of New Canaan: and strips Innocence the while, and comforts him." The executioner of their vengeance then went to work "in such manner that hee made Fairecloaths Innocent back like the picture of Rawhead and blowdy bones, and his shirte like a pudding wifes aperon. In this imployment Shackles takes a greate felicity, and glories in the practice of it." And "loe," Morton concluded, "this is the payment you shall get, if you be one of them they terme, without."

Morton rather deftly emphasized the sanctimonious cupidity of the Saints by having the punishment carried out in "the Counting howse," with the deacon expostulating with Ratcliff about being "so hasty for payment." More obvious still was Morton's denunciation of the nature of their sentence—its stupefying harshness created a stir in England and word even came back to Massachusetts, through one of Winthrop the younger's correspondents, that there had been "diuerse complaints against the severitie of your Gouernement especially mr. Indicutts and that he shalbe sent for ouer, about cuttinge off the Lunatick mans eares, and other greiuances." Morton also put before his readers Endicott's menacing conviction that the Puritans were God's chosen, so that to speak against them was to defile their holy mission; and as Hawthorne later had Endicott say, "woe unto them that would defile it." But Morton's analysis went beyond the economic level of pious acquisitiveness and the political level of religious nationalism to a psychological truth: his enemies were not coldly righteous monsters, however great their hypocrisy, but men who found their cruelty bloody good fun. Though they could hardly admit it, it gave them "greate felicity, and glories in the practice of it." Morton recognized in effect, again before Hawthorne, that the pleasures they took from the whipping post made it their equivalent of his Maypole.

3

As a Puritan, Endicott believed in a two-species theory of European humankind. Outside the true faith were the carnal men—men like the libertine Morton, whose carnality was palpable, the traducer Ratcliff, the blasphemous Quakers, and the Anabaptist trash from Rhode Island—all concupiscent, rational animals scarcely more worthy of consideration than the dancing bear Endicott ordered shot through the head in Hawthorne's story. Within the faith were the grace-endowed men, men redeemed from their sinful bodies by Jesus—men like Endicott, Winthrop, Dudley, Bradford, all instruments of God's purpose. But it was hard, devilishly hard as it were, for spiritual men not to ease down into their sinful bodies, not to think of sex for more than procreation, to avoid "impurity" of thought and act. Not to "fall back into nature" required no less than twenty-four-hour watches all the days of their lives.

Within this context Endicott naturally would have commanded, as Hawthorne had him do, that the young Lord of the May be shorn of his lovelock and curls. In fact as governor in 1649 he and the magistrates had sought to stop such adornment of wickedness:

> Forasmuch as the wearing of long haire after the manner of Ruffians and barbarous Indians, hath begun to invade new England contrary to the rule of Gods word, which saith it is a shame for a man to wear long hair, as also the Commendable Custome generally of all the Godly of our nation until within this few yeares Wee the Magistrates who have subscribed this paper (for the clearing of our owne innocency in this behalfe) doe declare and manifest our dislike and detestation against the wearing of such long haire, as against a thing uncivil and unmanly whereby men doe deforme themselves, and offend sober and modest men, and doe corrupt good manners.

Long hair took root in the lubricous skin, flaunted its origins, and seemed barbarously ungroomed, pubescent, suggestive of the unmentionable—of carnality, mortality, beastliness. The magistrates, with Endicott at their head, entreated the elders to manifest their zeal in ensuring that the members of their "respective Churches bee not defiled therewith." At the very least Endicott had cleared his name and witnessed his "own innocency" of such defilement.

Women were the opposite sex and as such a threat to purity. It was admittedly better to marry than to burn, so Endicott, before he emigrated and then nearly forty, married Anna Gover, a cousin of Matthew Cradock. She died shortly after their arrival at Naumkeag, however—Morton unkindly suggested that it was through the good offices of Samuel Fuller, butcher turned Plymouth physician, that Endicott had been cured "of a disease called a wife." In August 1630 he married Mrs. Elizabeth Gibson, a widow, of whom little more is known than of the first Mrs. Endicott, save that she bore him two sons and survived him.

The obscurity of his wives was not fortuitous, for Endicott would have considered it unmanly to let a "weaker vessel" take the lead or share authority and responsibilities. During the examination of Anne Hutchinson, for instance, he revealed keen interest in whether she had presumed to teach at meetings of *men*, found her defiance intolerable, hoped "the court takes notice of the vanity of it and heat of her spirit," and joined his brothers in finding her "unfit for our society." For a woman to presume to act as an enlightened individual with a conscience of her own would subvert the patriarchal family, church, and state, and lead to all the anarchic evils of the Antinomians, the Anabaptists, the Familists, "that filthie Sinne of the *Comunitie of Weomen*," as John Cotton defined the sect for Anne Hutchinson, "all promiscuus & filthie cominge togeather of men & Weomen. wthout Distinction or Relation of Marriage [all of which] will necessarily follow. . . ." Endicott, too, knew very well "where the foundation of all these troubles among us lies." He frankly believed, as Hawthorne had him say, that the female sex required the stricter discipline and supported Paul's injunction that woman not pray "with her head uncovered," which he interpreted to mean that she had to wear not only a bonnet but also a veil to meetings.*

Her reproductive flesh being more accessible, woman was the

* John Cotton successfully opposed the veil requirement. It does not follow, however, that Endicott's patriarchism, though characteristically forceful and forthright, was aberrant. Winthrop berated Anne Hutchinson for transgressing the Fifth Commandment (to honor parents) by casting reproach on "the Fathers of the Commonwealth." When she contested his assertion, he expressed reluctance "to discourse with those of your sex." In her church "trial" John Cotton admonished her sisters in the congregation not to be misled by her, "for you see she is but a Woman." In 1637 the Cambridge Synod actually allowed that women might meet, "some few together," but condemned larger meetings of women as "disorderly, and without rule."

devil's gateway to sin. Endicott had occasion to reflect on this commonplace of the times, for he had left a bastard son behind in England. The product of "an amorous episode in earlier years," in the words of biographer Mayo, the first John Endicott, Jr., was put out to a collier and provision made to apprentice the "poore boy" in "some good trade." Though Endicott provided for his support, under no circumstances did he want the boy with him as a standing reminder of a lustful past he was "myselfe ashamed to write of." As he wrote his agent, "onely I would not by any meanes have the boy sent over." He made no mention, of course, of the person who had given birth to this shame, the boy's mother.

4

Eternal vigilance against the body so as not to fall into former lusts required inordinate energy and was not always possible. "Marvelous it may be to see and consider how some kind of wickedness did grow and break forth here, in a land where the same was so much witnessed against and so narrowly looked unto, and severely punished when it was known," Bradford puzzled. He listed, among the "sundry and notorious sins" that had become surprisingly common, drunkenness and uncleanness, incontinence between persons unmarried and even married, and, worse, "sodomy and buggery (things fearful to name)." Perhaps the devil was moved to greater spite by the greater holiness of the New England churches; perhaps the close examination of church members simply exposed sins that would have remained hidden elsewhere; and, closer to the marrow of the matter:

> Another reason may be, that it may be in this case as it is with waters when their streams are stopped or dammed up. When they get passage they flow with more violence and make more noise and disturbance than when they are suffered to run quietly in their own channels; so wickedness being here more stopped by strict laws; and the same more narrowly looked unto so as it cannot run in a common road of liberty as it would and is inclined, it searches everywhere and at last breaks out where it gets vent.

Flash floods of backed-up life left behind absurd deposits of the sodomitic Saints: in this remarkable passage Bradford directly foreshadowed Freud, even down to the dam simile, on the discontents of repressive "civilization." What Bradford could not see, understandably, was that the pent-up natural impulses could find opening not only in forbidden venery but also, alternatively, in Puritan virtue.

Endicott's outbursts against the unchosen were partial returns of this suppressed sexuality. In the role of judge and executioner he gave socially sanctioned outlet to his hatred and fear of carnality, struck at his own by branding and whipping it in others, demonstrated all the while his own innocency, preserved the purity of church and state, and therewith extended their sway. He could measure his spirituality by the number of carnal men and women stretched out prostrate behind him. Unlike the poor buggers who channeled their frustrations into sex crimes, Endicott gained the acclaim of all right-minded people.

5

Specific biographical details aside, the exemplary religious life of the seventeenth century followed a set pattern. Wild youth ran into the misgivings of early maturity and a rising awareness of the tasks to which God put the soul of man. Yet, since "the flesh would not give up her interest" but shook off "this yoake of the law," "Secrett Corruptions" led to intermissions. The real conversion came toward age thirty, and then only after the man had been laid low by a realization of the emptiness of his spiritual pretensions. He then had revealed unto him the Lord's free mercy in Christ and felt his soul filled "with joy unspeakable." Yet this estate too declined by degrees "as worldly employments and the Love of Temporall things did steal my heart. . . ." Thus down through the years to the present, 1637, "I have gone under continuall conflicts between the flesh and the spirit, and sometimes with Satan himself (which I have more discerned of late then I did formerly) many falls I have had," confessed John Winthrop, whose illuminating "Relation of His Religious Experience" I have been quoting and paraphrasing. So Winthrop summed up his spiritual odyssey on his forty-ninth birthday; but so might his age-mate John Endicott, and so might many other Saints: "Oh, I lived in and loved darkness and hated light," said Oliver Cromwell of his unregenerate life as another country gentleman. "I was a chief, the chief of sinners."

Continual conflicts between the flesh and the spirit were the trademarks of Puritanism on both sides of the Atlantic. The Saints, as Christopher Hill and others have established, came generally from the industrious middle classes of town and country. Country gentry such as Winthrop and Cromwell, soldiers such as Endicott newly returned from the Lowlands, ministers such as Cotton, merchants such as Cradock—these and other "middling men" fought to control their own lusts within while resentfully witnessing what they took to be the triumphant advance of the flesh without, at both ends of the social scale. It flaunted itself in court debauchery, of course, and in the voluptuous theater and the sensuous arts. It taunted from bypath and lane in the voices of "masterless" men turned out upon the countryside by enclosure. It spread with the rogues, beggars, unemployed artisans, peddlers, strolling players, and vagabonds, all of whom increased alarmingly in number with the breakup of traditional society and the first measures of death control, piled up in London and other cities, and menaced from the slums as that many-headed monster, the mob. It even lurked in the fens and highlands and in such forests as Dean, Arden, and Sherwood, where lived analogues of New World "savages"—woodspeople said by one sober witness to be "as ignorant of God or of any civil course of life as the very savages amongst the infidels."

Diligence, thrift, sobriety, and prudence were the sharpest weapons in the Puritan arsenal: "The most urgent task," wrote Max Weber, was "the destruction of spontaneous, impulsive enjoyment, [and] the most important means was to bring order into the conduct of its adherents." In his classic *Protestant Ethic and the Spirit of Capitalism* (1904–5) Weber noted not only the fundamental antagonism to sensuality but also Puritan fascination with the idea of total control: "The Calvinist was fascinated by the idea that God in creating the world, including the order of society, must have willed things to be objectively purposeful as a means of adding to His glory; not the flesh for its own sake, but the organization of the things of the flesh under His will. The active energies of the elect, liberated by the doctrine of predestination, thus flowed into the struggle to rationalize the world." These energies did in fact flow into efforts to drain the fens of water and their lewdness, to deforest Dean and other shelters for those without the law, and shortly to raise the ax against hiding places of New World infidels. They flowed also into the Royal Society, a large majority of whose original members were Puritans, and into experimental science generally, attracted by the correspondence of values and perhaps above all pushed along, as Robert K. Merton concluded, by "the active ascetic drive which necessitated the study of Nature that it might be controlled." Puritans were not, therefore, merely responding to current footlooseness. That they abhorred, admittedly; but they were driven against it by a goal that shot so far ahead of their times as to make them almost modern: They sought nothing less than to master the masterless "natural man" and, for good measure, the rest of nature. Their total victory would have meant the world defleshed, its bones picked clean.

Fortunately the Protestant ethic never got quite that far, though it did take the seventeenth century a long stride in our direction and over the intervening centuries did shape, or misshape, lives and our notions of what is possible and desirable. At the time, however, there were other, more libertarian responses to the experience of masterlessness, such as that of the "vulgar royalist libertine" discussed in the preceding chapter and such as those of the men and women Christopher Hill lovingly discussed in *The World Turned Upside Down: Radical Ideas during the English Revolution* (1972).

"What produced alarm and anxiety in some was an opportunity for others," Hill pointed out, "though not an opportunity for climbing up the normal social ladder. A masterless man was nobody's servant: this could mean freedom for those who prized independence more than security." Familism, so hated by Cotton and Winthrop, was a case in point. It held that "Christ was in every believer" and referred to its adherents as members of the Family of Love who held property in common and thought all things come by and from nature, including bodily gratifications; as an underground tradition it had been spread by wayfaring traders since Elizabeth's reign. It was accused of begetting others: Antinomians, Seekers, Levellers, Diggers, Ranters, and more quietly, Quakers. Ranters shocked George Fox, the founder of the Society of Friends, with their boisterous scenes of singing, whistling, and dancing; the name they accepted for themselves was "My one flesh," a designation that took them to the opposite pole from the Saints. Their pantheism in fact had close parallels with Native American reverence for our Mother Earth. They sought to keep Dionysus from being banished, celebrated the body, and warned against the power or bondage so dear to the Puritans. In 1649 Abiezer Coppe, a truly extraordinary man, formulated the Ranter goal very nearly in Morton's words: men should go quickly to "spiritual Canaan (the living Lord), which is a land of large liberty, the house of happiness, where, like the Lord's lily, they toil not but grow in the land flowing with sweet wine, milk and honey . . . without money." As we know, this threatened countercultural revolution never happened: the land of large liberty was neither Puritan nor Restoration England. Cromwell lumped together milder men than Coppe as a "despicable and contemptible generation," as "persons differing little from beasts."

Yet Cromwell and his associates never freed themselves completely of counterculture pockets among "the people poor and seditious," not even after the suppression of the Levellers in 1649 and the scattering of the Diggers the following year. Nor did they ever work themselves clear of the ongoing conflict with traditional aristocratic culture, not even at the height of the revolution. And therein lies the fact of critical importance for understanding the initial basic difference between the Puritans who stayed behind and those who withdrew to fight the war against the flesh on the other side of the Atlantic.

The Great Migration of Puritans sailed in the conviction that the fight for purity had already failed in England. In his canvass of reasons for emigrating, John Winthrop recorded that the land itself seemed to have grown "wearye of her inhabitants"; proof was to be seen in the throngs of sturdy beggars in his native Suffolk. The Massachusetts Bay offered economic opportunities and land acquisition. Then too there were always the Indians to convert. But a careful reading of the current situation in Britain led him to the most pressing reason: "All other Churches of Europe are brought to desolation and it cannot be, but the like Judgment is comminge upon us." In fine, a relatively safe shelter for purity lay off their bow; there were no safe havens astern. Once across the ocean the emigrants no longer had to maintain their version of the Protestant ethic against other versions and other ideologies. By movement through space, this transplanted Puritanism had become the one true faith, and with no one—or no one recognizable—either above or below them, the emigrant middling men had become the only true men.

To their minds they were the only carriers of the last best hope of the Protestant Reformation. Hence, while the English Puritan version of the Protestant ethic was harshly intolerant, the New English variation was more so. The tempers of Winthrop and Endicott were more hair-trigger sensitive than Cromwell's when someone threatened to defile their holy mission. Abroad, to be sure, Cromwell hunted down the Irish; but he still fell short of the unyielding fury the Saints turned on the New England specimens of those "differing little from beasts," those "savages" who had both a different culture and a different color.*

* In the 1640s and 1650s countermigration blurred this distinction but perhaps underlined its validity. George Cooke, for instance, returned from Massachusetts to take command of a regiment in Ireland, became governor of Wexford, and put his experience with Indians to use fighting guerrilla bands. Two repatriates discussed below were also part of this sizable exodus.

6

Precisely because he lacked Winthrop's introspective power, facility with language, and inner complexity, John Endicott more directly symbolized the New England Saints. His rough, soldierly exterior stood for their will to harness all human energies to the chariot of their church and state or, better, could more transparently express their drive to bring the flesh with "her interest" under the "yoake of the law." In this sense he was Hawthorne's Puritan of Puritans, a pure embodiment of their lust for total power; in this sense he was a representative or pattern Puritan, clearly cast in the same mold as all the other "Johns"—Winthrop, Cotton, and as we shall see, Oldham, Gallop, Underhill, and Mason; even Stoughton, though the last's Christian name was "Israel."

Of course their eternal opposite was Thomas Morton. After Morton's return from his second deportation, Governor Endicott issued a warrant for his arrest, as we have noted, and wrote Winthrop for "speedy advice" on what else to do about the sinister movements of their old enemy: "It is most likely that Jesuits or some that way disposed have sent him over to do us mischief, to raise up our enemies round about us both English and Indian." But Morton was merely the most visible link between subversive forces within the colony and savages lurking on its borders. There were others. Endicott waged war on the lot.

In the arresting words of biographer Mayo, Endicott had an interest "in the welfare of the better type of American Indian." Unfortunately he had no opportunity to show more of this side of his character, for that sort was evidently in short supply, short-lived, or both. In 1629 Matthew Cradock, head of the Massachusetts Bay Company in England, had instructed him to bring Indians to the Gospel "but to be cautious and distrustful." Endicott never worked out how to proselytize distrustfully and in any event was more at home dealing with avowed enemies. In 1636 he was

---

Israel Stoughton, who returned in 1643, became a lieutenant colonel in the regiment of Colonel Rainborow. But the most famous fighting chaplain in the Roundhead forces was the Reverend Hugh Peter, who became a fiery exhorter of troops before battle and a recruiter without peer. He too commanded a regiment in Ireland, and his valor there, according to a contemporary, "General Cromwell himself so highly extols as to reckon this one preacher worth a hundred soldiers: always the first in attacking a rampart, he is followed by the rest so punctually that already he has taken several towns in Ireland by his sheer alacrity"—quoted by William L. Sachse, "The Migration of New Englanders to England, 1640–1660," *American Historical Review*, LIII (January 1948), 271.

the natural choice to command a punitive expedition against the Indians. He had reportedly seen service in the Netherlands against the Spanish, still bore the title of "captain," and was the last man to question his explicitly genocidal orders. As Winthrop summarized them in his *Journal*, Endicott was

> to put to death the men of Block Island, but to spare the women and children, and to bring them away, and to take possession of the island; and from thence to go to the Pequots to demand the murderers of Capt. Stone and other English, and one thousand fathom of wompom for damages, etc., and some of their children as hostages. . . .

CHAPTER IV

The Pequot War

. . . their quarrell being as antient as Adams time [was] propagated from that old enmity betweene the Seede of the Woman, and the Seed of the Serpent, who was the grand signor of this war in hand.

—EDWARD JOHNSON,
Wonder-Working Providence, 1653

TO ALL appearances it began innocently enough with a first victim. On his return from a voyage to the mouth of the Connecticut, John Gallop was skirting Block Island when he spied a pinnace belonging to John Oldham, another trader and an old planter. Made apprehensive by the cluster of Indians on her deck, Gallop fired on them with small arms and rammed his twenty-ton bark into the quarter of the lighter craft. Six Indians then leaped into the sea and drowned. Another ramming induced almost as many to plunge to their death. With most of the natives over the side, Gallop and his crew, consisting of one man and two boys, boarded the stricken pinnace, "whereupon," John Winthrop noted in his *Journal*, "one Indian came up and yielded; him they bound and put into [the] hold. Then another yielded, whom they bound. But John Gallop, being well acquainted with their skill to untie themselves, if two of them be together, and having no place to keep them asunder, he threw him bound into [the] sea." Under an old seine he found John Oldham's body, head cleft to the brains, legs slashed almost off. "The blood of the innocent," John Underhill declared, "called for vengeance."

In late August 1636 vengeance sailed from Boston in the person of Captain John Endicott, who commanded three pinnaces containing some four score men, two shallops, and two Indians. The fleet made Block Island at dusk a week later, landed through heavy surf, and easily dispersed forty or so Indians who sought to contest the invasion—their arrows bounced harmlessly off English armor that consisted of back, belly, thigh, and headpieces and of gorgets. Posting sentinels, the landing party bivouacked for the night, prepared to carry the war to the enemy in the morning.

The sole surviving eyewitness account of the first phase of the campaign that followed was John Underhill's invaluable *Newes from America . . . Containing a True Relation of Their War-like Proceedings These Two Years Last Past*, published in London in 1638. One of Endicott's captains, Underhill had also been a professional soldier in the Lowlands before his conversion to Puritanism, came to the Bay early and was one of the first deputies from Boston to the General Court, imprudently took Anne Hutchinson's side in the Antinomian controversy (and was to suffer for that later), and was withal a promoter of New England lands and opportunities. Indeed, Underhill's eccentric feeling for the wilderness made him resemble Morton a little, though he did not of course share the latter's friendly relations with its native inhabitants.

For two days Endicott's men hunted natives to kill or capture. The only reason they did not kill every man on Block Island, Underhill made clear, was because no one would stand up and fight, "the Indians being retired into swamps, so as we could not find them. We burnt and spoiled both houses and corn in great abundance; but they kept themselves in obscurity." The first day the invaders spent "in burning and spoiling the island," and so the next. In all they burned the wigwams of two villages, threw Indian mats on and burned "great heaps of pleasant corn ready shelled," "destroyed some of their dogs instead of men," and staved in canoes. But with "the Indians playing least in sight," the attackers could claim a body count of no more than fourteen—the Narragansetts later related they had killed only one man.

From Block Island they went down to Saybrook Fort at the mouth of the Connecticut and thence to the mouth of the Pequot, "the Pequeats having slain one Captain Norton, and Captain Stone, with seven more of their company." As they sailed along the shore, Indians came down to the water's edge, crying, "What cheer, Englishmen, what cheer, what do you come for?" Hoping to "have the more advantage of them," Endicott's men answered them not, so the Indians persisted: "What, Englishmen, what cheer, what cheer, are you hoggery, will you cram us? That is, are you angry, will you kill us, and do you come to fight?" Windbound at Saybrook, the expedition lay over four days before being able to make its way back up the coast to the mouth of the Pequot.

Once Endicott's party had cast anchor, the Pequots sent on board "a grave senior" to demand "of us what the end of our coming was." Endicott told him he came for the heads of the murderers of Captains Stone and Norton. The elder statesman explained that their chief sachem, Sassacus, was away, denied any Pequots had knowingly slain any Englishmen, and asked them to stay on their vessels while he conferred with his tribesmen. Endicott refused, ordered his men ashore, threatened to "beat up the drum, and march through the country, and spoil your corn," and announced the particulars of his commission, including the demands for wampum and hostage children. In return "these devil's instruments," as Underhill called them, offered to parley unarmed: They would leave their bows behind if the English soldiers would lay down their muskets. Endicott rather chose to "bid them battle. Marching into a champaign field we displayed our colors; but none would come near us, but standing remotely off did laugh at us for our patience." Angered by this further show of insolence, Endicott and his men fired on any Indians they could come near, burned their wigwams, spoiled their corn, "and many other necesaries that they had buried in the ground we raked up, which the soldiers had for booty."

"Thus we spent the day," Underhill again recorded, "burning and spoiling the country." That night they sailed for Narragansett Bay, where in the morning "we were served in like nature, no Indians would come near us, but run from us, as the deer from the dogs." So there too they "burnt and spoiled what we could light on" before setting sail for Boston. And so began what has been called the Pequot War.

But Captain John Mason, hero of the events that followed, put it well in his own *Brief History of the Pequot War:* "If the Beginning be but obscure, and the Ground uncertain, its Continuance can hardly perswade to purchase belief: Or if Truth be wanting in History, it proves but a fruitless Discourse." The truth is that the beginning of the Pequot War was both less innocent and more obscure than it appeared, and Mason unfortunately went on to perpetuate its uncertainties. Reconsider the first victim, John Oldham: Would not the dozen or so Indians John Gallop made pay with their lives satisfy any normal appetite for vengeance? Was Oldham all that innocent? He was in fact a grasping, contentious man who picked quarrels with everybody. Banished from Plymouth, Oldham was angry, according to Bradford, and "more like a furious beast than a man." Even Morton, whom Oldham had conducted to England in 1628, you will recall, found him passionate and moody, "a mad Jack in his mood." Oldham was more than capable of provoking the quarrel that led to his death. At an absolute minimum an inquiry into the particulars of the case might have allocated responsibility for his death and pointed the way toward the real criminals, dead or alive. Why then undertake to kill every man on Block Island? Above all, why undertake it if the natives there were not Pequots but Eastern Niantics, allies of the Narragansetts?

Of course rulers eager to make war can make do with almost any first victim, so long as his death will infect everyone with the feeling of being threatened and provide basis for belief that "the enemy," broadly defined, is responsible. If this minimal foundation be laid, every other reason for his death may be ignored or suppressed, as Elias Canetti observed, save one, the victim's "membership of the group to which one belongs oneself"—in the present instance a functional racism determined this group membership. But if every war requires a John Oldham, the Puritans overdid it in the Pequot War by having at least two earlier known first victims.

In 1633 Captain John Stone had come north from Virginia with a cargo of cattle. In Boston he drank a good deal, according to Winthrop's *Journal*, was found upon the bed in the night with one Barcroft's wife, and after using "braving and threatening speeches" against one of the assistants, was clapped in irons, fined a hundred pounds, and "ordered upon pain of death to come here no more." In Plymouth he attempted to stab Governor Prence with his dagger. On the Connecticut River he and Captain John Norton, who had sailed with him from Plymouth, quarreled with Indians and were killed aboard ship. Ill-cast always for the role of an innocent, Stone died under circumstances forever obscure—John Mason, for example, maintained that the Indians responsible "were not native Pequots, but had frequent recourse unto them." Whosoever responsibility, the incident did not become an issue until 1634, when the Pequots sought an alliance with the English and were told they would first have to give up the murderers of Stone and Norton. But the following year the possible use of their deaths became clearer with the arrival of John Winthrop, Jr., who had been commissioned by Lords Say and Brook to plant a new colony on the Connecticut and appointed to serve as its governor.

The narrative of Lion Gardiner demonstrated beautifully the commonplace that authorities are not infrequently better known by what they do than what they say. Gardiner had served the Prince of Orange as a professional army engineer before he was hired and brought over to take charge of the construction of Saybrook Fort. On his arrival in November 1635 he remonstrated with Winthrop the younger, George Fenwick, Hugh Peter, and other employers over their willingness to "make war for a Virginian [Stone] and expose us to the Indians, whose mercies are cruelties."

He entreated his associates to rest awhile until
Hunger" had been met, "till we get more st
that we hear where the seat of war will be,
provide for it . . . Mr. Winthrop, Mr. Fenwi
promised me that they would do their utmost en
suade the Bay-men to desist from war a year or two, t
be better provided for it." That time had not arrived t
summer, in Gardiner's estimation; the sudden descent of Endi
and his men on Saybrook was to his "great grief, for, said I, you come hither to raise these wasps about my ears, and then you will take wing and flee away." The Endicott expedition came against his will and left against his will, after having killed one Pequot, "and thus began the war between the Indians and us in these parts." And thus were all the improbable first victims shown to be largely incidental to the expansionist plans of the colonists.

2

Naturally the Pequots had their own victims, and a basic lack of concern for the tribe as persons has made the history of their time the "fruitless Discourse" Mason warned against. They were an Algonquian-speaking division of the Mohegans, who in turn have traditionally been considered a branch of the Hudson River Mahicans—the name of the last two tribes meant "wolves," and both were sometimes called Mohicans.

The European invasion of the Atlantic seaboard in the early 1600s led to a series of conquest wars among coastal tribes such as the Pequots, as whites claimed lands and pushed their trade up the river valleys. The Pequots established effective control of the Connecticut, of the coast from Saybrook (or Niantic) to Pawcatuck, and off shore of Eastern Long Island. But by 1632 they could no longer keep whites down river and were forced to yield a trading site to the Dutch at what is present-day Hartford. The following year, over the objections of both the Pequots and the Dutch, who had signed a treaty for the first trading post, the Plymouth planters established a post at present-day Windsor. This intrusion stirred up Indian recollections of old wrongs, as when Thomas Hunt, erstwhile associate of Captain John Smith, had seized twenty-seven natives under pretence of trading, pushed them into the hold, and in 1614 had carried them to Spain where a number of them wound up as slaves. Sassacus, who emerged as their chief sachem, was further enraged in 1633 when the Dutch killed his

Wopigwooit and several of his warriors. This context per-
helps explain the deaths of Stone and Norton, for the Pequots
that they did not distinguish between Dutch and English and
t Stone had attempted to abduct two braves. Nothing in any
this or elsewhere, to my knowledge, supports the colonists' contention that the Pequots were especially "treacherous and perfidious" or made them, as the Reverend William Hubbard said, "the Dregs and Lees of the Earth, and Dross of Mankind."

The Pequots were "insolent" only if it be insolent to resist subjugation and dispossession. They rightly saw the alleged murders as pretexts to those ends. At this point, after the unprovoked attack on them for Oldham's death, Sassacus and his council made a daring move to strike an alliance with their old rivals and enemies, the Narragansetts. As Bradford summarized their proposals, the Pequot ambassadors pointed out to the Narragansetts that the English strangers were overspreading the country and would in time deprive them of their lands; if they helped the English subdue the Pequots, "they did but make way for their own overthrow, for if they [the Pequots] were rooted out, the English would soon take occasion to subjugate them." But if the Narragansetts would hearken to the Pequots, they need not fear English firepower, for the Indians

> would not come to open battle with them but fire their houses, kill their cattle, and lie in ambush for them as they went abroad upon their occasions; and all this they might easily do without any or little danger to themselves. The which course being held, they well saw the English could not long subsist but they would either be starved or be forced to forsake the country.

This sophisticated plan for guerrilla warfare more than half persuaded the Narragansetts. Winthrop made a *Journal* entry that "Mr. Williams wrote, that the Pequods and Naragansetts were at truce, and that Miantunnomoh told him, that the Pequods had labored to persuade them, that the English were minded to destroy all Indians."

Historians have since marveled at the magnanimity of Roger Williams, symbol of the American liberal tradition. Overlooking his recent banishment, he not only kept Winthrop informed, as above, but sought to protect all the Massachusetts colonists from the dangers inherent in Indian combination. He energetically opposed the Pequot plan, and at very considerable personal risk, as he wrote John Mason decades later when they were both old men:

> When, the next year after my banishment, the Lord drew the bow of the Pequot war against my country . . . the Lord helped me immediately to put my life in my hand, and scarce acquainting my wife, to ship myself all alone in a poor canoe, and to cut through a stormy wind with great seas, every minute in hazard of my life, to the sachem's house. Three days and nights my business forced me to lodge and mix with the bloody Pequot ambassadors, whose hands and arms reeked with the blood of my countrymen, murdered and massacred by them on [the] Connecticut River, and from whom I could not but nightly look for their bloody knives at my own throat also.

Ultimately his friends the Narragansetts, who had given him shelter and granted him land, listened to his counsel against the confederation. In fact, Williams acted as a one-man Office of Strategic Services, watching out for "any perfidious dealing," warning Winthrop that the Pequots had heard of his preparations for war, proposing a Narragansett hit-and-run raid that would "much enrage the Pequts for euer against them, a thing much desirable," and sending a sketch of the Pequot forts and a Narragansett battle plan against them, including night assault—a plan that was later followed in all its essentials save one: "That it would be pleasing to all natives, that women and children be spared, etc."

On their part the Pequots, exasperated by the Endicott expedition and forced to go it alone, effectively harried Gardiner and his men at Saybrook and launched guerrilla raids that killed some thirty persons in the up-river plantations. In response, as their substitute for the detested Maypoles and other festivities celebrated elsewhere on May Day (O.S.) 1637—since regarded as "the proper birthday of Connecticut"—the General Court met at Hartford, declared offensive war against the Pequots, levied ninety men, and appointed Captain John Mason their commander.

3

Mason thus took up where Endicott left off. Also a veteran of service in the Dutch Netherlands, Mason had first migrated to Dorchester before moving from Massachusetts to Windsor, "to lay the Foundation of Connecticut Colony." A century later his editor, the Reverend Thomas Prince, held him to be of the same mettle as Captain Miles Standish, "who spread a Terror over all the Tribes of Indians round about him," except that Mason became "the equal Dread of the more numerous Nations from Narragansett to Hudson's River. They were Both the Instrumental Saviours of this Country in the most critical Conjunctures." Though a plain military man of action, Mason spoke the same language in his important narrative of the war. He shared the conviction of Endicott and other Saints that they were the proper recipients of God's special providences: "the Lord was as it were pleased to say unto us, The Land of Canaan will I give unto thee though but few and Strangers in it." He also remembered the rather more direct prophecy of the Reverend Thomas Hooker as they set out to fight Pequots, "that they should be Bread for us."

So they could eat their fill, Mason shunned Endicott's tactics of open challenges and awkward lunges at the natives, only to have them "fly away and hide in their swamps and thickets." Instead, after a skirmish established his presence on their western borders, Mason picked up Underhill and some Massachusetts militia at Saybrook and with a force of nearly eighty men sailed up the coast as though he were afraid to challenge Pequot power by landing from the sea. At Narragansett Bay he disembarked and marched his men back overland from the east to come "upon their backs." The surprise was complete. The Pequots were delighted that danger had sailed by them, and those in their fort near the Mystic River had just held a festival of gladness and were deep in sleep when the English attacked on Friday, June 5 (May 26, O.S.), 1637.

Just before dawn Mason stormed one entrance of the fort and Underhill the other. The sounds themselves must have been terrifying, with Pequot shouts of alarm, "Owanux! Owanux!" ("Englishmen! Englishmen!"), mixed with war whoops, screams of women and children, musket shots, barked orders. Warriors within the wigwams pelted the English with their arrows so effectively they made "the fort too hot for us," Underhill admitted: "Most courageously these Pequeats behaved themselves." Mason, reaching the same conclusion, declared to his men, "We must Burn them." Immediately stepping into a wigwam he "brought out a firebrand, and putting it into the Matts with which they were covered, set the Wigwams on Fire." Underhill from his side started a fire with powder, "both meeting in the centre of the fort, blazed most terribly, and burnt all in the space of half an hour." The stench of frying flesh, the flames, and the heat drove the English outside the walls: Many of the Pequots "were burnt in the fort, both men, women, and children. Others [who were] forced out . . . our soldiers received and entertained with the point of the sword. Down fell men, women, and children," Underhill observed, and all but a half-dozen of those who escaped the English fell into the hands of their Narragansett and Mohegan allies in the rear.

A short, quick step had thus taken the English from burning and spoiling the country to burning and spoiling some four hundred persons in little over an hour. Mason rejoiced that God had "laughed his Enemies and the Enemies of his People to Scorn, making them as a fiery Oven: Thus were the Stout Hearted spoiled, having slept their last Sleep, and none of their Men could find their Hands: Thus did the Lord Judge among the Heathen, filling the Place with dead Bodies!" Underhill granted that the "most doleful cry" from within the fort might have moved the English to commiseration, "if God had not fitted the hearts of men for the service. . . . But every man being bereaved of pity, fell upon the work without compassion, considering the blood they had shed of our native countrymen, and how barbarously they had dealt with them. . . ." When he was later asked, "Why should you be so furious?" he referred questioners to the Scriptures and their justification for killing women and children: "When a people is grown to such a height of blood, and sin against God and man, and all the confederates in the action, there he hath no respect to persons, but harrows them, and saws them, and puts them to the sword, and the most terriblest death that may be." More than a bad pun then was his complacent assertion: "We had sufficient light from the word of God for our proceedings."

"Our Indians," as Underhill called them, came up afterward "and much rejoiced at our victories, and greatly, admired the manner of Englishmen's fight, but cried Mach it, mach it; that is, It is naught, it is naught, because it is too furious, and slays too many men." Watching these Narragansett and Mohegan allies as they engaged the Pequots, shooting their arrows compass or at an elevation and then waiting to see if they came down on their targets, Underhill had already reached the professional conclusion that Indians might fight seven years and not kill seven men: Their "fight is more for pastime, than to conquer and subdue enemies." Mason concurred that their "feeble Manner . . . did hardly deserve the Name of Fighting." * Though the Pequots were the first

* Indeed Indian violence fell short of the systematic ferocity of Europeans. Underhill's and Mason's authoritative judgments that it was hardly more than a pastime have been ignored by those who have stressed, down through the centuries, Indian blood lust. Perhaps I should make clear that in these pages I have *not* suggested that the woodland Indians were pacifistically inclined, incapable of atrocities, as perhaps in the case of John Oldham (see p. 35), or lacking sadistic impulses of their own—after all, they too were human. But the land was theirs, in the beginning, and their violence must be placed within its cultural context and within the context of the European invasion. Prior to that they were not engaged in insane, unending war —as Francis Jennings has emphasized in *The Invasion of America: Indians, Colonialism, and the Cant of Conquest* (Chapel Hill: University of North Carolina Press, 1975), pp. 146–70. He rightly

of the New England tribes to sense the genocidal intentions of the English and the implications of their different style of battle, the fury of the attack on their fort still was demoralizing. Almost certainly they did not anticipate the relentless pursuit that followed.

In late June, well after the slaughter at the fort, Captain Israel Stoughton arrived at the Pequot River in command of one hundred and twenty Massachusetts militiamen. Later to become a colonel in Cromwell's army (and the father of Governor William Stoughton, of Salem witchcraft-trial fame), Stoughton was then a principal man in the Bay Colony and like Endicott an assistant. He came from Dorchester, whose inhabitants had early complained of "lack of room" and many of whom like Mason had moved overland to "hive" at Windsor in 1635. His own very concrete interest in Pequot lands had already been made explicit in a letter (ca. June 1636) to Winthrop the younger: Stoughton asked the new governor of Connecticut for land on which to build a warehouse to store the goods of his former neighbors from Dorchester and then closed with "a Motion to you in particular for my selfe, for some small portion, resoluing if you would shew me that fauor to count my selfe no small debtor to you for euer."

Shortly after his arrival Stoughton was guided by Narragansetts to a swamp twelve miles from the mouth of the Pequot, where about a hundred refugees had taken cover. When they were surrounded, the Pequots surrendered without a fight. Stoughton turned the men over to Skipper John Gallop, whom we first met off Block Island during the Oldham naval action. No doubt remembering his earlier exploit and enlarging on it, Gallop took twenty or so captives a little beyond the harbor and threw them bound into the sea or, as a Puritan historian exulted, fed "the fishes with 'em." Stoughton sent the women and children to the Bay and informed Winthrop the shipment was on its way. In his letter, dated late June 1637 and also reproduced in the *Winthrop Papers*, Stoughton revealed intriguing esthetic preferences in his personal request for one "that is the fairest and largest that I saw amongst them to whome I haue giuen a coate to cloath her: It is my desire to haue her for a servant if it may stand with your good likeing: ells not. There is a little Squa Stewart Calacot desireth. . . ."

---

noted that Indians were more peaceful internally than European nations and no more ferocious externally—his evidence suggested that the Pequots, for instance, were startled to learn their Puritan enemies considered women and children fair game. And "That all war is cruel, horrible, and socially insane is easy to demonstrate, but the nationalist dwells upon destiny, glory, crusades, and other such claptrap to pretend that his own kind of war is different from and better than the horrors perpetrated by savages. This is plainly false. The qualities of ferocity and atrocity are massively visible in the practices of European and American powers all over the world" (p. 170).

Stoughton was then joined by Mason with forty Connecticut troops and together they set off in pursuit of the starving families of Pequots, who, as Mason put it, could make "but little haste, by reason of their Children, and want of Provision: being forced to dig for Clams, and to procure such other things as the Wilderness afforded." The chase stretched out over sixty miles to Quinnipiac (present-day New Haven) and beyond to three miles westward of Fairfield, where three hundred of the quarry were literally run to ground. Many of those killed were tramped into the mud or buried in swamp mire. "Hard by a most hideous swamp, so thick with bushes and so quagmiry, as men could hardly crowd into it," Winthrop wrote, "they were all gotten."

On his return to the Bay, Israel Stoughton stopped off where Endicott had started, killed one or two, burned some wigwams, and forced the Block Island Niantics to become tributaries. Miantonomo, sachem of the Narragansetts, Winthrop also noted with pleasure, "acknowledged that all the Pequot country and Block Island were ours, and promised he would not meddle with them but by our leave."

CHAPTER V

The Legacy of the Pequot War

The devil would never cease to disturb our peace, and to raise up instruments, one after another.

—JOHN WINTHROP,
Journal, December 1638

THUS WAS God pleased to smite our enemies, said John Mason, "and to give us their Land for an Inheritance." Thus was God a mercantilist, he could have said, for on the economic level that is exactly what the Pequot War was about: the acquisition of Block Island and Connecticut. The Niantics and the Pequots were so insolent as to possess land the English wanted, and the story of their dispossession was repeated, with significant variations, up and down the Atlantic seaboard. Earlier the 1622 attack by Opechancanough and the Powhatan Confederacy had led whites on the Chesapeake Bay to establish an open hunting season on Indians, a precedent well-known in New England—indeed the attitudes of John Stone and John Oldham toward natives were probably influenced by this climate of extermination to the south, for Stone shipped out of Virginia and Oldham spent time there. Later John Underhill put his experience to good use for the Dutch, whose mercenary he became, in their wars with the Indians located just beyond what had been Pequot country.* In these and other instances the Indians were on the receiving end of European imperialism, their lands furthering the objectives of mercantilism and eventually of market capitalism. In this sense the Pequots were early victims of a process Marxists would later call "primitive accumulation."

Instruments of this larger historical process, the New England settlers also had their own very concrete individual interests. Stoughton sought land and a fair captive. If Mason sought fame

* With his company of Dutch troops Underhill surrounded an Indian village outside Stamford, set fire to the wigwams, drove back in with saber thrusts and shots those who sought to escape, and in all burned and shot five hundred with relative ease, allowing only about eight to escape—statistics comparable to those from the Pequot fort.

and fortune, he got them in full measure: he was made a major general of all the Connecticut forces, the staff of his post being handed to him by the Reverend Hooker, who, as Thomas Prince observed, "like an ancient Prophet addressing himself to the Military Officer," delivered "to him the Principal Ensign of Martial Power, to Lead the Armies and Fight the Battles of the Lord and of his People." Mason enjoyed the psychic income accruing to an instrumental savior and tangible assets as well, including an island of five hundred acres at the mouth of the Mystic that he claimed by right of conquest. Gardiner had a bay and an island named after him. The name of the skipper of "*Charons* Ferryboat" was perpetuated by Gallop's Island in Boston Harbor. And in recognition of "his great services to this country," the Massachusetts General Court later fittingly granted John Endicott a quarter of Block Island.

At the time, Endicott passed on a more modest request for a share of the spoils. In July 1637 his pastor at Salem, Hugh Peter, wrote Winthrop:

> Mr. Endecot and my selfe salute you in the Lord Jesus etc. Wee haue heard of a diuidence of women and children in the bay and would bee glad of a share viz: a yong woman or girle and a boy if you thinke good: I wrote to you for some boyes for Bermudas, which I think is considerable.

Roger Williams had already put in his claim in a letter to Winthrop of June 30, 1637:

> It having againe pleased the most High to put into your hands another miserable droue of Adams degenerate seede, and our brethren by nature: I am bold (if I may not offend in it) to request the keeping and bringing vp of one of the Children. I haue fixed mine eye on this litle one with the red about his neck, but I will not be peremptory in my choice but will rest in your loving pleasure for him or any etc.

Unlike the Reverend Hugh Peter, Williams had no part in the mania for selling boy captives into slavery in the West Indies but was merely an early seeker of a war "orphan" on the calm assumption that he knew and could promote the boy's "good and the common [good], in him," as he put it in his letter of thanks to Winthrop (July 31, 1637). Unfortunately an analytical history of the forcible relocation of native children in white homes, from Williams's day to our own, remains to be written. At its worst, as in Endicott's case, it seems to have meant the collection of trophies of the chase—as though one hunted wolves and brought home a pup. At its best, as in Williams's case, it generously bestowed upon the native, even though he was thought of as one of "Adams degenerate seed," a chance to live like whites. But this takes us beyond the economic benefits of the war.

On the political level the Pequot War was about extending English rule and laws, and about pacification of the countryside. Block Island was opened up for settlement and Connecticut launched as another colony. In a very real sense the Saints made the Pequot War do for the Connecticut Valley what the plague of 1616–17 had done for the Massachusetts Bay: It removed natives from the premises. Another clear political gain also came by design, with "the Indians in all quarters so terrified," as Winthrop wrote Bradford, that former friends of the Pequots were afraid to give them refuge. The war established the credibility of the English will to exterminate, lessened the likelihood of "conspiracies" to resist their rule, and established a peace based on terror that lasted more or less for four decades until the outbreak of what was called King Philip's War.

2

But the Saints were more than an armed band on the edges of empire and nascent capitalism, more than unconscious or half-conscious agents of a process men would come to call "modernization." They were there as "this Army of Christ," as Edward Johnson called them. A mere enumeration of their economic and political motives will not put us within sight of others that were complementary but not necessarily less fundamental. These English settlers were also Calvinist Christians. And to answer the question put to Underhill, "Why should you be so furious?" we must first try to sort out what the Indians meant to him, to Endicott, and to their comrades in arms.

Certainly before they left England their prejudices had already commenced to harden. They knew of the Portuguese and Spanish experience with nonwhites. Though they could compliment themselves on their presumed moral superiority to the Spanish, whose cruelty had already become proverbial through the writings of Bartholomé de Las Casas, that did not mean they themselves were not wondering about the issues debated at Valladolid in 1550–51, namely, whether the Indians had rational souls and deserved the name of men, nor of course did it mean that as Protestants they had to subscribe to the bull "Sublimus Deus" of Pope Paul III: "the Indians are truly men." Plenty of English pamphlets and books denied just that. Thus before he left Leyden, Bradford knew that the vast countries of America were "unpeopled": there were "only savage and brutish men which range up and down, little otherwise than the wild beasts of the same." On the other hand, as Christians they shouldered the awful responsibility of bringing the clear sunshine of the Gospel to those "savage and brutish men." In 1623, after Captain Standish had slaughtered eight friendly Indians at Wessagusset, the colonists' failure to resolve this dilemma in practice led to the famous lament of their former pastor John Robinson: "Oh, how happy a thing it had been, if you had converted some before you had killed any!"

Genesis made clear that all men were the progeny of Adam. Indians might be "Adams degenerate seede," as Williams said, perhaps descendants of the wandering Ham, but they still were of Adamitic origin and as such "our brethren by nature." This was the light side of Indians, one that showed them to be men and, no matter how heathen, susceptible to conversion to the one true faith. On their dark side, however, they were barbarous, brutish savages, much otherwise than the men who "peopled" Europe, and as "devil's instruments," beyond reach of the Word. At all times the realities of dispossession and subjugation threatened, notwithstanding the theological stakes, to superimpose the Indians' dark side over their light. In times of crisis the Saints could hardly see beyond the shadows of the Indians' darkness of origin and being.

The Pequot War was the Puritans' first decisive answer to the question of whether the Indians were truly men. The nature of that answer and the release it provided from profound ambivalences of attitude were vividly illustrated by Philip Vincent's *A True Relation of the Late Battell Fought in New-England,* published in London in 1638. A veteran of the Antinomian controversy in Massachusetts, Vincent's distaste for heretical views probably had much to do with his quarrel with John Underhill, whom he went out of his way to accuse of cowardice in the assault on the Pequot fort. Vincent was a Cambridge-trained scholar and former rector of Stoke d'Abernon in Surrey. In addition he brought to his analysis of the war experiences in other colonies in the Western Hemisphere, including Guiana and probably the West Indies and Virginia.

Vincent started out with a view of the natives that was mandatory if he were to avoid the heresy of entertaining the possibility of their separate origin:

> Their outsides say they are men, their actions say they are reasonable. As the thing is, so it operateth. Their correspondency of disposition with us, argueth all to be of the same constitution, and the sons of Adam, and that we had the same matter, the same mould. Only art and grace have given us that perfection which yet they want, but may perhaps be as capable thereof as we.

Yet within five pages, in a contrast between the excessive "lenity of the English towards the Virginian salvages" before 1622 and the resolute killings at Mystic in 1637, Vincent gave the natives a fixed characteristic of which they were more capable than whites: "These barbarians, ever treacherous, abuse the goodness of those that condescend to their rudeness and imperfections." * And a half-dozen pages farther along the natives had become not so much brutish as quite simply *brutes*. The triumph over the Pequots had assured the New England colonists

> of their peace, by killing the barbarians, better than our English Virginians were by being killed by them. For having once terrified them, by severe execution of just revenge, they shall never hear of more harm from them, except, perhaps, the killing of a man or two at his work, upon advantage, which their sentinels and corps-du-guards may easily prevent. *Nay, they shall have those brutes their servants, their slaves, either willingly or of necessity, and docible enough, if not obsequious.* [my italics]

The fires at the Pequot fort had consumed Vincent's reluctant conclusion that, as sons of Adam, the Indians were men. If their outsides said so, appearances were often deceiving. The loud and clear answer of the war was that the Indians were truly animals that could be killed or enslaved at will.

Vincent could close his narrative serene in the confidence that colonists held the whip hand over natives and that as Englishmen they enjoyed a biological edge over other Europeans through their capacity "to beget and bring forth more children than any other nation in the world," a capacity that, combined with their honoring "of marriage, and careful preventing and punishing of furtive congression, giveth them and us no small hope of their future puissance and multitude of subjects." This close he came to a fully articulated racism but felt no need to go farther. Through the

* Vincent was a pioneer exponent of the "one bloody good lesson" view of teaching natives how to behave, a view that was still being articulated in nineteenth-century Australia by colonials engaged in killing aborigines—see C. D. Rowley, *The Destruction of Aboriginal Society* (Canberra: Australian National University Press, 1970).

simple expedient of denying the humanity of the Indians he sidestepped the issue of heresy and ignored the theological havoc his denial in fact left behind. The received views on the unity of mankind were presumably left untouched if Indians were not men but brutes.* Not for another four decades were the implications of this tactic confronted on the theoretical level. Then Sir William Petty, physician and member of the Royal Society, tried to establish *The Scale of Creatures* (1676–77) by assigning "savages" an intermediate or "midle" place between man proper and animals in the chain of being; but this was just another instance of theory catching up with practice.

Yet the practice had a theory of sorts all along, which is to say that racism as a phenomenon came into existence long before the nineteenth-century ideology, with which it is usually identified, was concocted to justify overseas expansion and exploitation. Even then, as a theory, it was always left "very bare and rudimentary," as Otare Mannoni observed, "and those who believe it most firmly are precisely those who would be the least capable of explaining it." Whatever its ultimate definition, *racism* as I use the term in these pages consists of more than theory: it consists in habitual practice by a people of treating, feeling, and viewing physically dissimilar peoples—identified as such by skin color and other shared hereditary characteristics—as less than persons.

White racism developed in precisely those terms in early seventeenth-century New England. It was embodied in the prohibition against selling guns to the Indians and in "the new creede" Morton said the Saints imposed on every newcomer, that "the Salvages are a dangerous people, subtill, secreat, and mischeivous." It engendered the reasonable Pequot fear "that the English were minded to destroy all Indians." It was embedded in Endicott's orders to kill every man on Block Island. It led Bradford to speak of North America as "unpeopled." In May 1637 it prompted

* In fact Vincent's "solution" of the theological problem had been anticipated by Saint Augustine in the *City of God* (413–27). Augustine guardedly responded to accounts of dwarfs, giants, and wild men by holding that "either these things which have been told of some races have no existence at all; or if they do exist, they are not human races; or if they are human, they are not descended from Adam"—quoted by Penelope B. R. Doob in *Nebuchadnezzar's Children: Conventions of Madness in Middle English Literature* (New Haven: Yale University Press, 1974), p. 136. Vincent could not deny the existence of Indians, but he could and did deny they were a human race: Though with human outsides, they had brute souls that allowed Christians to enslave them guiltlessly. European images and attitudes toward Indians owed much to ancient and medieval fantasies of the "Wild Man" and madness. In addition to Doob's first-rate study, see Richard Bernheimer's *Wild Men in the Middle Ages* (Cambridge: Harvard University Press, 1952); and Edward Dudley and Maximillian E. Novak, eds., *The Wild Man Within* (Pittsburgh: University of Pittsburgh Press, 1972).

C[aptain John] Smith Taketh . . . [a] Prisoner, 1608, 1624. This engraving suggests that the Reverend Philip Vincent tended to exaggerate the "lenity" of the first English on Chesapeake Bay. Assuredly after 1622 they exterminated all the Indians they could reach, or in Captain Smith's words, all the subanimals that "put on a more unnaturall brutishness than beasts." (Courtesy of the Library of Congress.)

Winthrop to remind Bradford and his associates at Plymouth that the Bay colonists "conceiue that you looke at the pequents, and *all other Indeans,* as a common enimie" [my italics].

Only Morton and Roger Williams seemingly did not share these racial attitudes. The proposition that all Indians were a common enemy was directly attacked in *A Key into the Language of America* (1643):

Boast not proud English of thy birth and blood,
Thy brother Indian is by birth as good.
Of one blood God made him, and thee, and all,
As wise, as fair, as strong, as personal.

Elsewhere in that remarkable work Williams discussed the Narragansetts with respect and drew favorable contrasts between them and the colonists. As he was fond of saying, "it was not price nor money that could have purchased Rhode Island. Rhode Island was purchased by love." Yet by act and word Williams illustrated a painful truth: A belief in the brotherhood of man could share psychic space, in the same man, with a belief in its negation, the nascent white racism of the day. In a crisis, as we have seen, Williams instantaneously took the part of his "countrymen" against "the Barbarous." Though he saw some sparks of true friendship in Miantonomo, he wrote Winthrop, the Narragansetts lacked all fear of God and thus "all the Cords that euer bound the Barbarous to Forreiners were made of Selfe and Covetuousnes" (July 1637). In another letter he went beyond saying the Narragansetts were heathens by affirming that he "would not feare a Jarr with them yet I would fend of[f] from being fowle, and deale with them wisely as with wolues endewed with mens braines" (July 10, 1637).* When the Indians were not Narragansetts, beast metaphors replaced the simile: The Mohawks were "these mad dogs" (July 3, 1637). To stop "a Conglutination" or league between them and the Pequots, Williams urged Winthrop "that better now then hereafter the Pursuit be continued" of the fleeing bands of Pequot survivors (June 30, 1637). He returned to the need to track them all down in a letter of July 15, 1637:

> The generall speech is, all must be rooted out etc. The body of Pequin men yet liue, and *are onely remooved from their dens:* the good Lord grant, that the Mowhaugs and they, and the wh[ole] of the last vnite not. For mine owne Lot I can [not be] without suspicions of it.

The phrase I have italicized echoed down through the decades and centuries of American history, from Williams's day to Andrew Jackson's and beyond: In times of trouble natives were always wild animals that had to be rooted out of their dens, swamps, jungles.

* *Wolves* were quite unfairly held to be quintessential beasts, or as Williams put it in the *Key*, "the wolf is an emblem of a fierce, blood-sucking persecutor."

Some Meditations

Concerning our HONOURABLE

Gentlemen and Fellow-Souldiers,

In Purſuit of thoſe

Barbarous NATIVES in the NARRAGANSIT-Country;

and Their Service there.

Committed into Plain Verſe for the Benefit of thoſe that Read it. By an Unfeigned Friend.

[1]
THeſe *Indians ſtrong* have *waited long*
but now intend to ſee,
If that they can ſo play the man,
that they themſelves may free.

[2]
When I do Muſe, I cannot Chuſe
but mind Gods hand therein,
How he at firſt, *Indians* diſperſt,
that *Engliſh* might begin,

[3]
The Ground to Till, the Land to Fill,
with Men and eke with Beaſts.
This fifty year, ſome have been here,
And have Enjoyed Peace.

[4]
The *Indians* then, were as ſtrong men,
as now they ſeem to be,
And yet *two men* wou'd chace *twice ten*
and make them for to flee.

[5]
They ſtood in dread and were afraid,
the *Engliſh* to offend,
When they did meet 'em in the Street
their word was *VVhat Cheer Friend.*

[6]
It did appear that when they were
in Number many more
Than now they are, both near and far,
abroad, and at our door.

[7]
Then *Engliſh* men, there was not Ten,
where Thouſands now appear,
And yet alas, it comes to paſs,
of them we ſtand in fear.

[8]
Our Captains ſtout, they have gone out,
to fight them in the Field:
But in the Wood they were *withſtood*,
theſe Indians will not yield·

[9]
They had a Fight, in Heavens ſight,
kept up as in a Pound,
Men ſtrong & tall, were forc'd to fall,
and left upon the Ground.

[10]
Six Score and Ten of Valiant Men,
was wounded in that Fight,
And Fifty more, if not Three Score,
was ſlain thereby out-right.

[11]
And then at night, they left the Fight,
in Snow up to the knees,
And to their Tent, away they went,
almoſt ready to freeze.

[12]
And as ſome ſay, they loſt their way,
and Twenty Miles to go,
Marching all Night, till clear *Day light*,
up to the knees in Snow.

[13]
Some froze to Death, have loſt their Breath
with fierceneſs of the Cold,
Some Froze their Feet, they being wet.
'twas ſad for to behold.

[14]
It grieves my heart, & makes it ſmart
I cannot chuſe but weep,
To think on thoſe, who on the Snows
are forc'd to take their Sleep,

[15]
In Wilderneſs, in great Diſtreſs,
in Winters fierceſt Cold,
'Tis grief to me to hear or ſee,
'tis ſad for to behold.

[16]
Brave Gentlemen, the which have been
brought up moſt tenderly,
Are now come out, Ranging about,
and on the Ground do ly.

[17]
Open your Eyes, and Sympathize,
with thoſe brave Gentlemen,
A heart of Flint, it would Relent,
to think how it hath been·

[18]
All who are wiſe, will Sympathize,
with this our Engliſh Nation,
And ſorely grieve, and help relieve
them there in this ſad ſtation·

(19)
O Lord, ariſe, open the Eyes,
of this our Engliſh Nation,
And let them ſee, and alſo be,
ſaved with thy Salvation·

(20)
Caſt thou a dread, and make afraid
theſe Indians ſtrong and ſtout,
And make 'em feel the Sword of Steel,
that ſo they may give out.

(21)
Leſt they do boaſt of their great Hoſt,
and praiſe their god the Devil,
From whom indeed, this doth proceed
the Author of all Evil.

(22)
O *New-England*, I underſtand,
with thee God is offended:
And therefore He doth humble thee,
till thou thy ways haſt mended.

(23)
Repent therefore, and do no more,
advance thy ſelf ſo High,
But humbled be, and thou ſhalt ſee
theſe Indians ſoon will dy·

(24)
A Swarm of Flies, they may ariſe,
a Nation to Annoy,
Yea Rats and Mice, or Swarms of Lice
a Nation may deſtroy.

(25)
Do thou not boaſt, it is God's Hoſt,
and He before doth go,
To humble thee and make thee ſee,
that He His Works will ſhow.

(26)
And now I ſhall my Neighbours all
give one word of Advice,
In Love and Care do you prepare
for War, if you be wiſe.

(27)
Get Ammunition with Expedition
your Selves for to Defend,
And Pray to God that He His Rod
will pleaſe for to ſuſpend.

(28)
And who can tell? but that He will,
be Gracious to us here?
Thoſe that him ſerve He will preſerve
they need it not to fear.

(29)
Though here they dy immediately,
yet they ſhall go to Reſt,
And at that day the Lord will ſay,
happy art thou, and bleſt.

(30)
But unto thoſe, that be His Foes,
the Lord to them will ſay,
Depart from Me, even all ye
that would not Me Obey.

(31)
My Friends I pray, mark what I ſay,
and to it give good heed,
And to the SON, ſee that you Run,
and Him Embrace with ſpeed.

(32)
And now my Friend, here I will End,
no more here ſhall I write,
Or leſt I ſhall ſome Tears let fall,
and ſpoil my Writing quite.

December 28, 1675. W. W.

Re-printed at *N-London, April* 4. 1721

Some Meditations Concerning . . . Barbarous Natives, 1675. First printed to celebrate the decisive Swamp Fight during what was called King Philip's War, these verses were pressed into use during other Indian "troubles." This broadside has the distinction of being the oldest American news ballad to survive; it shows that from the beginning Indians had a bad press. They were *natives,* not *men* and emphatically not "Honourable

3

The Pequots were "rooted out" as a tribe. Winthrop put the body count at between eight and nine hundred. In 1832 one observer counted only "about forty souls, in all, or perhaps a few more or less" living in the township of Groton; in the 1960s the official figure was twenty-one Pequots in all of Connecticut.* And there would have been no known living members of that tribe had the colonizers had their way. They sought, as Mason said, "to cut off the Remembrance of them from the Earth." After the war, the General Assembly of Connecticut declared the name extinct. No survivors should be called Pequots. The Pequot River became the Thames, and the village known as Pequot became New London, in "remembrance," the legislators declared, "of the chief city in our dear native country."

The Saints were an Armed Band of Christ with a need to unpeople the country and make it into *New* England, and that they did; they also had a keen need to engage the forces of darkness. The "stirs at Boston" over Anne Hutchinson and her followers had already put their nerves on edge, and battle with the Pequots, Larzer Ziff pointed out, offered an outlet for their sense of baffled righteousness. After they had exterminated the Pequots and banished the Antinomians, the Reverend Thomas Shepard, Hooker's successor at Newtown, made the connection explicit in his thankful observation that the Lord had delivered the country from both "the Indians and Familists (who arose and fell together)." To their minds the Antinomians beckoned men to licentiousness, "all

* In the 1970s the Connecticut Indian Affairs Coordinator estimated there were less than six hundred people in the state who could claim descent from the Pequot nation. *Mystic* (Conn.) *Compass Comment*, November 27, 1975.

Gentlemen," even had they not worshipped "their god the Devil" (verse 21). Like Captain John Smith's earlier classification of them as unnatural beasts, they appear here as vermin, a swarm of flies, "Yea Rats and Mice or Swarms of Lice" (verse 24). Full blown, this racism was to shower epithets on such "wild varments" as they came to be called—on the Indian-hater's "nits and lice" of the nineteenth century and on the overseas empire-builder's "termites" of the twentieth. Fittingly, Wait Still Winthrop, grandson of the first Governor John Winthrop, was almost certainly the author W. W. and thus through this broadside a worthy legatee of the tradition of looking at all Indians "as a common enimie." (Courtesy of the Massachusetts Historical Society.)

promiscuus & filthie cominge togeather of men & Weomen," as John Cotton warned, and threatened an anarchic breakdown of lawful order and regimented reason. But the bestial Indians already *were* the nature that colonists would fall back into should they yield to such impulses. That is, the Antinomians were carnal men like the libertine Morton; the Indians were carnal beasts—concupiscent, but not rational, animals. They were *the flesh* with "her interest" outside the yoke of law. They flew to the thickets and swamps of their female wilderness the way dark inner "Secrett Corruptions" eluded spiritual authority.

Now, as an Armed Band of Christ, the Saints took pleasure in executing God's will. *Vengeance is mine, saith the Lord, I will repay it:* Endicott, Mason, and the others naturally were pleased to act as His agents. But their racial attitudes made certain that repaying Indians would heighten their pleasure. On the clear moonlight night before the assault on the Pequot fort, Mason recorded, "we appointed our Guards and placed our Sentinels at some distance; who heard the Enemy Singing at the Fort, who continued that Strain until Midnight, with great Insulting and Rejoycing, as we were afterwards informed." The Pequot festival of gladness, with the singing, rejoicing, and no doubt dancing, along with other animal pleasures, made "our Assault before Day" all the sweeter to contemplate. The reality even in retrospect moved the stern old soldier to rhapsody:

> I still remember a Speech of Mr. Hooker at our going aboard; That they should be Bread for us. And thus when the Lord turned the Captivity of his People, and turned the Wheel upon their Enemies; we were like Men in a Dream; then was our Mouth filled with Laughter, and our Tongues with Singing; thus we may say the Lord hath done great Things for us among the Heathen, whereof we are glad. Praise ye the Lord!

Why should they be so furious?—at Mystic the Saints' suppressed sexuality at last broke out and found vent in an orgy of violence. Like men in a dream they burned and shot *the flesh* they so feared and hated in themselves: the breaking of the dam filled them with delight, their mouths with laughter, their tongues with singing.*

* After the slaughter at Fairfield swamp, John Winthrop wrote William Bradford of "the lords greate mercies towards vs, in our preuailing against his, and our enimies; that you may rejoyce, and praise his name with vs." His concluding remarks suggested that those who threw themselves into such violent delights were refreshed but likely to become addicts, vengeance junkies:

Praise ye the Lord!—Morton never had a better time around his Maypole.

4

By contrast Morton's unrepressed joys at Merry Mount had hurt no one. True, as Bradford charged, he and his friends drank, erected an eighty-foot Maypole, and danced "about it many days together, inviting the Indian women for their consorts, dancing and frisking together like so many fairies, or furies, rather; and worse practices." But the dancing and "worse practices" were between and among consenting adults, you might say, and helped free the participants from that oppressive starvation of life that Hawthorne maintained "could not fail to cause miserable distortions of the moral nature." To paraphrase Bradford a little, Morton's "wickedness" was not stopped "by strict laws" and thus did not flow destructively back upon the self; nor did it flow outward to find projective "vent" in negating nature and nature's children, the Indians. Unlike Endicott, Morton experienced no need to scourge "bestial pagans" and to burn and spoil their country. By contrast he indulged in good clean fun.

"Of the Revells of New Canaan" suggests that Morton indulged in much more. Filled with humor, this fascinating chapter of *New English Canaan* poked fun at the Saints for impoverishing their lives by "troubling their brains more then reason would require about things that are indifferent." He thumbed his nose at them with a drinking song to illustrate the "harmless mirth" of his young men—Hawthorne's sons of Pan—who had invited "lasses in beaver coats [to] come away" and made them "welcome to us night and day." The chorus went

---

> Our people are all in health (the lord be praised) and allthough they had marched in their armes all the day, and had been in fight all the night, yet they professed they found them selues so fresh as they could willingly haue gone to such another bussines. [July 28, 1637]

Just this kind of refreshment had been promised by the Reverend Thomas Hooker's anthropophagous send-off: "they should be Bread for us." With the slaughter of Indians, the Saints swallowed at one sitting the carnality of wilderness creatures and their own inadmissible fantasies and yearnings. They rose up with new powers that would help make these Englishmen into Americans, as Richard Slotkin has argued persuasively in *Regeneration through Violence: The Mythology of the American Frontier* (Middletown, Conn.: Wesleyan University Press, 1973).

Drinke and be merry, merry, merry boyes;
Let all your delight be in the Hymens ioyes;
Jo[y] to Hymen, now the day is come,
About the merry Maypole take a Roome.

"The Song" was a relatively straightforward celebration of drink and sex, though one had to know that Hymen was the Greek god of marriage rites, a beautiful youth who carried his "bridal torch" to a virgin's vagina.

"The Poem" at the head of the chapter, on the other hand, was really a riddle, and it demands full quotation:

Rise Oedipeus, and, if thou canst, unfould
What meanes Caribdis underneath the mould,
When Scilla sollitary on the ground
(Sitting in forme of Niobe,) was found,
Till Amphitrites Darling did acquaint
Grim Neptune with the Tenor of her plaint,
And causd him send forth Triton with the sound
Of Trumpet lowd, at which the Seas were found
So full of Protean formes that the bold shore
Presented Scilla a new parramore
So stronge as Sampson and so patient
As Job himselfe, directed thus, by fate,
To comfort Scilla so unfortunate.
I doe professe, by Cupids beautious mother,
Heres Scogans choise for Scilla, and none other;
Though Scilla's sick with greife, because no signe
Can there be found of vertue masculine.
Esculapius come; I know right well
His laboure's lost when you may ring her Knell.
The fatall sisters doome none can withstand,
Nor Cithareas powre, who poynts to land
With proclamation that the first of May
At Ma-re Mount shall be kept hollyday.

All this was "enigmattically composed," Morton chuckled, and "pusselled the Seperatists most pittifully to expound it." Twitting "those Moles" for their blind presumption in thinking "they could expound hidden misteries," he grandly expounded it himself in terms that left the reader no less pitifully puzzled.

Better at explaining why Morton should not have traded guns to the Indians, Charles Francis Adams, when preparing his annotated edition of *New English Canaan* in the late nineteenth century, gave up on the verse in disgust, denouncing it for its "incomprehensibility." And its Sphinxian inscrutability was baffling to all save an Oedipus or perhaps a Renaissance Platonist. To unfold its hidden meaning the reader had to reach for the understanding that pagan gods and goddesses of fertility functioned as metaphors for universal forces in nature. Tides of fecund seas flowed in "so full of Protean formes that the bold shore" presented the yearning Scylla with "a new parramore." To apprehend Morton when he "[did] professe, by Cupids beautious mother," the reader had to know that Cupid (or Eros) was born of Venus (or Aphrodite), the goddess of beauty and love, who in turn rose from the sea's white foam where Uranus' sex organs had fallen. The creative power and physical passion then swept in with such force no one could withstand them. In the final three lines the alert reader could be expected to perceive that May Day was to be kept a *holy* day but would have to be quick and something of a classicist to know that by "Cithareas powre" Morton invoked the Ionian Island of ancient Cythera, site of a center of the cult of Aphrodite, and still quicker to understand that Morton was proposing another such center at Merry Mount.

This reading of Morton's allegory makes it closely resemble Botticelli's as depicted in *The Birth of Venus* (Uffizi, Florence): In the painting the goddess of love rises up newborn from the white foam of the waves and is blown by spring winds on the shore as a symbol of generative power. Once deciphered, "The Poem" confirms Morton's deep interest in what Edgar Wind called *Pagan Mysteries in the Renaissance* (1958). Like Sir Thomas More in *Utopia* (1551), Morton defended pleasure; and like Marsilio Ficino, the Florentine Neoplatonist who influenced Botticelli, he manifestly believed *voluptas* should be reclassified a noble passion. Like other Renaissance men with a taste for bacchic mysteries, he spoke in riddles and deliberately composed "enigmattically" because he believed, with Dionysus, that "the divine ray cannot reach us unless it is covered with poetic veils." These poetic veils were what Pico della Mirandola, another fifteenth-century Florentine Neoplatonist, called "hieroglyphic" imagery: All pagan religions had them, he maintained, to protect divine secrets from profanation. These were the "hidden misteries" the host of Merry Mount flaunted before the Saints.

Morton's use of hieroglyphics also ironically brought him closer in feeling and thought to the "Salvages" he invited to the revels. Old World and New World pagan entered into communion with the beyond through figurative speech, rites, songs, and dances, and not through denotative language, ratiocination, and three-hour sermons. Above all the dance, hated by the Saints as barbarous madness, brought Morton close to his Indian neighbors. The fusion he sought between Old and New World paganism became visible around the Maypole, as hand reached out to hand and rhythms of breath and blood within matched vibrations from the wilderness without—suddenly man and his environment were a harmonious unit.

5

But the dance was over and so was this great opportunity to see if whites and reds could live with themselves, each other, and the land. As Morton smilingly confessed, "hee that playd Proteus, (with the helpe of Priapus,) put their noses out of joynt, as the Proverbe is." According to tradition, Proteus could change himself into any shape he pleased, but if caught and held, he would foretell the future. Caught and held, Morton warned carnal white and carnal Indian alike that the Puritans would cut off their ears and worse, "if you be one of them they terme, without." So it came to pass that when the future marched forward with Endicott's remorseless stride, men and countryside were burned and spoiled.

Some presentiment of this destiny led Miantonomo, after having listened too long to the kindly Roger Williams, to repeat the Pequot proposal for an Indian alliance. According to Lion Gardiner, Miantonomo told the Indians on Long Island:

> So are we all Indians as the English are [all English], and say brother to one another; so must we be one as they are, otherwise we shall be all gone shortly, for you know our fathers had plenty of deer and skins, our plains were full of deer, as also our woods, and of turkies, and our coves full of fish and fowl. But these English have gotten our land, they with scythes cut down the grass, and with axes fell the trees; their cows and horses eat the grass, and their hogs spoil our clam banks, and we shall all be starved.

In 1643 the commissioners of the United Colonies of New England decreed that Uncas, the Mohegan sachem, might kill Miantonomo, their former ally because, as Winthrop recorded, he was head of "a general conspiracy among the Indians to cut off all the English," because he was "of a turbulent and proud spirit, and would never be at rest," and because of lesser reasons. So it was that Uncas's brother slipped up behind the great Narragansett sachem and "clave his head with an hatchet, some English being present."

Lion Gardiner wrote his account of all these events fifteen years before King Philip's War. By then an old man, preoccupied with Indian "plots" and in fear to see and hear what he thought would "ere long come upon us," he nevertheless still believed that "although there hath been much blood shed here in these parts among us, God and we know it came not by us." But it was as if even in death Miantonomo would never be at rest, for Gardiner confided in a moment of self-realization: "I think the soil hath almost infected me. . . ."

Part Two

Founding Fathers and Merciless Savages

Indulge, my native land! indulge the tear,
That steals, impassion'd, o'er a nation's doom:
To me each twig, from Adam's stock is near,
And sorrows fall upon an Indian's tomb. . . .

—Timothy Dwight,
"The Destruction of the Pequods," 1794

CHAPTER VI

Timothy Dwight of Greenfield Hill

> I am clear in my opinion, that policy and oeconomy point very strongly to the expediency of being upon good terms with the Indians, and the propriety of purchasing their lands in preference to attempting to drive them by force of arms out of their Country; which . . . is like driving the wild Beasts of ye forest . . . when the gradual extension of our settlements will as certainly cause the savage, as the wolf, to retire; both being beasts of prey, tho' they differ in shape.
>
> —GEORGE WASHINGTON
> to James Duane,
> September 7, 1783

IN THE YOUNG republic, critics and admirers alike agreed the Reverend Timothy Dwight was a giant of Puritan values and attitudes. He preached an unyielding Calvinism that went back to his grandfather Jonathan Edwards and beyond, at least to his forefather John Dwight, who emigrated from England to Massachusetts in 1635. His religious doctrine and political views would have been accounted sound by John Endicott and John Mason. He was, as Vernon Louis Parrington said in *The Connecticut Wits* (1926), "a walking repository of the venerable *status quo.*" Chaplain of the Connecticut Continental Brigade during the Revolution, he inspired troops with the old-time religion and composed patriotic songs to cheer them on. After Yorktown he was ordained pastor of the Congregational Church at Greenfield Hill, Fairfield Township, Connecticut. And of course after his election in 1795 he became the legendary president of Yale College.

As a young man Dwight strove to become the American Homer. In "America: Or, a Poem on the Settlement of the British Colonies," circulated at Yale in manuscript in the early 1770s, he anticipated the American centuries to come by envisioning a vast empire "of light and joy! thy power shall grow

> Far as the seas, which round thy regions flow;
> Through earth's wide realms thy glory shall extend,
> And savage nations at thy scepter bend.
> Round the frozen shores thy sons shall sail,
> Or stretch their canvas to the Asian gale.

His patriotic songs also fused the Puritan sense of mission with the messianic nationalism of the Revolution. Everywhere people sang his "Columbia":

> Columbia, Columbia, to glory arise,
> The Queen of the world, and child of the skies!

In America's first epic poem, *The Conquest of Canäan* (1785), Dwight decisively identified the causes of ancient Israel and the new United States of America—and therewith repudiated the promise of Thomas Morton's *New English Canaan*. *The Conquest of Canäan* was in fact an expanded and allegorized version of the Book of Joshua's account of the heroic struggle to bring the Children of Israel to the Land of Canaan: Dwight dedicated the epic to Washington, the Mason-like "Saviour of his Country," and made him the American Joshua who triumphed over fiendish, wolflike Canaanites with their "childish rage." Parallels with the Puritan justification of the destruction of the Pequots were drawn tighter in the poem's justification of the killing of lovers and babies (later prone to sodomy and unspeakable acts), so "God be witness'd sin's unchanging foe": "Should then these infants to dread manhood rise,/ What unheard crimes would smoke thro' earth and skies!" America was the last stop on the westward march of empire, in fine, the sole heir apparent of Israel's mission "To found an empire, and to rule a world."

2

The pastoral *Greenfield Hill: A Poem in Seven Parts* (1794) was Dwight's last major poem. From the eminence of "the sweet-smiling village" of which he was minister, the poet looked out on a landscape that had over the generations become "my much-lov'd native land!" The villagers flourished, and in the surrounding countryside men owned the land they worked, fields where "every farmer reigns a little king." A visitor would see no nobles or serfs here, for "one extended class embraces all." To be sure, "yon poor black" singing gaily reminded the poet of slavery, but the "laughing mind" of that son of Africa bore witness he was "kindly fed, and clad, and treated" and took "his portion of the common good." As for savages, "Through the war path the Indian creep no more" but "peace, and truth, illume the twilight mind." In Part Four, "The Destruction of the Pequods," Dwight looked from his vantage point toward Long Island Sound a few miles away and could see with pleasure "a Field called the Pequod Swamp, in which, most of the warriors of that nation, who survived the invasion of their country by Capt. Mason, were destroyed. . . ."

Lion Gardiner notwithstanding, the land had not been infected but fertilized by all the Indian bodies: "Even now, perhaps, on human dust I tread," Dwight mused, perchance on that of some ambitious chief who roused his followers to deeds sublime "and soar'd Caesarean heights, the Phoenix of his age." The sleeping nations below were America's vanished "Ancient Empires." They gave the land a past:

> And, O ye Chiefs! in yonder starry home,
> Accept the humble tribute of this rhyme.
> Your gallant deeds, in Greece, or haughty Rome,
> By Maro sung, or Homer's harp sublime,
> Had charm'd the world's wide round, and triumph'd over time.

Moreover, without Indians the forefathers would have had no way to prove themselves in trials by ordeal as they wrested the wilderness from stiffening red hands. "Glowing Mason," later by "generous Stoughton join'd," claimed "the fiends of blood" in battle and in the pursuit that came to an end just down the hill: "Mason, Mason, rung in every wind." Nor should a grateful posterity forget "hapless Stone," "gallant Norton," and "generous Oldham!" Finally even the Indians, as men "from Adam's stock," made claims on the poet's faith and generous humanity that led him to "indulge the tear" over their savage "woes severe."

But Dwight took pleasure in treading on human dust at unreckoned peril. On his body-strewn pages, not so much the land as truth itself "in blood was drown'd." To establish that the Puritans had fought the Pequots in self-defense he had to engage in sheer projection: Pequots "resolved to attempt the destruction of the

English, with the strength of their own tribes only." His glorification of their slaughter smashed head-on into his insistence that the heroic forefathers were gentle followers of Christ. From the wreckage he retrieved only a Sunday-school story: In the beginning meek Christians had abandoned Europe, "that land of war, and woe" to come "o'er th' Atlantic wild, by Angels borne":

> The pilgrim barque explor'd it's western way,
> With spring and beauty bloom'd the waste forlorn,
> And night and chaos shrunk from new-born day.
> Dumb was the savage howl. . . .

The "dusky mind" of the savage was lit with wisdom's beam, transformed to that of "the meekly child." Even the Pequots received kind ministrations:

> Even Pequod foes they view'd with generous eye,
> And, pierc'd with injuries keen, that Virtue try,
> The savage faith, and friendship, strove to gain:
> And, had no base Canadian fiends been nigh,
> Even now soft Peace had smil'd on every plain,
> And tawny nations liv'd, and own'd Messiah's reign.

In the last analysis outsiders, Canadian fiends no less, had forced the hands of innocent Puritan killers.

"Wolves play'd with lambs," Dwight said of the Indians after they had been instructed by missionaries John Eliot and Thomas Mayhew, but the Indians at his Pequot Swamp were the beasts of old. On the level of statement, Indians were as Genesis dictated, brother offshoots of Adam; on the level of action, and the words and images flowing from acts that mattered, Indians were New World Canaanites. They were animals of the forest: "murderous fiends" who often disemboweled captives in their "dark abodes"; "sable forms" who crept along "snaky paths" to aim their "death unseen" and scream "the tyger-yell." They were the tawny animal side of creatures, incarnations of the night and chaos that shrank from the Saints' newborn day:

> Fierce, dark, and jealous, is the exotic soul,
> That, cell'd in secret, rules the savage breast.
> There treacherous thoughts of gloomy vengeance roll,
> And deadly deeds of malice unconfess'd;
> The viper's poison rankling in it's nest.

"Yet savages are men," Dwight reminded himself, only a Spenserian stanza later, but still too late—the Christian creation myth had already sunk beneath the surface with all hands aboard.

3

Like John Adams, to whom *Greenfield Hill* was dedicated, Dwight had inherited his racist attitudes intact. His bad poetry provided good evidence of the continuity of Indian-hating. Perhaps "one extended class embraces all," as he said, but that did not mean it embraced such nonpersons as yon poor black and yon poor Indian.

Indeed the one class could barely embrace radical whites who contended they were as good as their betters. Like John Endicott, Dwight came to see parallels and linkages between savages lurking on the frontiers and the rebellious commons at home. In his prose and verse he fought the "banded hellhounds" who favored the French Revolution with wrathful and by now familiar invective. The Indians' "snaky paths" became "snaky anarchy, the child of hell" beloved by champions of personal freedom. In his tract *The Duty of Americans at the Present Crisis* (1798), he put the issue bluntly: "Without religion we may possibly retain the freedom of savages, bears, and wolves; but not the freedom of New England." Here the animality of Indians was unquestionable, but the white enemies of Christ were not much better when they generated "the cruelty and rapacity of the Beast." And like Endicott, Dwight saw close ties between politics and sex, and the dangers of physical passion given vent by song and dance. Again the dance: Would the present crisis "change our holy worship into a dance of Jacobin phrenzy, and that we may behold a strumpet personating a *Goddess* on the altars of Jehovah?"

But for all his bitterness, Dwight was involved here in a family quarrel with fellow citizens of the new republic. As "the Pope of Federalism" he fought Deists and advocates of the Enlightenment with the same fervor Endicott had fought Anabaptists from Rhode Island. The advancing Jeffersonianism provoked his hatred precisely because to his mind it freed men to act like savages and other animals. Within the family of whites, on the other hand, there was not much dispute over the nature of Indians. After all, one of the more heinous crimes charged against George III in the Declaration of Independence was that he had instigated "the merciless Indian Savages."

CHAPTER VII

John Adams

> The Indians are known to conduct their Wars so entirely without Faith and Humanity, that it will bring eternal Infamy on the Ministry throughout all Europe if they should excite these Savages to War. The French disgraced themselves last War by employing them. To let loose these blood Hounds to scalp Men and to butcher Women and Children is horrid.
>
> —JOHN ADAMS
> to James Warren,
> June 7, 1775

> It is said they are very expensive and troublesome Confederates in war, besides the incivi[li]ty and Inhumanity of employing such Savages with their cruel, bloody dispositions, against any Enemy whatever. Nevertheless, such have been the Extravagancies of British Barbarity in prosecuting the war against us, that I think we need not be so delicate as to refuse the assistance of Indians.
>
> —JOHN ADAMS
> to Horatio Gates,
> April 27, 1776

JOHN ADAMS served with Thomas Jefferson on the Committee of Five appointed to draft a declaration, and collaborated effectively with him in the diplomatic service following the Revolution. A period of estrangement blew up out of the political passions and controversies surrounding their presidencies, but their friendship took root again in the tranquillity of retirement. They were brought back together by another signer of the Declaration of Independence, Benjamin Rush, a mutual friend who had a keen sense of their historic importance as "fellow laborers in erecting the great fabric of American independence!" Not unaware themselves of their seminal role as Founding Fathers and personifications of the republic, they wrote in the full knowledge that posterity was already peering at their every word. Their correspondence from 1812 to their deaths in 1826 was nevertheless rich and, on occasion, revealing.

Close upon their reconciliation Adams announced (October 12, 1812) "an event of singular Oddity." After he had himself searched for Thomas Morton's *New English Canaan* for half a century, John Quincy Adams had finally found and purchased a copy at an auction in Berlin. His father, an acquisitive bibliophile, exulted in the find, in the likelihood "the Berlin Adventurer is I believe the only one in America," and in the whimsical turn of events "that this Book, so long lost, should be brought to me, for this Hill [that Morton called Merry Mount] is in my Farm. There are curious Things in it, about the Indians and the Country. If you have any Inclination, I will send you more of them." For openers he summarized the goings-on at Merry Mount:

> On May Day this mighty May Pole was drawn to its appointed Place on the Summit of the Hill by the help of Savages males and females, with the Sound of Guns, Drums, Pistols and other Instruments of Musick. A Barrel of excellent Beer was brewed, and a Case of Bottles, (of Brandy I suppose) with other good Chear, and English Men and Indians Sannups and Squaws, danced and sang and revelled round the Maypole till Bacchus and Venus, I suppose were satiated.

But Maypole dancing and singing and reveling had little appeal to Adams's angular, Apollonian correspondent, who was most fond of reading he could apply "immediately to some useful science." Jefferson's reply (December 28, 1812) showed him more interested in bibliography than in Bacchus and Venus—he even declined the invitation to discuss interracial sex, a topic presumably of concern to him, as we shall see.

Adams had naturally hoped for more. He half-apologetically thanked Jefferson for digging out data for him on a matter that was "selfish, personal and local"; still he doggedly pursued it by reporting that, after "a Philipic against the Puritans and a Panegiric on the Indians," Morton had maintained there was evidence Indians had heretofore worshipped the great god Pan, and that, along with other proofs, showed them to be descendants "of the scattered Trojans." This time (January 26, 1813) Adams struck a spark. Jefferson had long pondered the question of whence descended the Indian tribes and had derided the bootless speculation of those who traced them back to the ancient Hebrews, the

Tartars, the Persians, and others: "Moreton's deduction of the origin of our Indians from the fugitive Trojans," he wrote, "and his manner of accounting for the sprinckling of their Latin with Greek is really amusing" (May 27, 1813). One might as well assume a separate creation of the Indians, he observed, or "no creation at all, but that all things have existed without beginning in time, as they now exist, producing and reproducing in a circle without end. This would very summarily dispose of Mr. Morton's learning."

Such serene philosophizing did not really advance the discussion, however, for Adams had already seized the occasion to make fun of Morton's erudition and critical judgment. An uncompleted draft of a letter (dated February 2, 1813) further revealed the passing absorption with Morton he was never quite able to communicate to the Virginian: Morton's *New English Canaan*, Adams wrote, was "infinitely more entertaining and instructive to me, than our Friend Condorcets 'New Heaven' was almost 30 years ago, or than Swedenborgs 'New Jerusalem' is now." Though for Adams, as for his Puritan ancestors, Morton remained "this incendiary instrument of spiritual and temporal domination," he was a firebrand who was safely extinguished and whose *New English Canaan* had thereby become more curious than wicked—it needed no longer to be immediately consigned to the flames as the devil's notebook. Like Swedenborg with his direct intercourse with angels and like Condorcet with his prophecy of the ultimate perfection of man, Morton had been a drunken enthusiast, drunk (as Adams saw him) on his vision of "English Men and Indians Sannups and Squaws" dancing and reveling together in joyous New Canaan. Adams was entertained by the extravagance of the notion and by the dead rebel generally: As he wrote Jefferson (January 26, 1813), "The Character of the Miscreant . . . is not wholly contemptible. It marks the Complextion of the Age in which he lived. How many such Characters could You and I enumerate, who in our times have had a similar influence on Society!"

No more than Adams had Jefferson tolerated live firebrands—indeed the latter's recent vendetta against Aaron Burr suggested a lower tolerance level. Adams's personal experience was sufficient, at all events, to keep him from suspecting Jefferson of the libertarianism imputed to him by posterity: In the critical year 1776 he and Jefferson had drafted another document, the first United States Articles of War. Their draft passed with a provision punishing "traitorous or disrespectful words" against Congress or the state legislatures—as Leonard Levy has pointed out, they drew on the British model, simply substituting legislatures for king and royal family as the bodies against whom it became a crime for militiamen and Continentals to speak contemptuously. Adams could be confident that Jefferson felt as he did about "such Characters" as Morton: those who defied well-ordered states put themselves outside civil society and were the proper recipients of governmental wrath.

2

Adams had the satisfaction of knowing that was exactly what had happened to Morton, and the added pleasure of owning both the miscreant's book and his Merry Mount. His was the delight of being a surviving heir of the victors. As he outlined the family saga for Benjamin Rush (July 19, 1812), the first Adams in America was named Henry, "a Congregational dissenter from the Church of England, persecuted by the intolerant spirit of Archbishop Laud, [who] came over to this country with eight sons in the reign of King Charles the First." The Adamses acquired Merry Mount, or Mount Wollaston, and were the original proprietors of the town of Braintree, incorporated in 1640: "This Henry and his son Joseph, my great-grandfather, and his grandson Joseph, my grandfather, whom I knew, tho' he died in 1739, and John, my father, who died in 1761, all lie buried in the Congregational churchyard in Quincy, half a mile from my house." Each of the five generations in this country had reared numerous children, "multiplied like the sands on the seashore or the stars in the milky way, to such a degree that I know not who there is in America to whom I am not related. My family, I believe, have cut down more trees in America than any other name! What a family distinction!" His family had multiplied and prospered, he believed, through "industry, frugality, regularity, and religion," but especially through the last:

> What has preserved this race of Adamses in all their ramifications, in such numbers, health, peace, comfort, and mediocrity? I believe it is religion, without which they would have been rakes, fops, sots, gamblers, starved with hunger, frozen with cold, scalped by Indians, &c., &c., &c.

The faithful shall inherit the earth.

What a family distinction! Even when allowances are made for Adams's half-facetiousness, he still provided backing for Hans Huths's contention in *Nature and the American* (1957) that the ax was "the appropriate symbol of the early American attitude toward nature." So did another Henry Adams—author of the great *History of the United States* (1889–91) and great-grandson of John Adams—who wrote of white Americans in 1800: "From Lake Erie to Florida, in long, unbroken line, pioneers were at work, cutting into the forests with the energy of so many beavers, and with no more express moral purpose than the beavers they drove away." Over the eight generations, including those of the immigrant Henry and the historian Henry Adams, the trees came crashing down, with any higher goal progressively less visible. And in his discussion of Indian affairs during the administrations of Jefferson and Madison, the historian penned some lines with wry bearing on the preservation of "this race of Adamses": He invoked "the law accepted by all historians in theory, but adopted by none in practice; which former ages called 'fate,' and metaphysicians called 'necessity,' but which modern science has refined into the 'survival of the fittest.' " His great-grandfather knew nothing of this latest refinement, of course, and would have been nonplussed by the note of skepticism in the historian's tone: John Adams had no thought of questioning the providential necessity that led to a withering away of scalping Indians and such rakes as Morton.

Despite his notorious temper and vanity, John Adams was by no means an unkind man, uninterested in the Indians and unconcerned over their plight. When Jefferson wrote that his early familiarity with the Indians had given him a lifelong attachment to them (June 11, 1812), Adams replied that "I also have felt an Interest . . . and a Commiseration for them from my Childhood" (June 28, 1812). When he was a boy, neighboring Indians used to call at his father's house and he himself had visited their wigwam nearby "where I never failed to be treated with Whortle Berries, Blackberries, Strawberries or Apples, Plumbs, Peaches, etc., for they had planted a variety of fruit Trees about them." But that was seventy years ago and now "we scarcely see an Indian in a year."

The New England tribes had so withered away that Adams had to turn to Jefferson as the expert on current Indian affairs. He questioned him in particular about their religion and about the "Wabash Prophet," as he called Tenskwatawa, Tecumseh's brother (February 10, 1812). What Jefferson told him enabled him to come up with an insight of startling originality: "The Opinions of the Indians and their Usages, as they are represented in your obliging letter of 11. June, appear to me to resemble the Platonizing Philo [of Alexandria], or the Philonizing Plato, more than the genuine system of Judaism" (June 28, 1812). Though Adams confessed himself "weary of contemplating Nations from the lowest and most beastly degradations of human life, to the highest Refinement of Civilization," he at least took the trouble to inquire into Indian thought, to sense that the profound inwardness of its embrace of the Great Mystery established affinities with the "civilized" shamanism called Platonism, and to note as well parallels with modes of Egyptian and Oriental thought.

3

The trouble was that the Indians were beastly degradations of human life. They scalped men, butchered women and children, and were "by disposition" cruel and bloody-minded. Furthermore, they had no capacity or potential for citizenship, not to mention civility: "Negroes, Indians, and Kaffrarians," Adams wrote Rush, "cannot bear democracy any more than Bonaparte and Talleyrand . . ." (April 11, 1805).* Dictatorial Frenchmen aside, Adams's association of peoples unable to bear democracy in fact offered support for Chief Justice Roger Taney's famous dictum decades later in *Dred Scott* v. *Sandford* (19 Howard 393, 1857): When the Constitution was adopted, the chief justice was to hold, Negroes with slave ancestry were considered unfit for citizenship "as a subordinate and inferior class of beings, who had been subjected by the dominant race, and whether emancipated or not, yet remained subject to their authority, and had no rights or privileges but such as those who held the power and the government might choose to grant them." Red tribesmen were no more fit for citizenship than black chattels: Adams obviously did not have either in mind when he informed Rush, in the letter quoted earlier, that with the increase and scattering of his family, "I know not who there is in America to whom I am not related." He exaggerated, of course, but not to the extent of positing relationship with slaves or savages. By definition there were no nonpersons in the Adams lineage.

* By *Kaffrarians* Adams meant the inhabitants of the Transkeian Territories, then called Kaffraria, of South Africa. Now called Kafirs or Kaffirs, they are a Bantu-speaking Negro people. Adams's anthropology thus slipped a bit in distinguishing them from Negroes, though he was right on the mark in using the term as the epithet since common in racist references to all Negro Africans. In the context of his letter to Rush, Adams was manifestly seeking a term of ultimate bestiality; his purpose would have been better if not more accurately served by *Hottentots*, then thought by Europeans to be "the most appallingly barbarous of men, though other people were darker and lived in worse climates" (Winthrop D. Jordan, *White over Black* [Baltimore: Penguin Books, 1969], pp. 226–27).

Notwithstanding the whortleberries of his youth, Adams's commiseration toward the Indians never stretched very far. His racism joined with a messianic nationalism to make Indians at best obstacles to the rise of the American Republic. From the beginning, he believed, their claims to the lands they inhabited were questionable. In a letter to Judge William Tudor (September 23, 1818), he granted they had Lockean rights "to life, liberty, and property in common with all men" but then demanded what rights they had beyond those: "Shall we say that a few handfuls of scattering tribes of savages have a right of dominion and property over a quarter of this globe capable of nourishing hundreds of millions of happy human beings? Why had not Europeans a right to come and hunt and fish with them?" When happy European hunters and fishermen did come, they generously respected the tribes' shadowy rights to the land, "entered into negotiations with them, purchased and paid for their rights and claims, whatever they were, and procured deeds, grants, and quitclaims of all their lands, leaving them their habitations, arms, utensils, fishings, huntings, and plantations." Ordinarily honest with himself, Adams allowed himself to be comforted here with a fable no less fanciful than Timothy Dwight's reading of the Puritan past: "In short," Adams concluded, "I see not how the Indians could have been treated with more equity or humanity than they have been in general in North America."

In his exchanges with Jefferson, Adams had expressed distrust of the mystical, Platonizing tendency of Indian religion and of such of their prophets as Tenskwatawa:

> I think these Prophecies are not only unphilosophical and inconsistent with the political Safety of States and Nations; but that the most sincere and sober Christians in the World ought upon their own Principles to hold them impious, for nothing is clearer from their Scriptures than that Their Prophecies were not intended to make Us Prophets. [February 10, 1812]

Now in his letter to Judge Tudor, Adams stated his position still more forcefully:

> What infinite pains have been taken and expenses incurred in treaties, presents, stipulated sums of money, instruments of agriculture, education, what dangerous and unwearied labors, to convert these poor, ignorant savages to Christianity! And, alas! with how little success! The Indians are as bigoted to their religion as the Mahometans are to their Koran, the Hindoos to their Shaster, the Chinese to Confucius, the Romans to their saints and angels, or the

> Jews to Moses and the Prophets. It is a principle of religion, at bottom, which inspires the Indians with such an invincible aversion both to civilization and Christianity. The same principle has excited their perpetual hostilities against the colonists and the independent Americans.

In this remarkable paragraph Adams fused racism, nationalism, and Enlightenment bigotry into a whole that hardly commiserated with the Indians: If that was kindness, it was the kindness that kills.

CHAPTER VIII

Thomas Jefferson

> You congratulate her [the U.S.A.] on the Indians' becoming somewhat civilized: and on the increase, instead of the dwindling, of several of their tribes, due to their increased knowledge of agriculture. The inhabitants of your country districts regard—wrongfully, it is true—Indians and forests as natural enemies which must be exterminated by fire and sword and *brandy*, in order that they may seize their territory.
>
> They regard themselves, themselves and their posterity, as collateral heirs to all the magnificent portion of land which God has created from the Cumberland and Ohio to the Pacific Ocean.
>
> —PIERRE SAMUEL DU PONT DE NEMOURS
> to Thomas Jefferson,
> December 17, 1801

"OF ALL historical problems," declared Henry Adams, "the nature of a national character is the most difficult and the most important." Of all the revered fathers who shaped that character, he might have added, none was more important, not even his great-grandfather, than Thomas Jefferson, immortal author of the hallowed Declaration of Independence, symbol of liberalism, philanthropist, idealist of the practical, philosopher of immediately useful ends, whether of buildings, moldboards of plows, or origins of Indians—to many of his generation and to many since, Jefferson represented the ideal American self. His ambition went beyond mere nationality to embrace, in Henry Adams's words, "the whole future of man."

The Sage of Monticello was a monument of cool rationality, devoted to order and discipline, temperate in habit, measured in speech, proverbially lucid in writing. No one wrote more about his opinions on more topics. His life was an open book or rather, given the volume of his papers, an open library, a repository of data on the American character. Yet the curious truth is that subsequent "intimate" biographies succeed best in establishing his enormous elusiveness and leave the inmost man unexplored and uncharted. In Jefferson's literary remains, ringing enunciations of abstract ideals are easily come by; perhaps no less a national characteristic, hints of living relationships between these ideals and the man are most difficult to find. There were thus more than echoes of family quarrels in the cautionary words of John Quincy and Charles Francis Adams in their *Life of John Adams* (1871):

> The character of Thomas Jefferson presents one of the most difficult studies to be met with by the historian of these times. . . . More ardent in his imagination than his affections, he did not always speak exactly as he felt towards either friends or enemies. As a consequence, he has left hanging over a part of his public life a vapor of duplicity, or, to say the least, of indirection, the presence of which is generally felt more than it is seen.

This timely warning has largely gone unheeded. That politicians and scholars have so easily fixed a different Jefferson image in the American mind, however, says more about the eager receptiveness of their countrymen than about the persuasiveness of the historians' arguments. Parrington, Henry Steele Commager, Dumas Malone, Julian P. Boyd, and Merrill D. Peterson have presented a hero who was the "voice of imperishable freedoms," the man who wanted posterity to remember: "I have sworn upon the altar of God, eternal hostility against every form of tyranny over the mind of man." But this eternal hostility falls under the weight of evidence assembled by Leonard W. Levy in his brilliantly iconoclastic *Jefferson and Civil Liberties* (1963): the ideologue of freedom came out for dictatorship in times of crisis, supported loyalty oaths, favored prosecution—at the state level—for "seditious" libel, accepted concentration camps for the politically unreliable, adopted censorship, and indulged in other authoritarian acts that led Levy to find "a strong pattern of unlibertarian, even antilibertarian thought and behavior extending throughout Jefferson's long career." When it came to slavery, Jefferson early spoke out against the institution and confided to correspondents he hoped for its end. Yet the careful analysis of David Brion Davis in *The Problem of Slavery in the Age of Revolution* (1975) demonstrated that Jefferson consistently trimmed and equivocated about emancipation, had only a theoretical interest in it, and never in fact transcended the loyalties of Virginia's wealthiest families—instead he had "an extraordinary capacity to sound like an enlightened reformer while upholding the interests of the planter class."

Winthrop D. Jordan had already reached a comparable conclusion in *White over Black* (1968): Jefferson's "vigorous antislavery pronouncements . . . were always redolent more of the library than the field." In his view, Jefferson did not implement his theoretical commitment to abolition because his hand was stayed by his own unrealized prejudices against blacks. Though he shared these prejudices with other planters, along with their interests as a class, he had his personal cross to bear as the slaveholding philosopher of freedom. Thus in *Notes on the State of Virginia* (Paris, 1785) he tried to extricate himself by depicting blacks as creatures of the body and sensation rather than of the mind and reflection, and doubted their fitness for freedom; but he disguised this "anti-Negro diatribe" by casting it as a scientific hypothesis subject to further verification.

Jordan took us below a generalized class analysis to Jefferson's great personal aversion to miscegenation or, as he put it, "to the mixture of colour here." On November 24, 1801, shortly after he became president, he wrote James Monroe, then governor of Virginia, his rhapsodical reflections on the destiny of American white men: they would multiply and "cover the whole northern, if not southern continent, with a people speaking the same language, governed in similar forms, and by similar laws; nor can we contemplate with satisfaction either blot or mixture on that surface." So that no black would ever blot that surface and stain "the blood of the master," the great revolutionary proposed to rid the country of them all, those previously freed and those newly emancipated: through "expatriation" to Africa or the West Indies—he wavered on "the most desirable receptacle"—the problem would disappear with the last shipload.

By going beneath the manifest content of Jefferson's rhetoric about blacks, Jordan went deeper into his racism than anyone before or since. At the risk of appearing singularly ungrateful for this remarkable contribution, I must add that it is a pity he did not submit Jefferson's statements on the Indians to the same kind of skeptical, probing analysis. Instead he embraced as "heartfelt" the passages in *Notes on Virginia* defending the Indian against the aspersions of Buffon, the famous French naturalist—with no less validity Jordan might have accepted as heartfelt Jefferson's nationalistic defense of New World quadrupeds, those "other animals of America," as he put it in his famous letter to General Chastellux

(June 7, 1785, V, 6).* Confronted by three races in America, Jordan asserted, Jefferson "determinedly turned three into two by transforming the Indian into a degraded yet basically noble brand of white man." Though Jefferson stressed the color of blacks, "he denied the Indian's tawny color by not mentioning it," favored intermixture of whites and Indians over removal of the latter, and in an 1813 letter to Baron von Humboldt welcomed this mixture of "their blood with ours." Hence Jordan's conclusion: "Amalgamation and identification, welcomed with the Indian, were precisely what Jefferson most abhorred with the Negro. The Indian was a brother, the Negro a leper."

But it was precisely when Jordan himself turned from blacks to reds that he suffered his vision to cloud by vapor rising off Jefferson's lines. In one of the passages from the *Notes* he quoted, it was made quite clear that three races had not been turned into two: "To our reproach it must be said," Jefferson had admitted, "that though for a century and a half we have had under our eyes the races of black and red men, they have never yet been viewed by us as subjects of natural history." Nor in the very letter to Humboldt quoted by Jordan did Jefferson deny Indians their color, but on the contrary predicted that the "confirmed brutalization, if not extermination of this race in our America is . . . to form an additional chapter in the English history of [oppression of] the same colored man in Asia, and of the brethren of their own color in Ireland" (XIV, 20–25). As to whether the Indian was a brother to Jefferson, we shall see.

2

Color identified the players in the game. According to Jefferson's notes of a conversation with George Hammond on June 3, 1792, the young British minister asked him what he understood as the right of the United States in Indian soil. The secretary of state replied that it was a preemptive right, that is to say, the right to buy to the exclusion of others, including the English: "We consider it as established by the usage of different nations into a kind of *Jus gentium* [law of nations] for America," Jefferson declared, "that a white nation settling down and declaring that such and such are their limits, makes an invasion of those limits by any

* For an explanation of the abbreviated references, see p. 478.

other white nation an act of war, but gives no right of soil against the native possessors." But in the contest with American whites for their lands, Hammond saw the native possessors as the big losers: The British believed, he told Jefferson, that it was the intention of the United States "to exterminate the Indians and take the lands." Such directness was hardly diplomatic finesse, nor was it true, apparently, for Jefferson assured him that "on the contrary, our system was to protect them, even against our own citizens: that we wish to get lines established with all of them, and have no views even of purchasing any more lands of them for a long time" (XVII, 328–29). Less than two decades later, before Jefferson handed over the presidency to Madison, white America had acquired through purchase and otherwise 109,884,000 acres of Indian land.

Notwithstanding Hammond's grasp of the motives of his American cousins, it was true that Jefferson had long thought of himself as a protector of the Indians. There were those early associations and attachments he mentioned to John Adams. "Our Indians," as he revealingly called them, were products of the environment, of "our America," and therefore proper subjects of the natural history he reproached his countrymen for neglecting. In the *Notes* his ranking of Logan's speech with the best of Demosthenes or Cicero was testimonial sufficient of his appreciation of native eloquence. The need to collect vocabularies of their languages became equally obvious with his lament "we have suffered so many of the Indian tribes to extinguish" without having previously collected "the general rudiments at least of the languages they spoke." With his own vocabularies, mysteriously stolen as he left the White House in 1809, he had planned a work to demonstrate the Asiatic origins of American Indians or, in his more exact and more vivid imagery, "the probability of a common origin between the people of color of the two continents" (April 18, 1806, XI, 102).

To the small group of French savants for whom he drafted the *Notes*, Jefferson half-apologized for the absence of any Indian monuments in his native state, "for I would not honor with that name arrow points, stone hatchets, stone pipes, and half-shapen images. Of labor on the large scale, I think there is no remain as respectable as would be a common ditch for the draining of lands; unless it would be the barrows, of which many are to be found all over this country." His reply to a query about Indian remains contained an account of his own perpendicular slash through a burial mound near his home, his thorough examination of the bones, and his conjecture that they might have made up a thousand skeletons. Whatever the occasion of their interment, they continued to be "of considerable notoriety among the Indians." Some thirty years previously a passing band had gone directly through the woods to the mound and "having staid about it for some time, with expressions which were construed to be those of sorrow, they returned to the high road . . . about a half a dozen miles . . . and pursued their journey."

The absence of imposing remains did not keep Jefferson from affirming directly to General Chastellux that the Indian was, "in body and mind, equal to the white man." Unlike the black, whom he supposed not so, the Indian was or should be a member of the family. Thus as president he welcomed visiting tribesmen to the White House as though they had come to a reunion of the clan: On January 7, 1802, for instance, he saluted "our brothers" from the Miami, Potawatomi, and Wea nations: "Made by the same Great Spirit, and living in the same land with our brothers, the red men, we consider ourselves as of the same family; we wish to live as one people and to cherish their interests as our own" (XVI, 390). The following year he observed to Colonel Benjamin Hawkins, the Creek agent:

> In truth, the ultimate point of rest and happiness for them is to let our settlements and theirs meet and blend together, to intermix, and become one people. Incorporating themselves with us as citizens of the United States, this is what the natural progress of things will, of course, bring on, and it will be better to promote than to retard it. . . . And we have already had an application from a settlement of Indians to become citizens of the United States. [February 18, 1803, X, 363]

To this end he placed his famous plan before one delegation after another: They should give up the chase, dispose of lands needed only for hunting, become tawny yeomen farmers, and intermix with the white population. Since they were "now reduced within limits too narrow for the hunter's state," as he said in his second inaugural address, "humanity enjoins us to teach them agriculture and the domestic arts" (March 4, 1805, III, 378).

But by this point the dissociation between Jefferson's word and act had become very great. In early 1805, while he proposed *incorporation* in his inaugural on March 4, he suggested *removal* to a Chickasaw delegation on March 7: "We have lately obtained from the French and Spaniards all the country beyond the Mississippi called Louisiana, in which there is a great deal of land unoccupied by any red men. But it is very far off, and we would prefer giving you lands there, or money and goods, as you like best, for such parts of your land on this side of the Mississippi as you are disposed to part with" (XVI, 412). But those far-off lands on which Jefferson wanted the Chickasaws were as "unoccupied" as the wilderness Bradford and the Pilgrims came to was "unpeopled" —as Annie Heloise Abel commented on this passage, Jefferson spoke "without an unnecessary regard for the truth." And earlier, barely five months after he had written Colonel Hawkins so eloquently on the desirability of intermixture, he drafted a constitutional amendment (July 1803) to validate the Louisiana Purchase and provide for the removal of Eastern tribes to upper Louisiana. His abortive amendment was the first formal advocacy of colonizing Indians across the Mississippi.*

Color defined citizenship. In the second draft of his proposed amendment Jefferson provided that Louisiana be made a part of the United States and "its white inhabitants shall be citizens, and stand, as to their rights and obligations, on the same footing with other citizens of the U.S. in analogous situations. . . . Florida also, whenever it may be rightfully obtained, shall become a part of the U.S., its white inhabitants shall thereupon be Citizens & shall stand, as to their rights & obligations, on the same footing with other citizens of the U.S. in analogous situations." Though just five months earlier Jefferson had proudly told Hawkins of an application for citizenship from an Indian settlement, location unspecified, now he took not the slightest step, in these drafts or elsewhere, toward making red inhabitants citizens—not those presently in the states and presumably to be colonized west of the Mississippi, nor those presently there, nor those in Florida, "whenever it may be rightfully obtained." The reality is that Jef-

* For a discussion of this proposed amendment and its background I am indebted to what is still far and away the best study of it—"The History of Events Resulting in Indian Consolidation West of the Mississippi," *Annual Report of the American Historical Association for the Year 1906* (Washington: Government Printing Office, 1908), I, 233–450, by Annie Heloise Abel. She argued that the idea of removal originated with Jefferson and first came to him *after* the acquisition of Louisiana (the treaty of cession was signed May 2 but antedated April 30, 1803). I think she erred slightly by missing the import of Jefferson's earlier letter (February 27, 1803, X, 370) to William Henry Harrison: "Our settlements will gradually circumscribe and approach the Indians, and they will in time either incorporate with us as citizens of the United States, *or remove beyond the Mississippi*. The former is certainly the termination [N.B.!] of their history most happy for themselves; but, in the whole course of this, it is essential to cultivate their love" (my italics). Though he then still favored incorporation, Jefferson was toying with the idea of removal, therefore, *before* the purchase provided "unoccupied" lands to dump them on —colonizing reds in the West was the internal counterpart of his external "expatriation" scheme for blacks. Finally, I must note that Jefferson's world is not as lost to our time as some have contended: His use of *termination* here directly anticipated the official policy of that name a century and a half later. In the 1950s Presidents Truman and Eisenhower were still seeking love and an end of Indian history through detribalization and incorporation.

ferson never abandoned his goal of removing red noncitizens. After the Chickasaws had been invited to far-off lands in 1805, the Choctaws and the Cherokees were pressured to leave in 1808. On May 4 of that year a delegation of Cherokee farmers from Tennessee asked Jefferson to assign their lands to them in severalty and to make them citizens. He put them off, spoke of the attractiveness of lands beyond the Mississippi "where game is plenty," but for a moment brought his celebrated lucidity to bear on their problem: Only "our Great Council, the Congress," he kindly instructed them, could make Indians citizens (XVI, 432–35). Needless to add, he put no such proposal before the legislature.

Jefferson's views on physical amalgamation were no less beguiling. As early as 1691, to prevent "abominable mixture and spurious issue," Virginia had passed an act banishing any free white who intermarried "with a Negro, mulatto, or Indian man or woman, bound or free." In subsequent reenactments Indians were dropped but that did not mean that their intermarriages no longer suffered legal disabilities, for Indians had come to be included within the term *mulatto:* as a 1705 act read, the child of an Indian was to be "deemed, accounted, held and taken to be a mulatto." Several other assemblies also prohibited red and white marriages —John Adams's Massachusetts, for instance, had done so in 1692 and renewed the prohibition in 1786. All along the seaboard, laws and mores made legal marriages exceedingly rare. Moreover, those who advocated a contrary course were themselves infected by the underlying racism. Historians have been fond of citing William Byrd's famous advocacy of 1728 without, seemingly, scrutinizing it closely. Byrd's argument was that the Englishman should have stomached the Indian's darkness of skin for his own security and good: "Nor would the shade of skin have been any reproach at this day," he added reassuringly, "for if a Moor might be washed white in three generations, surely an Indian might have been blanched in two." But that reflection demonstrably did little about present antipathies. The colonel did not himself put his theory to the test, of course, nor did Jefferson. The latter in truth took not one practical step toward that end, not as the key member of the General Assembly's Committee of Revisors—charged in 1776 with a thorough overhaul of Virginia's laws—and not as president later when his executive leadership might have done something to reverse institutionalized discrimination and the free-floating prejudice against "persons of color." At best Jefferson's championship of white–red miscegenation has to be put down as rhetoric, more of the head than the bed.

3

Color, alas, also conditioned candor. The student of Jefferson's indirection might well begin by holding up his "Indian Addresses" alongside his confidential messages and letters. On November 3, 1802, for instance, he wrote Handsome Lake, the great Seneca prophet, that his fears for the territorial security of the Iroquois were groundless:

> You remind me, brother, of what I said to you when you visited me the last winter, that the lands you then held would remain yours, and shall never go from you but when you should be disposed to sell. This I now repeat and will ever abide by. We, indeed, are always ready to buy land; but we will never ask but when you wish to sell. . . . Go on then, brother, in the great reformation you have undertaken. Persuade our red brethren then to be sober, and to cultivate their lands; and their women to spin and weave for their families. [XVI, 393–96]

Abide by this assurance he did, repeating it throughout his administrations. On January 18, 1809, for instance, he met the complaints of a delegation of chiefs with the reassurance that the United States had no right to purchase lands "but with your own free consent" and "we will never press you to sell, until you shall desire yourselves to sell it" (XVI, 466–70).

Yet in 1802, less than two months after his comforting words to Handsome Lake, a working paper "Hints on the Subject of Indian Boundaries" (December 29, 1802, XVII, 373–77) contained Jefferson's covert suggestions for extinguishing titles to lands they refused to sell; the following month a confidential message shared his disingenuous hints with Congress—minus some awkward details—on how to undermine Indian leaders who persisted "obstinately in these dispositions." Though duty required him to submit his views to the legislature, he warned that, "as their disclosure might embarrass and defeat their effect, they are committed to the special confidence of the two houses" (January 18, 1803, III, 489–92). Therewith he secretly launched a systematic campaign of psychological warfare against the tribes: it deprecated or denied their own agriculture by urging them "to abandon hunting" and adopt white techniques of tillage, stock raising, and domestic manufactures; it entrapped their leading men into running up debts at government trading posts so they would have to sell their lands to pay; and it bought the complaisance of others through "largesses" or, in plain language, bribes. Jefferson's letters to lieutenants in the field filled in the details.

"With candor" Jefferson outlined his policy for General Andrew Jackson, who had his eyes on Creek and Cherokee holdings, and guaranteed him he would pursue "unremittingly" its two main objects: "1. The preservation of peace; 2. The obtaining lands" (February 16, 1803, X, 357–60). He rallied Colonel Hawkins, whom Jackson and the Georgians accused of being "more attached to the interests of the Indians than of the United States," by emphasizing the necessity of the Creeks "learning to do better on less land." He proposed a painfully revealing figure of speech to illustrate the lesson: The Creeks should have "the wisdom of the animal which amputates and abandons to the hunter the parts for which he is pursued . . . with this difference, that the former sacrifices what is useful, the latter what is not" (February 18, 1803, X, 360–65). The trapper's imagery also surfaced in his "private and confidential letter" to Governor William C. C. Claiborne of the Louisiana Territory: Claiborne should give the Southern tribes presents of arms, ammunition, and other essentials "to toll them in this way across the Mississippi, and thus prepare in time an eligible retreat for the whole" (May 24, 1803, X, 390–96).

But Jefferson's letter to Governor William Henry Harrison of the Indiana Territory revealed most candidly how the trap should be sprung, leaving the Indians dependent on the market economy and relieved of their extensive forests:

> To promote this disposition to exchange lands, which they have to spare and we want, for necessaries, which we have to spare and they want, we shall push our trading uses, and be glad to see the good and influential individuals among them run in debt, because we observe that when these debts get beyond what the individuals can pay, they become willing to lop them off by a cession of lands.

To keep tribes from stiffening "against cessions of lands to us," he directed that their minds "be soothed and conciliated by liberalities and sincere assurances of friendship"—Jeffersonisms that roughly translated as bribes and philanthropic commiseration—and suggested a secret agent be sent to one chief, "as if on other business," to inculcate the advantages of land cession. In his view, all of this made up the system that would best promote the interests "of the Indians and ourselves, and finally consolidate our whole country to one nation only. . . . The crisis is pressing; what ever can now be obtained must be obtained quickly." Finally,

Jefferson knew he could trust his fellow Virginian to realize how "sacredly" this letter was to be "kept within your own breast, and especially how improper to be understood by the Indians. For their interests and their tranquillity it is best they should see only the present age of their history" (February 27, 1803, X, 368–73).

Though Jefferson had responded to Handsome Lake's egalitarian "Brother" in kind, and sometimes so addressed other Indians during his first months in the White House, he soon settled into saluting delegations as "Children," "My Children," "My son." * Listen to our first truly imperial president admonish "My son Kitchao Geboway" on February 27, 1808:

> My son, tell your nation, the Chippewas, that I take them by the hand, and consider them as a part of the great family of the United States, which extends to the great Lakes and the Lake of the Woods, northwardly, and from the rising to the setting sun. . . . You see that we are as numerous as the leaves of the trees, and that we are strong enough to fight our own battles, and too strong to fear any enemy. When, therefore, we wish you to live in peace with all people, red and white, we wish it because it is for your good. . . . [When the English] shall be driven away, my son, what is to become of the red men who may join in their battles? Take the advice then of a father, and meddle not in the quarrels of the white people, should any war take place between them; but stay at home in peace, taking care of your wives and children. [XVI, 425–27]

Here the gap between word and act narrowed a bit, with the advice to red men not to meddle in the affairs of white people. If in the family, and their inclusion was by no means always certain, then they were there as permanent children who had to be sent to bed when adults discussed serious matters. Their paternal benefactor would protect them in their absence by a curious calculus: what procured gratifications for white citizens, namely Indian lands, was also in the best interest of his red children. Anyhow, as noncitizens they could not interact as equals with any of the whites who counted when the votes were in. They were nonpersons in a civil sense or, in an adaptation of Chief Justice Taney's famous opinion again, they had a subordinate and inferior status

* There was a harbinger of this shift even in the exchanges with Handsome Lake. In March 1802 the prophet opened their correspondence by addressing him directly as "Brother." Jefferson's response was conveyed by Secretary of War Henry Dearborn, who began with a salutation to his "Brothers" Handsome Lake and the other Iroquois delegates but then swerved to tell them their needs were being well considered by their "father and good friend, the President of the United States"—the letters were discussed by Anthony F. C. Wallace in *The Death and Rebirth of the Seneca* (New York: Vintage Books, 1972), pp. 266–72.

that gave them, like the blacks, "no rights or privileges but such as those who held the power and the government might choose to grant them." * The Indian was not, therefore, Jefferson's brother. Under the best of circumstances he was merely "my son" who had to be protected despite himself.

Of course circumstances were not always at their best so that Jeffersonian philanthropy might take effect. It was that side of his Indian policy that made it "essential to cultivate their love." There was another side, however, as Jefferson went on to impress upon Harrison:

> As to their fear, we presume that our strength and their weakness is now so visible that they must see we have only to shut our hand to crush them, and that all our liberalities to them proceed from motives of pure humanity only. Should any tribe be foolhardy enough to take up the hatchet at any time, the seizing the whole country of that tribe, and driving them across the Mississippi, as the only condition of peace, would be an example to others, and a furtherance of our final consolidation. [February 27, 1803, X, 370–71]

* In *Dred Scott* v. *Sandford* (19 Howard 393, 1857) Taney differentiated sharply between Negroes and Indians. The latter had their own political communities, he pointed out, and were simply "foreigners not living under our Government. It is true that the course of events has brought the Indian tribes within the limits of the United States under the subjection of the white race. . . . But they may, without doubt, like the subjects of any other foreign Government, be naturalized by the authority of Congress, and become citizens of a State, and of the United States. . . ." In the event, however, citizenship of Afro-Americans was affirmed by the Fourteenth Amendment, ratified July 28, 1868, while Indians, not covered by that amendment or any other, had to wait over a half-century until Congress passed the Indian Citizenship Act of June 2, 1924. (The major exceptions were those Indians who received land title under the Dawes Act of 1887—for the germane bills, see Charles J. Kappler, ed., *Indian Affairs, Laws and Treaties* [Washington: Government Printing Office, 1903–29], I, 33–36; IV, 420.) Despite Taney's words to the contrary, then, his judgment in fact applied to Indians and did so long after blacks had ceased to be excluded formally from the political compact.

In 1884 the Supreme Court, in the case of *John Elk* v. *Charles Wilkins* (112 U.S. 94), upheld the decision of a lower court that denied Elk, a Native American separated from his tribe and seemingly otherwise qualified, the right to register to vote on the grounds that he was not a citizen of the United States. Two dissenting justices denounced the decision in language that distinctly echoed Taney's dictum about Afro-Americans. The majority opinion, said the dissenters, showed that "there is still in this country a despised and rejected class of persons, with no nationality whatever; who, born in our territory, owing no allegiance to any foreign power, and subject as residents of the States to all the burdens of government, are yet not members of any political community nor entitled to any of the rights, privileges, or immunities of citizens of the United States." In 1913 the Supreme Court, in *U.S.* v. *Sandoval* (231 U.S. 39), ruled in favor of Pueblo land claims in an opinion grounded on an assumption of the racial inferiority of the Native American claimants. In Associate Justice Willis Van Devanter's words, the Pueblos were "essentially a simple, uninformed, and inferior people"—quoted by Wilbert H. Ahern, "Assimilationist Racism: The Case of the 'Friends of the Indian,' " *Journal of Ethnic Studies*, IV (Summer 1976), 29.

CHAPTER IX

Jefferson, II: Benevolence Betrayed

> There are thousands and tens of thousands, who would think it glorious to expel from the Continent, that barbarous and hellish power, which hath stirred up the Indians and the Negroes to destroy us.
>
> —THOMAS PAINE,
> *Common Sense*, 1776

"NOTHING could be more embarrassing to Jefferson," unkindly observed Henry Adams, "than to see the Indians follow his advice." The Shawnee prophet Tenskwatawa was a prime case in point.

Given the name Laulewasika, or Loud Mouth, at his birth in 1775, he was seven years younger than his brother Tecumseh. He arrived in hard times. A few months earlier a band of white hunters had murdered his father as an "insolent savage." When he was two years old other whites shot down Cornstalk as the celebrated old Shawnee chief stood peaceably at the door of his cabin. When he was five George Rogers Clark led an army of Kentuckians across the Ohio into Shawnee country, burned the village of Old Requa to the ground, and leveled five hundred acres of green corn. When he was twenty the fighting paused with the signing of the Treaty of Fort Greenville, by which the Northwestern tribes gave up the southern half of present-day Ohio and part of Indiana. But by 1795 he did not much care anymore; the whiskey kegs at the treaty grounds interested him more than the losses, humiliation, and poverty of his tribespeople. Unlike Tecumseh, already an outstanding warrior who showed statesmanship in refusing to come to Greenville, Laulewasika stayed there to lap up the white man's firewater and to retreat further from the collective hell into his despondency and alcoholism. For the next half-dozen years and more he wandered through the Old Northwest, drinking, quarreling, and doctoring a little with herbs.

But then, shortly after the turn of the century, Laulewasika became Tenskwatawa, "the Open Door." This sudden transformation was truly remarkable, but no more so than the events leading up to it. The Great Kentucky Revival was in full swing. Among the participating sects were the Shakers, who had arrived in America in 1774 from their native Lancashire, where their roots went back a century and more to the seething religious radicalism we considered in chapter 3. Shakers believed they could live without sin, advocated nonviolence and complete honesty, sought revelations and delivered prophecies, and had been contemptuously named for giving their faith physical expression through shakings, jerkings, stampings, reelings. They defended the dance as an ancient form of worship and drew on Jeremiah 31:4–13, to justify turning "their mourning into joy" by going forth "in the dances of them that make merry." Their ecstatic movements gave them entry to Native American devotion through dance rhythms and perhaps provided connections at midcentury with Smohalla's Dreamer cult along the Columbia and still later with the Indian Shakers of the Puget Sound area. Tenskwatawa, at all events, could not have understood their words and thus could only have been influenced by their physical movements. When he later learned what those expressed, he gave the Shakers credit for the great change in his life: he stopped drinking, had revelatory visions of his own, and worked out a "Code" that fused their ideas, save for celibacy, with Algonquian mysticism and traditions.

Like Handsome Lake, another former drunkard, Tenskwatawa followed the turnabout in his life with a heroic effort to restore the shattered culture of his people. His apocalyptic message raced through the Shawnees and beyond to other tribes; in 1808, at Tecumseh's urging, he established headquarters of an intertribal confederacy on the banks of the Tippecanoe in Indiana. That summer, while trying to reassure William Henry Harrison of his peaceful intentions, he obligingly outlined his goals for the apprehensive governor:

> That we ought to consider ourselves as one man; but we ought to live agreeably to our several customs, the red people after their mode, and the white people after theirs; particularly, that they should not drink whiskey; that it was not made for them, but the white people, who alone knew how to use it . . . that they must always follow the directions of the Great Spirit, and we must listen to him, as it was he that made us: determine to listen to nothing

> that is bad: do not take up the tomahawk, should it be offered by the British, or by the long knives: do not meddle with any thing that does not belong to you, but mind your own business, and cultivate the ground, that your women and your children have enough to live on.

Under influence of his teaching, Indians refused offers of whiskey, to Harrison's surprise, and seemingly adopted the whole Jeffersonian program, including the raising of stock and cultivation of the rich Wabash River bottoms near Prophet's Town. Their insistence upon racial solidarity and their angry refusal to consider land cessions were no doubt largely Tecumseh's contribution to the Code—the land belonged to all the Indians, Tecumseh maintained, and could not be sold off tract by tract by village chiefs who had been pressured or bought off by government officials and agents. But even these views were not necessarily hostile to white interests or in conflict with Jefferson's professions—after all, he had repeatedly promised the tribes to buy their lands only with their own free consent.

Tenskwatawa and Tecumseh had inspired and directed a "revitalization movement," moreover, a movement defined by Anthony F. C. Wallace as "a deliberate, organized, conscious effort by members of society to construct a more satisfying culture." Handsome Lake's *Gaiwiio*, or Old Way, was just such an effort among the Iroquois, one that Jefferson had apparently given his benediction, you will recall, by urging the Seneca prophet to carry on "the great reformation you have undertaken." Hence, why did he not now welcome a comparable effort by the Shawnee brothers to restore cultural integrity within the Northwestern tribes? No doubt part of the answer was that Jefferson bestowed his good wishes on the Iroquois because they no longer constituted a major threat to white interests, no longer held vast tracts of desirable land, and now held a more conciliatory attitude toward white goods and customs. But that was not the whole answer.

2

The truth was that Jefferson had little understanding of Native American culture and less concern for its integrity. To be sure, he was a collector of Indian words, but these vocabularies were only the bones of their living languages. He literally collected Indian physical remains, but these bones said more about Jefferson than they did about live Indians. As he had to know, he would have been guilty of grave robbing, *pro confesso*, had he entered a Charlottesville cemetery, dug up skeletons, counted the bones, and published his findings in *Notes on Virginia*. Yet he had no hesitation in excavating the mound near his home and putting the results before his readers and before members of the American Philosophical Society. Had he respectfully studied the ways of his red neighbors, he would have known that they considered burial grounds sacred and that the descendants of the Temple Mound builders, who had some thirty years previously visited the site he cut into, had done more than, as Jefferson put it in the *Notes*, "staid about it for some time, with expressions which were construed to be those of sorrow." As we have seen, Thomas Morton, whom he derided, had taken the trouble a century and a half earlier to discover that Indians held it impious to deface graves and were accustomed to gather around them at certain times to bewail the dead.

"But the dead have no rights," declared Jefferson (XV, 43). His contempt for the past was as profound as the Indians' reverence. In his second inaugural he listed the powerful obstacles that stood in the way of his administration's benevolent endeavors to induce Indians "to exercise their reason, follow its dictates, and change their pursuits with the change of circumstances." His endeavors were

> combated by the habits of their bodies, prejudice of their minds, ignorance, pride, and the influence of interested and crafty individuals among them, who feel themselves something in the present order of things, and fear to become nothing in any other. These persons inculcate a sanctimonious reverence for the customs of their ancestors; that whatsoever they did, must be done through all time; that reason is a false guide, and to advance under its counsel, in their physical, moral, or political condition, is perilous innovation; that their duty is to remain as their Creator made them, ignorance being safety, and knowledge full of danger; in short, my friends, among them is seen the action and counteraction of good sense and bigotry; they, too, have their anti-philosophers. [III, 379]

The sly diversionary assault here on white conservatism should not deflect our attention from the main thrust of Jefferson's attack: No less imperiously than the Puritans did he command the Indians to negate themselves. The Saints had demanded they enter Calvin's world to become tawny Christians; Jefferson demanded they enter Newton's to become Enlightenment philosophes. In both sets of demands they first had to accept "civilization"—private property, haircuts, European dress and language, white domination, the lot—before they could reach true faith or right reason. And the grim joke was that white racism made it impossible for all but a handful to achieve full church membership or full citizenship.

"When once we quit the basis of sensation, all is in the wind," Jefferson wrote Adams (August 15, 1820). Like his friend, Jefferson despised the "whimsies, the puerilities, and unintelligible jargon" of Platonism (July 5, 1814) and believed firmly that to speak of "*immaterial* existences is to talk of *nothings*." Hence Tenskwatawa's ecstatic trances and claims to have reached the beyond had to be nonsense or worse. The Shawnee prophet was "more rogue than fool," Jefferson replied to Adams:

> His declared object was the reformation of his red brethren, and their return to their pristine manner of living. He pretended to be in constant communication with the great spirit, that he was instructed by him to make known to the Indians that they were created by him distinct from the Whites, of different natures, for different purposes . . . that they must return from all the ways of the Whites to the habits and opinions of their forefathers. . . . I concluded from all this that he was a visionary, inveloped in the clouds of their antiquities, and vainly endeavoring to lead back his brethren to the fancied beatitudes of their golden age. I thought there was little danger of his making many proselytes from the habits and comforts they had learned from the Whites to the hardships and privations of savagism, and no great harm if he did. We let him go on therefore unmolested. [April 20, 1812]

Let him go on: Was it not remarkable that the prototypical liberal had ever thought of molesting the free exercise of religious beliefs? And that on the Indians' own lands!

To say that the prophet had been allowed to go his way unmolested, in any event, was not entirely accurate. In 1807 Charles Jewitt, an Indian commissioner in the Northwest, proposed that he be seized and imprisoned. In response Jefferson allowed that if the Indians were to handle the prophet "in their own way, it would be their affair." This read just a bit as if the president were suggesting the prophet's assassination at the hands of red enemies, a suggestion that was in fact considered in Washington after Jefferson left the White House.* Now Jefferson hastened to in-

* In 1811 the Ohio Indian agent, John Johnston, made the suggestion in more inclusive form by proposing two assassinations to the secretary of war: "If it was agreeable to the Government," he wrote, "the Indians could be easily prevailed on to kill the Prophet and his Brother [Tecumseh]." Quoted in Robert F. Bauman, "Review of *John Johnston and the Indians . . . ,*" *Ethnohistory*, V (Fall 1958), 392–96.

struct Secretary of War Dearborn, however, that "we should do nothing towards it," with the "it" relating, presumably, to the arbitrary jailing of the prophet. "That kind of policy is not in the character of our government," Jefferson continued, "and still less of the paternal spirit we wish to show towards that people." On the other hand, bribes were not out of character for the United States: "But could not Harrison gain over the prophet," he asked in a question verging on an order: Tenskwatawa was "no doubt a scroundrel, and only needs his price?" Anyway, he added reassuringly, other unspecified Indian "operations we contemplate" would, if successful, "put them entirely in our power" (August 12, 1807, XI, 325).

3

We hold these truths to be self-evident. . . . In his formative statements of white American liberalism, Jefferson had committed himself to the proposition that all men had a natural right to adult independence. Yet he obviously relished being the omnipotent and omniscient father of red children, and his paternalism had the objective we just noted of bringing those not already removed "entirely in our power." Like fatherly Prospero in *The Tempest* (1611), he pretended to have no selfish interests of his own and acted toward his wards only out of his upswelling benevolence. But also like Shakespeare's prefiguration of all the paternal colonizers to come, he in fact indulged hidden psychological cravings for absolute power by keeping Indians in a state of childish dependence. "Good" children were like the Shawnee Blackhoof, who had resigned himself to the wishes of white officials and in turn lived off their largesses. "Bad" children were Calibans like Tecumseh and Tenskwatawa, who refused to be bribed into perpetual submissiveness. They were the ones who repaid their white father's unfailing kindness with an obstinate defiance that threatened to unmask his possessive domination. And that is not to say this was yet another manifestation of the "vapor of duplicity" hanging over Jefferson's public life—to my mind John Quincy Adams was closer to the truth when he observed that in deceiving others, Jefferson "seems to have begun by deceiving himself." He really seems to have considered it more than rank ingratitude when Indians resisted being brought entirely under his power: it was a betrayal of their too-lenient father, for which they deserved a good thrashing. And if they persisted, they would get what was coming to them—and more. In his next letter to Dearborn he instructed him to "prepare for war in that quarter" and "to let the Indians understand the danger they are bringing on themselves": "We too are preparing for war against those, and those only who shall seek it; and that if ever we are constrained to lift the hatchet against any tribe, we will never lay it down till that tribe is exterminated, or driven beyond the Mississippi." They should be told plainly: "In war, they will kill some of us; we shall destroy all of them" (August 28, 1807, XI, 342–46).

Nestled under Jefferson's philanthropy was an ominous will to exterminate. Indians *were* children, red shadows dancing through the trees, naughty boys and girls who combated him, as he said in his second inaugural, with "the habits of their bodies." And if they would not listen to reason, would not bury the dancers in themselves, Jefferson would bury them. Their open resistance, moreover, provided a delightful release from the burdens of his insincere love. Still, who would willingly acknowledge delight in contemplating infanticide? No one, least of all Jefferson—and he could draw on two time-tested ways of hiding from himself and others the meaning of Indian extermination.

One way was to absolve or seemingly absolve the Indians of responsibility for opposition to his "operations." Eternal children, they were too immature themselves to launch a campaign of resistance; hence they had to have been manipulated or tampered with by crafty adults, most likely renegade whites or foreign agents. Thus he had allowed Tenskwatawa to proceed as a harmless visionary, Jefferson wrote Adams, "but his followers increased till the English thought him worth corruption, and found him corruptible. I suppose his views were then changed; but his proceedings in consequence of them were after I left the administration, and are therefore unknown to me; nor have I ever been informed what were the particular acts on his part which produced an actual commencement of hostilities on ours" (April 20, 1812). Someone would have had a hard time informing Jefferson of particular hostile acts of the prophet, for the reference was to General Harrison's unprovoked and unauthorized invasion of Indian country in 1811 and to the so-called Battle of Tippecanoe that stopped the movement toward Indian confederacy.* Nor did Jefferson say why the

* The great Tecumseh fell at the Battle of the Thames in October 1813. Tenskwatawa died in Kansas in 1837. When George Catlin encountered him across the Mississippi and painted his picture, he found this once "very shrewd and influential man" destroyed by circumstances: "he now lives respected, but silent and melancholy in his tribe" (*North American Indians* [London: Chatto and Windus, n.d. (1841)], II, 133–35).

seductions of the English succeeded where the bribes of the United States had not, but because the "backward" Northwest tribes had gone over to the enemy, he wrote Adams in another letter, "we shall be obliged to drive them, with the beasts of the forest into the Stony mountains. They will be conquered however in Canada. The possession of that country secures our women and children for ever from the tomahawk and scalping knife, by removing those who excite them" (June 11, 1812). Adams quite agreed: "Another Conquest of Canada will quiet the Indians forever and be as great a Blessing to them as to Us" (June 28, 1812). And these Canadian seducers or exciters played exactly the same role for Jefferson and Adams as the "base Canadian fiends" who stirred up Pequots had played for Timothy Dwight: It is permissible to think of this mask for motive as the "outside agitator" refrain and desirable to remember that it appeared in the Declaration of Independence joined to the second major justification.

Jefferson called his bill of particulars against George III—surely the busiest outside agitator of all time—the "black Catalogue of unprovoked injuries." The alleged injury involving Indians appeared for the first time—that is, before June 13, 1776—on the first page of Jefferson's first draft of the Virginia constitution, where it followed a dark crime: The Crown had prompted "our negroes to rise in arms among us; those very negroes whom he hath from time to time refused us permission to exclude by law." Jefferson then charged George III with "endeavoring to bring on the inhabitants of our frontiers the merciless Indian savages whose known rule of warfare is an undistinguished destruction of all ages, sexes, & conditions of existence." This charge survived two other Jefferson drafts, as well as committee deletions and amendments, and was printed substantially unchanged in the Virginia constitution adopted June 29, 1776. Meanwhile, Jefferson transferred it to his "original Rough draught" of the Declaration of Independence, where it survived the corrections, additions, and deletions of John Adams, Benjamin Franklin, and the two other members of the Committee of Five. In the Continental Congress debate there was one slight addition and one slight deletion, to make the charge against George III read as we know it in the parchment copy:

> He has excited domestic insurrections amongst us, and has endeavoured to bring on the inhabitants of our frontiers, the merciless Indian Savages, whose known rule of warfare, is an undistinguished destruction of all ages, sexes and conditions.

I rehearse all these details since this rationale for extermination came from Jefferson and was embedded in the great charter of white American liberties. The two engrossed justifications were both racist: first, childish Indians were especially prone to manipulation by tyrannous outside forces; second, Indian savages were especially merciless. And if their violence was especially merciless, could whites not exterminate them in good conscience?

Like Dwight and Adams, Jefferson answered with a resounding *yes!* "You know, my friend," he wrote Baron von Humboldt, "the benevolent plan we were pursuing here for the happiness of the aboriginal inhabitants of our vicinities." The United States had spared nothing to keep them at peace with one another, to teach them agriculture, to have their blood mix "with ours," but all for naught:

> the interested and unprincipled policy of England has defeated all our labors for the salvation of these unfortunate people. They have seduced the greater part of the tribes within our neighborhood, to take up the hatchet against us, and the cruel massacres they have committed on the women and children of our frontiers taken by surprise, will oblige us now to pursue them to extermination, or drive them to new seats beyond our reach. [December 6, 1813, XIV, 23]

Jefferson, believing the long-awaited day of extermination or pursuit over the horizon had finally arrived, made "necessary to secure ourselves against the future effects of their savage and ruthless warfare." Freed at last from the weight of their benevolence, he and his compatriots might shatter cultures, kill people, destroy homes and crops, and still be in no wise responsible for the genocidal process of which they were instruments. The useful English were guilty of it all, of "the confirmed brutalization, if not the extermination of this race in our America. . . ."

CHAPTER X

Driving Indians into Jefferson's Stony Mountains

> Next to the case of the black race within our bosom, that of the red on our borders is the problem most baffling to the policy of our country.
>
> —JAMES MADISON
> to Thomas L. McKenney,
> February 10, 1826

THE DECLARATION OF INDEPENDENCE, Jefferson rightly observed, was "an expression of the American mind." "The merciless Indian Savages" therein were of a piece with Timothy Dwight's "murderous fiends," George Washington's "beasts of prey," and John Adams's butchering "blood Hounds." In fact, while Jefferson's phrase met with the approval of the other Founding Fathers to a man, it simultaneously established massive continuity with the colonial past. Edward Waterhouse, historian of the 1622 attack on the Virginia colonists, characterized the Indians there as "by nature sloathful and idle, vitious, melancholy, slovenly, of bad conditions, lyers, of small memory, of no constancy or trust. . . ." No less were the natives in the Declaration the New England "ravening wolves" of old, those that slunk through the pages of the Reverends Cotton and Increase Mather, William Hubbard, and those of that other exemplary liberal, Roger Williams, with his "mad-dog" Mohawks and wild-animal Pequots in their "dens." The "new creed" Thomas Morton had outlined with contempt in the 1630s, namely, that the "Salvages" were "a dangerous people, subtill, secreat and mischeivous," had by the 1770s become an axiom so universally accepted it was writ large on the birth certificate of the United States of America.

The past was present during the debate on the Declaration from the second to the fourth of July. It was there in that very embodiment of capitalism, Benjamin Franklin, who sat quietly through the discussions, resting comfortably on his maxim: "*God gives all things to industry*." With the proverbial practicality of *Poor Richard* (1732–57), he represented, as Max Weber saw, the identical attitudes of Puritan worldly asceticism, "only without the religious basis, which by Franklin's time had died away." Not all utilitarian seriousness, however, he proved as capable as John Adams of a half-playful outline of the imperial mission of their countrymen: "By *clearing America* of Woods," he once pointed out, they were "*Scouring* our Planet. . . ." And in a famous sentence, he anticipated an America cleared as well of the children of the woods: "If it be the design of Providence to extirpate these savages in order to make room for the cultivators of the earth, it seems not improbable that rum may be the appointed means." As a speculator in Western holdings, he was by no means a playful or disinterested student of this particular providential windfall. Linked to the past and a shaper of things to come, Franklin illustrated in his writings and speculations a truth destined for future neglect: If agricultural products later provided the wherewithal for the rise of industrial capitalism in the New World, the whole process commenced on Indian lands acquired by pressured purchase, fraud, and violence.

Still, it was not entirely accurate to say that the religious basis of the Protestant ethic "had died away" by the 1770s. It lived on full-blown in such anachronisms as Timothy Dwight. It was physically present at the Continental Congress in the person of John Adams, who argued strongly for the Declaration on the floor. His list of reasons for the flourishing house of Adams contained the timeworn virtues: "industry, frugality, regularity, and religion." His hostility toward the dancing and singing and reveling—of the Indians, of Morton, or of other "rakes, fops, sots"—was a swift-flowing current, powerful as Winthrop's and Endicott's. If it had occurred to John Adams that the Revolution's liberty trees and poles were lineal descendants of Morton's Maypole, he would not have been amused by this little ruse of history. Instead he championed sumptuary laws and thumped his chest with seemly masculine seriousness: "Vanities, levities, and fopperies," he proclaimed in "Thoughts on Government" (1776), "are real antidotes to all great, manly, and warlike virtues."

By contrast, Jefferson's roots went down by the Chesapeake and across to Newton and Locke. We have been taught to remember the vast difference between the Calvinist view of man's depravity and Jefferson's Enlightenment view of man's possibilities. This stress on difference—however important some of the dissimilarities—has obscured connections and identities, some of which were suggested by the important common grounds Jefferson and

Death of Jane McCrea, John Vanderlyn, 1804. Originally intended to illustrate Joel Barlow's flawed American epic, *The Columbiad* (1807), Vanderlyn's painting helped set the pattern for an endless series of pictorial indictments of Jefferson's "merciless Indian Savages." Always the epic contrast was between dusky evil and fair innocence, between maddened red cruelty and helpless white virtue. "Helpless women have been butchered," said Andrew Jackson in 1818, "and the cradle stained with the blood of innocence." In images or words, this Declaration theme traveled well: The Chinese Boxer, said Theodore Roosevelt in 1900, had "his hands red with the blood of women and children." (Courtesy of the Wadsworth Atheneum, Hartford, Connecticut.)

Adams discovered in their correspondence. Both Puritanism and the Enlightenment made contributions to "the specific and peculiar rationalism of Western culture," to quote Weber again, with the Enlightenment the surprising but true "laughing heir" of Puritanism. Though he did so under different auspices, Jefferson's attempt to rationalize the world was no less determined and no less fantastic than the Puritan's. Jefferson's hostility to mysticism, impulse, and the pleasures of the body was scarcely less than John Adams's and John Winthrop's. "The Indian became important for the English mind," Roy Harvey Pearce noted in 1953, "not for what he was in and of himself, but rather for what he showed civilized men they were not and must not become." This was true for the Founding Fathers as well, except that to them the Indian was no longer so much an anti-Christ, a "skulking heathen," a dusky "bond-slave of Sathan." For Jefferson the Indian was rather the face of unreason. If he chose to remain an Indian, in the face of all paternalistic efforts to the contrary, then he confessed himself a madman or a fool who refused to enter the encompassing world of reason and order. The red child then had to be expelled from the landscape of pastoral tranquillity—or be buried under it. Like the Puritan, Jefferson regarded the Indian's culture as a form of evil or folly. Indeed Jefferson's admonitions to exercise reason were more disorienting, for they were accompanied by benignant smiles at "My children." Yet from the Indian point of view, the end result was pretty much the same: death, flight, or cultural castration—the lopping off and abandoning to the hunter, as Jefferson proposed, of "the parts for which he is pursued."

Just as the Indian never quite became one of the elect among the Saints, so did he not quite become a man among the revolutionists. Many historians have discussed the significance of the deletion of Jefferson's so-called philippic against slavery and the fact that while putting their case before a "candid world," the revolutionists said not one word about being slaveholders themselves. But few have gone on to observe the significance of the *inclusion* of the "merciless Indian Savages" charge—Carl Becker, for instance, discussed it not at all in his good book on the Declaration. Yet by virtue of what was there about Native Americans and not there about Afro-Americans, the Framers manifestly established a government under which non-Europeans were not men created equal—in the white polity, as Louis Hartz has shown, they were nonpeoples. Racism pervaded the Revolution.

The republic's natal day marked the decisive emergence of the white American character out of European culture. From the first plantings the Indian was of critical importance for the colonist's understanding of who he was *not* and for his definition of self that evolved; therefore, the Founders, and in particular Jefferson, were of critical importance in setting forth the characteristics of that self. To Jefferson we have turned for primary data on the hypocritical national commitment to the equal rights of all men everywhere and the concurrent, daily, fundamental denial of equality to people of color as, in Chief Justice Taney's terms, a "subordinate and inferior class of beings." It is only fair to add that at the time the dissociation between revolutionary word and racist act troubled other than Loyalists and foreign critics. Abigail Adams, for instance, wondered out loud about the "passion for liberty" of Jefferson and the Virginians, since "they have been accustomed to deprive their fellow creatures of theirs." Sadly, she and her husband failed to follow this up with the realization they themselves were part of ongoing national efforts to deprive the Indians of theirs—they failed, surely, because like other white Americans they saw the Indians not as fellow creatures but as merciless savages.

2

"And our John has been too much worn to contend much longer with the conflicting factions," John Adams wrote Thomas Jefferson on January 22, 1825. "I call him our John, because when you was at Cul de sac at Paris, he appeared to me to be almost as much your boy as mine." Though the old days in France were long past, John Quincy Adams was at the very least Jefferson's godson and, before he became president, proved it by his handling of the most urgent problem he confronted as secretary of state under James Monroe.

At stake was Florida, East and West. In 1803 Jefferson had drafted his proposed constitutional amendment to apply to it, you will recall, "whenever it may be rightfully obtained." Such scruples and his former mentor's policies to that end seemed weak and vacillating to John Quincy Adams, then an expansionist-minded U.S. senator from Massachusetts. Shortly after he became secretary of state, Adams commenced negotiations that led to the acquisition of Florida as part of the Transcontinental Treaty of 1821; in the meantime the negotiations had clearly revealed Spain's willingness to cede the area in any general settlement of boundaries and claims. But in the spring of 1818 it seemingly had already been obtained, however rightfully or wrongfully: Major

General Andrew Jackson, at the head of an overwhelming force of whites and Indian auxiliaries, had invaded the Spanish possession, crushed red and black resistance, summarily executed red and white "exciters" thereof, captured the town of St. Marks near the Atlantic and the post of Pensacola on the Caribbean, and commanded General Edmund P. Gaines to take St. Augustine. Word of his successful filibuster created consternation in Washington. As Adams put it in a July 15 entry in his *Memoirs* (1875):

> Attended the Cabinet meeting at the President's, from noon till near five o'clock. The subject of deliberation was General Jackson's late transactions in Florida, particularly the taking of Pensacola. The President and all the members of the Cabinet, except myself, are of opinion that Jackson acted not only without, but against, his instructions; that he has committed war upon Spain, which cannot be justified, and in which, if not disavowed by the Administration, they will be abandoned by the country.

Adams argued strenuously that Jackson had not really violated his orders: the violation was apparent only and dictated by the necessity of the case and the misconduct of Spanish officials in Florida; all his actions might be justified as "*defensive* acts of hostility."

By his lone stand Adams blocked open repudiation of Jackson, though the order to take St. Augustine was countermanded and arrangements worked out to return St. Marks and Pensacola to the Spanish authorities. To counter congressional attacks on the administration and the international outcry against the execution of two British subjects, Adams wrote a lengthy communication, addressed to George W. Erving, the U.S. minister in Madrid, but intended for wide circulation. Dated November 28, 1818, this important document amounted to a White Paper in which Adams attempted "a succinct account of the late Seminole War" that would completely "justify the measures of this Government relating to it, and as far as possible the proceedings of General Jackson. The task is of the highest order: may I not be found inferior to it!" It was a tall undertaking.

Jackson's letters from the field revealed just how tall. Southwesterners had secured the best of the Creek lands under the treaty dictated by Jackson in 1814, but to their dismay Florida still eluded their grasp after the Battle of New Orleans. Clashes followed on both sides of the border as white Americans moved in to "pacify" and settle the disputed area. The First Seminole War commenced on November 21, 1817, when a force under the command of General Edmund P. Gaines attacked and burned the frontier Indian village called Fowltown. Later Creek agent David B. Mitchell would testify that "before the attack on Fowltown[,] aggressions . . . were as frequent on the part of whites as on the part of Indians." Jackson had no such balanced view, of course; he approved Gaines's report and wrote Secretary of War Calhoun (December 16, 1817) that "this check to the Savages, may incline them to peace; should it not, and their hostility continue, the protection of our citizens will require that the Wolf be struck in his den."

With orders from Calhoun (December 26, 1817) that permitted him to pursue the enemy to "his den," Jackson marched his men into Florida, hit the Mikasuki towns in the Seminole country, burned three hundred houses, and carted away "the greatest abundance of corn cattle etc."; then he moved beyond his orders to occupy St. Marks, from which he wrote Rachel Jackson:

> I may fairly say, that the modern Sodom and Gomorrow are destroyed . . . the hot bed of the war. Capt McKeever [of the U.S. Navy] who co[o]perated with me, was fortunate enough to capture on board his flotilla, the noted Francis the prophet, and Homollimicko [Himollemico], who visted him from St marks as a British vessell the Capt having the British colours flying, they supposed him part of [Captain George] Woodbines Fleet from new providence coming to their aid, these were hung this morning, I found in St marks the noted Scotch villain Arbuthnot who has not only excited but fomented a continuance of the war I hold him for trial. [April 8, 1818]

Far from being "the noted Scotch villain," Alexander Arbuthnot was a middle-aged trader from the British island of New Providence. At home in Nassau he was an established merchant with a reputation of fair-mindedness and humaneness. His presence in Florida was for trade, not political intrigue, though he sympathized with the Seminoles, on occasion acted as their spokesman, and angrily denounced "the wanton aggressions of the Americans." For such villainy Jackson had him hanged. James Parton, the general's nineteenth-century biographer, characterized the execution as "of such complicated and unmitigated atrocity, that to call it murder would be to defame all ordinary murders." Also convicted by a drumhead court-martial on almost equally flimsy evidence, Robert Ambrister, a British adventurer, died before a firing squad.

But what of "the noted Francis the prophet"? Hillis Hadjo or Josiah Francis, as he was known to traders and other whites, was indeed a noted warrior and religious leader of his people. The son of a Creek woman and a white man, he became a follower of Tecumseh and an inveterate foe of United States expansion. Tecumseh named him "the great prophet of the Creeks" at Tukabatchi in 1812 and took him along on a trip across the Mississippi to enlist Western tribes in their confederacy. Hillis Hadjo fought Jackson in the War of 1812, survived, and afterward visited England where he was feted, commissioned a brigadier general, and received by George IV. Like Tenskwatawa among the Northwestern tribes, Hillis Hadjo was a revitalization prophet among the Creeks and Seminoles, and in a sense his fate was a logical extension of the earlier proposals to imprison or assassinate the Shawnee prophet. The village of Hillis Hadjo and his comrade Himollemico lay about three miles from St. Marks. Seeing a gunboat conspicuously flying British colors, they paddled out to it, received a cordial invitation to "partake of hospitalities," and once aboard found themselves in the jaws of a sprung trap.* The chiefs were then bound and hauled to St. Marks, where Jackson unceremoniously ordered they be hanged.

Jackson had already made up his mind about their guilt, anyhow, along with that of Arbuthnot and Ambrister. James Parton quoted orders to Captain McKeever *before* the seizure of St. Marks and hence before the drumhead courts-martial: "It is all important that these men be captured and made examples of." Evidence that Hillis Hadjo and Arbuthnot, for example, were not "exciting" the Seminoles to war was thus largely beside the point. Nevertheless, Jackson emphasized to Calhoun (May 5, 1818) the legality of the proceedings by which Arbuthnot and Ambrister were convicted "as exciters of this savage and negro War." Testimony at their courts-martial, he maintained, more than justified their condemnation by presenting "scenes of wickedness, corruption, and barbarity at which the heart sickens and in which in this enlightened age it ought not scarcely to be believed that a christian nation would have participated, and yet the British government is involved in the agency." Fortunately George IV was beyond Jackson's reach, but "the modern Sodom and Gomorrow" was not. The homespun Prospero declared to Calhoun that "the Savages . . . must be made dependant on us" and kept out of the reach of "the delusions of false prophets, and the poison of foreign intrigue."

Jackson's "Proclamation on Taking Possession of Pensacola"

* The false flag of the gunboat that decoyed them aboard anticipated the white flag of truce under which General Thomas S. Jesup's men captured Osceola two decades later during the Second Seminole War—but one war at a time.

(May 29, 1818) tightened the rationale: "The Seminole Indians inhabiting the territories of Spain have for more than two years past, visited our Frontier settlements with all the horrors of savage massacre—helpless women have been butchered and the cradle stained with the blood of innocence." As existing treaties obliged them, Spanish authorities should have moved decisively to put an end to these atrocities. They had not and "the immutable laws of self defence, therefore compelled the American Government to take possession of such parts of the Floridas in which the Spanish authority could not be maintained." Freed by alleged precedent and immutable law, Jackson could play Death as executioner.

I imagine Jackson riding across Florida looking like Albert Pinkham Ryder's *Death on a Pale Horse* (1888–1910). Michael Paul Rogin thought that during the recent Creek War he must have had the appearance of Albrecht Dürer's *King Death on Horseback* (1505). Old wounds, fevers left over from smallpox, recurrent malaria, and chronic dysentery had given his skin a yellowish tint and made his tall frame emaciated, almost skeletal. Rogin's analysis compellingly established that Jackson warred against his body as he warred against the Indian: he sought to make both dependent on his will, to transform aggression and death into paternal authority, and to make himself and his men efficient "engines of destruction." Indians were "Ruthless Savages," to be sure, but inefficient ones since they had no comparable control over their passions—Rogin noted that at New Orleans Jackson complained to Monroe, then secretary of war, that his Indian auxiliaries were unreliable "because after every battle they wanted to return home for a dance." The general's pleasures were of a different order.

It was as though barely imaginable "scenes of wickedness, corruption, and barbarity" really were concentrated in the shadowy recesses of the Florida swamps. It was not given to Everyman to destroy a modern "Sodom and Gomorrow." It was furthermore a racial war or, as Jackson put it, a "savage and negro War." Mixtures of Native Americans and Afro-Americans had become so numerous in the lower South, whites adopted the term *griffes* to distinguish these offspring from other mulattoes. Their presence and that of blacks and reds—runaway slaves and Seminoles and Creeks, hated and feared in part for different reasons—made their common resistance especially menacing and therefore their subjugation and destruction especially satisfying.*

* The First Seminole War had among its origins conquest designed to secure Indian lands and therewith deny sanctuary to runaway slaves—see Joshua Giddings, *The Exiles of Florida* (Gainesville, Fla.: University of Florida Press, 1964), pp. 35–56.

Behind his Old Hickory facade Jackson's spirits soared as the triumphs of his campaign mounted. From the Suwanee he wrote Rachel Jackson (April 20, 1818) that he had occupied Chief Bowleg's town and "here we obtained a few cattle, and about three thousand bushels of corn . . . and this was a providential supply, the truth is that we have been fed like the Iseralites of old in the wilderness." Like John Mason and others before him, he saw himself in Old Testament terms as the instrument of an avenging God: "The hand of heaven has been pointed against the exciters of this war, every principle villain has been either killed or taken." And this New Israelite could boast to his wife: "I think I may say that the Indian war is at an end for the present, the enemy is scattered over the whole face of the Earth, and at least one half must starve and die with disease."

Jackson's imperiousness kept pace with his elation. In a summary report to Monroe (June 2, 1818) he declared: "The hords of Negro Brigands must be drove from the bay of Tennessee and possess ourselves of it, this cutts off all excitement . . . by keeping from there all foreign agents." Later he told Monroe (August 10, 1818) it was essential for Gaines to take St. Augustine: "The innocent blood of our aged matrons and helpless infants demand it, as by this alone can we give security and peace to our country." To George W. Campbell, U.S. minister to Russia, he confessed regret at not having stormed the fort at Pensacola so he could have hanged the Spanish governor (October 5, 1818). And all this at a time when the triumph over his own body was hardly less gloriously complete: "My health is greatly impaired," he wrote Monroe (August 10, 1818), "my constitution broken."

Unlike Jackson in the field, John Quincy Adams was denied the pleasure of acting out fantasies of violence. The nascent bureaucratization of the United States meant division of labor and the mechanization of violence—men as "engines of destruction"—and put more distance, physical and occupational, between Adams and Jackson than between Winthrop and Endicott. Nevertheless, Adams could participate vicariously in Jackson's campaign, see if his individualistic rampage could not be bent to serve national interests, and therewith overcome both his and the administration's enemies. Adams's whole-hearted embrace of Jackson's actions in fact created a problem for those modern historians who have contended that the enlightened Indian policy of statesmen in Washington was frustrated and ultimately defeated by fiercely acquisitive frontiersmen—in 1818 the division of labor required for efficient subjugation and dispossession of the Indians was neatly illustrated by the working partnership of Adams and Jackson.

The audacity of his White Paper proved Adams fully equal to the challenge raised by the general's spectacular invasion. The root principle of his dispatch to Ambassador Erving (November 28, 1818) was that *everything* Jackson did was defensive. By not keeping their treaty obligations, Spanish officials had allowed Florida to become a "derelict" province, open to every enemy, "civilized or savage." In President Monroe's name, Adams demanded that the Spanish government discipline the governor of Pensacola and the commandant of St. Marks for their aid to "these hordes of savages in those very hostilities against the United States which it was their official duty to restrain." They deserved punishment for that and for the "hostile spirit" they manifested when Jackson, under "the necessities of self-defence," was forced to invade their province, wage war against its inhabitants, and occupy their posts. Arbuthnot and Ambrister, "by the lawful and ordinary usages of war," might have been executed on the spot "without the formality of a trial"; instead Jackson had generously granted them "the benefit of trial by a court-martial of highly respectable officers." Of their guilt there was no doubt: Arbuthnot was "the firebrand by whose touch this negro-Indian war against our borders had been re-kindled." No sooner did Arbuthnot "make his appearance among the Indians, accompanied by the prophet Hillis Hadjo, returned from his expedition to England, than the peaceful inhabitants on the borders of the United States were visited with all the horrors of savage war—the robbery of their property, and the barbarous and indiscriminate murder of women, infancy, and age." In chastising these deluded hordes of "wretched savages" and in executing their exciters, therefore, Jackson had acted on motives "founded in the purest patriotism."

The White Paper was a virtuoso performance that added luster and reach to the theme of merciless savages and outside agitators. As recently as 1949, in an account embellished by his own twentieth-century references to "fleeing savages" and "murderous Seminoles," historian Samuel Flagg Bemis called it "the greatest state paper of John Quincy Adams's diplomatic career." It spiked congressional guns, quieted international protests, helped bring Spain around on the Western boundary question, and contributed to the boom that would eventually take Jackson from the field to the White House.

It was of small importance that the key plea of self-defense was denied by one of Jackson's letters (to Monroe, January 6, 1818) showing that his aim was conquest, that Adams said not a word about the attack on Fowltown that touched off the war, that the defenseless state of Seminole villages proved the Indians had not prepared for war, that St. Marks was not in imminent danger of falling into the hands of "negro-Indian banditti," that no evidence established Arbuthnot as the "firebrand" who had rekindled the war, that the hanging of Hillis Hadjo mattered so little it was ignored, and the like. The stunning success of the White Paper rested on ignorance of these facts, to be sure; but it depended still more on widely shared prejudices. Its success demonstrated that Adams, the rest of the Monroe administration, and the country shared Jackson's operating assumptions and attitudes.

"In this narrative of dark and complicated depravity," Adams laid bare the racism he shared with his countrymen. With two hundred years of conditioning behind them, his readers had little need for the reminder that, cruel as war was in its mildest forms, it was "doubly cruel when waged with savages; that savages make no prisoners, but to torture them; that they give no quarter; that they put to death, without discrimination of age or sex." Though Adams took the "ordinary ferociousness" of the Indians as a given, he also believed they were childishly incapable, by themselves, of bringing it to bear on white enemies: "It so happened, that, from the period of our established independence to this day, *all* the Indian wars with which we have been afflicted have been distinctly traceable to the instigation of English traders or agents." Retributive justice therefore demanded the lives of leading warriors taken in arms, "and, still more, the lives of the foreign incendiaries, who, disowned by their own governments, and *disowning their own natures,* degrade themselves beneath the savage character by voluntarily descending to its level" (my italics). Why should anyone then protest over "the rights of runaway negroes, and the wrongs of savage murderers"?

Adams's prosecutor's brief was enthusiastically received by a hanging jury of readers, foreign and domestic. The assumption the United States had to intervene in the "derelict" province of Florida, articulated five years before the Monroe Doctrine was announced, looked forward eighty-six years to Theodore Roosevelt's so-called corollary of preventive intervention in the affairs of any South American or Caribbean country where there was "chronic wrongdoing, or an impotence which results in a general loosening of the ties of civilized society." In 1818, assuming the same international police power, Adams directly anticipated the tumid virility of Roosevelt's message: It had to be irresistibly demonstrated to Spain that "the right of the United States can as little compound with impotence as with perfidy." More ominous still was Adams's laying of a groundwork in international law for genocidal acts against "a ferocious nation which observes no rules." Quoting Vattel's *Droit des Gens* (1758), he calmly observed that if a commander " 'has to contend with an inhuman enemy . . . he may take the lives of some of his prisoners, and treat them as his own people have been treated.' The justification of these principles is found in their salutary efficacy for terror and example."

All the surviving ex-presidents, and that meant all the presidents but Washington, approved Adams's position. We may safely assume his father's response. James Madison complimented Monroe (February 13, 1819) on the handling of Jackson's invasion and noted that Adams had "given all lustre to the proof that the conduct of the General is invulnerable to complaints from abroad." But it must have been still more gratifying to Adams that his putative godfather placed a benediction on the great effort. On January 18, 1819, Jefferson wrote Monroe recommending that it and three other Adams papers be translated and widely distributed abroad:

> The paper on our right to the Rio Bravo [in Texas], and the letter to Erving of November 28 are the most important and are among the ablest compositions I have ever seen, both as to logic and style. . . . It is of great consequence to us, and merits every possible endeavor, to maintain in Europe a correct opinion of our political morality. These papers will place the event with the world in the important cases of our western boundary, of our military entrance into Florida, and of the execution of Arbuthnot and Ambrister.

3

White American racism not only followed the moving frontier into Indian nations; Florida in 1818 demonstrated the ease with which it spilled over internationally recognized boundaries. There, the baffling problem "of the black race within our bosom" and "of the red on our borders," in Madison's formulation, was compounded by their physical intermixture and their union in what was called the Indian and Negro War. Rebellions in South and Central America also raised the color question to the foreign policy level. What should be the response of the United States to those new republics where the intermixture of black, red, and white had been under way for centuries?

At home, color was a critical constitutional issue in the last phase of the struggle over the admission of Missouri to the Union. In 1820 the territory submitted a constitution to Congress that contained a clause requiring the new state legislature to pass a law banning free Negroes from entering the commonwealth. This apparent violation of the equal privileges and immunities section (Art. IV, Sec. 2) of the Constitution raised another storm in Congress. When Monroe turned to Madison for advice, the latter wrote (November 19, 1820) that he did not have the territorial document at hand and could not be sure, therefore, "but a right in the States to inhibit the entrance of that description of coloured people, it may be presumed, would be as little disrelished by the States having no Slaves as by the States retaining them." Madison had an acute sense of the disrelish, on both sides of Mason and Dixon's line, for "free people of color."

Madison was justly known as the Father of the Constitution and as a hardheaded political economist who realized that government existed to protect property. In the celebrated tenth number of *The Federalist* (1787) he based the clash of faction, party, or class—the stuff of European politics—on the unequal division of property and traced that back to the unequal faculties of men. But reds and blacks were conspicuously absent from this classic statement of liberal ideology because they were outside the politics of private property. Apart from the anomalous few, blacks did not own property; they were property. Reds did not own property individually; they held transient collective claims to lands already possessed or coveted by whites. Hence both Afro-Americans and Native Americans fell outside Madison's categories and outside the system framed by the Anglo middle-class fragment. Their de facto exclusion, however, had not been spelled out in the Constitution; by 1820 their cloudy status had become an urgent issue with threatening implications. In his letter to Monroe, Madison obviously regretted the earlier omissions and the resultant obscurities: "There is room, also, for a more critical examination of the constitutional meaning of the term 'citizens' than has yet taken place, and of the effect of the various civil disqualifications applied by the laws of the States to free people of color."

Madison's language became uncharacteristically theological when he discussed skin color: "All these perplexities," he wrote General Lafayette (November 25, 1820), "develope more and more the dreadful fruitfulness of the original sin of the African trade." Plainly aware that his Lockean liberalism based on property politics could not account for such "original sin," Madison probed for its meaning in "public sentiment," as he called the irrational underpinnings of what were no less plainly racist attitudes. If only an "asylum" in Africa could be found for the "unhappy race" of blacks, he declared to Lafayette in another (undated) letter, that might help take "out the stain." Somehow they had to be colonized elsewhere:

> The repugnance of the whites to their continuance among them is founded on prejudices, themselves founded on physical distinctions, which are not likely soon, if ever, to be eradicated. Even in States, Massachusetts for example, which displayed most sympathy with the people of colour on the Missouri question, prohibitions are taking place against their becoming residents. They are every where regarded as a nuisance, and must really be such as long as they are under the degradation which public sentiment inflicts on them.

Yet owners would be distressed by the "blank" in laborers created by colonization, he wrote Frances Wright (September 1, 1825), since "there would not be an influx of white labourers, successively taking the place of the exiles"; blacks could not stay on as free workers, an economically sensible solution, since "the physical peculiarities of those held in bondage . . . preclude their incorporation with the white population." Possibly there was a ray of light across the Caribbean in those South American states where physical peculiarities had less force, "owing, in part, perhaps, to a former degradation, produced by colonial vassalage; but principally to the lesser contrast of colours. The difference is not striking between that of many of the Spanish and Portuguese Creoles and that of many of the mixed breed." It was baffling, at all events, as *pigment* came to figure explicitly with *property* in the political economy of James Madison.

That the United States was on the move made certain the problem would grow still more perplexing. On the eve of war with Britain John Quincy had written John Adams (August 31, 1811):

> The whole continent of North America appears to be destined by Divine Providence to be peopled by one *nation*, speaking one language, professing one general system of religious and political principles, and accustomed to one general tenor of social usages and customs.

This echoed virtually word for word Jefferson's 1801 prediction to Monroe, quoted in an earlier chapter (p. 80), except Jefferson raised the possibility that "our rapid multiplication" might conceivably be throughout the hemisphere: it would "cover the whole northern, if not southern continent." The destiny of both men was manifestly white and English-speaking, with Muskhogean-speaking people such as the Seminoles, for example, having no part in it. But what about Spanish-speaking peoples?

The problem was further complicated by the lingering presence of European powers. In 1823 the threat of their intervention to suppress revolts in Spanish America led Foreign Secretary George Canning to propose that the United States join Great Britain in pledging not to take any of the former colonies and not to permit any other power to do so. When consulted, Jefferson and Madison were inclined to go along with the proposed joint declaration, though there was the obvious danger of embroilment in European rivalries. In the background correspondence leading up to the famous presidential message of December 2, 1823, Jefferson reminded Monroe:

> I have ever deemed it fundamental for the U.S. never to take active part in the quarrels of Europe. Their political interests are entirely distinct from ours. Their mutual jealousies, their balances of power, their complicated alliances, their forms and principles of government, are all foreign to us. They are nations of eternal war. All their energies are expended in the destruction of labor, property and lives of their people. On our part, never had a people so favorable a chance of trying the opposite system of peace and fraternity with mankind. [June 11, 1823, XV, 435–36]

Unfortunately the elder statesman limited his system of peace and fraternity to white American kind: Another passage in his letter read that Cuba alone was a problem—he had for the moment forgotten Texas—that the United States should not immediately go to war over it, "but the first war on other accounts will give it to us, or the Island will give itself to us, when able to do so." Jefferson's principle of the two spheres was not so much a contrast between European despotism and American freedom, then, as it was a declaration of hemispheric imperialism.

Precisely because he did not want to block off the longed-for acquisition of Spanish possessions, Cuba and others, John Quincy Adams responded coolly to Canning's overtures and stood out successfully in the cabinet against any joint declaration. The doctrine that bore Monroe's name thus studiously avoided the proposed self-denying ordinance, for the very good reason that the United States hoped to acquire Spanish-American territory. In operational terms the Monroe Doctrine declared to the world that the United States and only the United States was free to expand in the Western Hemisphere. It was, Samuel Flagg Bemis noted approvingly, "a voice of Manifest Destiny."

On December 7, 1824, almost exactly a year to the day after his famous message on foreign policy, Monroe presented a plan to Congress on "civilizing the Indians" by removing them to the West. He returned to the topic in January 1825, asking Congress to enact "a well-digested plan" that would move the Indians across the Mississippi, establish a government to protect them and their property, and provide means to instruct them in "the arts of civilized life and make them a civilized people." Though "experience had shown" the impossibility of whites and Indians becoming one people in their present state and though the tribes could "never be incorporated into our system in any form whatever," Monroe rejected forcible removal as "revolting to humanity and utterly unjustifiable." On the other hand, if they would only consent to be removed, white Americans would "become in reality their benefactors."

One of the curiosities of American scholarship has been the failure of historians to explore the connections between Monroe's doctrines on foreign policy and on the Indians. They were mirror images of each other. Indian removal at home reflected a policy abroad addressed to "natives" (and potential European colonizers) throughout the New World. Both came not, in any important sense, from the pen of any one author, Monroe or Jefferson or John Quincy Adams, but were collective creations, expressions of white American nationalism. And both provided *prospective* justification for the rapid multiplication of citizens of the United States and for their expansion onto the lands of nonwhites.

Of course Monroe's removal proposal was indigestible. It functioned principally as a benchmark of pressures rising toward the "utterly unjustifiable"—forcible removal—and of formal recognition that the United States was a white man's country: no more than blacks could reds in fact be incorporated into the larger polity and society without "either blot or mixture on that surface." The most notable characteristic of Monroe's "plan" was the same yawning vagueness that had characterized the African colonization schemes of Jefferson, Madison, and Monroe. No one knew how officials should go about getting the "consent" of those to be expatriated: What about those who clung stubbornly to "the soil of their nativity"? If what little the Eastern tribes still had was being taken away from them, why should they feel secure in their new homes? And was it not presumptuous to ask them to regard those who sent them packing as their benefactors? But such questions fell on the deaf ears of those who believed in ethnic magic.

Following in his predecessors' footsteps, Monroe merely asked the Native Americans, officially and still cordially, to become the Vanishing Americans. Unlike the blacks, who could more fittingly be sent "back" to some such place as Sierra Leone, the Indians should move "on" beyond the horizon of the Great Desert—or be driven, as Jefferson had prophesied, "with the beasts of the forest into the Stony mountains."

Part Three

Philanthropists and Indian-Haters

The carnage was complete: about six hundred Indians, men, women, and children, perished; most of them in the hideous conflagration. . . . The remnants of the Pequods were pursued into their hiding-places; every wigwam was burned, every settlement was laid waste. . . . The vigor and courage displayed by the settlers on the Connecticut, in this first Indian war in New England, struck terror into the savages, and secured a long succession of years of peace. The infant was safe in its cradle. . . . Under the benignant auspices of peace, the citizens of the western colony resolved to perfect its political institutions, and to form a body politic by a voluntary association. The constitution which was thus framed was of unexampled liberality.

—George Bancroft,
"Extermination of the Pequods," 1834

CHAPTER XI

Westward Ho! with James Kirke Paulding

> The first year of his arrival he was only the lord of a wilderness, the possession of which was disputed equally by the wild animals and the red men who hunted them. By degrees, however, the former had become more rare, and the latter had receded before the irresistible influence of the "wise white man," who, wheresoever he goes, to whatever region of the earth, whether east or west, north or south, carries with him his destiny, which is to civilize the world, and rule it afterwards.
>
> —JAMES KIRKE PAULDING, *Westward Ho!*, 1832

TWO CENTURIES of expropriation and extermination left their mark on the citizen of the United States. They helped shape his political institutions and shored up Paulding's conviction that his destiny had become irresistible. In 1624 an engraver had shown John Smith capturing a Pamunkey at pistol point by grabbing his hair, but the red man towered over the intrepid captain, and the mortal combat in the background threatened to go badly for the outnumbered English. By 1853, when Horatio Greenough's *Rescue Group* was erected at the national Capitol, such a reversal had long been almost literally unthinkable.

From the center of Greenough's famous sculpture arose a giant backwoodsman to grapple implacably with the merciless savage who stood with his back toward him. Pinioned by the massive arms of his white nemesis, the Indian appeared a mere stripling, although the tomahawk in his upraised hand, now arrested forever, was proof aplenty of his premeditated savagery. A helpless woman and her baby, white society's most tender and vulnerable members, still cowered to the rear, as yet unaware of their deliverance—thanks to the intervention of this super-Boone, at least this infant's cradle would not be stained with the blood of innocence. And as if this were not enough, to the left of his master snarled an Indian dog—symbol of nature domesticated and the white man's best friend—ready to spring on command. Reportedly Greenough himself neatly summed up his realized objective: He had wanted to show "the peril of the American wilderness, the ferocity of our Indians, the superiority of the white-man, and why and how civilization crowded the Indian from his soil."

This was not the first visual representation of terror on the frontier. Countless crude illustrations in the captivity narratives—the penny dreadfuls and shilling horrors that channeled hatreds against particular tribes and chiefs—depicted a dusky but never completely nude figure always lifting his trade hatchet to commit "Horrible and Unparalleled Massacre!" upon the proverbial prey: "Women and Children Falling Victims to the Indian's Tomahawk." On a different esthetic and intellectual level were some paintings of frontier conflict. Charles Deas's two spectacular horsemen entangled in *The Death Struggle* (1845), for instance, depicted their combat with unmistakable bias and yet came close to expressing the intertwined fates of red man and white since the invasion of North America. But it was Greenough's genius to capture in one stone knot of figures, with the putative victims present both as witnesses and as justification, the defeat of the Indian at the hands of avenging white "civilization."

Even in the *Rescue Group*, however, the diminutive Indian had human form. By itself Greenough's graphic message could not erase the likelihood that the white rescuer would shortly kill his red brother. Hence the statue had to be interpreted in the light of recurrent denials, some of which we have already considered, of the very possibility of brotherhood. The reasoning behind the denials was transparently self-protective and self-celebratory, constantly tripped up those who advanced it, but for all that seemed to offer a seductive way out of an intolerable impasse: Brotherhood could exist only between persons; not Greenough's Indian nor any other qualified for that ideal relationship because he was not a person, not one of the "people."

Just as William Bradford had observed in the early 1600s that he was immigrating to an "unpeopled" wilderness, so nine generations later John G. Nicolay and John Hay observed in their monumental *Abraham Lincoln* (1890) that the army of pioneers who first immigrated to their subject's native state had been impelled "by an instinct which they themselves probably but half comprehended. The country was to be peopled, and there was no other way of peopling it but by sacrifice of many lives and fortunes."

Rescue Group, Horatio Greenough, 1853. Originally erected on the Capitol's East Front steps, Central Portico, this marble statuary group was removed during renovations in 1958 and never replaced. Government officials and others found the symbolism embarrassing; Native Americans and their friends found it an affront. After more than a century of proud display, Greenough's classic depiction of Indian-hating was thus in a sense locked away in the nation's memory closet. (Courtesy of the Library of Congress.)

The Death Struggle, Charles Deas, 1845. (Courtesy of Shelburne Museum, Inc., Shelburne, Vermont.)

And then Daniel Boone himself, according to John Filson in *Kentucke* (1784), unwittingly tied all the generations together with his observation that on his first trip (1769) the Dark and Bloody Ground, as white settlers called the country beyond the mountains, was no more than a howling wilderness inhabited by "savages and wild beasts": among the latter were buffalo more numerous than cattle in the settlements and no less "fearless, because ignorant, of the violence of man." Seemingly even the North American bison refused to recognize the Indians as men, as *people* with their own peculiar violence.

Yet the ultimately unblinkable truth was that the wilderness had never been unoccupied. Thus, shortly after the landing at Plymouth, Miles Standish found it necessary or desirable to slaughter eight of its friendly inhabitants. Boone himself remarked, again according to Filson: "My footsteps have often been marked with blood." And Nicolay and Hay faced the awkward necessity of accounting for the death of the pioneer Abraham Lincoln in 1784 at the hands of persons unknown—unknown but undeniably sojourners in supposedly vacant Kentucky. Indeed the lucky shot of one of the fallen man's sons brought down a red assailant and thereupon President Lincoln's Uncle Mordecai became a dedicated Indian-hater: "Either a spirit of revenge for his murdered father" motivated Uncle Mordecai, concluded Nicolay and Hay, "or a sportsmanlike pleasure in his successful shot, made him a determined Indian-stalker, and he rarely stopped to inquire whether the red man who came within range of his rifle was friendly or hostile." But if Indian-stalking was people-stalking, as it surely was, then what about their contention, a few paragraphs earlier, that white pioneers had to "people" the country?

For writers dedicated to the creation of a national literature, the critical vehicle for a national mythology, the problem was perhaps more pressing than for sculptors and painters: Did real people lurk in the woods just beyond the line of white settlement? If the answer was yes, then how could one maintain the dream of American innocence versus European evil? "It is of great consequence to us, and merits every possible endeavor," as Jefferson said, "to maintain in Europe a correct opinion of our political morality." And if the answer was an unhappy yes, then what happened to the Christian principle of the brotherhood of man and the Enlightenment principle of the equality of man? This painful prospect, along with the Indian's reputed mercilessness, may have had something to do with the first American novelist's response: Through *Edgar Huntly* (1799), Charles Brockden Brown declared that "I never looked upon or called up the image of a savage without shuddering." Of course, James Fenimore Cooper's Natty Bumppo seemed somehow to straddle this fearsome gap between the red world and the white, but his mythic stature makes it convenient for us to consider the Leatherstocking Tales later. James Kirke Paulding (1778–1860), William Gilmore Simms (1806–

70), and Robert Montgomery Bird (1806–54), three of Cooper's contemporaries without his mythopoeic reach, wrote "border" romances with images of the Indian that disclosed as much or perhaps even more about the crystallizing national style.

2

James Kirke Paulding was a man of the people. "For myself," he wrote historian George Bancroft (November 21, 1834), "I never have, and never mean to desert the side of the People. I was born in the ranks of Democracy." He was born in a Dutch family that suffered financially for supporting the Revolution; he grew up hating the British and the exclusive privileges he assumed they stood for. In the 1790s he moved to New York City from the Hudson village of Tarrytown, found work as a clerk, joined the literary circle around Washington Irving, and became known for his witty collaboration with Irving in the *Salmagundi Papers* (1807–8). In 1815, after enlisting with Timothy Dwight and other patriots in the literary war against Britain, he fired off a salvo against the English *Quarterly Review* that earned him President Madison's friendship and an appointment in Washington as secretary of the board of navy commissioners. When Monroe appointed him navy agent for New York in 1824, he stepped into a sinecure that left him free to pursue his literary career. In 1838 his loyal service in the ranks of democracy, his written encomiums of Andrew Jackson, and his friendship with Martin Van Buren lifted him into the cabinet of the latter as secretary of the navy.

In *The Backwoodsman* (1818) Paulding marched heroic couplets into the Ohio Valley in search of the saga of pioneering. Like Timothy Dwight, Paulding sought materials for the epic of white America. His most significant discovery in this disjointed poem was a despairing Indian prophet, perhaps modeled on Tenskwatawa, who defied the Great Spirit with his dying breath:

> If thou hadst power, why then refuse thine aid?
> If not, then have thy vot'ries idly prayed;
> Thou art a cheat that in the heav'ns dost dwell,
> Take my defiance, and so fare thee well.

Or goodbye, and go to hell? More than loss of faith, Paulding's prophet clearly suffered from a bad case of plethora.

In *The Dutchman's Fireside* (1831)—loosely connected, Fielding-like character sketches set in upstate New York before the Revolution—Paulding assumed readers would not find it farfetched that Mohawks would want to drink the blood of Sybrandt Westbrook, his bashful and possibly autobiographical hero—the author himself, like Westbrook, did not marry till he was forty. He also assumed they would understand why these merciless savages would rave as "yelling fiends" under the influence of rum and their "deep and never-dying hatred . . . for the white man." Readers would know what the hero found out along the way, namely, that Indians were never more untrustworthy than when they were most friendly. Timothy Weasel, the most memorable character in the gallery, became a local legend acting on this home truth. A Vermonter, Weasel had lost his entire family and all his neighbors to a raiding party of Canadian Indians, an experience that changed his life. As Paulding had Sir William Johnson, the famous Indian commissioner, explain to Westbrook, Weasel afterward became a genuine Indian-hater; he had already killed almost a hundred and withal pursued "this grand object of his life" with style and ever-fresh enjoyment in killing "those tarnal kritters," as he called them: "It is inconceivable with what avidity he will hunt an Indian; and the keenest sportsman does not feel a hundredth part of the delight in bringing down his game, that Timothy does in witnessing the mortal pangs of one of these 'kritters.' " Lest he seem too enthusiastic in his vicarious sharing of these inexpressible delights, Sir William immediately added his conventional disapproval of Weasel's monomania: "It is a horrible propensity." Nevertheless, Indian-hater that he was, the commissioner employed him and found "his services highly useful."

While imitating Henry Fielding and Oliver Goldsmith in his own verse and prose, Paulding demanded an original national literature of his countrymen. It was most likely to develop in the West, he granted Daniel Drake, chronicler of that region, "far distant from the shores of the Atlantic where every gale comes tainted with the moral, political, and intellectual corruption of European degeneracy" (January 1, 1835). Bowing to Drake's presumed firsthand knowledge and to his criticism, he admitted that in writing *Westward Ho!* (1832) he had himself relied "on the ideas of others, rather than my own."

Westward Ho! was indeed derivative, much of it based on the writings of Timothy Flint, another chronicler of the West, but it was truly awful in its own right. It was redeemed in part, however, by its title and by its historical function as a carrier of widely shared ideas and values. The plot itself was a sentimental commonplace: Shortly after the turn of the century, Colonel Cuthbert

Dangerfield, an impecunious Virginia gentleman, packed up his family and moved to the Dark and Bloody Ground, where he became sharp and tough fighting Indians; there, by the end of the second volume, his daughter—"the sweetest sap that ever was boiled into maple sugar"—made a man and a husband of Dudley Rainsford, an obliging if slightly deranged young heir whom she cured of religious frenzy and presentiments of evil. A broadhorn captain who said he was "half horse, half alligator"; a "blackey" who preferred to be thought of as "um gemman of choler"; a black housekeeper named "Snowball"; "Black Warrior," a tame Indian who occasionally emerged from his hut to orate in pidgin English —these and other odd types crowded the pages of the novel. But none was fit to hold a candle to the old hunter Ambrose Bushfield, prime exemplar of the new land.

From the vast regions of the Mississippi Valley, Paulding saw rising up "a new and primitive race." Bushfield represented this new breed, with a bit of Mike Fink and Davy Crockett in his makeup, like Nimrod Wildfire in Paulding's popular comedy, *The Lion of the West* (1830). A collateral ancestor was Cooper's Leatherstocking. His direct descent, however, was from the patriarch of them all. Indeed he was said to have joined Daniel Boone during the early days in "Old Kentuck"; he also had a passion for hunting and a need for "elbow room"; following the original, he soon sought solitude across the Mississippi, beyond white settlement; there he too was to die after his last shot, "sitting upright against a tree, his rifle between his legs, and resting on his shoulder."

Not a lover, Bushfield was a hater. In his world there were white men, "Ingens" and wild animals, and "niggers." The last was of course the epithet of ultimate contempt. When "Snowball" insisted she had seen black ghosts, he erupted: " 'A black ghost!' cried Bushfield, breaking into a loud laugh. 'I'd as soon think of a white nigger.' " Not much higher in the scale of creatures were the "varmints" called Indians. Even the Dangerfields' tame "Black Warrior" struck him as less than human: "I'm a nigger if I think this copper-washed man is a right clean, full-blooded feller cretur."

It was in character for tenderfoot Dudley Rainsford to say he pitied the Indians. Bushfield thereupon gave him a lesson in Kentucky realities:

> When I first remember this country, nobody could sleep of nights for fear of the Ingens, who were so thick you couldn't see the trees for them. There isn't a soul in all Kentucky but has lost some one of his kin in the Ingen wars, or had his house burnt over his head by

> these creturs. When they plough their fields, they every day turn up the bones of their own colour and kin who have been scalped, and tortured, and whipped, and starved by these varmints, that are ten thousand times more bloodthirsty than tigers, and as cunning as 'possums. I, stranger, I am the last of my family and name; the rest are all gone, and not one of them died by the hand of his Maker.

Confronted by such passionate Indian-hating, Rainsford "could not help acknowledging, that to judge rightly of the conduct of mankind in all situations, we should know the necessity under which they laboured, and the provocations to which they were exposed." Yes, Rainsford helped us see, surely we should know the circumstances before sitting in judgment on those who have suffered so terribly . . .

Timothy Flint, Paulding's mentor on the West, had already used this device in his "Indian Fighter," published in the *Token* for 1830. There the Harvard-educated missionary related the life of another Indian-hater: The Hermit of Cap au Gris had lost his loved ones to merciless savages, vowed the usual vengeance, and spilled such "copious libations of Indian blood" the afflicted tribes called him Indian Fighter. Rather daringly for a reverend gentleman, Flint implicitly invited readers to celebrate with him the bloodletting of his secular backwoodsman, warm their inner selves with the thought of Indian villages he had burned, and then draw back with the narrator to a cool Christian distance by having the Indian-hater return to the fold: "Say to those that come after me," Flint's Hermit besought, "that it is wise as well as Christian to stay the storm of wrath and leave vengeance to Him who operateth by the silent and irresistible hand of time, and will soon subdue all our enemies under his foot." Weasel and Bushfield underwent no such transformation, of course, but Paulding placed their stories within the sensibilities of Sir William Johnson and Dudley Rainsford, and had these cultivated figures share Flint's circumspect taste for slaughter: Like the devil, the Indian-hater made a fine instrument of the Christian lust to punish enemies eternally. More specifically, Rainsford's misplaced pity and forced acknowledgment provided a "civilized" frame for understanding and accepting the otherwise unacceptable. Yes, the reader was asked to reflect, is it not a little too easy to be virtuous at a distance? A little cheap to forgive merciless savages when we ourselves have not suffered, as Bushfield and all the others had, at their hands?

But who were "we"? Obviously we were WE THE PEOPLE of

the United States. . . . And that Paulding was a man of the people did not mean for him, any more than for the Kentuckians, that the people included those not of his "own colour and kin." The racism in his fiction had its counterpart in his egregious defense of the peculiar institution in *Slavery in the United States* (1836). Later in "Dred Scott," a manuscript fragment on "this doughty gentleman of colour" (1845), he warned abolitionists against preaching that the slave had no obligation to his master, for the proposition cut the other way as well: "The master being thus left to the mercy of wild hearts in the Shape of man" might justifiably deal with "Slaves as with a horde of Savages ignorant or regardless of all these obligations imposed by the Laws of man or the Laws of God."

While Bushfield and Weasel endlessly brought down "varmints" in his novels, Paulding himself actively Indian-hated on the congenial terrain of Jacksonian politics. Old Hickory's program to remove the diminished "horde of Savages" still east of the Mississippi received his early and energetic support. Thus on August 16, 1830, he congratulated Congressman Richard Henry Wilde from Georgia for a speech in the House advocating removal: "If I had not been already convinced, it would have satisfied me of the necessity as well as humanity of removing the Indians from the bosom of Georgia and Alabama." He impatiently awaited the several Trails of Tears and was angered by Indian obduracy and resistance. Finally, during the Second Seminole War, he exploded in a letter to his brother-in-law Gouverneur Kemble (June 15, 1838): Paulding hoped Joel R. Poinsett, secretary of war and thus the official responsible for the forcible removal of the Seminoles and others, would "be soon rid of the Indian heroes":

> I say heroes, for it seems only necessary for a bloody Indian to break his faith; plunder and burn a district or two; massacre a few hundred soldiers; and scalp a good number of women and children; to gain immortal honour. He will be glorified in Congress; canonized by philanthropists; autographed and lithographed, and biographied, by authors, artists, and periodicals; the petticoated petitioners to Congress will weep, not over the fate of the poor white victims, but that of the treacherous and bloody murderer.

"The treacherous and bloody murderer" reference was a bit of projection. Paulding probably had in mind Osceola, who had been treacherously captured the preceding October as he stood under a white flag of truce. The great Seminole warrior had since died in a dungeon at Fort Moultrie, South Carolina.

Paulding's contempt for the "petticoated petitioners" who protested such violation of trust underlined his sexual politics. To his mind those effeminate bleeding-hearts—the abolitionists and other reformers—were the sort who also opposed the annexation of Texas and presented petitions to Congress against slavery. But Paulding still had hopes, he wrote President Van Buren, "hopes, that the *men* of the Country, will get the better of the *women*" (September 22, 1837).

3

Paulding's stand on abolition and removal made his appointment as secretary of the navy acceptable to Southerners and Westerners. One of his first duties was to forward instructions (August 11, 1838) to Lieutenant Charles Wilkes, who had been appointed commander of the United States Exploring Expedition. The Wilkes, or South Sea, Expedition was to explore and map the Southern Hemisphere, the Pacific Ocean, and the Northwest Coast of America, and to sail back to the Orient via the great circle route for Japan, the Straits of Sunda, and the port of Singapore—Wilkes eventually spent five years at sea, sailing his flotilla around the world and "discovering" the Antarctic, a "new" continent, along the way. Though Wilkes was explicitly told "the Expedition is not for conquest but discovery," it was in fact ambitious, government-sponsored, seagoing Manifest Destiny, with commercial conquest a major objective. It was part of the thrust of United States power, economic and political, south into Latin America and west into the Orient and the China trade. It involved mapping the South Pacific, a farther West than any of the continental explorers would ever reach.

Yet our primary interest here is not the "door" Wilkes knocked on, Commodore Matthew C. Perry stuck his foot in at midcentury, and John Hay supposedly threw "open" at the turn of the century. Nor need we pause over the fact that Joel Poinsett, on the scene before Paulding's appointment, had selected Wilkes, put the expedition together, and very probably wrote most of the instructions Paulding handed on. Wilkes's instructions were issued over Paulding's name and reflected his views, those of Poinsett—no less of a racist and imperialist—and in fact those of all onward- and outward-looking citizens in the ranks of democracy. The expedition had scientific pretensions, with a group of experts aboard, reports of the various learned societies, including the

American Philosophical Society, and Albert Gallatin's list of English words for Indian vocabularies. The undertaking represented the best wisdom Jacksonian America could muster in the fields concerned, including the fledgling science of ethnology.

Paulding's instructions sent the underlying assumptions and attitudes of Indian-hating around the world. Wilkes and his men were warned that, "among savage nations, unacquainted with, or possessing but vague ideas of the rights of property, the most common cause of collision with civilized visitors, is the offence and punishment of theft." Wilkes and his men should therefore keep their eyes on their property. They should use forbearance in punishing offenders, beware of arrogance, and exercise "that courtesy and kindness towards the natives, which is understood and felt by all classes of mankind." They should "appeal to their good-will rather than their fears," but be especially on their guard against the former:

> You will, on all occasions, avoid risking the officers and men unnecessarily on shore at the mercy of the natives. Treachery is one of the invariable characteristics of savages and barbarians; and very many of the fatal disasters which have befallen preceding navigators, have arisen from too great a reliance on savage professions of friendship, or overweening confidence in themselves.

That read as if it were the lesson Sybrandt Westbrook learned in *The Dutchman's Fireside*. It was, however, the received wisdom around the fireside of white America.

In his five-volume narrative of the expedition, Lieutenant Wilkes concluded that indeed the Pacific slope was destined to be "possessed as it must be by the Anglo-Norman race." Thus did life in the United States imitate art as Wilkes acted out the theme of *Westward Ho!* quoted in the epigraph to this chapter.

CHAPTER XII

An American Romance in Color: William Gilmore Simms

> "The Yemassee" is proposed as an *American* romance. It is so styled as much of the material could have been furnished by no other country.
>
> —WILLIAM GILMORE SIMMS, preface to *The Yemassee*, 1853

"THUS, in the beginning all the World was *America*," said John Locke in the second treatise *Of Civil Government* (Sec. 49; 1690). In America the beginning occurred anew each time "civil society" gained another western vantage point over "the state of nature." Just as once upon a time the Atlantic was the Western Sea and North America lay beyond the western edge of consciousness, so in the mid-eighteenth century the Alleghenies were the western mountains and Kentucky lay beyond in the shadows, a dark wilderness awaiting discovery and the beginning.

What was just about to end appeared hauntingly in George Caleb Bingham's famous *Daniel Boone Escorting Settlers through the Cumberland Gap* (1851–52). The painting, as Henry Nash Smith once observed, "showed the celebrated Kentuckian leading a party of settlers with their wives and children and livestock out into a dreamily beautiful wilderness which they obviously meant to bring under the plow." Boone stood out from the group, as in Greenough's statue, but behind and above him, mounted on a horse whiter than snow, was Woman, repository of "civilization" —according to William Gilmore Simms, Boone had naturally exulted that his "wife and daughter were the first white women that ever stood on the banks of the Kentucky river." In the painting Boone naturally had his rifle on his shoulder, but behind him came settlers armed with axes against that distant reverie of sycamores and oaks and chestnuts. So long as those primeval forests remained standing, the land would never be ready for the plow.

Yet an earlier and still more dramatic moment in the very beginning was suggested by the subtitle of Simms's 1845 essay on Boone: "The First Hunter of Kentucky." In the lines that followed, Simms gave Boone credit for being a dead shot and for possessing a keen eye, great elasticity of muscle, and great powers of endurance—for having in short all the physical and moral characteristics that made him first in rank among hunters. But by "first" Simms meant primarily foremost in time, the hunter who preceded all the others and found an untouched part of the continent. Thus Simms commenced and concluded with Boone *alone*, the great discoverer: "Standing upon Cumberland mountain, and looking out upon the broad vallies and fertile bottoms of Kentucky, he certainly thought of himself so. We have no doubt he felt very much as Columbus did, gazing from his caraval on San Salvador . . . or Vasco Nuñez [Balboa], standing alone on the peak of Darien, and stretching his eyes over the hitherto undiscovered waters of the Pacific." Now, on the peak of Darien Balboa did not in fact stand alone: Indians guided and accompanied him on his epic march across the isthmus to "discover" the true Western Sea and to claim "all the shores washed by it." Nor in truth was Boone ever alone in Kentucky, and that was part of his problem, Simms's, and ours.

"But Boon ceased to be alone in this march of discovery into the Kentucky wilderness," wrote Simms. "There were other spirits like himself, destined to open the way for the thronging multitudes that began to cry aloud for homes of their own." Had we tarried at the Gap with the figures in Bingham's painting, it followed, the forests would have become a legend before our eyes, the ground we stood on would have become the Cumberland Gap National Historical Park (20,184.20 acres; established 1955), and Kentucky's recent boom in tourism would have brought in still more multitudes to make us even less alone. By then the dreamy beauty would have gone from the view. In his essay, at all events, Simms evinced more interest in the march of discovery than in the march of improvement.

Without knowing the author's name and background, Simms based his essay on John Filson's *Discovery, Settlement, and Present State of Kentucke* (1784) or, more exactly, on the Boone narrative therein that was republished in Brooklyn under the title *Life and Adventure of Colonel Daniel Boon*, Written by Himself (1823). A surveyor and land speculator and quondam schoolteacher, Filson had supposedly written the narrative at Boone's dictation. In what was to become a tradition, Simms denounced the writer's clumsy

Daniel Boone Escorting Settlers through the Cumberland Gap, George Caleb Bingham, 1851–52. (Courtesy of Collection, Washington University, Saint Louis.)

attempts at eloquence and poetry, the flourishes and "marks of ambitious composition quite unlike our hunter," but proceeded to follow his emphasis on the *discovery* of Kentucky, quoted him extensively, and revealed his own enchantment over what the writer had made of Boone.

To be sure, Filson was guilty of looking at "the ample plain" of Kentucky as a series of "beauteous tracts," each chock-full of potentially marketable "sylvan pleasures." Yet his realtor's rhetoric was relatively artless and his archaic images in the end left even the critics more captivated than irritated. Criticisms bounced off the narrative, anyhow, as it proceeded with the unimpeachable authority of a dream: "It was on the first of May, in the year 1769," Filson's Boone commenced, "that I resigned my domestic happiness for a time, and left my family and peaceable habitation on the Yadkin River, in North Carolina, to wander through the wilderness of America, in quest of the country of Kentucke. . . ." But why had he temporarily resigned his domestic happiness? He just had, that was all, the way you and I just do something in a fantasy. Agents for the Loyal Land Company had made their way through the Cumberland Gap in 1750, and white settlers had long since defied the royal proclamation of 1763 by moving into Kentucky—but to interrupt with these points of historical fact would have been ungracious and largely irrelevant. In 1769 Boone wandered, no more and no less, in quest of "Kentucke."

Simms picked up the sequence to record that when one of Boone's companions was killed by the Indians, he became "the first anglo-norman victim of the red man, in the lovely wilds of Kentucky." Daniel Boone's brother, Squire, unexpectedly strolled into their camp accompanied by a stranger, but the stranger soon left and others disappeared, leaving Daniel and Squire Boone the only remaining members of the quest. An obsessive concern for the exact place name and precise date of the goings and comings formed a patina of narrative realism. Thus one year to the day after the quest had begun, that is, "on the first of May, 1770," according to Daniel Boone's reckoning, "my brother returned home to the settlement by himself, for a new recruit of horses and ammunition, leaving me by myself, without bread, salt or sugar, and without company of my fellow creatures, or even a horse or dog." Like Robinson Crusoe on his island, at first Boone was uncomfortable, almost melancholy as "a thousand dreadful apprehensions presented themselves to my view"; but then "the diversity and beauties of nature I met with in this charming season, expelled every gloomy and vexatious thought. . . . I had gained the summit of a commanding ridge, and, looking round with astonishing delight, beheld the ample plains, the beauteous tracts below."

Yet the wilderness, for all its green charms, "concealed the painted and ferocious savage," as Simms put it sympathetically, "and he who hunted the deer successfully through his haunts, might still, while keenly bent upon the chase, be unconscious of the stealthy footsteps which were set down in his own tracks." And there they were in all this solitude, the American cousins of Friday's unforgettable Footprint in the Sand.

The supposition that the Kentucky wilderness was Boone's desert island became less fanciful a few sentences later when Simms had the First Hunter gaze "over the ocean waste of forest which then spread from the dim western outlines of the Alleghanies, to the distant and untravelled waters of the Mississippi." Never was Crusoe more alone in his own ocean wastes as he peered vainly for the dim outlines of sails on his untraveled waters. And like Boone he had to contend with those surprising, frightening, stealthy footsteps.

"Thence emerged the story of Robinson, in the way a dream might occur," observed Otare Mannoni, an acute student of the psychology of colonizers. "When this dream was published, however, all Europe realized that it had been dreaming it. For more than a century afterwards the European concept of the savage came no nearer reality than Defoe's representation of him." Certainly Filson's and Simms's concept came no nearer reality than Defoe's, but I do not suggest either man necessarily had the English writer in mind when he wrote about Kentucky. I do suggest that Filson's nearly identical dream of a world without people came no less directly from his unconscious, a suggestion that helps explain the appeal of his narrative for imaginative writers, for Simms, and for such major figures as Byron and Cooper. And since white Americans already shared the dream and had lived it for more than a century, Filson's formulation, for all its clumsiness, helped make a folk hero of Boone in his own lifetime.

In the dream, a European—or in our case a Euro-American—flees from repressive "civil society," all that everyday domestic happiness, to the ocean island of his imagination where his desires rule uncontested by other people. This "flight from mankind" Mannoni traced in Defoe to "a misanthropic neurosis" and found the origin of this widely shared condition in "the lure of a world without men" and in a lack of sociability "combined with a pathological urge to dominate." Simms himself saw something of the sort in Boone, anticipating Mannoni's phrase almost word for word as he pictured the First Hunter "flying, as it were, from his kindred and society." Moreover, Boone's misanthropy became part of American folklore, as in the popular account of his complaint even after moving to Missouri: "I had not been two years at the licks before a d——d Yankee came, and settled down *within an hundred miles of me!!*" But the psychiatric label cannot be sustained by any detailed evidence from Boone's early life, misty as that was, and in any event it diverts attention from Mannoni's richly suggestive metaphor.

The metaphor links the emotional satisfactions of colonizing North America to the appeal of colonial life elsewhere—in darkest Africa, Asia, the South Seas. It suggests that Boone's desert island of Kentucky was America writ small, just as that white America was the archetypal desert island writ large. It establishes connections over time to the earliest colonizers along the seaboard, to men like Winthrop and Endicott who fled the Old World, where all the churches were "brought to desolation," for their own "howling wilderness," where their wishes became laws.* And it takes us, part way at least, from the individual level to the collective and helps us see that this repudiation of the world of parents carried with it certain guilts. Simms went out of his way to assuage these by interlacing Boone's flight with that of all the American revolutionists from the treacherous mother. While Boone was building his fort on the banks of the Kentucky:

> The revolutionary war had begun, and our benign mother of Great Britain had already filled the forest with her emissaries, fomenting . . . [the Indians'] always eager jealousy, and the common appetite for war and plunder. While Congress were making the declaration of independence, at Philadelphia, Boon was already waging the conflict.

While Jefferson was drawing up his indictment of the malevolent father George and mother Britain, Boone was locked in combat with the merciless savages the king had fomented in far away Kentucky. Hence the fugitives, one and all, had good reason for their flight.

Or so it seemed. And as the Footprint in the Sand, the Indian was either Jefferson's mindless child, a dependent Friday, or his merciless savage, a cannibal. Frequently he was both at the same time or at least in the same essay. Simms rightly noted that Indian

* That many of these first colonizers fled European mankind in groups or companies did not make Robinson Crusoe any less of a paradigm of their flight—as Mannoni pointed out, the fact that *several* children run away to the bottom of the garden to play at being Robinson Crusoe "makes no difference—it is still a flight from mankind, and intrusion must be guarded against."

violence had been almost pacific: "His wars were seldom bloody until he encountered the Anglo-Norman, and then he paid the penalty of an inferior civilization. The loss of a warrior was a serious event—the taking of a single scalp was a triumph. To gain but one shot at a foe, an Indian would crouch all day in a painful posture." Yet Simms also had him raging in thickets, a "painted and ferocious savage" with all "the terrors and vicissitudes" of his merciless warfare. In both halves of this image the Indian was seen as semihuman, and thus his presence could never stop Boone from being absolutely alone. Simms actually had the Boone brothers wander through the unsung countryside, "bestowing names, like other founders of nations, upon heights, and plains, and waters." * But he had to know the rivers and ridges had names, for he knew the object of the Boone quest already had the Indian name *Kentucky*.

To recognize the implications of discovering what was already known to others, of being alone in the presence of Indians thick as leaves on the trees, of renaming the named—to have recognized what he was doing would have obliged Simms to recognize the existence of Indians as *others*, other people.

On occasion he seemed as far from that level of self-awareness

* Not only had these natural delights been unnamed, they had been unseen (by "civilized" eye), according to Daniel Bryan's still more revealing lines in his epic *The Mountain Muse: Comprising the Adventures of Daniel Boone; and the Power of Virtuous and Refined Beauty* (Harrisonburg, Va.: Davidson and Bourne, 1813). Boone toured the opening world

> He form'd! bestowing *names* on streams and founts,
> On plants and places yet anonymous,
> And yet unvisited by other eye
> Emiting Civilizement's softened beams,
> Than the Adventurer's own.

But the fantasy was still more antique. In *The Faerie Queene* (1590) Edmund Spenser rhetorically asked Elizabeth:

> Who euer heard of th'Indian *Peru?*
> Or who in venturous vessell measured
> The *Amazons* huge riuer now found trew?
> Or fruitfullest *Virginia* who did euer vew?
> Yet all these were, when no man did them know.

From before "fruitfullest *Virginia*" was properly planted, then, the Indian was "no man"—a nonperson, who could neither view nor know. One of the threads we have pursued over the succeeding centuries was his continuing status as a nonperson. He apparently remains so in surprising quarters. In an illuminating essay on the new geographical literature of Columbus, Vespucci, Oviedo, and others, David Beers Quinn concluded that with the circulation of their writings, "America became part of human experience; it was assimilated, even if imperfectly, into European concepts"—"New Geographical Horizons" in *First Images of America*, ed. Fredi Chiappelli et al. (Berkeley: University of California Press, 1976), II, 654. *O poor brave new world:* previously unnamed, unmeasured, unknown, nonhuman! That this lethal sixteenth-century delusion lives unwittingly on in a man who has written perceptively of Elizabethan empire-building in Ireland is all the more astonishing—and discouraging.

as Paulding. When Simms was editor of the *Charleston City Gazette,* for instance, he summarized the Indian character "as it is" (April 30, 1831): "We say every thing of the North American . . . when we call him a mere savage, like all the others, and no better than any savages, but a few degrees removed from the condition of the brute." Yet the same man was capable of a brilliant insight a few years later in his essay on Indian literature and art. Simms then held that the general insensibility of whites to the power of Indian oratory could only be accounted for

> by reference to our blinding prejudices against the race—prejudices which seem to have been fostered as necessary to justify the reckless and unsparing hand with which we have smitten them in their habitations, and expelled them from their country. We must prove them unreasoning beings, to sustain our pretensions as human ones—show them to have been irreclaimable, to maintain our own claim to the regards and respect of civilization.

This was on the level of—and may have been drawn from—Montesquieu's observation about slaves: "It is impossible for us to suppose these creatures to be men, because, allowing them to be men, a suspicion would follow that we ourselves are not Christians." From whatever source, Simms's insight took him beyond self-indictment to a dawning apprehension of the national web of self-deceptions.

In the same collection, *Views and Reviews* (1845), Simms began his essay on Boone with an epigraph from the eighth canto of *Don Juan* (1822). Delighted by the depth of Boone's plunge out of society into "wilds of deepest maze," Lord Byron had seen the backwoodsman of Kentucky as quite literally alone in nature,

> happiest among mortals any where;
> For killing nothing but a bear or buck. . . .

Simms had long admired Byron but felt constrained to point out his mortal error here: "Boon's rifle occasionally made free with nobler victims than bear and buck. He was a hunter of men too, upon occasion." Yet this gave away too much at the very beginning, so Simms quickly added: "Not that he was fond of this sport. His nature was a gentle one—really and strangely gentle—and did not incline to war." Strange indeed, for a First Hunter who looked for ground where he would have to smite "the savage man, as well as the scarcely more savage beast of the same region. . . . He could take a scalp with the rest, and might feel justified in the adoption of a practice which, when employed by whites, had its very great influence in discouraging the Indian appetite for war." It was a pity the practice, when employed by reds, did not discourage the white appetite for war. Still, might not Indians be drawing on the same psychology of deterrence to protect what was after all their country?

But, for Simms, Boone was all by himself in the wilderness. For him to have consistently regarded Indians as persons with a psychology of their own would have upended his world. It would have meant recognizing that "the state of nature" really had full-fledged people in it and that both it and the cherished "civil society" had started out as lethal figments of the European imagination. Only later did these delusions enter American political theory through John Locke, drafter of *The Fundamental Constitutions of Carolina* (1669) and tutelary saint of Jefferson's Declaration and of Madison's Constitution.

2

Like his friend Paulding, William Gilmore Simms sought the epic of America. Paulding was "one of the earliest pioneers in the fields of American letters," read Simms's dedication in *The Damsel of Darien* (1839), his historical romance of Balboa in Spanish America. Simms saw himself continuing the struggle of the older writer, of Cooper, and of all the others who worked for a cultural declaration of independence that would be a fit complement to the political. A serious literary critic and historian, he probed the past for usable themes and found in such figures as Pocahontas and Boone precious national resources. But he was best known as a writer of some thirty volumes of fiction in many of which he celebrated the rise of the colonies and of the new Union. Simms rejected the realism of the domestic novel, or novel of manners, in the tradition of Richardson and Fielding, and adopted instead the "romance," a form he defined as "the substitute people of the present day offer for the ancient epic." His chosen instrument, in short, was the modern epic in which heroes and scenes wild and wondrous might give rise to a truly American mythology.

The Yemassee (1835) was Simms's most widely read work and his most important romance of Indian warfare—it was, as he said, "an *American* romance." Unlike Paulding, Simms could tell a story, and this was one that appealed to readers with a taste for melodrama and a touch of sentiment for the passing of a once powerful and gallant race. It centered on an actual event in colonial Carolina, the resistance in 1715 of the Yemassee and associated tribes to further encroachments upon their lands. The colonists of course triumphed, led by a dashing but mysterious young cavalier named Gabriel Harrison, who prowled about to see what dirty work was afoot, claimed the damsel in the final pages, and then turned out to be Governor Charles Craven in disguise! Moreover, Simms followed his master Cooper in allowing the Indians some space to tell their side of the story and to have a secondary plot of their own. Their doomed cause was led by Sanutee, a philosophical patriot in the mold of Tamenund in *The Last of the Mohicans* (1826), and supported by his faithful wife, Matiwan; it was betrayed by their son, Occonestoga, a once upright young warrior who had been corrupted by white ways and firewater. On both sides there were secondary figures, including Sanutee's lieutenant Ishiagaska, the sinister medicine man Enoree-Mattee, Harrison's body servant Hector, and the bloodhound Dugdale, famous for hunting Indians.

Now, more interesting than the content of the double plot was the show of lights and darks, the choreography of colors as they advanced on each other, retreated, whirled, and finally collided. As Tocqueville observed the year Simms published this tale, "the danger of a conflict between the white and black inhabitants of the Southern states of the Union (a danger which, however remote it may be, is inevitable) perpetually haunts the imagination of the Americans like a painful dream." That was characteristically acute, save that the great French observer should have added *red* to the interplay of colors, for conflict with the Indians had been perpetual and was not some dreadful probability on the horizon. Like James Madison, Simms had a color-haunted imagination, haunted by "the black race within our bosom" and "the red on our borders." In an important sense the "border" romances were Simms's attempts to work out in fiction the baffling problem of the latter, while paying passing attention to the former.

In *The Yemassee* even Sanutee had a sense of the importance of color castes:

> He was a philosopher not less than a patriot, and saw, while he deplored, the destiny which awaited his people. He well knew that the superior must necessarily be the ruin of the race which is inferior—that one must either sink its existence in with that of the other, or it must perish. He was wise enough to see, that, in every case of a leading difference betwixt classes of men, either in colour or organization, such difference must only and necessarily eventuate in the formation of castes.

Shades of color seemed to shape or reflect relationships even among the Yemassees. The subtle subchief who was Sanutee's lieutenant appeared "a dark, brave, collected malignant, Ishiagaska, by name." Certainly the hue of none of the Indians, not even Sanutee, could compare with that of Gabriel Harrison, who had "a rich European complexion, a light blue eye, and features moulded finely, so as to combine manliness with as much of beauty as might well comport with it." His hair hung down in long clustering ringlets, and his English cavalier garments, combined with those of the American forests, set off his shining good looks marvelously.

Imagine the bed chamber of his beloved Bess Matthews, the damsel, one night at the height of the troubles, with the dark Ishiagaska creeping up on the half-slumbering girl:

> The dress had fallen low from her neck, and in the meek, spiritual light of the moon, the soft wavelike heave of the scarce living principle within her bosom was like that of some blessed thing, susceptible of death, yet, at the same time, strong in the possession of the most exquisite developments of life. Her long tresses hung about her neck, relieving but not concealing its snowy whiteness. One arm fell over the side of the couch, nerveless, but soft and snowy as the frostwreath lifted by the capricious wind.

The hand of "the wily savage" was not stayed, needless to add, by the helplessness of that "sweet white round," the Snow Maiden, but had to be stopped by more powerful arms. And beside her chilly charms, what chance had the "dusky" matron Matiwan or Occonestoga's former love, "the dusky maiden" Hiwassee?

Simms manipulated these symbols of light and darkness for all they were worth, with unintended and unperceived consequence: His clash of colors had in awkward fact highlighted the colonists' dark side. There was the sturdy black Hector, "my Hector" or "my snow-ball," as Harrison sometimes called him, who was as much an extension of his owner's ego as Boone's rifle was of his reach. Indeed when Harrison offered the slave freedom for saving his life, Hector refused to be separated: " 'I d——n to h——ll, maussa, ef I guine to be free!' roared the adhesive black, in a tone of unrestrainable determination. 'I can't loss you company, and who de debble Dugdale guine let feed him like Hector? 'Tis onpossible, maussa, and dere's no use for talk 'bout it. De ting aint right; and enty I know wha' kind of ting freedom is wid black man?' " So did Hector refuse to be severed from his master and stubbornly placed his black hand back upon the leash of a still darker side of Harrison.

Dugdale may have looked a little like the dog in Bingham's *Rescue Group*, though the famous bloodhound probably resembled more the Irish wolfhound. To remind Dugdale that in Carolina he was an Indian hound, Harrison trained him by using a figure stuffed and painted to resemble a redskin and made more enticing "by deer's entrails hanging around his neck, [while] Hector, holding back the dog by a stout rope drawn around a beam, the better to embarrass him at pleasure, was stimulating at the same time his hunger and ferocity." With a snap of his fingers, Harrison could send the dog springing for the jugular of a merciless savage. Locked in combat with one during the hostilities, he had only to shout "Dugdale!"

> The favorite, with a howl of delight, bounded at the well-known voice, and in another instant Harrison felt the long hair and thick body pass directly over his face, then a single deep cry rang above him. . . . He partially succeeded in freeing himself from the mass that had weighed him down; and looking up, saw the entire mouth and chin of the Indian in the jaws of the ferocious hound. . . . Dugdale held on to his prey, and before he would forego his hold, completely cut the throat which he had taken in his teeth.

Dugdale was to Harrison what Paulding's Timothy Weasel was to Sir William Johnson: a highly useful Indian-hater. Not the least of his uses was to allow his master—and Simms and his readers—to experience the howl of delight as a dark surrogate self sprang on bestial prey. Well might Harrison embrace Dugdale, then, to attest "the deep gratitude which he felt for the good service of his favorite."

Even Dugdale's thirst for this blood had had to be whetted. As Ishiagaska said, "The dog must smell the blood, or he tear not the throat." So it was with the Yemassees. They sang their famous scalp song of taking the path of the enemy,

> I hear him groan, I see him gasp,
> I tear his throat, I drink his blood.

The capture and sacrifice of a heroic Irishman gave them their first taste. The hydrophobic convulsions of prophet Enoree-Mattee inflamed them as he continued

> those fearful incantations to the war-god, which seemed to make of himself a victim. He was intoxicated with his own spells and incantations. His eye glared with the light of madness—his tongue hung forth between his clinched teeth, which seemed every moment, when parting and gnashing, as if about to sever it in two, while the slaver gathered about his mouth in thick foam, and all his features were convulsed.

And the Dionysiac underpinnings of this drama of song and seizure became explicit when Simms likened Indians to the early Greeks and had their war dance make of them one frenzied horde in the thick night,

> a thousand enemies, dark, dusky, fierce savages, half-intoxicated with that wild physical action which has its drunkenness, not less than wine. Their wild distortions—their hell-kindled eyes—their barbarous sports and weapons—the sudden demoniac shrieks from the women—the occasional burst of song, pledging the singer to the most diabolical achievements. . . .

So in the final analysis Dugdale did not cut a person's throat but a mad animal's, just as the white Carolinians had cut down a pack of rabid dogs, "these cursed bloodhounds," as one put it, these "redskins that hunt for nothing but our blood!" Or as the poor Irishman cried out, "Ye bloody red nagers!"

3

The salient fact about William Gilmore Simms's birthplace is that it was in Charleston, South Carolina, but outside the planter class that ruled the city, the state, the region. His father had emigrated from Ireland shortly after the American Revolution, failed as a Charleston merchant, moved west to Tennessee, where he became an officer in Andrew Jackson's campaigns against the Creeks and Seminoles, and ultimately came to rest in Mississippi as a planter. Like Davy Crockett and other Tennessee Volunteers who served under Old Hickory, he was a "discontented & forever wandering man," or so at least his son remembered him. Left behind to grow up with his maternal relatives and under the particular care of Grandmother Singleton, the younger Simms later recalled his fascination with her tales of Indian warfare and heroic frontier days in Virginia and the Carolinas. As a young man he made two long trips to the current frontier, observing with interest the circumstances of white settlers and Indians along the way, but he refused his father's invitation to join him in Mississippi. Instead he returned to Charleston, was admitted to the bar, became a newspaper editor, a minor politician, and, from the 1830s to the Civil War, an extraordinarily prolific man of letters.

In 1836 Simms finally married into a marginal family of the ruling class. His father-in-law, Nash Roach, who owned only about seventy slaves, bequeathed him the hand of his daughter Chevillette Eliza and the plantation of "Woodlands" for their family home. Thereafter the sweat of black brows permitted Simms to remain in the library at "Woodlands" writing his romances and essays. In *The Yemassee* he had taken pains to point out that the blacks had sided with the white Carolinians in the racial war with the Indians and that sensibly adhesive Hectors would not choose to leave their too-generous masters. Now a direct beneficiary of slavery, he defended the institution with rising stridency, arguing with John C. Calhoun that it was a positive good or, in his own words, "an especially and wisely devised institution of heaven." By frequently identifying himself so fully and so vehemently with the peculiar institution, Simms became the leading literary symbol of the Old South.

Such commitment to slavery and southern aristocracy, argued Richard Slotkin in *Regeneration through Violence* (1973), left Southern writers, "William Gilmore Simms, James K. Paulding, Joseph P. Kennedy, and Robert M. Bird, among them," incapable of responding to "the West as a psychological problem of immense national significance." Now, there were important regional variations in the response to the West, some of which Slotkin traced sensitively; there were particular problems facing Southern writers who tried to translate their dedication to a stratified society from the seaboard to the more egalitarian border. But the idea of the West was as much the South's as the North's, and the physical reality no less so: After all, it was a Southerner who acquired much of it through the Louisiana Purchase.

Part of the problem stems from the difficulty of locating the sections with some specificity. Daniel Boone lived his adult life in slave territories and states, in North Carolina, Kentucky, and Missouri: Was he therefore a Southerner, a Westerner, or both? Slotkin cited Davy Crockett, who was born near Greeneville, Tennessee, as a symbol of "the contemporary West": Was he therefore more of a Westerner than the elder Simms, who wound up farther south in Mississippi, another slaveholding state? Say his son had joined him there: Could William Gilmore Simms then have become a Western writer like the Eastern missionary Timothy Flint, who traveled up and down the Mississippi Valley and who lived for some time in the slaveholding state of Missouri? The truth was that slaves and a stratified society had long since moved into the "West" of Slotkin's folk hero Boone and of that other frontier hero, Old Hickory.

This prevailing confusion about the West, where it began—did it ever end?—and where the South and the North left off, made tracing regional variations chancy. The Virginia and Carolina borders of Simms's Grandmother Singleton were once the West. His father's service under Jackson helped clear out the Indians and extend it. Simms's trips to visit his father gave him direct knowledge of a farther West. And he remained throughout his life a political partisan of his father's old commander—according to one of his best critics, Simms was "a committed Jacksonian and a consistent Jeffersonian." He even stood by Old Hickory during the Nullification controversy, at considerable personal cost and risk for a Charleston newspaper editor, and remained until 1848 an ardent member of the party of Jackson and Van Buren. Actually, his politics and his response to the West as a national issue were virtually identical with those of his friend Paulding.

In *The Yemassee*, Indians functioned as white America's own link with the ancient epic, and that meant a national tie to a past as remote as that of all those foreign visitors who superciliously spoke of the "newness" of the United States. But now, to allow for the continued rise of the Union and of the superior race, they had to be cleared from the land. Society had a place for blacks in the bottom caste but no place where reds could be put and kept. Hugh Grayson, the youthful rival of Harrison and a common man with leadership potential, undoubtedly spoke for Simms in singling out their color as the major reason why they would have to go:

> It is utterly impossible that the whites and Indians should ever live together and agree. The nature of things is against it, and the very difference between the two, that of colour, perceptible to our most ready sentinel, the sight, must always constitute them an inferior caste in our minds. . . . They lose by our contact in every way; and to my mind, the best thing we can do for them is to send them as far as possible from communion with our people.

This was pretty powerful stuff in 1835 on the eve of all the Trails of Tears. Very shortly Simms's fellow Carolinian Poinsett would forcibly send the Indians as far away as possible, to the cheers of Paulding, the *Northern* writer.*

* Slotkin's major misidentification, or mislocation, in a generally important work bears mentioning because it is both understandable and instructive: Paulding did write *Letters from the South* (1817) and *Slavery in the United States*, a fervid defense already mentioned; he also contributed to the *Southern Literary Messenger* (Richmond) and otherwise left evidence an observer might mistake as proof of Southern origin. He was, however, born in Putnam County, New York, and except for two trips to the South and West, spent his entire life in the North and in Washington. But for an able writer to mislocate him in the South helps remind us that racism did not grow exclusively below Mason and Dixon's line and that the response to the major psychological problem of the West, namely the Indians, was indeed *national* in scope.

CHAPTER XIII

Nicks in the Woods: Robert Montgomery Bird

> The whole object was here to portray the peculiar characteristics of a class of men, very limited, of course, in number, but found, in the old Indian days, scattered, at intervals, along the extreme frontier of every State, from New York to Georgia; men in whom the terrible barbarities of the savages, suffered through their families, or their friends and neighbors, had wrought a change of temper as strange as fearful.
>
> —ROBERT MONTGOMERY BIRD,
> preface to *Nick of the Woods*, 1853

RETURN FOR a moment to the pretty scene of *Daniel Boone Escorting Settlers through the Cumberland Gap* (p. 133) and imagine the trailblazer wearing a sweat shirt with TRANSYLVANIA COMPANY emblazoned across the front. The anachronism grates, shatters the edenic bliss, and wreaks esthetic havoc, I grant; but historical truth is bound to seem an intruder in George Caleb Bingham's deathless interpretation: under the handsome yellow buckskins of his hero beat the heart of a land company agent.

Actually Boone was a business associate of Colonel Richard Henderson, chief promoter of the Transylvania Company. He located and later surveyed tracts of rich lands, supervised the dubious treaty with the Cherokees that granted Henderson and his fellow speculators some twenty million acres beyond the Gap, blazed the Wilderness Trace at the head of an armed band of thirty men, and later that season, 1775, guided into Kentucky the party of colonizers Bingham made immortal. Boone built Boonesboro on a "commanding site" along the Kentucky. Henderson provided the capital for pushing this salient into Indian country, helped with logistics and support, and guided additional settlers to the fort. Moreover, he shared his own grandiose schemes for a Western empire at least to the extent of allowing Boone to stake out vast claims for himself.

In 1776 the Virginia legislature repudiated the claims of the Transylvania Company, and subsequent court proceedings adjudged Boone's title entries imperfect. Stripped of his holdings, the First Hunter crossed the Mississippi at the turn of the century in disgusted flight from "civilization." But the essential point should not be lost in transit: however he disliked the litigious quarrelsomeness of the white settlements and whatever his yearnings for intimate communion with nature, he was still an ambitious land speculator. And as a leading property owner, before all the ejectment suits, he served several terms in the Virginia legislature and became a sheriff and county lieutenant.

The legend-makers abstracted Boone from this economic and political context. But capital and organization were necessary to establish permanent beachheads in Kentucky, as Boone discovered when he unsuccessfully made the attempt on his own in 1773. Only after repeated assaults by land companies and government forces did the Cumberland Mountains yield the pass to the army of colonizers celebrated by Daniel Bryan in *The Mountain Muse* (1813):

> Swift on, o'er the rude-featured Wilderness
> The sinewy sons of Enterprise proceed.

An advance agent of Transylvania, Boone also had held a series of official positions dating back to his participation in General Edward Braddock's disastrous campaign against Pittsburgh early in the Seven Years War (1755). As part of Virginia's expansion into the Northwest, Governor Dunmore sent surveyors into the region and in 1774 appointed Boone to guide a party back from the falls of the Ohio River. He carried out this mission with such celerity, the governor appointed him commander of three garrisons during what was called Lord Dunmore's War against the Shawnees and Ottawas. When Virginia repudiated Henderson's land claims and made Kentucky a county of Virginia, Boone was commissioned a captain of militia. Now, all this meant that he had sterling credentials as a professional empire-builder. If he was "an instrument ordained to settle the wilderness," as John Filson had him say, then he was an instrument forged in the seaboard settlements. He carried the stamp of their values and ideas, including their enterprising expansionism, and cut into the wilderness to further their objectives as well as his own. Far from being an isolated First

Hunter, he was as much an agent of collective purpose as those two other famous captains, Lewis and Clark, and had sponsorship almost as direct and official.

In this light *what* Boone intended to do after he passed through the Cumberland Gap stood forth in bold relief, for he was by then a veteran of the war against the wilderness and its inhabitants. But *how* he waged the war remained the critical ambiguity in Bingham's painting and in reality. That is, how would he use that rifle on his shoulder? Was he the strangely gentle empire-builder Simms saw in him, by necessity alone "a hunter of men too, upon occasion"?

Boone's vaguely Quaker origins in Pennsylvania suggested he was. John Bakeless, his best modern biographer, concluded he was, finding "always a faintly chivalrous note in Daniel Boone's warfare with the Indians." It was this Boone who became the prototype for a series of American folk heroes who were disinclined to fight but driven to *High Noon* confrontations by dire circumstances.*

On the other hand, just how reluctant a killer was Boone? The keen First Hunter would have been scarcely human had he not been tempted, on occasion, to share the passion for Indian-stalking of Uncle Mordecai Lincoln and other Kentuckians. Adam Poe, for instance, saw them as the noblest game in the woods: "I've tried all kinds of game boys!" the imitable Colonel Frank Triplett had Poe say in *Conquering the Wilderness* (1883): " 'I've fit bar and painter (panther) and catamount, but,' he added regretfully, and with a vague, unsatisfied longing in the plaintive tones of his voice, 'thar ain't no game like Ingins—No, sir! no game like Injins.' " The gentlemanly Boone may well have caught such game in his fabled sights: In his enormously popular nineteenth-century biography, Timothy Flint proudly recounted how Boone had once picked off two Indians with a single bullet.

John Filson's Boone was less a sporting killer and more a grim soldier in the state of siege just lifting after a decade and more of intense fighting. Indeed, while Filson included Boone's Crusoe-

* The barbarity of the villain(s), and the hero's sense of honor, overcame the promptings of conscience or surrogate conscience, as the schoolmarm functioned in Owen Wister's *The Virginian* (1902). In real life Alvin C. York, the World War I hero who grew up on the Wolf River in the Cumberland Mountains, was a more recent exemplar. When Sergeant York died in 1964, the Associated Press obituary noted that he had served in the 82nd "All-American" Division, that "Onward Christian Soldiers" would be sung at his funeral, and that his strong religious convictions were part of his legend: "Once a conscientious objector, he went on to answer his country's call to World War I and killed 25 Germans and captured 132 more virtually single handed in the battle of Argonne Forest." From Boone's day to York's, men who heroically overcame principle to kill for their country frequently enjoyed all-American status.

like adventures and captivity episodes, the body of his famous appendix was in the tradition of the colonial war narratives and Boone in the tradition of the Indian-war heroes of old. Like John Underhill, Boone went in for body counts. After one dreadful siege of Boonesboro, for example, "we had two men killed, and four wounded, besides a number of cattle. We killed of the enemy thirty-seven, and wounded a great number." Following the defeat of the Kentuckians at Blue Licks in 1782, Boone joined the punitive expedition across the Ohio led by General George Rogers Clark, conqueror of the Old Northwest. They raced to heart-warming vengeance through the villages where Cornstalk and Tenskwatawa once lived: "We . . . burnt them all to ashes," Boone recounted, "entirely destroyed their corn, and other fruits, and every where spread a scene of desolation in the country." And then the count: "In this expedition we took seven prisoners and five scalps, with the loss of only four men, two of whom were accidentally killed by our own army." Boone wasted precious little sympathy on the "barbarous savage nations" of Shawnees and others who appeared as predictably merciless savages. I can detect no "faintly chivalrous note," as Boone's biographer would have it, in his taut references to "the tender women, and helpless children, [who] fell victims to their cruelty."

Simms's contention that Boone was not fond of the sport of hunting men was hence not entirely supported by the record. Probably he was a less enthusiastic Indian-killer than Mordecai Lincoln or Simon Kenton, that other hero of the Ohio–Kentucky frontier. Beyond that the record is spotty or nonexistent. It is complicated by the fact that, as Richard Slotkin has pointed out, after Boone read Filson's book, he "went so far as to model his public statements on Filson's characterization of him." This was yet another instance of life imitating art but one that probably justifies us in assuming that the attitudes toward Indians Filson attributed to Boone met with Boone's approval.

In what has always been a print culture, American folk heroes from Boone to Paul Bunyan pose the vexing problem of sifting out in what measure they were originally concocted by journalists and public relations men. Just as Davy Crockett leaped onto the national stage with a boost from Whig writers and politicians, so Boone became an international hero through a collective dream articulated by a promoter to sell Kentucky to settlers and land speculators. But our concern is primarily the Boone legend, and for that the real Boone became increasingly irrelevant, especially after he began to talk like Filson. Anyway Boone was from the beginning sufficiently shadowy so that writers said more about themselves than about him in their characterizations. A relatively amiable man, Simms saw in him a great discoverer and strangely gentle empire-builder. Less amiable, Paulding cast him as Ambrose Bushfield, an unrestrained Indian-hater. Then a novelist with still greater inner tensions picked up the legend of the good Quaker hunter and created from it one of the nearly great characters in American fiction, a mutilated Boone who became, as we shall now see, a demoniacal killer, an Old Nick loose in the woods.

2

By a latitudinarian definition of Southern, Robert Montgomery Bird was a Southern writer, born in New Castle, Delaware, the sixth of seven children in a substantial farm family of English–Dutch stock—he grew up tall and fair and was said to be "of commanding presence." As a young man he crossed over to another border state and studied medicine at the University of Pennsylvania. When he was thirty-one years old he married the daughter "of a Lutheran minister of a good Philadelphia family." By then he had already established himself as a writer. More interested in letters than in general practice, he gained early recognition as a dramatist, wrote novels with exotic settings in Spanish America before he turned to the Kentucky frontier for his most famous work, and toward the end of his life, discouraged and ill, desultorily tried his hand at history. Since his active years were spent in Pennsylvania, literary historians refer to him as "probably the ablest man of letters that Philadelphia produced" or as a "Philadelphia . . . physician of restless temperament with a flair for storytelling." Notwithstanding this Middle-State identification and his party affiliation—he once planned to stand for public office as a Whig candidate—Doctor Bird shared with Paulding and Simms a sympathy for the South and its institutions.

In 1829, 1833, and 1835 Bird visited the West, was particularly taken by Kentucky's Mammoth Cave, and later wrote that he had found his proper field "in the history of early Western colonization." Kentucky was the scene and Indian warfare his principal theme. Among his papers at the University of Pennsylvania, editor Cecil B. Williams found a prospectus indicating that in the 1830s Bird had originally planned a series of tales of Kentucky. His manuscript notations for the first indicated it was to be a romance, "The Forest-Rover," with Boone the central figure:

Opening—Boone, in Solitude—Cabin—
Canebrake, &c. &c.
A Scene of Wolves

Next were to come tales of the Battle of Blue Licks and of Simon Kenton; the series was to conclude with "The Fighting Quaker—Last of his line." In the event Bird wrote only the last, but it contained elements of the first three projected novels. In the composite Boone may have been the model for Bird's "honest Colonel of the Militia." More critically, though Boone disappeared as the named central figure, his Quaker parentage and the moral problem it posed for a Kentuckian survived in Nathan Slaughter, Bird's title character. In his 1853 preface to the novel he professed to have drawn Nick from an obscure legend—the source of which he had forgotten—that touched "the existence of just such a personage, whose habitat was referred to Western Pennsylvania, [and that] used to prevail among the contemporaries, or immediate successors, of Boone and Kenton." But the novelist quite properly claimed the right to write within the limits of "poetical possibility." Those limits gave him plenty of room, I venture, to shuffle Boone's gentle origins behind the most grotesque Indian-hater of them all.

Nick of the Woods or the Jibbenainosay: A Tale of Kentucky (Philadelphia, 1837) enjoyed immediate and lasting public favor. Two London editions appeared the same year, and over the following century others appeared in English, German, Dutch, and Polish, for a total of twenty-nine editions in all—in the early 1970s it was still available as a reasonably priced hardbound and paperback.

The time was 1782, tail end of the Revolution and year of the Battle of Blue Licks. The place was the Salt River Valley, through which ran an extension of Boone's Wilderness Trace. From Virginia had come two of the gentry, Captain Roland Forrester and his cousin Edith, in love but deprived of their inheritance by a villainous lawyer, who had left them little more than the option to wander like Boone in quest of "Kentucke." Moreover lawyer Braxley sicked the Shawnees after them, and that meant captivities and rescues too numerous to mention before they were finally reunited in each other's arms. But beneath this melodrama was Nathan Slaughter's struggle to the death against the Shawnees, Bird's real story and the source of his book's enduring popularity.

With contemptuous sarcasm frontiersmen at Colonel Tom Bruce's fort called Slaughter "Bloody Nathan" because, as the honest militia commander explained to Roland Forrester, "he's the only man in all Kentucky that *won't* fight! and thar's the way he beats us all hollow. Lord, Captain, you'd hardly believe it, but he's nothing more than a poor Pennsylvany Quaker; and what brought him out to Kentucky, whar thar's nar another creatur' of his tribe, that's no knowing." Bruce tolerated him as an old fool, but he and everybody else in Kentucky believed it the duty of all able-bodied men to fight "the murderers of their wives and children." While staggering under the weight of their reproach and abuse, Nathan Slaughter was the very model of the other-cheek ethic, larding his conversation with "thees" and backing away from quarrels with a depth of humility that convinced real fighting men he was a "no-souled crittur!" Little did they know that hidden under his pious pacifism was the butchering Jibbenainosay, as the Shawnees called their nemesis, the "Spirit-that-walks" through the district killing all Indians unlucky enough to get in his way, scalping them, hacking their skulls open, and slashing his fearful mark into their breasts—a brace of deep knife cuts in the shape of a cross, Forrester learned, was the way he marked "all the meat of his killing."

Through the trackless forest fled the terrified young couple, pursued by the Shawnees, who in turn were pursued by Old Nick, alias Nathan Slaughter. A born captive, Edith fluttered—helpless white womanhood in the clutches of Indians determined to deliver her to lawyer Braxley. "Hapless Edith! wretched Edith!" cried Roland Forrester. "Was ever wretch so miserable as I?" Such extravagance of grief led his singular companion to think back to true wretchedness and ask severely: "Is thee wretched because thee eyes did not see the Injun axe struck into her brain?" Thereupon Nick told the young man how, a scant ten years before, Nathan Slaughter had been happy in the mountains of Pennsylvania far away, surrounded by the wife of his bosom, his "gray old mother," and the children of his body, *five* "little innocent babes."

Nick's story was that of Paulding's Timothy Weasel and Ambrose Bushfield, that of Timothy Flint's "Indian Fighter," and that of Judge James Hall's still earlier Samuel Monson, "The Indian Hater" (*Western Souvenir . . . for 1829*), but more vivid and made immeasurably more gripping by the fact that Nathan Slaughter was a man of peace when the Shawnees came upon him. So they might know he was a friend, he handed their chief his hunting rifle and knife:

> With my own knife he struck down my eldest boy! with my own gun he slew the mother of my children!—If thee should live till thee is gray, thee will never see the sight I saw that day! When thee has

> children that Injuns murder, as thee stands by,—a wife that clasps thee legs in the writhing of death,—her blood, spouting up to thee bosom where she has slept,—an old mother calling thee to help her in the death-struggle:—then friend, *then* . . . thee may call theeself wretched, for thee will be so!

The passions awakened by these dreadful recollections brought on overpowering agitation of mind: Suddenly Nick dropped to the earth in an epileptic fit, "his mouth foaming, his eyes distorted, his hands clenched, his body convulsed. . . ." As Roland Forrester raised him up his cap fell off baring a horrible scar, years old, where "the savage scalping-knife had done its work on the mangled head." If ever a man had grounds for declaring eternal war against a race . . .

So had Nathan Slaughter, the man of peace and goodwill, become Nick, the man of war and vengeance. With this story off his chest, Nick recovered rapidly, breathed more freely out from under his professions of philanthropy, and began to speak easily, like all white Kentucky, of the Shawnees as "thieves and dogs." He made ready to kill five who were torturing Captain Ralph Stackpole "with an alacrity of motion and ardor of look that indicated anything rather than a distaste for the murderous work in hand." Across the Ohio in the Miami country this ardor became more explicitly sexual as Nick finally placed his hands on the chief who had slain his family. Disguised as an Indian, he crept up on Wenonga as the "wretched squalid sot" lay passed out before the entrance to his own wigwam, looked down on his "dark grim visage" with "a feeling not less exciting, if less unworthy, than fear" and with no one

> at hand to gaze upon his own, to mark the hideous frown of hate, and the more hideous grin of delight, that mingled on, and distorted his visage, as he gloated, snake-like, over that of the chief. As he looked, he drew from its sheath in his girdle his well-worn, but still bright and keen knife,—which he poised in one hand, while feeling, with what seemed extraordinary fearlessness or confidence of his prey, with the other along the sleeper's naked breast, as if regardless how soon he might wake. But Wenonga slept on, though the hand of the white man lay upon his ribs, and rose and fell with the throbs of his warlike heart. The knife took the place of the hand, and one thrust would have driven it through the organ that had never beaten with pity or remorse; and that thrust Nathan, quivering through every fibre with nameless joy and exultation, and forgetful of everything but his prey, was about to make.

Like Boone, let us boldly bestow a name: Nick's tremulous joy was the orgiastic delight of violence. It was merely preliminary this time, however, for "the wail of a female voice" broke in upon the two men and staggered Nick back to recollection of the captive damsel and his mission of rescue.

Consummation was not long in coming. At the very first opportunity Nick buried a tomahawk in Wenonga's brain, ripped off his gray scalp lock, and with two thrusts of the chief's own knife, delivered with the utmost strength, laid open his naked breast, "dividing skin, cartilage, and even bone, before it, so sharp was the blade and so powerful the hand that urged it." With his ardor now fully aroused, Nick gave forth "a wild, ear-piercing yell, that spoke the triumph, the exulting transport, of long-baffled but never-dying revenge."

Shortly thereafter all the captives were rescued by General George Rogers Clark—or Colonel Bruce, as Bird had it—at the head of an army of Kentuckians aching to avenge their recent defeat at Blue Licks. Nick burst upon them, no longer the inoffensive Nathan. Wenonga's scalp, "yet reeking with blood," hung at his belt; the chief's ax, "gory to the helve," dangled from his bloodstained hand: "His eyes beamed with a wild excitement, with exultation, mingled with fury; his step was fierce, active, firm, and elastic, like that of a warrior leaping through the measures of the war-dance." Like Filson's Boone, whom we have already placed on this expedition into the Shawnee country (p. 150), Nick helped spread a scene of desolation: "He was among the most zealous in destroying the Indian village, applying the fire with his own hands to at least a dozen different wigwams, shouting with the most savage exultation, as each burst into flames."

Indeed, Nick was a Boone with a secret infection. As Bird put it in his 1853 preface, " 'Indian-hating' (which implied the fullest indulgence of a rancorous animosity no blood could appease) was so far from being an uncommon passion in some particular districts, that it was thought to have infected, occasionally, persons, otherwise of good repute, who ranged the woods, intent on private adventures, which they were careful to conceal from the public eye." Thus, put the other way round, had Forest-Rover Boone had a secret infection and such a common private passion, he would have been Old Nick.

At Bruce's fort the reproach and abuse heaped on nonviolent Nathan changed overnight to cheers and hearty hurrahs for the valiant Nick. The special traits prized there had been made clear earlier, in an admiring discussion of Tom Bruce the younger, who came from "the best stock for loving women and fighting Injuns in all Kentucky!" The boy had taken the scalp "of a full-grown Shawnee before he war fourteen y'ar old." When he was kidded for not having grabbed the chance to kill the Indian's companion, "he blubbered all night, to think he had not killed them both." Nick met and surpassed such standards of excellence.

The character of Ralph Stackpole was a particularly interesting case in point. Captain of the horse thieves, a ring-tailed roarer, and champion brawler—"Down then, you go, war you a buffalo!" —Stackpole had sentimentally dedicated himself to Edith's cause or, as he said, to "angeliferous madam's fighting against all critturs, human and inhuman, Christian and Injun, white, red, black, and party-colored." With considerably greater success than Paulding, Bird had drawn on the blossoming Southwestern humorists for Stackpole, modeling him after Mike Fink, with a dash of Simon Kenton's fame as a horse thief thrown in for good measure.

"Well," Mike Fink had said, "I walk tall into varmint and Indian" and did so to the amusement of white folks everywhere. For instance, in the *Western Souvenir . . . for 1829*, edited by the indefatigable James Hall, appeared the tall tale of "Mike Fink, the Indian, and the Deer": Hunting in the Pennsylvania woods, the legendary scout spied an Indian aiming at the deer he had just stalked. Fink caught the Indian in his own sights, and the moment the latter fired, "the bullet of Fink passed through the red man's breast." That was good parsimony as well as good shooting, for it brought down two beasts with one bullet. Another of Fink's famous shooting feats, though it appeared in print after *Nick* was published, was his symbolic castration of a red man silhouetted against the morning sky ("The Disgraced Scalp-Lock," by Thomas Bangs Thorpe), and still another his marvelous shot at the heel of a black ("Trimming a Darky's Heel," by John S. Robb). Stackpole shared this chuckling contempt for the victims of white cruelty and like Fink believed it to be "as praiseworthy to bring in the scalp of a Shawnee, as the skin of a panther." As had the original alligator–horse, Stackpole illustrated what Benjamin Botkin once characterized as "the essential viciousness of many of our folk heroes, stories, and expressions." To Stackpole's mind Indians were merely "red niggurs." Nick's skill as a super slaughterer clearly won his wholehearted admiration.

This admiration cut across class and regional lines. After Nick had told Roland Forrester his story, what the Shawnees had done to him and his family, he demanded eagerly, "had they done so by *thee,* what would thee have done to them?" The young Virginian's reply made the butcher's monomania a perfectly reasonable response to his terrifying experience:

> "Declared eternal war upon them and their accursed race!" cried Roland, greatly excited by the story; "I would have sworn undying vengeance, and I would have sought it,—ay, sought it without ceasing. Day and night, summer and winter, on the frontier and in their own lands and villages, I would have pursued the wretches, and pursued them to the death."

Well might "a grin of hideous approval" have lighted up Nick's face, for a "civilized" frame had just been placed around his unappeasable fury. Roland Forrester was no backcountry braggart like Stackpole but a genteel Easterner who would shortly return to his ancestral plantation, with Edith on his arm and his slave Emperor, as adhesive as Simms's Hector, attending him. Before they left, Forrester spelled all this out by thanking Nick for all he had done: " 'We owe you life, fortune, everything,' he cried, extending his hand; 'and be assured neither Edith nor myself will forget it.' " But Nick refused his invitation to go with him to Virginia for a share of their wealth and a new life. At the novel's end he had disappeared, "going no man knew whither." No one knew, but the reader was left with the likelihood Nick had followed his game —"the meat of his killing"—farther into the West where other new "civil societies" would depend on his Indian-hating and owe him everything for villages burned and scenes of desolation spread.

Moreover, when Roland Forrester called Indians an "accursed race," spoke of the Shawnees as "yonder crawling reptiles," and voiced approval of Nick's murders, he was the genteel mask of his creator, the urbane Philadelphia physician. Like Simms, Bird worked with color contrasts, Edith's "pale visage" and the "dark grim visage" of the red man. "Black Vulture" was the translation of the name of Wenonga, Nick's deadly enemy; and the color of this carrion-feeder was not fortuitous—in his posthumous frontier tale "Ipisco Poe" (1889), Bird had a drunken Shawnee chief named "Niggurnose." And like Paulding, Bird presented parchment-thin merciless savages: the Philadelphian's Indians were "the most heartless, merciless, and brutal of all the races of men." With surprising directness he likened children "and savages, who are but grown children, after all." But his Indians were more bestial than infantile, and his novel was still about "The Forest-Rover in Solitude—A Scene of Wolves."

Bird went further than Simms and Paulding in revealing his own delight in the destruction of forest brutes. In scenes reeking with latent homosexuality he exultingly carved crosses in red corpses with his pen while Nick thrust his keen knife into their naked breasts. Penetration of red bodies presumably destroyed such temptations to instinctual gratification by butchering their embodiments: Bird was visibly upset that Indians, grown children at best, swung easily from laughter to anger and back, expressing desires and venting tensions on the way. He did cynically grant, in his 1853 preface, that "the Indian is a gentleman; but he is a gentleman with a very dirty shirt." "Civilization" was what the Indian did not have, and "civilization" meant for Bird a real gentleman, himself, with sterilized or subdued urges and with proper order imposed on his life, even after his wedding to the Lutheran minister's daughter. Doctor Bird's shirt was always spotless, and this starched self-control probably made his not "uncommon passion" of Indian-hating especially rancorous. That rigidity, and some secret psychic wound of his own, may have given him a sense of the dammed-up aggressions underlying Nathan Slaughter's Quaker professions of philanthropy. Apart from his published work, however, Bird the man remains as obscure as the reference to his "restless temperament." What is clear is that in his violent pornography he verged on explicit acknowledgment that white Americans were the real devils in the woods, proud of it, and, to his mind, justifiably so.

Whatever the source of his inner demons, Bird wrote out of his flawed genius a revealing allegory. Through a Quaker man of peace he demonstrated the horrifying consequences that would have awaited other Anglo-Americans had they acted on such principles of nonviolence and goodwill. As it was, what Nathan Slaughter had suffered personally, in extreme form, the whole country had suffered collectively when their wives and children and aged mothers fell before the Indians' indiscriminate barbarity. In return, the United States might expropriate and kill and even scalp them, for "such is the practice of the border, and such it has been ever since the mortal feud, never destined to be really ended but with annihilation, or civilization, of the American race, first began between the savage and the white intruder." Appearing in 1837, Bird's allegory helped citizens believe they might proceed in good conscience with "emptying"—in the elegant euphemism of a modern historian—the Eastern states of those merciless savages still ambulatory.

3

In 1837 Bird had sought to forestall such criticism by prefacing his work "with some apology for the hues we have thrown around the Indian portraits in our picture,—hues darker than are usually employed by the painters of such pictures." But the single fact that the Indian waged war "upon women and children, whom all other races in the world, no matter how barbarous, consent to spare," accounted for the dark colors on his palette and the faithful likenesses they shaded in. When *Nick* appeared, a reviewer in the *Southern Literary Messenger* found it a welcome corrective to contemporaneous romanticizations of the Indian. A century later Vernon Louis Parrington agreed, noting that in his depiction of Indians, "Bird has put all his romanticisms aside. There is no sentimentalizing of the noble red man in the brisk pages of *Nick of the Woods;* the warriors are dirty drunken louts, filled with unquenchable blood-lust, whom the frontiersman kills with as little compunction as he would kill a rattlesnake." So for unsentimental "stirring action . . . certain to end in blood-letting," see Nick. In *The Indian in American Literature*, published in 1933, Albert Keiser called his chapter on Bird "Stark Realism on Kentucky's Dark and Bloody Ground."

In stark reality all but three of the Indians in *Nick* ranged through the brush an undifferentiated pack of "thieves and dogs." Of those singled out, one was a ridiculous old chief who in a drunken rage shot his own horse. He watched over captive Roland Forrester with "a miser's, or a wild cat's affection"; a truly unrepressed brute of "dark visage," he swung wildly from rage to laughter, a grown child in whom "we find malice and mirth go hand in hand." Feral ferocity also characterized Wenonga's "old crone," "the old hag" who was as heartless as her mate: "No ray of pity [for fluttering Edith] shone even for a moment from her forbidding, and even hideous countenance." But the distillation of all this hideousness was Wenonga himself, introduced to the reader as "an old Indian of exceedingly fierce and malign aspect, though wasted and withered into the semblance of a consumptive wolf." Later, "with a wild and demoniacal glitter of eye, that seemed the result of mingled drunkenness and insanity," he gabbled in his own tongue, "being perhaps too much enraged to think of any other,—'I am Wenonga, a great Shawnee chief. . . . I love white-man's blood.' " But that sort of love was not romantic, and these were surely not sentimental portraits.

Bird's vilified Indians were one and all blackened monsters, drawn not from reality but from the unconscious. In 1853 he prefaced a new edition with another denial, maintaining that "our own Cooper" had wrapped Indian character in "poetical illusions" as imaginary and contrary to nature "as the shepherd swains of the old pastoral school of rhyme and romance." Bird had instead confined himself "to real Indians" and professed

> it was with no little surprise he found himself taken to account by some of the critical gentry, on the charge of entertaining the humane [*sic!**] design of influencing the passions of his countrymen against the remnant of an unfortunate race, with a view of excusing the wrongs done to it by the whites, if not of actually hastening the period of that "final destruction" which it pleases so many men, against all probability, if not possibility, to predict as a certain future event.

But as we noted a couple of paragraphs earlier, Bird had himself confidently predicted the final annihilation of the Indians, barring their improbable "civilization." His slip in using the word *humane* in the extraordinary sentence just quoted, whether genuine or mere editorial carelessness, suggested a flinch of the unconscious away from even stating the profound inhumanity of what he had in fact done: added his mite to hastening the "final solution" of the Indian "problem."

Attacked so often as "Noble Savages," James Fenimore Cooper's Indians always had their irony: From one end of the Leatherstocking Tales to the other, from *The Pioneers* of 1823 to *The Deerslayer* of 1841, all but a handful were stock merciless savages. Most were Iroquois, or Mingoes, as Natty Bumppo called them, reptiles such as Magua in *The Last of the Mohicans* (1826), who was every bit as inhuman as Bird's Shawnees: "Iroquois—devil—Mingoes—Mengwes, or furies—," was Natty's summary in *The Pathfinder* (1840); "all are pretty much the same. I call all rascals Mingoes." On the high Plains their counterparts in *The Prairie* (1827) were the Sioux "demons rather than men." In his other writing and political activity, moreover, Cooper was not an impassioned advocate of Indian rights. In *Notions of the Americans* (1828), for instance, he vilified those still in the East as "a stunted, dirty, and degraded race," sympathetically discussed the national government's attempts to help them, and said scarcely a word about the ongoing expropriation of their lands and extermination of their

* In this context Bird would hardly have used the word *humane* ironically or cynically, for that would have verged on avowal of a premeditated atrocity of the spirit.

peoples. The *real* Indians were anyhow in the West: "A few degraded descendants of the ancient warlike possessors of their country are indeed seen wandering among the settlements, but the Indian now must be chiefly sought west of the Mississippi, to be found in any of his savage grandeur." Moral sanction for removal of the decayed relics of the tribes east of the Mississippi grew lushly in such prejudice and misstatement.

Yet the same man created a handful of exceptional red men, the Delawares of Chingachgook and Uncas, and the Pawnees of Hard-Heart. Many critics have remarked that for these red heroes the novelist had turned primarily to the Reverend John Heckewelder, whose *History, Manners, and Customs of the Indian Nations* (1819) celebrated his beloved Delawares and maligned their enemies. As a source, the book was useless to Paulding and Bird, for unlike Cooper they could make do with background from writers hostile to *all* Indians—Bird, for instance, drew heavily on John A. McClung's legendary Indian-killers in *Sketches of Western Adventure* (1832) and on James Hall, Timothy Flint, and others for model Indian-haters. But Cooper had a particular need for the Moravian missionary's work, and that need provides a key for understanding his Indians and the hostility to them.

With the build-up of accretions the legendary Boone had become more definitely a double: half empire-builder and half white-Indian. Natty Bumppo was Boone II, the hunter and fugitive from the "settlements." In his famous 1850 "Preface to the Leather-Stocking Tales," Cooper explained that "in a moral sense this man of the forest is purely a creation." He had given him a character natural to a man who possessed "little of civilization but its highest principles as they are exhibited in the uneducated, and all of savage life that is not incompatible with these great rules of conduct." Natty Bumppo thus lived on the border of "civilization" and had his character all but completely shaped by "savage life." Cooper's critical problem was that for Natty to have any authenticity, Indians had to have real space, if not equal time, in his novels. And to keep them from turning out as travesties of human beings, like all of Bird's Indians and most of his own, he desperately needed someone who could knowledgeably and sympathetically discuss tribal culture from within his own experience. That very rare white was John Heckewelder, who provided him with indispensable data on Delaware myths, manners, and customs and with a sense of their spiritual oneness with nature.

But as Simms was to discover when Sanutee and his tribe nearly toppled the sentimental plot of *The Yemassee*, once Indians entered novels with a measure of their living complexity, they took on fictional life of their own. Only through these savages in residence did Natty learn to live in his wilderness home. And even then doing so depended on his having the strong spiritual sense Simms attributed to Boone: "He felt the union between his inner [nature] and the nature of the visible world, and yearned for their intimate communion." In the Leatherstocking Tales, that yearned-for communion was established through the mythopoeic friendship of Natty Bumppo and Chingachgook. With the Indian reverence for brother animals, Leatherstocking sharply criticized white settlers in *The Pioneers* for their "wasty ways" of slaughtering fish and passenger pigeons. In *The Last of the Mohicans* he still more subversively questioned the slaughter of nature's human inhabitants: "Nothing but vast wisdom and unlimited power should dare to sweep off men in multitudes." No, nothing, not even their color, implied the novel's epigraph from Shakespeare: "Mislike me not for my complexion. . . ." Nearly as obsessed by color as Simms, Cooper had his own profound ambivalences; he made Leatherstocking a stickler for white blood "without a cross," yet allowed Cora Munro to say of Uncas: "Who, that looks at this creature of nature, remembers the shades of his skin!" In *The Prairie* he allowed the noble Hard-Heart open violation of the color taboo: "Such as were abandoned to the worker of evil could never be brave or virtuous, let the colour of the skin be what it might." Natty made the Pawnee his spiritual heir, anyhow, and continued to think of the red men as "the rightful owners of the country." And in *The Deerslayer,* last of the series but first in the chronology of Natty's life, the young hunter stated anarchistic principles with breathtaking implications for white–red relations: "When the colony's laws, or even the king's laws, run a'gin the laws of God, they get to be onlawful, and ought not to be obeyed." A colony law in point was that establishing scalp bounties for Indian men, women, and children—the two professional bounty hunters in the novel, Hurry Harry March and Floating Tom Hutter, were in themselves effective answers to Bird's insistence that Indians alone preyed on the helpless.

Now, Leatherstocking no doubt represented the subversive side of Cooper himself as well as the white-Indian side of Boone. But Cooper's hero needed the Indians as much as they needed him, I have ventured, and through him they found enduring means to pose the westward course of empire as a moral problem: How could a system of justice in the clearings be built upon a record of injustice in the wilderness? A society of Christian brotherhood erected upon its denial? Respect for law and order based on broken treaties, bribery and debauchery, and a thoroughgoing contempt for the natural rights of natives? Finally, there was the singular allure of forest freedom set against the oppressive conformism of the settlements. As Leatherstocking remarked in *The Pioneers*, "I lose myself every day of my life in the clearings."

Such good questions and attitudes naturally evoked the wrath of Indian-haters. General Lewis Cass, soon to become Jackson's secretary of war, brusquely dismissed Cooper's Indians as "beings with feelings and opinions, such as never existed in our forests." Historian Francis Parkman, for whom the slaughter of Indians was on a level with that of buffalo, understandably tended to accept the monstrous characterization of Magua while rejecting Uncas as a phony Indian, as one of Cooper's "aboriginal heroes, lovers, and sages, who have long formed a petty nuisance in our literature." Bird thus merely followed in their footsteps in censuring Cooper for his "poetical illusions." Bird simply could not tolerate the intrusion into a novel of a red man who seemed human. Moreover, he picked up the Indian-fighter side of Boone and rejected out of hand the white-Indian side. For Bird, wildness could never be tonic nor could nature ever be sweetly refreshing green woods and clear waters. Nature was instead the black unconscious within—the psychic counterpart of Kentucky's Mammoth Cave that had so impressed him—and the Dark and Bloody Ground without. If white men like Natty sought intimate communion with it, therewith they became "renegades from the States, traitors to their country and to civilization." As he had a white apostate say in *Nick*, "I'm a white Injun, and there's nothing more despicable." Hence questioning the beneficence of westward expansion was treasonous. Like that of so many of his contemporaries, Bird's nationalism made the eagle scream, as a passage on Roland Forrester suggested: The young Virginian had the integrity of spirit of

> the mighty fathers of the republic [who conducted the country] through the vicissitudes of revolution to the rewards of liberty, [and who] would not stoop to the meanness of falsehood and deception even in that moment of peril and fear.

In the end, alas, even Cooper recanted. The last of the Leatherstocking Tales, with their special structural and ideological requirements, had been published in 1841. Tired, bitter, increasingly religious, he created one of the most egregious Uncle Tomahawks ever in his last Indian story, *The Oak-Openings* (1848).

Scalping Peter started as a Caliban, a Tecumseh; but Cooper converted him to the Gospel of Progress so he could become a good Indian:

> Injin don't own 'arth. 'Arth belong to God, and he send whom he like to live on it. One time he send Injin; now he send pale-face. *His* 'arth, and he do what he please wid it. No body any right to complain. Bad to find fault wid Great Spirit. All he do, right; nebber do anyt'ing bad. His blessed Son die for all color, and all color muss bow down at his holy name.

That all color must bow down at the holy name of "westward the course of empire" should have calmed even a Bird's wrath. When Cooper exterminated his own white-Indian tendencies in this wretched pidgin-English, literature was the loser. But for a novelist of his stature to wind up in the camp of Paulding, Bird, and all the other enemies of his Leatherstocking Tales is a valuable indication of the power and pervasiveness of American Indian-hating.

CHAPTER XIV

Friend of the Indian: Colonel McKenney

Onward, and yet onward, moved the bands, clothed in winter in the skins of beasts, and in summer free from all such encumbrance. The earth was their mother, and upon its lap they reposed. Rude wigwams sheltered them. Hunger and thirst satisfied, sleep followed—and within this circle was contained the happiness of the aboriginal man.

—THOMAS L. MCKENNEY,
Memoirs, 1846

IN THE EARLY summer of 1826 Colonel McKenney marked time in Detroit while Governor Lewis Cass readied the expedition they would lead through the wilderness to negotiate a treaty with the Ojibwas at the far end of Lake Superior. Only three weeks away from Washington and his office in the War Department, McKenney already felt a bit melancholy. "Yesterday," he recorded on June 21, "I felt an unaccountable loneliness; and could by the aid of a little fancy . . . have imagined myself on the island of *Juan Fernandez,* another Selkirk; and as much a solitary as he" (T, 125).* Undeniably, he was surrounded by society "of the most agreeable and polished sort"; indeed that very evening he had been invited to "Col. H——'s; the Governor and his family, I learn, are to be there, and in general, the beauty and fashion of the city." But the charm of Colonel Hunt's party merely spiced the delicious melancholy he experienced playing Alexander Selkirk and that is to say playing Robinson Crusoe; for Defoe's hero was modeled on Selkirk, a castaway who lived alone for years (1704–9) on an islet in the South Seas. A later note at Sault Ste. Marie made certain that this and no other was McKenney's fantasy: Further last minute details had to be attended to, he wrote, "preparatory to the

* For an explanation of the abbreviated references, see p. 490.

step we are about to take, six hundred miles beyond the limits of civilization, and where we shall be alone among the mountains, and forests, and lakes" (T, 179). An oneiric Crusoe, McKenney imagined himself on the brink of a plunge into a world without people.

As it happened, the commissioners had arrived at Sault Ste. Marie on the Fourth of July, or "this beloved 4th of July," as McKenney put it. "We, of the present generation, I know, love this day and reverence it, but we cannot feel as do our patriot Fathers—as Jefferson, and Adams, and Carroll feel—those three surviving signers of that glorious instrument which lies at the foundation of our liberty" (T, 174–75).* At the school within the fort McKenney approvingly noticed a scroll:

NATIONAL JUBILEE
Fiftieth Anniversary
of American Independence

From a feeble infancy she has grown to
a giant size, and a giant's strength. . . .
Our agriculture has reduced the
wilderness to submission.

From then through July 6, drilling of troops at the fort, inspections, "manoeuvrings and firings, &c." provided visible intimations of this giant's strength. Nearby Indians watched the exercises and "it was easy to see that they had yielded the contest for supremacy," McKenney thought. "They looked as if they believed the white man had got the ascendancy" (T, 180).

As white men, the commissioners exerted themselves to maintain this apparent ascendancy. Before their expedition pulled up to the American Fur Company post at Fond du Lac on July 28 they formed the barges and canoes into a squadron that contained their staffs, a detachment of soldiers, a company band, and accompanying Indians—all stretched out over a quarter of a mile, all in order, and "all with flags flying, and martial music." "Hail Columbia" "rose over the mountain tops for the first time since their formation," recorded McKenney; "Yankee Doodle" rang out while the commissioners landed (T, 274). Ashore the troops drilled each morning and were inspected by General Cass and Colonel McKenney, with the latter in his militia uniform. This exhibition was good for the Indians, they believed, and would "prove a safeguard

* It was not until their return from the treaty council that the commissioners learned of the deaths of Jefferson and Adams on that day of days, passings of the Fathers that happened not by chance, McKenney concluded, but rather by appointment of "the supremely wise" (T, 368).

to many a trader and traveller, who, but for the remembrance that troops can be marched into their country . . . would feel the gripe of an Indian's hand, and the incision of his knife about the crown" (T, 295).

The red men danced to a different drummer. They greeted the commissioners with "a dance of ceremony," or festival dance. Naked bodies appeared all black or half red and half black; their faces were painted with the same colors and with green and yellow in circles, stars, points, lines; feathers and bells hung from their hair; bird skins, knives, and looking glasses dangled at their waists; and furs or tails trailed from their heels:

> Round and round, they went, with a kind of double short step, first with one foot, and then with the other; but the motion throughout was up and down. When they had gone twice or thrice round the circle, the drums would give the signal, when they would scream and whoop, and clap their mouths with their hands—then stand. I could see from their breathing—for they were all naked (except the *auzeum* [or loincloth],) and painted,—that their dancing was a severe exercise. [T, 285]

Another band greeted the white treaty-makers with even greater physical abandon:

> Young men, naked and painted as before, formed themselves on the ground into a circle of twenty feet in diameter, and two, and sometimes three entered the ring, and keeping time to the drums, exhibited the most violent contortions of body. Nothing can exceed the variety of figures into which these people can throw their bodies. [T, 288]

All this left McKenney unwilling to hear further of the happiness of the Indians, of their independence and freedom from restraint. They had simply demonstrated "the superior excellence of civilized, and polished and Christian society, over that of the savage" (T, 287).

Or so he said. Actually the experience left him shaken, less certain of the superiority of his society and its military exercises than these conventional opinions would suggest. He had witnessed "pure play," as Johan Huizinga characterized such dances as these, "the purest and most perfect form of play that exists." No longer *Homo Ludens*, or Man the Player, McKenney must still have felt the stirrings of some faint memory of the lost child in himself as he stood or sat, held stiffly in place by his uniform and a lifetime of repressions, and outwardly disapproved of these

Bear Dance, Preparing for a Bear Hunt, George Catlin, 1832. At the spiritual center of their cultures, Indian dances were prayers in motion—critically important means of communing through their rhythmic bodies with the beloved Mother Earth. A most exceptional Anglo-American, Catlin respectfully painted such dances and even sensed in them, long before Johan Huizinga, forms of sacred play. (Courtesy of National Collection of Fine Arts, Smithsonian Institution.)

physical excesses—and on Sunday, no less! Certainly the extraordinary passage that followed was shot through with ambivalences and revealed how threatened he felt:

> Such a sight presents a wide field for moral reflections; and furnishes a dark foreground to the picture I have just sketched, of the repose, of the peace of the Sabbath! No one can witness such a scene, and look upon bodies of the finest mould, for they are all such, and one especially the most perfect I ever beheld, and would in Italy be worth its thousands for a model, without feeling anxious for the arrival of the time (but how slow have been its advances!) when all these unmeaning and barbarous customs shall give place to the refinements of civilized life, and the sensual object which led to this, be changed to the nobler one of which their faculties are so manifestly capable. [T, 287]

All those naked, "mahogany colored," beautifully moving bodies, and that most perfect one especially, shook his rigidity and spurred his anxiety to wrench them out of "all these unmeaning and barbarous customs." *

The council began on August 2, 1826, and a treaty was duly drawn up and signed or marked. It gave the United States "metals or minerals from any part" of the Ojibwa or Chippewa country, and that turned out to mean the Mesabi iron range and deposits of metal riches even beyond the commissioners' wildest reckoning (T, 479–83).

At the outset McKenney had felt fortunate in having "no ordinary instructor" at his side, for "few men have so intimate a knowledge of Indian character as Governor Cass. . . . I wrote him before I left home, that I should attend as a pupil—and, therefore, would expect him to conduct the whole proceedings. I do not, therefore, expect to open my lips on this occasion" (T, 312–13). But that was too much to expect of a man who could rarely choke down his good intentions or restrain his own sweeping generalizations about the sameness of "Indian character" everywhere. After the treaty had been accepted "without a dissenting voice" on August 5, McKenney addressed the Indians and demanded they produce the alleged murderers of four whites. "It is a serious matter," he declared, and unless they obeyed "you will be visited with your great father's heaviest displeasure. No trader shall visit you—not a pound of tobacco, nor a yard of cloth, shall go into your country. This is not a thing to pass away like a cloud." Spokesmen for the band under suspicion replied it was difficult to speak for their absent tribesmen. But this would not do, so on the Sabbath, August 6, McKenney announced that if the wanted men were not delivered in the spring, the storm would come:

> If they are not surrendered then, *destruction will fall on your women and children*. Your father will put out his strong arm. Go, and think of it. Nothing will satisfy us but this. [T, 471; my italics]

Not surprisingly a brave told these stern fathers they might expect to see the young men the following spring.

McKenney even managed to have the last word. In the farewell address he admonished chiefs "no more to do bad actions," war-

* McKenney's breathless reaction to that body of perfect mold calls to mind Benjamin West's more esthetic response when he first saw the Belvidere Apollo in Rome (1760): "My God, how like it is to a young Mohawk warrior!" The Italian virtuosi present were mortified the young Quaker painter should compare their most perfect masterpiece with a "savage" but were mollified by his explanation of how like the pagan god were the Mohawks: " 'I have seen them often,' added he, 'standing in that very attitude, and pursuing with an intense eye, the arrow which they had just discharged from the bow' "—John Galt, *Life and Studies of Benjamin West* (London: T. Cadell and W. Davies, 1820), pp. 103–6.

riors to keep "the word of a warrior, and not of a dog," and young men "not to turn dogs." He also saluted them all as

> *Friends and Brothers,*—We will have good things to tell your great father who lives towards the rising sun. We will tell him his Chippeway children are men, and great men; that during this Council they behaved well; that they listened like good children to his counsel. [T, 473–74]

Red men who behaved themselves like good children were always dear to his heart.

And that was McKenney's philanthropy in action. I have quoted him copiously just as it reached Indians out beyond the frontier, for his descriptions based on notes from there provided a clarity of outline lacking in benevolent letters from his office and in his retrospectively defensive *Memoirs.* The location itself and his feeling about it mattered. Surrounded by people during the council, this nineteenth-century Selkirk still felt cut off, surrounded "by solitude and the lakes," still felt himself "somewhere, in the neighbourhood, at least, of *the end of the world*" (T, 320; McKenney's italics).

2

While at Fond du Lac the trill of a bluebird reminded Thomas Loraine McKenney (1785–1859) of home and "of the death scene of my beloved mother," who had died when he was ten: *"But she was such a mother!"* (T, 278–79). He did not say why, in his *Memoirs, Official and Personal,* for his autobiography started out with his first appointment to public office and left the reader free to assume he had had a mother and a preofficial existence. Other sources reveal he was the son of a Maryland family of "sturdy and consistent Quakers" (H, 25) and throughout his life considered himself in spirit, if not membership and attendance, one of the Society of Friends (V, 3). In 1823 he made fleeting reference to his father in testimony before the House Committee on Indian Affairs: "I was bred a merchant, and had all the advantages of information arising out of a large business, and frequent intercourse with our principal cities, as well since as during my initiatory progress in the counting-house of my father" (M, I, 301). He was twenty-one and still in his father's business on Maryland's East Shore at the time of his marriage to Editha Gleaves: His bride was, he said, an "exemplary and truly estimable lady" (V, 3). Little more is known of her save that she bore him a daughter who died early and a son, William, who merits mention later in this chapter. McKenney's Quaker spirit and rearing did not deter him from military service in the War of 1812 or from later rising perhaps to the rank of colonel in the Washington militia. For a fair while after the war he was disappointed in his search for public office. When Madison appointed him superintendent of Indian Trade in 1816, he had been subsisting as a partner in a foundering dry-goods store on Pennsylvania Avenue.

Colonel McKenney looked a little like Jackson. He was tall and had a military carriage and a shock of red hair that was prematurely turning white—once in the field, he recounted, a contretemps with several Indians ended when they mistook him for Old Hickory and abruptly retreated (M, I, 156). He was perennially in debt. On the basis of his annual salary of $2,000 and his prospects, one assumes, he induced the Bank of Columbia to lend him $15,000 to buy "Weston," a Georgetown estate where he kept three gardeners busy on the grounds and two black servants and four slaves at work in the mansion (V, 21, 36). Emotionally he was a loquacious man of quick sympathies, a windstorm of declamations for missionary societies, Sunday schools, and the Bible; patriotic addresses that fixed the United States "in the plans of the Eternal" as "the empire of freedom, and of mind"; and denunciations of whiskey, slavery, cruelty to dumb animals, and mistreatment of his charges, the Indians. Politically he became a lieutenant of Secretary of War John C. Calhoun and, after his own dismissal from office, an active campaigner for the Whigs. Intellectually he knew little or nothing about Indian culture and affairs but brought to the War Department the same enthusiasm he would have shown in the Senate, had he succeeded in his earlier effort to be named secretary of that body.

From 1816 to 1830, McKenney later gloried, no other man had the power he exercised over the North American Indians (H, 22). A recent biographer agreeably calls him *Architect of America's Early Indian Policy* (V). Once in office he took over and administered the government trading posts, or factories, as they were called, and pronounced this system "one of pure humanity, embracing a supply of the wants of the Indians without reference to profit" (M, I, 18). Yet he soon used his position to become heir apparent of Jeffersonian philanthropy, and that tradition, as we have seen, went beyond mere trade.

Exactly when McKenney became a champion of Indian reformation is not clear. It was not before his appointment and, given his vagueness about dates, how long thereafter remains obscure. One Fourth of July in 1817 or 1818, as he remembered, he addressed the citizens of Washington and Georgetown on "the glory of our institutions" and then went home to reflect he had not even thought "of those to whose country we had succeeded" (M, I, 32). The next day he ran across the letter of a Moravian missionary to the Cherokees, opened a correspondence with him, and soon became aware of the network of mission stations and schools among the tribes. Most were Protestant, with Quakers and Moravians perhaps the most experienced and accomplished, but with other denominations and sects close behind—the Congregationalists and Presbyterians had the strongest single organization in the American Board of Commissioners for Foreign Missions, established in 1810. These propagators of the Gospel and McKenney spoke the same language. By temperament and upbringing he was ready to welcome them as co-workers and valuable auxiliaries; they came to regard him as a strategically placed ally in the War Department.

Unlike the Indian-haters, McKenney insisted the red man was, "in his intellectual and moral structure, our *equal.*" Good men in charge of the various mission activities had already proved that, he related in his *Memoirs:* "I did not doubt then, nor do I now, the capacity of the Indian for the highest attainments of civilization" (I, 34). A case in point, if not conclusive proof, was James Lawrence McDonald, the seventeen-year-old son of a Choctaw woman and a white man, who walked into his office one day in 1818 accompanied by two members of the Baltimore Yearly Meeting of Friends. Since 1813 Quakers had educated the youth on War Department funds; two had brought him to Washington to find a suitable job and had been passed on by Calhoun with the suggestion something might be found for him in Indian Trade. McKenney snapped up the suggestion, thereby making McDonald probably the first Native American to find a place in the Washington bureaucracy, made arrangements to take him into his own home to live at "Weston," at government expense, and used the young Choctaw to exhibit the Indian's capacity for improvement (M, II, 109–19; V, 40–41).

McKenney also lobbied on the hill for legislation that would produce a host of McDonalds through more Indian schools, circularized benevolent associations, mission societies, and churches asking them to memorialize Congress to the same end, and tried to keep knowledge of his own role in the campaign as secret as possible, so the memorials would seem a great upsurge of popular sentiment. Petitions did stream in to both Houses and produced

McKenney's supreme achievement, the passage in 1819 of a bill that appropriated $10,000 annually "for the civilization of the Indian tribes adjoining the frontier settlements." The bill in effect established *incorporation* of Indians as the national goal and was the institutional expression of Jefferson's old hope of teaching them farming and the domestic arts. State and church also came together as McKenney worked out with the missionaries how the funds were to be spent in support of existing schools, in building new ones, and in furthering the advance of Indians in the English language, "in the various incipient branches of learning, in agriculture, and the mechanic arts" (M, I, 35–36). Everywhere, it seemed, "the wilderness was to blossom as the rose." By 1825 McKenney pronounced the Cherokees "civilized." By 1830, he emphasized proudly, "there were over *eighteen hundred Indian children* in these schools. . . ."

In the meantime pressures had mounted against the factory system. Private fur traders, led by John Jacob Astor and the American Fur Company, manufacturers, and merchants attacked attempts by the government to control commerce with the Indians, leveled charges of favoritism and corruption against McKenney, and in 1822 persuaded Congress to abolish the system and his office.

Again out of a job, McKenney shortly found funds, somewhere, to establish the semiweekly *Washington Republican and Congressional Examiner*. All the evidence unearthed since indicates that John Quincy Adams was close to the mark in his *Memoirs* entry of July 28, 1822: "I doubt much, however, whether Mr. McKenney's paper will be independent. I think it originated in the War Office, and will be Mr. Calhoun's official gazette as long as it lasts." It did champion Calhoun's candidacy for presidential nomination, favor the American Colonization Society, publish communications from Indian missions, and the like; but it failed to get sufficient government patronage and private circulation to keep going. In 1823 McKenney turned the paper over to other hands and retired from journalism deeper in debt than ever (V, 89–90).

On March 10, 1824, after much searching about and an abortive effort to become an assistant postmaster general, McKenney finally collected his political debts when Calhoun appointed him to head up the new Bureau of Indian Affairs (BIA). Technically his appointment was not as superintendent but only as a clerk with an annual salary of $1,600, but subsequent legislation would supposedly—though it never did—confirm his status and make his salary retroactively commensurate. Over the next half-dozen years he built up and tightened the organization of the Indian service bureaucracy and zealously served Secretaries of War Calhoun,

James Barbour, and John Eaton. In 1826 he joined Cass in the Fond du Lac expedition and treaty negotiations discussed earlier. In 1827 the two men again traveled into the Northwest, this time to a point near Green Bay where they negotiated a treaty (Butte des Morts) with the Menominees and Winnebagoes. Leaving his mentor behind, McKenney descended the Mississippi and crossed the South to negotiate agreements on his own with the Chickasaws, Choctaws, and Creeks, as part of the continuing national effort to move these tribes across the Mississippi. But as it happened McKenney was forcibly "removed" before they were: In 1830 Jackson ousted him from office on the grounds he was "not in harmony" with administration Indian policy (M, I, 262).

And that was McKenney's official philanthropy in outline. But what had happened to his great program of "civilization"? When and why had he shifted from *incorporation* to *removal?* This "dramatic shift in attitude," as Francis Paul Prucha characterized it in *American Indian Policy* (1962), has led apologists of the colonel's career into a swamp of their own digging. One recent biographer, for instance, tied the shift to the death of McKenney's Choctaw ward, who fell or threw himself off a cliff after a white woman rejected his marriage proposal: "The tragedy shocked McKenney into the realization that civilizing the Indians involved far more than simply teaching them English and making them into farmers" (V, 199). But James Lawrence McDonald's probable suicide was in the summer of 1831 and that would date McKenney's realization of the failure of Indian reform and his turn "to the removal program as the last best hope for the Indians" the year *after* he left the War Department.

In his much-noted chapter "Civilization and Removal," Father Prucha leaned heavily on McKenney's philanthropy for his depiction of government Indian policy as always humane: "That men as knowledgeable in Indian ways and as high-minded as Thomas L. McKenney, Lewis Cass, and William Clark were long-time and ardent promoters of Indian removal should give us pause in seeing only Jacksonian villainy behind the policy." After a pause to refresh ourselves with the reflection that few critics, if any, have seen only one man's villainy behind the policy, we may go on to a convenient, concurrent summary of Prucha's position in his "Thomas L. McKenney and the New York Indian Board," *Mississippi Valley Historical Review*, XLVII (March 1962): McKenney was "perhaps the best informed man in the United States on Indian affairs"; he produced letter after letter that showed "a deep and genuine concern for bettering the conditions of the Indians and protecting them in their rights"; and he had great hopes for the program "to civilize and educate the Indians," so great he did not doubt "the speedy incorporation of the Indians into the states and territories where they dwelt":

> But McKenney had changed his mind after an extensive tour of the frontier in 1827. What he had seen convinced him of the degradation of large numbers of Indians; and he could no longer believe in the imminence of their pulling abreast of whites in civilization. [637]

This formulation has the strength of precisely dating McKenney's "dramatic shift." Its weakness is, alas, rather more fundamental: It is untrue.

Well before McKenney visited a single tribe on the frontier he advocated "colonizing" Indians on lands "it is proposed to make theirs, and *forever*" (T, 69–70). On his return from Fond du Lac in 1826 he got into an argument with a Wyandot about removal and then firmly maintained it was "the only measure, in my opinion, that will preserve them"—removal promised the final solution of their problem, as it were, for it looked "to a last and permanent home for our Indians" (T, 426–27). Still earlier, when Monroe proposed voluntary removal in his annual message of 1824 (see p. 115) and Calhoun endorsed the plan, McKenney of course fell in line. Indian degradation and extermination would be inevitable without removal, McKenney agreed. Monroe had therefore spoken "the language of humanity, dictated by wisdom and experience" (M, I, 240). To be sure, McKenney himself had just weeks earlier spoken grandly of the imminent reformation of the Indians, *where they were*, but there was little or no drama in his shift. Like changing synchromesh gears, McKenney smoothly moved to an emphasis on *removal* as the essential precondition of what he called *civilization*.

And this truth is complemented by two others: McKenney was not knowledgeable about Indian ways and not always high-minded in his dealings with them. In fact, on occasion he was not even knowledgeable about Indian existence. On his trip through Utica in 1826, for instance, he noted that "no longer does the council fire of the Mohawks burn on Tribes' hill, nor their war-whoop echo along the shores of their own river. . . . *Not a single Mohawk lives!*" (T, 55). The Mohawks had indeed been pushed out of the valley that bears their name, but fortunately some lived on—and live on today—at Akwesasne, or St. Regis, opposite Caughnawaga on the St. Lawrence. They were currently one of the tribes McKenney had responsibility for as chief government administrator of their affairs. Though he thought them all dead, he did know of their reservation and vacuously imagined it occupied by a nonexistent tribe, "The St. Regis Indians" (T, 433–34).

Contemporaries assumed Colonel McKenney to be knowledgeable about Indian ways simply by virtue of his office. Publication in 1827 of his *Tour to the Lakes* seemingly confirmed this official status. But before he took office he knew next to nothing about Native Americans; thereafter he simply shared the prejudices and ignorance of General Lewis Cass, whom he acclaimed his tutor, and other old hands in the Indian service. "Of all men," McKenney declared, "an Indian is the most improvident, and furnishes the most painful example of a reckless disregard to the impoverishing and life-consuming effects of intemperance" (M, I, 20). Two hundred years earlier, when Europeans first arrived among them, "there was scarcely an Indian on the continent, who could comprehend an abstract idea; and at this day, the process is neither common nor easy" (M, II, 39). The "disgusting habits" they had then they have still:

> Indians, in their uneducated and unimproved state, appear to be the same every where, and to have nearly the same habits, and customs, and manners. They *powowed,* or conjured, it seems, in Massachusetts colony, near two centuries ago, and they do the same to this day on Lake Superior.—They howled then, and greased their bodies, and adorned their hair, and the same practices are yet maintained by the Chippeways of the lakes, and by all other Indians of whom I have read, whose improvement has not been studied, and who have never been taught the lessons of morality, and cleanliness, and industry. [T, 379]

"Hear! Hear!" one imagines John Endicott crying from the shadows. But McKenney thought only the Indians were unchanging, "*stationary*" as he italicized their condition, fixed as in past ages "in this state of helpless ignorance and imbecility" (M, II, 39–40).

A man who threatened the Ojibwas with destruction of their women and children, as we witnessed McKenney threaten at Fond du Lac, is perhaps better described as high-handed rather than high-minded. At all times in the field he expressed utter contempt for Indian culture and obviously relished exposing their shamans as "crafty impostors" (e.g., M, I, 169–72; ITNA, II, 407). In 1827 at Butte des Morts he and Cass took it upon themselves to appoint a head chief for the Menominees. McKenney promoted a warrior named Oshkosh to that station by placing a medal around his neck and admonishing him: "You are now the great Menominee Chief. You will take care & act like a man & not like a dog"—it was also in character for the commissioner later to deny thus interfering in Menominee tribal affairs (V, 161–62). But another incident at Butte des Morts still more graphically revealed McKenney's ignorance of Indian ways and his overbearing attitudes.

As the council was breaking up a warrior slashed the arms of his mother-in-law for trying to keep him from visiting a whiskey peddler. "What shall we do with this man?" McKenney asked Cass, who "answered promptly, *'Make a woman of him.'* And so we did." McKenney commenced the symbolic castration by taking the Indian's knife, breaking it off at the hilt, and placing the handle back in his hand. Holding the warrior's hand aloft, he declared:

> No man who employs his knife as this man employs his, has a right to carry one. Henceforth, this shall be the only knife he shall ever use. Woman, wherever she is, should be protected by man, not murdered. She is man's best friend. The Great Spirit gave her to man to be one with him, and to bless him; and man, whether red or white, should love her, and make her happy. [M, I, 91]

McKenney made the man strip and put on the petticoat of "an old squaw . . . [one that was] stiff with the accumulated grease and dirt of many years." Prepared for his execution, the brave was caught off guard by this mortification before an audience of a thousand reds and whites. After he was rushed from the scene of his humiliation, an interpreter caught his first words: "I'd rather be dead. I am no longer a brave; I'm a WOMAN!" (M, I, 92). And if this was not exactly what he said, it was what McKenney and Cass wanted to hear.

The triumphant commissioners remained serenely unaware they had exposed themselves when they acted out their assumption that the ultimate degradation for man was to be made a woman. And McKenney did not grasp just how self-revelatory he had been in assigning to woman the role of canine devotion as "man's best friend." He grasped or thought he grasped the fact that the tribes had to be "civilized" to liberate Indian women from their endless, less-than-dog's-life drudgery for their lords and masters. But this gratuitous assumption, lifted almost directly from Jefferson's *Notes on Virginia,* again was more self-revealing than descriptive of the sad lot of Native American women. Among the misplaced Mohawks, for example, descent had been matrilineal, and in their "matriarchate"—or quasi-matriarchy—women had more social freedom, more economic autonomy, and more political power than were currently enjoyed by the "exemplary and truly estimable" Editha Gleaves McKenney and her white sisters of the republic. Iroquois women took and left mates almost at will, selected chiefs, participated in decision-making, and started wars; men hunted and trapped, acted as diplomats, and fought wars. But this sort of delicate balance of roles and powers was as much an unknown to McKenney as the current whereabouts of the Mohawks—he would have been astounded to learn that among them *woman* was not a term of abuse.

So did the incident say more about the commissioners' ignorance of Indian ways and their imperious prejudices than they could have intended. It disclosed as well the underside of McKenney's philanthropy:

> Now this mode of punishment was intended to produce moral results, and to elevate the condition of women among the Indians. It was mild in its physical effects, but more terrible than death in its action and consequences upon the offender. Henceforth, and as long as I continued to hear of this "brave," he had not been admitted among his former associates, but was pushed aside as having lost the characteristics of his sex, and doomed to the performance of woman's labor, in all the drudgery to which she is subject.

This loving parent manifestly drew psychic income from chastising the offending ex-brave.

It was always unfair to expect McKenney to have been consistently high-minded. After all, he was a birthright Quaker in the War Department, its philanthropist in residence. By political heritage and personality he was a scaled-down or miniature Jefferson and thus even more likely to fail abysmally in fusing the rhetoric of benevolent intentions with the realities of national expansion. As a flywheel attached to the principal government engine of that expansion, he naturally countenanced bribery and employed secret agents, advocated taking hostages and approved spirited military movements to "chastise" refractory tribes, and resorted to those threats to kill women and children (e.g., V, 159, 210, 214). Above all, he dealt with Indians from the center of a web of deceptions and manipulations.

An address McKenney delivered to the Chickasaw chiefs during his southern tour of 1827 showed just how tangled his philanthropy became:

Brothers: Give me your ears, and, what is of equal importance, give me your confidence. If you think I am come to do you wrong, or give you bad counsels, you do me great injustice. I am not come but as your friend, and if there is a chief present who doubts this, let him speak, and I will not say another word.

Brothers: I know well who you are that I am addressing. I know you are not children, but men, and men of experience, and men of wisdom. I know, too, that the smoke of this council-fire comes not of ashes, but of living fire—it rises out of our hearts, for we are friends. [M, I, 324]

Now, whatever McKenney's friendly feelings for Levi Colbert and the other chiefs, he stood before them as an emissary of the War Department on an official mission to get their lands and show them the path across the Mississippi. He had not come as a disinterested friend. As for knowing they were men of wisdom, he spoke very differently to his own white brothers about their and other Indians' infantilism. In his letters from the Indian Office he repeatedly declared them children, nothing but children, or as he wrote Chairman John Cocke of the House Committee on Indian Affairs on January 23, 1827, the same year as the above address, they were "only children, and require to be nursed, and counselled, and directed as such" (NA, RG75, M21, R3, 328). No confidence man ever worked harder to gain the trust of his childish marks, as this excerpt from a model speech (1829) suggests: "Try us this once. Do not distrust our object; it is your welfare, only, we seek" (M, I, 247).

"Our object" was only their all—all their remaining lands east of the Mississippi. The rush of events toward forcible removal, the "utterly unjustifiable" as Monroe had named it, can be followed in the *Memoirs of John Quincy Adams*. Adams noted, for instance, that at a cabinet meeting on December 22, 1825, the entire afternoon had been devoted to a discussion of the affairs of the Creeks and Georgia. Secretary of War Barbour, McKenney's superior, proposed to stop making treaties with the tribes altogether and simply declare them subject to U.S. laws. Secretary of State Henry Clay thought that impracticable:

> that it was impossible to civilize Indians; that there never was a full-blooded Indian who took to civilization. It was not in their nature. He believed they were destined to extinction, and, although he would never use or countenance inhumanity towards them, he did not think them, as a race, worth preserving. He considered them as essentially inferior to the Anglo-Saxon race, which were now taking their place on this continent. They were not an improv-

> able breed, and their disappearance from the human family will be no great loss to the world.

These opinions "somewhat shocked" Barbour, while Adams feared they had "too much foundation." A bare two years later McKenney returned from his southern tour with the triumphant report he had persuaded the Creeks to part with the remnant of their land in Georgia: "Mr. McKenney is very voluble, and magnified his office," Adams noted dryly. "He told me all the arguments that he used . . . and how he put down Ridge and Vann" (December 6, 1827).* During their conference Adams informed him that tribes in New York had protested the Butte des Morts treaty he and Cass had negotiated the past summer, alleging the lands had already been ceded by the government to them (cf. M, I, 83). Adams was clearly a bit taken aback by McKenney's response: He "made light of it, but gave me no satisfactory answer." At another White House meeting Barbour informed the president of the Cherokee attempt to frame a constitution and referred him to a report of Colonel McKenney "recommending that notice should be given them as early as possible that this cannot be permitted" (February 8, 1828). Obviously the colonel's protestations of heartfelt friendship for red men of wisdom rang more hollowly in Adams's office than they had over the council fires.

Calhoun-partisan McKenney had easily swallowed Georgia's states' rights argument that it had sovereignty over all the lands and peoples within its borders (M, I, 330–36). After the election of 1828 he could hope this position would help in his desperate effort to keep out of the way of Jackson's new broom. And when such bona fide missionary groups as the American Board of Foreign Missions rebuffed his entreaties to come out in support of administration policy, McKenney formed his own religious front. It had a name out of his files on humanitarianism—the Indian Board for the Emigration, Preservation, and Improvement of the Aborigines of America—and a paper membership made up primarily of clergymen he had rounded up in the New York Dutch Reformed Church—hence the short title, the New York Indian Board. But

* John Ridge and David Vann were leading Cherokee chiefs whose advice had been solicited by Opothle Yoholo and the Creeks. When they advised against parting with the land for any consideration, McKenney ordered Ridge from the council and threatened, in the name of the president of the United States, to sever relations with the Creeks unless they repudiated Ridge. This awful prospect was enough to bring the Creeks around. Opothle Yoholo did dare to tell the commissioner during their negotiations that "he talks too much"; McKenney replied "that the welfare and happiness of the Creeks was all that their Great Father at Washington sought in this interview." For an unwittingly revealing account, see ITNA , II, 22–30.

this paper tiger could hardly stand, let alone whip up support for emigration and all the rest. When the upsurge of popular sentiment did not materialize in 1829 as it had in 1819 for the "civilization" bill, the clerk in charge of the Bureau of Indian Affairs had failed to make himself indispensable to the Age of Jackson. For all these last-minute efforts to toady, McKenney's career in official philanthropy came to a close in August 1830.

Of course, he may have resigned of his own accord, as his *Memoirs* imply, when the enormity of the death marches became clear—Michael Rogin has estimated that of the 100,000 Indians pushed or transported across the Mississippi between 1824 and 1844, from one-fourth to one-third died on the way or shortly thereafter. Unhappily such principled protest was out of character for a man who always displayed a ready flexibility of convictions. A case in point was his fawning invitation to John Jacob Astor to join him on the treaty expedition to Butte des Morts (V, 154–55); in his *Memoirs* he put the fur baron's name in boldface and, despite the fact that Astor more than any other man was responsible for ending the "pure humanity" of the factory system, gave him a puff as "this sagacious and wonderful man" (I, 20). If he could bring himself to placate his old enemy Astor, surely he could have worked something out with Old Hickory. And in reality his and Jackson's positions on removal were not all that different. McKenney had warmly endorsed administration policy before he was fired. "We believe if the Indians do not emigrate, and fly the causes, which are fixed in themselves, and which have proved so destructive in the past, they *must perish!*" (M, I, 241). Furthermore, "in regard to the disposition of *the great body* of the Indians within our States, we speak advisedly when we say, they are *anxious to remove*" (M, I, 244; McKenney's italics). He believed they were held back by threats and opposition from within their own ranks, primarily from "half-breeds" like John Ridge and others—this flipped over the customary invocations of "outsiders" and may be thought of as the "inside agitator" refrain. Just as advisedly, Secretary of War Cass later knew full well "the great body of the Indians within our states . . . are anxious to remove," also placed the blame on inside agitators, and took the pretense right up to the infamous treaty of New Echota (1835), in the face of hard evidence from the field—in his files—that nine-tenths of the Cherokees would have, given the opportunity, rejected removal.

In all likelihood McKenney would have been there standing loyally beside his mentor, had he not been dismissed, working to implement what the great body of Indians so obviously wanted.

Indeed logic and compassion would have compelled him to be there. How could he have stood by while his wards perished? As he had noted on October 28, 1829:

> Seeing as I do the condition of these people, and that they are bordering on destruction, I would, were I empowered, take them *firmly* but *kindly* by the hand, and tell them they must go; and I would do this, on the same principle that I would take my own children by the hand, firmly, but kindly and lead them from a district of Country in which the plague was raging. [NA, RG75, M21, R6, 140]

To be sure there were his repeated assurances to the Indians they would not be forced to go—but McKenney had assured them of a lot of things. Sometimes children have to be taken in hand for their own good, firmly in hand, and was that not exactly what General Winfield Scott was to do? Their Great Father in Washington or his stand-in, as the colonel saw himself, could have done no less. And I believe he would have so acted with no conscious insincerity. Starting like Jefferson by spinning out deceptions for others, McKenney finished like his model by dangling from his own web.

3

James Madison had identified for McKenney the two most intractable problems facing the country in words we have already seen in an epigraph: "Next to the case of the black race within our bosom, that of the red on our borders is the problem most baffling to the policy of our country" (February 10, 1826). In McKenney's hands Madison's formulation underwent an interesting sea change before it surfaced as a misquotation in his 1829 speech to the New York Indian Board: "There are, to use the words of a distinguished citizen of Virginia, two problems yet to be solved, both having, so far, puzzled the ingenuity of the politician, and baffled the wisdom of the sage. 'One of these relates to the black population, which we carry in our bosom; the other to the red population which we carry on our back' " (M, I, 229). Under such awkwardly placed burdens, no wonder the white man fell into perplexity.

McKenney manifestly considered Afro-Americans hopelessly unsuited for "civilization." At one point he expressed sympathetic interest in a proposal that whites unbosom themselves of their black burdens by colonizing them "some where, far off in the West" (T, 72–73); his perennial remedy was to export "the evils" of their presence through the American Colonization Society (M, I, 229; V, 88). Yet support of their colonization elsewhere hardly amounted to passionate antislavery, and the depth of his opposition was further called into question by the slaves he himself owned.*

The red population constituted the second baffling problem. McKenney was sanguine he had the solution in his proposal to get them off the collective white back through "civilization." His "schemes to amend the heads and hearts of the Indians," in the hostile words of Thomas Hart Benton (V, 68), were first to transform them where they were and later (post–1824, as we have seen) to transform them once they had been colonized somewhere far in the West. Wherever they were, they had to be weaned from their merciless savagery. One of the colonel's interesting experiments to that end was with his Choctaw ward.

In 1818 McKenney put James Lawrence McDonald to work in his office and also took "this poor Indian boy" into his own home where, he noted proudly, he "made no distinction between him and my son [William], in dress or attentions" (M, II, 111). The War Department paid for McDonald's board and room, education and incidentals, and he in turn worked there copying his benefactor's letters, as the notation on one to Calhoun indicated: "This letter is copied *in haste* by our little Indian" (V, 41). He also enrolled the young man in an academy and delighted in introducing him to skeptics as gratifying proof the Indian was not irreclaimable. Great was his dismay one day, therefore, when the principal of the academy declared, " 'Really, Colonel, I do not know what I shall do with McDonald.' Instantly I feared some latent Indian quality had burst forth, and that all my high hopes were to be destroyed." Fortunately the Indian within McDonald was not the

* On his 1826 and 1827 expeditions this latter-day Crusoe took along his own Friday in the person of a black man he called his "faithful servant Ben" (M, I, 63; H, 65). Among his other services, Ben provided comic relief. On the way to Fond du Lac, for instance, McKenney mounted his back to be carried across a stream and was pitched headforemost into the water when the black man lost his footing. "Ben was before me in an attitude of horror!" but was reassured when his master proved understanding: "I have only proof that you have yet to learn how to ferry me over a stream. He insisted on another trial; so after I got into dry apparel, I gave him the opportunity, and he took me over most triumphantly" (T, 238–39). On their travels Ben stayed close to his master, "sleeping at my feet"—McKenney actually said that Ben "follows me like my shadow" (T, 322, 408). And like the fearful Friday, seemingly Ben had no last name: the reader of McKenney's books would never learn that his surname was Hanson (V, 140). To my knowledge, incidentally, McKenney never commented on the essential identity of his advocacy of black expatriation and of red removal.

issue at all, this time: He had so far outstripped the other students he had to be put in a class by himself. After about three years and at Calhoun's suggestion, McKenney made arrangements for McDonald to study law in the Ohio office of John McLean, the future associate justice of the Supreme Court.

But the experiment started turning sour even before McDonald left "Weston." When McKenney told "him of the profession I had chosen for him," he expressed the fear it would all be lost on him: "*I am an Indian.*" He handed McKenney a letter he had just received from his brother, who was apparently an army lieutenant with the given name Thomas Jefferson. The letter warned

> he had one of two things to do—either throw away all that belonged to the white race, and turn Indian; or quit being Indian, and turn white man. *The first, you can do; the last, it is not in your power to do.* The white man hates the Indian, and will never permit him to come into close fellowship with him. [M, II, 113]

McKenney brushed this aside, saying his brother "had greatly erred," and persuaded the distraught youth to go ahead with his career. When McDonald returned to Washington in 1824 as part of a Choctaw delegation, he had his own law practice in Jackson, Mississippi. McKenney soon discovered in the young attorney a tough negotiator, very nearly one of the "unruly sort," as McKenney called those who objected to his benevolence, almost a "bad bird" or troublemaker. "You know me well, and my feelings towards your people," he reminded McDonald in a personal note; "I am their friend" (V, 129). McDonald's drinking also disturbed him: "A conflict between his Indian caste and his hope of overcoming it, and rising above its effects upon his prospects, shook him from his balance, and he fell before the strife, into habits of intemperance" (M, II, 116). The latent Indian quality the colonel feared had seemingly burst forth—there was pathetic irony in the degree to which McDonald had been made to fear this in himself; as he had said, he wanted his career "to free the character of educated Indian Youth (with some degree of Justice cast upon it) of a proneness to relapse into Savagism" (V, 45). But the bitterest pill for the intelligent lawyer was the fact that in a number of states, including Mississippi, native peoples were in the predicament later made infamous by the Dred Scott case: "We are denied privileges to which, as members of the human family, we are of right entitled," read the Choctaw address to Congress of February 18, 1825. "However qualified by education we may be, we are neither permitted to hold offices, nor to give our testimony in courts of justice, although our dearest rights may be at stake. Can this be a correct policy? Is it just? Is it humane?" (M, II, 121–22n). The questions were McDonald's, and the absence of good answers dragged his despair ever deeper.

McKenney washed his hands of the intemperate Choctaw and refused to reply to his letters. McDonald's death did not keep his erstwhile benefactor from emotional lectures on "the tragic end of this gifted and accomplished youth" who had been turned down by a white woman:

> On making his proposal, it was rejected with promptness, and, as he thought, with scorn. In a moment his caste came before him. "You are an Indian, and degraded," rang in his ears. Hope fled—despair assumed dominion over him. All that his brother had written to him, was now seen by him to be reality. The spectre was too formidable for his power of resistance—he rushed to the river, sprang off a bluff, and drowned himself! [M, II, 118–19]

In this humus questions spring up like weeds. Then had Thomas Jefferson McDonald not warned his brother of at least a psychological truth? If so, what about the colonel's easy judgment that Thomas Jefferson McDonald had "greatly erred" and ought not to have expressed such sentiments? And if he had been right that it was impossible for native people to gain full acceptance in a caste society, then what about the colonel's own advice to James Lawrence McDonald or, for that matter, what about his overall "civilization" program to turn them into whites? But such queries would have left McKenney unmoved. He simply was not gifted with sufficient courage and insight to see himself as part of the problem.

About the same age as McDonald, William McKenney was only a slightly less dramatic disappointment. In seeking a loan to help straighten his son out, McKenney described him as "a noble youth . . . who is not so provident as I hope he will be" (V, 291), words that not by chance echoed the legendary improvidence of "Noble Savages." To the colonel's mind, *discipline* was the universal solvent for such waywardness. He was in fact a pioneer advocate of institutions of total control. On his way through New York in 1826 he had responded enthusiastically to the silent lockstep control and centralization of "punishing power" at Auburn Prison—the then new prototype of all the maximum security prisons to come—and had concluded that "this doctrine of responsibility is altogether orthodox, and is good every where, in prisons, as well as in church and state" (T, 80). He could not very well put his son in Auburn, however, so in 1831, the year of McDonald's probable suicide, he did the next best thing by putting him aboard the U.S.S. *Potomac* for a two-year cruise. Since he was shipping out as an unpaid assistant to the surgeon, the elder McKenney had to borrow money for his outfit. He had put his son aboard the warship, he wrote his creditor, "in the fondest hopes that William was prosecuting his voyage, and deriving benefit from the *system* of the ship, in all things pertaining to his habits" (V, 291). Abroad William McKenney jumped ship, however, returned to New York, and after his mother's death in 1835, apparently disappeared forever.*

At Detroit General Cass had called friendly auxiliaries his "pet Indians." McKenney's reference to McDonald as "our little Indian" was almost equally suggestive. In 1824 the son of Levi Colbert took McDonald's place at "Weston," and shortly thereafter McKenney passed on word to the chief that his son was doing well in school and would "lose his Chickasaw language in a few years—it is slipping away from him very fast." When Dougherty Colbert left to study surveying in 1827, McKenney gave him one final piece of advice: "Now, my poor boy, as we are about to part, I can do no more than urge upon you the lessons I have endeavored to teach you—and remember, *the world is wicked*" (V, 190–91). Meanwhile at Fond du Lac the colonel had tried to pick up another boy but Chief Jackopa of the Ojibwas refused to part with his fourteen-year-old son, indicating by a gesture that he would not be so dismembered (H, 210).

McKenney had more luck in the South. On his tour of 1827 he gained custody, he said at their parents' urging, of William Barnard, a Creek, and Lee Compere, a Yuchi: "My little Indian boys were about ten and thirteen years old, Lee being the youngest." A hundred miles down the road Lee Compere still did not want to leave his family, so McKenney had the stage stopped:

> Now, William, tell Lee he can go home, if he wishes to go. This was scarcely said, before the little fellow, who had learned some English at the missionary school, seized his bundle, and was, in a twin-

* Readers of McKenney's *Memoirs* will look in vain for mention of his spouse's death and of his son's disappearance (see instead V, 281), one reason among many his recollections were so much more official than personal. Those close to McKenney died or faded away, leaving him a patriarch without a family or at least without a close family. He had two brothers, but one was estranged and the other stationed at a distance, a chaplain in the U.S. Navy. For the last decades of his life the colonel was one of the city's homeless males, living in a series of rooming houses and hotels from Philadelphia to Boston. Without citing his source, Frederick Webb Hodge noted "the son, William, was still alive, and unmarried, in 1857 [two years before his father's death], after which all trace of him is lost" (ITNA, I, x).

> kling, out at the side of the stage, and going down over one of the fore-wheels; when seeing him determined to go, I told Ben to reach out and take him in. He was inconsolable, and remained so till we reached Augusta, in Georgia. [M, I, 188]

All this was designed "to show the fearlessness with which the young savage throws himself upon his own resources" (ITNA, II, 53).

In Augusta McKenney had Ben Hanson buy them clothes, and "I then had their hair cut. Ben took them into a chamber of the hotel, and gave them a thorough cleansing; when they were brought to me, dressed, not in a very handsome suit of clothes only, but in smiles. A couple of prettier boys could be found nowhere." Back in Washington, McKenney put Barnard and Compere in a school run on West Point principles, with strict discipline, martial exercises, and military police. Their uniforms alone would establish a tie to the school "of the most agreeable sort," he believed (M, I, 189n); elsewhere he expressed confidence the routine "was in unison with their naturally martial dispositions," though admittedly "neither of them liked the exact enforcement of strict rules" (ITNA, II, 54). When McKenney had to leave the War Department, Jackson refused to let him take Barnard and Compere. After they returned to their people, the former got into "an Indian quarrel" and fled to the Seminoles in Florida. "Of Lee, I have never heard anything" (M, I, 190n).

So did the experiments in "civilization" end with a whimper. They revealed all the essentials of what McKenney meant by the term. In the words of the epigraph at the head of this chapter, Indians had the earth for their mother and on her "lap they reposed." "Civilization" would lift them off, separate them from their forests, cut their hair, cleanse away all traces of intimate communion with their mother, and clothe them in the habits of restraint, obedience, punctuality, sobriety, and of course, industry. And this list of Puritan virtues was precisely the curriculum of the "Nurseries of Morality," as Robert F. Berkhofer, Jr., called the mission schools supported by McKenney's act of 1819. Discipline therein bore close likeness with that at Auburn, at West Point, and aboard the U.S.S. *Potomac*. In both the North and the South, Indian families complained bitterly over their children's peonage—at the Mackinaw school in Michigan children were bound by legal indenture to the mission so they could not leave or be removed by their families. Everywhere whippings were common, as they were in the white schools, and the emphasis was on repression: At the Brainerd school in Tennessee, for instance, Indian children practiced their handwriting in sweeping letters: "Command your temper" (V, facing p. 33). But the ultimate instrument of alienation—and empire—was the language they were forced to write and speak. "The Indian tongue," McKenney ruled, was the great obstacle to "civilization." He therefore insisted only English be taught in the schools and was even hostile to Sequoyah's written Cherokee (e.g., NA, RG75, M21, R3, 103). As had Dougherty Colbert's Chickasaw, Indian languages should be made to slip away fast.

"Civilization" triumphant thus meant the termination of everything Indian. No less than Jackson and novelist James Kirke Paulding, McKenney confidently expected the destruction of tribal cultures. He too was a patriot who looked "upon the American people as one great family" (M, I, 153) and in this sense Jefferson and Adams were also his "patriot Fathers." Like the others McKenney read the metaphor literally. Yet the United States of America was not a family but a nation state, and McKenney, who was so little more than his official role, was the instrument of that national society's war on family or kinship societies. He severed the family connections of his "little Indians," used mission schools to batter tribal relations, and by stripping away native languages sought to cut off all ties between generations. Like Paulding and other Indian-haters, he acted on his assumption that tribal man was a merciless savage. He differed, however, in wanting to pry the young tribesman loose from the state of nature wherein he was inevitably *"cruel, revengeful,* and *ungovernable"* (M, II, 74). He differed too in emphasizing that impatience of restraint was inherent in *all* men and might have cited his own son as an example. The inference he drew followed: *"the civilized* is not the state most congenial to man, and . . . his instincts and his tastes combine, whether he be white or red, to attach him to the repose and indolence of a state of nature" (M, II, 103). Therefore *necessity* alone, as he stressed in an unsigned contribution to the *National Journal,* could bring man to enter "a willing captive into the restraints of civilization" (April 26, 1825). In terms of his own analysis, then, McKenney's war was on human nature itself.

Though "civilization" would presumably be good for everyone, McKenney had in fact given up on adult Indians, old birds who would not shed tribal ways and languages, and for their obstinacy would perish sooner rather than later. He wisely directed his efforts to "the fledglings of the forest," as he put it in another metaphor, for these young birds could be "tamed, and blessed, and made to feel the blessed influence of Christian teaching, and of Christian hope" (M, I, 154). Snared young, the Indian was capable of the demands McKenney placed on him: "I do no more than assume that he is *human;* that physically, intellectually, and morally, he is, *in all respects,* like ourselves; and that there is no difference between us, save only in the color, and in our superior advantages" (M, II, 80). But skin color was no minor matter, as McDonald should have taught the colonel. How were McKenney's "little Indians" to gain superior advantages if the color-conscious white society would not have them? Actually color, and not only black color, mattered a great deal to McKenney. Witness his staging of the Light Show: In the beginning the red man had lived without light, he said: "Not one ray of that blessed light which comes from the Gospel, bringing with it *'life and immortality,'* had penetrated the darkness that brooded over his mind, wrapping the future in such dismal and appalling mystery!" (M, II, 33). McKenney's imagery was no more passing literary fancy than Simms's and Bird's. In the conflict between whites and reds, he later granted, "there was light on one side, and darkness on the other; there was education, knowledge, religion, against ignorance, superstition, and paganism" (M, II, 61). The red man was aboriginally a creature of darkness. He lurked on even within McKenney's "little Indians," clamoring to burst out. That this was so for the colonel became certain in his half-hidden identification of them as natural objects or creatures, "fledglings of the forest."

The Indian joined black Ben Hanson, his "faithful servant," as McKenney's shadow—the projection of all the fears and anxieties that followed him about, the animal within himself with whom he could never negotiate a treaty. At Fond du Lac he had sprung forth in idealized form, that dancing naked body of perfect mold. Had McKenney let down his guard, a dancing counterpart might have leaped out of him, joined the circle, chanted, copulated, and run off into the free and boundless forest. And insofar as he saw and felt "savagery" to reside in the special animality and mortality of dark-skinned peoples, his "civilization" was intrinsically racist.

On the lecture circuit in the 1840s McKenney asked audiences to fancy themselves on the continent before the arrival of the white man: no hamlets, villages, cities; no vessels on the rivers; no cultivated fields, no gardens. Presently "groups of savages should be seen." They war, return to their village [*sic*] smeared with blood, are greeted by "half-naked squaws, and children, and dogs," dance triumphantly, but, with the flash of lightning and peal of thunder, flee to wigwams and caverns in the hill-sides:

> We become alarmed—not at the thunder-storm, but at our isolated, wilderness-bound, and exposed situation; and we look instinc-

> tively around for our species. But the white man is nowhere seen! [M, II, 56]

So he had been king of his island all along, a Selkirk—though that famous castaway was said to have danced with his goats—a Crusoe, a Prospero. Without his own species to interfere, McKenney had played the kind, self-denying father while enjoying absolute dominion over the dark creations of his imagination.

CHAPTER XV

Professional Westerner: Judge Hall

> The frontispiece prefixed to this volume [II] exhibits a lively representation of the noblest sport practised upon this continent—the hunting of buffalo. These animals . . . retired as the country became settled by civilised men, and are now found only on the great prairies of the Far West, whose immense extent, with the scarcity of timber and water, renders them uninhabitable by human beings.
>
> —McKENNEY and HALL, *Indian Tribes of North America*, 1836–44

IN THE OLD War Department Building on Pennsylvania Avenue Colonel McKenney had turned his room on the second floor into what was literally a charnel house, that is, a depository for human remains. There the visitor could finger bones McKenney had found in a cave. In 1826 he had also extracted bones from a burial mound near Cass's farm outside Detroit, "some ribs, a bit of *os frontis*, and pieces of *vertebrae*," but they were decayed and had crumbled in his hands. Eager for something to add to his collection, he accepted the offer of a skull of "enormous dimensions" from Major Thomas Forsyth, agent at Rock Creek—the skull had come from that mound and, if the agent could only relocate it, would be of interest to phrenologists simply for its size alone (T, 122).* And still earlier McKenney had asked factors to send in a few scalps, with data on their former owners and on the wars in which they were lifted, while not revealing to the tribes how desirable they were as collectors' items: "these you had better get incidentally," he instructed, "as by telling the Indians they are disgraceful appendages & ought to be sent away" (V, 239).

* For an explanation of the abbreviated references, see p. 490.

Under McKenney's direction, the Indian Office had become more museum than charnel house, however, in the sense of anthropologist Stanley Diamond's definition of museums as "those repositories for the loot of the world and testaments to imperial power." Display cases bulged with loot, or what McKenney called Indian "curiosities": arrows, quivers, and bows; tomahawks, pipes, and baskets; moccasins, leggings, shirts, and headdresses; strings of wampum, fur capes, and skins. McKenney had started gathering these artifacts in 1817 and added to his stock at every opportunity. As General Cass wrote him almost a decade later, it was imperative to collect such materials "for the history of a most interesting portion of the human family, thrown upon this continent we know not how & hastening to Extinction—unless the work is done speedily, it can never be done. Many of the tribes have already disappeared, and many others have dwindled into insignificance, & all have lost much of their distinctive character. We can yet, however, rescue much from destruction" (NA, RG75, M234, R429, January 20, 1826). McKenney hardly needed the reminder. He had already pressed agents and missionaries to hasten the collection of Indian vocabularies—they should be especially attentive to any "isolated being known to you, as the 'last man' of his Tribe—to get from him the words called for. Such a man may be looked upon as the connecting link between time and eternity, as to all that regards his people; and which, if it be lost, all that relates to his Tribe is gone forever!" (V, 244).

A sense of the absolute evanescence of red lives prompted this "last man" theme and set the tempo for McKenney's acquisition of paintings for his famous gallery. The earliest dated back to 1821, when Petalesharro, the Skidi Pawnee brave, Eagle of Delight, the Oto noted for her beauty, and other Plains Indians visited the capital. Thereafter McKenney still more vigorously favored bringing in red deputations so they too might have the salutary experience of seeing for themselves the seat of white power: "This mode of conquering these people," he pointed out to Secretary of War Barbour, "is merciful and it is cheap in comparison to what a war on the frontier would cost, to say nothing of the loss of human life" (H, 82; cf. ITNA, II, 294). Yet not incidentally the arrival of these delegates almost automatically meant more portraits—painted principally by Charles Bird King, the Washington artist who was a former student of Benjamin West—and thus an enlarged collection for the Indian Office. Nicholas Biddle later recalled he first met McKenney there, "surrounded by uncouth portraits of savages of both sexes, whose merits he explained with as much unction as a Roman Cicerone—how nearly extremes touch when so civilized a gentleman was in contact with so wild & aboriginal a set" (H, 62). The financier's simile was apt but just slightly out of focus: McKenney was more the unctuous curator of a museum of the dead and dying, a head-hunter seeking to preserve "the exterior and appearance of these hapless People" so as to have visual answers to the question posterity would ask: *What sort of being was the red man of America?* Moreover, he wrote in this 1828 letter that appeared in the *National Intelligencer*, each head cost little: "The average cost of this collection, since 1821, is perhaps $3,000 for *one hundred and sixteen heads*, and the cost for each head, including the full length likenesses, of which latter there are five, is about $33" (ITNA, I, xxxiv). All these captured likenesses lined his walls as bargain trophies.

Of all the Indians in the United States, only those in his gallery were not potential targets for forcible removal. In early 1830 Jackson approved the petition of two entrepreneurs to copy the paintings for a countrywide exhibition but instructed Secretary of War Eaton to insist the copies be made "in the Portrait room." The originals "ought not to be taken from the room" (V, 254). McKenney put the petitioners off and ignored the president's instructions, for he had already entered into partnership with Samuel Bradford, the Philadelphia printer, to reproduce the collection and to provide a biographical sketch for each red subject —to this end the colonel circularized agents and missionaries for data on "the character & life of each person." But McKenney's dismissal later in the year threatened to end the project. No longer could he simply let Bradford walk out of his office with paintings under his arm, and too clearly Jackson would turn down his formal application to copy the pictures. Desperate, he turned to Secretary of War Cass, who had just joined the cabinet, and worked out an informal arrangement with his old associate whereby he was allowed to remove a half-dozen or so portraits at a time. In Philadelphia they were copied in oil by Henry Inman, an artist who had just joined a local lithographic firm. They were then quietly returned for rehanging and another lot removed. It was all quite informal and strictly confidential. Soon, McKenney exulted, *"all will be in my hands*—And I shall esteem this great gallery *a fortune"* (V, 262; H, 106; McKenney's italics).

"It is good to be shifty in a new country," said Simon Suggs, the frontier rogue in Johnson J. Hooper's 1843 book of that name. McKenney anticipated this good advice by cutting off competitors and by successfully spiriting the portraits out of the War Department for his own commercial venture. With the ease of a confidence man he shifted a few words back and forth and—presto!—

his language of benevolence became the language of acquisition. There was gold in those craggy red features, hundreds of thousands, a half-million, *a fortune!* His enthusiasm was infectious, in the beginning, carrying along a series of publishers and printers, artists and lithographers, politicians and journalists. With biographical data presumably coming in from the field and many of the portraits in hand, all that remained was for McKenney to find someone to put it all together. He offered half his share of the prospective riches to Jared Sparks, editor of the *North American Review*, if he would do the writing. Sparks nibbled at the bait but prudently refused to bite. He did introduce the colonel to James Hall, whom he had recently urged to write on Daniel Boone for his review (H, 107). Unwilling to write for what Sparks offered to pay, Hall snapped up the McKenney proposition when it came to him through the editor. Avarice was the bond of this improbable union of professional philanthropist and authority on Indian-hating.

2

"The Hon. Judge Hall," as he was identified on the title page of his first book, plunged in for the kill. In early 1836, indeed, he moved to corner the market by proposing to join forces with George Catlin, the artist and pioneer ethnologist. After a meeting in Cincinnati, where Hall had resettled, a long letter (ITNA, I, xxix–xxxii) from Philadelphia made quite clear why he was offering Catlin a once-in-a-lifetime opportunity: "Should you think proper to join us, we shall have in our hands a complete monopoly; no other work can compete with that which we could make." His own part was "to do the writing. Messrs. Key & Biddle [who had replaced Bradford] furnish all the funds, and attend to the labor of publishing, selling, etc." As plans stood, the work would include twenty numbers, each with six portraits and accompanying printed matter, and the whole would be known as "McKenney and Hall's *History of the Indian Tribes of North America.* A portion of the latter will be biographies of the distinguished men. My materials for this part of the work are very voluminous and of the most authentic character, having been collected from a great number of Indian agents and other gentlemen who are personally acquainted with the Indians." Catlin had therefore only to make some of his portraits available to share the "immense profit":

> Your object, I presume, will be to make money by the exhibition of your gallery, and it will doubtless be a fortune to you. But you could in no way enhance the value of your gallery more than by publishing a part of it in such a work as ours.

Hall badly mistook his man.

For the preceding half-dozen years Catlin had wandered among the Plains tribes, unsupported by government agency or private firm, gaining truly authentic materials at their tribal sources and painting a still neolithic world with a few lovely colors. Obviously he hoped to make a living from his work, did in fact later maintain himself in Europe by exhibiting it, but never looked on native peoples as so many red objects to be consumed or exploited for enormous personal gain. On the contrary, he had early on decided "the history and customs of such a people are themes worthy of the lifetime of one man, and nothing short of the loss of my life, shall prevent me from visiting their country and of becoming their historian." Accordingly, this great painter of the North and South American Indian disdainfully refused to be hanged in the necrological McKenney–Hall gallery.

So "I became editor," Hall wrote in his autobiography, "and set to work, with my usual ardor and energy (which were not small)." Since everything pivoted around the portraits, he was especially pleased with the work of Charles Bird King, "a good artist, who from taking so many, acquired the art of hitting off the savage expression, and the exact tint of the tawny complexion." He also counted on the "vast hoard" of official correspondence and other materials McKenney had said he would make available. "Do you know McKenney?" Hall asked Evert A. Duyckinck, editor of the *Cyclopaedia of American Literature*, in 1855:

> If you do, you will imagine the sequel—if not, and whether or not, what I write of him, is to be strictly *inter nos*. I found the 1st No partly written—and I had to begin there, in the middle of that No and finish it. With some most agre[e]able social qualities, my friend the Col. was as lazy a man as ever lived, and as unreliable a mortal as ever made big promises. His hoard of materials dwindled down to almost nothing, and after exhausting them, I could neither get him to furnish more, or to aid in writing. I went on alone. The labor was Herculean. Here were a long list of Indian heroes, to be supplied with biographies—of whom we knew nothing but the names. But I was compromised to the work—and determined to do it—and to make the work what was intended: an authentic National work.

Publishers came and went and "at last a couple of Yankees," Rice and Clarke, got hold of it and finished it. Hall had given the project all his spare time for eight years, but in the end, he complained, "McKenney & I were to have had half the profits, but got little or nothing." Happily, as Simon Suggs might have said, there was more than one way to skin a cat: During this time Hall was also cashier of the Commercial Bank of Cincinnati and later, after its charter expired, became president of a smaller bank of the same name, a lucrative position that beat hollow writing the lives of Indian heroes. As he wrote Duyckinck, "my circumstances are quite easy—thanks to banking—not to authorship."

So great were the strains of coauthorship they even split up this patriotic union of two men undertaking a great authentic national work. It was an unkind cut for Hall to call McKenney as lazy as any man who ever lived, for the latter cast the Indian in that role. The colonel expressed his attitude toward his partner by ignoring him on the title page of his *Memoirs,* where he listed himself alone as "Author of 'The History of the Indian Tribes of North America,' Etc., Etc." He also took his revenge, such as it was, by appearing as sole author of the octavo editions of 1848, 1850, 1856, etc. (ITNA, I, lix).

Probably Hall's account of their relationship was reasonably accurate in outline, but only the most dauntless of readers would accept his details at face value. Adjoining his autobiographical account of their collaboration, for instance, was his contention that his campaign biography of William Henry Harrison was not that at all but had been written with no reference to the election of 1836, a patent untruth easily refuted by his own correspondence at the time. He also maintained that apart from a few facts from explorers Pike and Long and Indian agent Schoolcraft, "nothing was compiled from books" for *Indian Tribes.* Actually he drew freely on McKenney's *Tour* and from his *Memoirs* (e.g., ITNA, II, 426–32) and indeed from a series of articles he had himself contributed to the *Illinois Monthly Magazine* (May, June, July, and August 1831) under the interesting title "On the Intercourse of the American People with the Indians." In fine, *caveat emptor:* The judge's facts were not all original, not all were "new and interesting," and not more than a handful were "strictly reliable."

Almost certainly his partner's contribution was a bit more substantial than Hall let on. Inspired perhaps by McKenney's example, Hall deepened the muddle about the misplaced Mohawks by making the tribe "the most powerful nation of New England" in the colonial period and then erred in the other direction by locating them later "in the *western* part of New York (III, 123, 153; my italics). Either or both authors may have been responsible for the egregious error of making over Metacom, or King Philip, the great *Wampanoag* statesman of the 1670s, into a principal chief of the long-battered and scattered *Pequots* (III, 28).

On the conceptual level the confusion became simply monumental, suggesting authorship by someone in an advanced stage of dissociation, part philanthropist and part Indian-hater. In some contexts Indians were "fierce, rapacious, and untamable" (I, 104), possessed of a seemingly innate "savage appetite for blood" (III, 153): "We find that the Indian, when seeking revenge, and especially when foiled in an attempt upon the primary object of his hatred, becomes possessed of an insatiate and insane thirst for blood, which impels him to feed his passion, not only with the carnage of the helpless of the human race, but even by the slaughter of domestic animals" (II, 112). Yet in other contexts the Indian appeared under all his cruelty as basically a tamable man and "the nursery stories that have left the deepest impressions on our memories" were called into question: "Indeed we have been taught to consider the Indian as *necessarily* bloodthirsty, ferocious, and vindictive." True, the Indian had all those characteristics as the merciless savage he was, "but it is no less true, that he has never been taught those lessons of humanity which have, under the guidance of civilization and Christianity, stript war of its more appalling horrors, and without which we should be no less savage than the Indians" (I, 201–2). One is tempted to attribute to McKenney this recognition of the Indian's unredeemed backwardness and to Hall the perpetuation of nursery stories of his unredeemable bloodthirstiness. As we saw in the preceding chapter, however, quick assumptions about the colonel's philanthropy are dangerous. Such reasoning would lead us, for instance, to identify Hall as the author responsible for depicting the Winnebago We-Kau as "meagre—cold—dirty in his person and dress, crooked in form—like the starved wolf, gaunt, hungry, and blood-thirsty." The rancor in this description flowed as if from the pen of Hall or of the Eastern Indian-hater Robert Montgomery Bird and for pure bile matched the novelist's likening of the Shawnee Wenonga to "a consumptive wolf." But this dehumanization of We-Kau in fact came from McKenney's *Memoirs* (ITNA, II, 431; M, I, 113).

In *Summa theologica* (1267–73) St. Thomas Aquinas had established certain criteria for the "just war": it had to be declared by the sovereign, had to be waged against those who deserved to be killed or hurt for their evil, and had to be waged by those who had rightful intentions, that is, those who sought to advance good and avoid evil. A little like Father Prucha in the following century, the authors of *Indian Tribes* sought to meet the final criterion head-on: "We have asserted that the policy of our Government, and the intentions of the American people towards the Indians, have been uniformly just and benevolent" (III, 202; cf. I, 9; II, 361). Yet attempts to demonstrate these perennial good intentions repeatedly floundered and finally collapsed in undeclared disaster (e.g., II, 176, 190).

The recent Black Hawk War, so-called, was a case in point. *Indian Tribes* provided most of the details (II, 58–91). Black Hawk, "whose unpronounceable Indian name we shall not attempt to repeat," on one occasion had been hunting on tribal land when "a party of white men seized him, charged him with having killed their hogs, and beat him severely with sticks." Treaties then in force gave him and the Sauks right to live and hunt on lands previously ceded to the United States, but whites ignored the legal position, and as with "the greater number of our Indian wars," this was "incited by the impatience of our own people to possess the hunting-grounds of the receding savage." In 1827 citizens burned down about forty lodges in the Sauk and Fox village at Rock Island, Illinois. In 1831, "under these circumstances, the Government required the removal of this nation from the ceded tract." In April 1832 Black Hawk openly recrossed the Mississippi with about a thousand of his people to visit their old home and mourn their dead. Their intentions were eminently peaceful: "Notwithstanding their merciless rule of warfare, which spares no foe who may fall into their hands, however helpless, they passed the isolated cabins in the wilderness without offering the slightest outrage to the defenceless [*sic*] inhabitants." Yet the alarm went out, the governor called out the militia, and an advance party slaughtered all but one of "five or six Indians who were approaching them with pacific signals." The "war" had commenced. Then living in the state, Hall witnessed citizens rally to the colors almost to a man, so great was their hatred of Indians, "and especially were all gentlemen who had any aspiration for political preferment, eager to signalise themselves in this field." Such was the benevolence of Hall's neighbors, the campaign became more of a springboard to political preferment than he could have anticipated. In the regular and militia forces that attacked and pursued the old brave and his band of Sauk families were two future presidents, Abraham Lincoln, nephew of the Indian-stalker, and Zachary Taylor, the army officer who subsequently achieved still greater recognition fighting Seminoles and Mexicans; the Whig presidential nominee in 1852, Winfield Scott; the next three governors of Illinois; and at least ten future members of Congress, state legislators, and lesser office holders.*

After a pursuit of hundreds of miles the Sauks were finally cornered near the mouth of the Bad Axe in Wisconsin "and nearly the whole party slain or captured."† Black Hawk escaped, later surrendered, and was paraded through the East as a prisoner of war:

> Thus ended a war instigated by a few individuals to forward their own sinister views, but which cost the Government more than two millions of dollars, besides needlessly sacrificing many valuable lives. But while we condemn the beginning of this contest, we would award credit to those who afterwards became engaged in it. However unjustly a war may be brought about, it becomes the cause of our country whenever hostilities have commenced, and

* Cecil Eby, *"That Disgraceful Affair," the Black Hawk War* (New York: Norton, 1973), pp. 18, 289–92, lists those who turned in their "war credits, at inflated value, for political preferment." With an eagerness for the fray that would have made his Uncle Mordecai proud, Lincoln enlisted "at the first tap of the drum," according to John G. Nicolay and John Hay, *Abraham Lincoln* (New York: Century, 1890), I, 87–100. The "war" brought the young rail-splitter his first public recognition. Afterward he returned to his humble home in New Salem, from which he was "soon to start on the way marked out for him by Providence."

† Hall sensibly did not go into the details of how the hundreds of Sauks were slain. For an unflinching account of how they were, of how "the two-hundred-year heritage of Indian hating" surfaced in the bloodletting at Bad Axe, see Eby's *"That Disgraceful Affair,"* pp. 243–61. The fugitive families were trapped at the edge of the Mississippi by pursuing soldiers on land and by a steambarge—wryly named the *Warrior*—in the water. When they tried to surrender, they were cut down under their white flag by a six-pounder mounted on the barge. Except for this technological advance, the ensuing slaughter was reminiscent of that at the Pequot Swamp. It was in this "battle" that one of the soldiers, John House, found, near the bank, an infant tied to a piece of cottonwood bark, deliberately shot the baby, and delivered himself of the immortal words: "Kill the nits, and you'll have no lice." Another volunteer shot a woman in the back after his bullet passed through the arm of her child strapped there. The mother died on the spot but the child was taken to headquarters where the injured arm was amputated, perhaps unnecessarily. Soldiers ranged through the woods and along the riverbank raping and shooting women, killing children, in fine, as one put it, "fast getting rid of those demons in human shape" or, as another put it, those "wretched wanderers [who] are most like the wild beasts than man."

Lincoln missed Bad Axe and the preceding "battle" of Wisconsin Heights and afterward made light of his tour of duty: "By the way, Mr. Speaker, did you know I am a military hero?" he later sarcastically asked in the House. "Yes sir; in the days of the Black Hawk war, I fought, bled, and came away. . . . [If General Cass ever] saw any live, fighting Indians, it was more than I did; but I had a good many bloody struggles with the musquetoes."—"Speech in the U.S. House of Representatives on the Presidential Question, July 27, 1848," in *Collected Works of Abraham Lincoln*, ed. Roy P. Basler (New Brunswick, N.J.: Rutgers University Press, 1953), I, 509–10. Intended to be very funny, the speech showed Lincoln to be a representative white man of his time and place—his celebrated compassion never quite made it across the color line. For Lincoln, as for Jefferson, Monroe, Clay, et al., the blacks had to be colonized elsewhere and the reds had to be cleared out, or killed, before citizens could truly enjoy having a white man's country. Lincoln could celebrate along the way, however, as he did in an 1838 speech to the Springfield Young Men's Lyceum, that "we, the American people . . . find ourselves in the peaceful possession, of the fairest portion of the earth, as regards extent of territory, fertility of soil, and salubrity of climate"—(ibid., 108.)

> honour should be awarded to the citizen who draws his sword to repel an armed foe from our borders. [II, 79–80]

So much for honor and the uniformly just intentions of citizens and their government: Out of this disaster all that could be salvaged was Judge Hall's doctrine of the "unjust war." It turned St. Thomas on his head but no doubt the "Angelic Doctor" was out of place near the mouth of the Bad Axe.

At the very beginning of all these killings and hurtings, I remarked (p. 5) that the Second Seminole War was dragging on in the 1830s much as the Vietnam War was to drag on thirteen decades later. Led by Osceola, or Asi-Yaholo, meaning Black Drink Singer, the Seminoles repudiated the fraudulent Payne's Landing Treaty (proclaimed in 1834) that cleared the way for their removal through the signatures or marks of a few chiefs, resisted heavy-handed efforts to secure their compliance, and finally, after being whipped, tortured, and locked up, in 1835 launched a determined guerrilla struggle to maintain their freedom. During the war Osceola was captured by the treachery of United States forces, and his wife was sold as a slave—that made the United States, in the words of one of Cass's biographers, "a trafficker in human flesh." Before hostilities ended, the United States sent some 30,000 troops "to root the Indians out of their swamps," along with half its Marine Corps and a number of Navy warships and transports. Of this conflict that raged while *Indian Tribes* came off the press Colonel McKenney later wrote in his *Memoirs:*

> It was little thought of, that the employment of . . . [force] to compel acquiescence to the terms of a treaty [Payne's Landing], which were never binding upon the Indians, would cost some thirty millions of dollars, and lead to as lavish a waste of as patriotic blood as ever the earth drank up, and that a train of disasters would follow, through nearly a seven years' war, reflecting anything and everything upon our arms and country, *but honor.* The war was alike unjust, inhuman, and inglorious. But the agony is over, and there remains no remedy but to study, in the future, to atone for the evils of the past. [I, 283; McKenney's italics]

This ringing denunciation was accurate and suggested McKenney did not accept the "unjust war" doctrine.

So one would assume Hall responsible for the tortured justifications that appeared in *Indian Tribes,* tortured justifications that may be compared with the earlier account of Jackson's invasion of Florida and his summary executions of reds and whites (pp. 105–108). Like the Sauks, according to *Indian Tribes,* the Seminoles had been abused and insulted, "the most scandalous outrages were perpetrated upon their persons and property, provoked often by their own ferocity and bad faith, but nevertheless wholly inexcusable," all of which led to "the political paradox . . . of a people practically oppressed by a magnanimous nation, entertaining towards them the kindest sympathies, and annually expending millions for their defence, support, and welfare" (II, 362–63). Still, ran the account in *Indian Tribes*, the dark Seminoles had been tampered with in the past by spies such as Arbuthnot and Ambrister, whom the gallant General Jackson had rightly executed; Osceola was himself an "aspiring demagogue"; and his people were "themselves intruders into a land previously occupied by the Europeans, from whom the American Government derived title by purchase" (II, 337, 374, 375). Thus "the United States, having the right as well as the power to remove them, resistance could only lead to a war, wholly unjustifiable because hopeless" (II, 375–76)—the Seminoles were like John Endicott's Quakers, that is, in having thrown themselves on the point of the sword. And thus, "knowing the perfidious character of these people," General Thomas Sidney Jesup had been entirely justified, nay, had only carried out his duty, in seizing Osceola and his associates under their flag of truce. They were intruders "and if it was lawful to remove them, there could be no moral wrong in taking them wherever they could be found" (II, 386–87).

So the political paradox—indeed the formulations in this absolutely staggering defense were all Hall's. Our confidence that McKenney was in profound disagreement oozes away, however, when we reflect that he repeatedly took sole credit, proudly, for these formulations and everything else in the three volumes. It disappears when we learn that in 1842 he proposed balloons be used to root the Seminoles out of their swamps (V, 3). His proposal came too late to be implemented but gave the colonel a niche in the military history of counterinsurgency tactics and placed him among the innovators who made the Seminole War look modern —the use of war rockets, rubber pontoon boats, and bloodhounds like Dugdale in *The Yemassee*, in anticipation of the K-9 Corps of WWII.

Hanging in the McKenney–Hall gallery were valuable portraits of men and women of dignity who have since vanished. The biographies, along with Hall's history of the Indian nations, to which we shall return, were grand receptacles for the racism, nationalism, imperialism, and religious bigotry of the age. And, however inactive McKenney's collaboration, they helped reveal the underlying affinities of Indian-hating and philanthropy. Judge Hall drew on Colonel McKenney and General Cass for data and interpretation; McKenney drew on Cass for attitudes and assumptions integral to Indian-hating. Both McKenney and Hall saw the Indian as a merciless savage, habitually indolent, perpetually infantile, incapable of abstract reasoning, a fit target for removal (e.g., II, 112, 246, 262, 269; III, 7, 21, 94). Above all, neither man saw anything wrong in writing of "the great prairies of the Far West" as, in the words of the epigraph at the head of this chapter, "uninhabitable by human beings" and then discussing Petalesharro, Eagle of Delight, and other Indians who lived there. But in this too they merely revealed the curious notions and prejudices of their time; for Major Stephen Harriman Long, the Dartmouth graduate, former West Point instructor, and explorer, who was one of their sources, had already declared the Great Plains unfit for human habitation and then had discussed the Indian tribes living in this "Great American Desert."

3

In *Flush Times in Alabama and Mississippi* (1853) Joseph Glover Baldwin presented Sargeant Prentiss, ideal gentleman of the Old South. Free from any taint of Yankee meanness or littleness, Prentiss was generous to a fault, magnificent even in his vices, as when he nonchalantly bet a fortune on the turn of a card, and chivalric in his courtesy; under this knightly exterior was a temper "instant in resentment, and bitter in . . . animosities, yet magnanimous to forgive." Yet Judge Baldwin's overdrawn and idealized portrait was actually of a native Yankee, as W. J. Cash pointed out, namely of Northerner Prentiss, "studying to get on as a politician in the deep South." It was good to be shifty in flush times, South or North. Exactly like Prentiss, Hall was a native Easterner studying to get on in new country, except he settled north of Mason and Dixon's line and thus had to adopt slightly different coloration in order to embody in himself his new neighbors' dreams of themselves.

James Hall (1793–1868) was born in Philadelphia. Like McKenney, he came from Quaker stock, though the Halls had long since abandoned the faith and made names for themselves in military service. Both his father and his grandfather served in the Revolution. He grew up in Philadelphia, commenced the study of law in 1811 but gave that up to enlist in the War of 1812, became a lieutenant of artillery under Winfield Scott, and after the Niagara campaign was officially commended for "brave and meritorious services." The young lieutenant liked the military, liked "the faithful precision of its movements, the subordination of strict discipline, and the steady coolness of determined courage." Fortunately, since he sought to stay in action, after the hostilities ended he landed an appointment as an artillery officer in Stephen Decatur's expedition to Algiers. Unfortunately that little war petered out when Decatur reached an agreement with the Barbary powers. So after a cruise of the Mediterranean "I returned at the close of the same year [1815], and was stationed at Newport, R.I. and afterwards at various other Posts until 1818 when I resigned, having previously resumed the study of law, at Pittsburgh, Pa., where I was then stationed, and been admitted to the bar." All that was neglected in this autobiographical account was the major crisis of his career: In 1817 he had been court-martialed for negligence and insubordination, had been dismissed from the service, but managed to have his punishment—the conviction stood—remitted by President Monroe. Then and only then had he been in a position to resign.

Things were only infrequently the same as Hall construed them. Whatever the merits of the case against him, he was in fact a professional ordnance officer who left the service under a cloud of charges and countercharges. Omit this crisis from his record as he omitted it from his autobiography: might he then have remained a career officer like Winfield Scott and had firsthand experience of the Indian wars he later merely chronicled? Very possibly. Up to that point his trajectory had been eastward. Say the conflict with the Barbary powers had dragged on: might he have become an interpreter of the Near East to the great republic of the West? Who knows? But this much is sure: His destiny was not manifest before he descended the Ohio on a keelboat to take up the practice of law in Shawneetown, Illinois. That was in 1820, so the first *twenty-seven* years of his life had been spent in the East. Over the next *eight* years he was to become "The Hon. Judge Hall," author of *Letters from the West* (1828), interpreter of the frontier to the East, propagandist for his adopted region, professional Westerner.

Like Melville's symbolic *Pequod* (a haunting name) Hall "was not so much bound to any haven ahead as rushing from all havens astern." Astern was failure, the abrupt end of his military career and lack of success at the Pittsburgh bar. Ahead was "new country," the new state of Illinois: "I resolved to go to a new country, to practice my profession where I could rise with the growth of the population—but allured in fact by a romantic disposition, a thirst for adventure, and a desire to see the rough scenes of the frontier." Almost straightaway he became district attorney for ten counties, and that meant he could look down the barrel of the law at counterfeiters, horse thieves, murderers, and other rough backwoodsmen. In 1825 he landed an appointment as circuit judge and served until 1827, when the legislature abolished the circuit system but not "The Hon. Judge" title—for that fell into place during these *two* years and stuck like flypaper for the next *forty*. His banking career commenced in 1827 when he was appointed treasurer of Illinois for a four-year term. In January 1833, after being out of office two years, after the death of his first wife, and after *twelve* years in Illinois, none of them on any identifiable frontier, he backtracked to Cincinnati. There he remarried, resumed banking, and was on hand to become McKenney's partner, in which role we first met him.

Betwixt and between his duties as soldier, jurist, and banker, Hall became a man of letters. He came from a family given to writing prose and verse, with a mother who achieved a measure of recognition for her prim *Conversations on the Bible* (1818) and brothers who published and edited the *Port Folio*, a Philadelphia literary magazine to which he himself frequently contributed. In Illinois he edited newspapers, founded a state historical society, and edited the first Western literary annual, the *Western Souvenir . . . for 1829* (1828), wherein the germinal "The Indian Hater" first appeared. In 1830 he established the *Illinois Monthly Magazine*, and when he retraced his steps eastward, he took the literary periodical with him to Cincinnati, where it became the *Western Monthly Magazine* (1833–35). In addition to *Letters from the West*, he had an armful of books to his credit, among them *Legends of the West* (1832), in which "The Indian Hater" reappeared; *Sketches of History, Life, and Manners in the West* (1834–35), wherein appeared his celebrated chapter, "Indian hating.—Some of the sources of this animosity.—Brief account of Col. Moredock"; *Tales of the Border* (1835), wherein "The Pioneer" continued the theme of Indian-hating; of course *Indian Tribes of North America*, in which he used his materials on Indian-hating for a number of passages (e.g., II, 78; III, 155–58); and *The Wilderness and the War Path* (1846), wherein "The Indian Hater" made a valedictory appearance.

As his titles indicated, Hall's subject was the West. Whatever their names, *Letters* or *Legends* or *Tales*, he had been writing the same book over and over, *The Romance of Western History* (1857), as he called his last published work, a sly reproduction of the 1834–35 *Sketches* and hence the post-valedictory appearance of

"Indian hating . . . Col. Moredock." This bibliographer's nightmare prompted a contemporary rhymester to accuse the judge of making new books out of old, and of selling again, "the present year,/The work I last year sold." But *caveat emptor*—and since when has a good story suffered in the retelling?

Even after he left the bench Judge Hall seemed to have his robes on. His portrait in the *Romance* looked out at the reader from appropriately grave eyes set deep under a broad, noble forehead and over heavy jowls. He was short, stout, *solid* looking, just the man to instill confidence in court or bank. To others he seemed "courtly," "dignified" in bearing, "genial," though the suggestion of geniality about the eyes was undercut by tightly closed, thin lips. In fact, as Edwin Fussell observed, he bore a close resemblance to Charlie Noble in *The Confidence-Man: His Masquerade*, also published in 1857. There, Herman Melville described Noble as exuding a "warm air of florid cordiality, contrasting itself with one knows not what kind of aguish sallowness of saving discretion lurking behind it." Hall's florid cordiality was everywhere in evidence, as in his fulsome compliment to his new Western neighbors: Their contemplation of "boundless plains" and other "majestic features of their country [had] swelled their ideas."

Such obliging rhetoric made the judge much in demand as a speaker and raconteur. At Fourth of July orations he extolled the swelling grandeur of the American empire and declared to pleased audiences that nothing could stop their westward expansion short of the Pacific. In addresses to the Erodelphian Society, the Young Men's Mercantile Library, and other bodies, he was wont to denounce the abolitionists and the "pernicious dogmas of Garrison," invite support of the American Colonization Society, take a stand against Jackson and for a firm fiscal policy, champion the literary nationalism of Bird and compliment Paulding for his "sterling Americanism," unmask all British travelers as "hired agents" of their government, and anathematize vulgarity and depravity in literature: "Fielding and Smollet[t] were men of genius, but we cannot believe that any man of delicacy ever read their novels without disgust, and we should be sorry to accuse any lady of having read them at all." He had picked up his "frontier" lore in swings around the circuit of county courthouses, where he pumped his brethren of the bar for yarns and usable oddities; but unlike Lincoln, who later shaped his style in just this milieu, the judge never let on that a smutty story had fetched his fancy. Officially he disapproved of any writing that could not be shared with readers of his mother's reflections on the Bible.

In *Blackwood's Magazine* of February 1825, Yankee novelist John Neal recalled his own precocity as a sharper in his mother's store: There as a boy he had learned "how to sell tape—lie—cheat—swear, and pass counterfeit money—if occasion required—as it would, sometimes in a country, where that, which was counterfeit, and that, which was not, were exceedingly alike, not only in appearance, but in value." If that was true of Portland, Maine, and Neal was to be trusted on untrustworthiness, then how much more true was it of the backcountry out toward the frontier, that meeting place of nature and fraud?

At the other end of the seaboard another judge, Augustus Baldwin Longstreet, could claim his *Georgia Scenes* (1835) were mint pieces, since they were "By a *Native* Georgian." In conscience or caution Hall could make no such claim, of course, so success in passing off his writing as the real thing depended on a little friendly connivance from readers. Those in the West—for the most part also newcomers none too certain about their identities—readily winked at the judge's habit of writing *as if* he were a *native* Westerner, a native by adoption, as it were. The masquerade required delicacy, as in his introduction to the *Western Souvenir*, published a few years after his arrival in Illinois. All the contributors were old settlers, he shyly implied, for the annual had been "written and published in the Western country, by Western men, and is chiefly confined to subjects connected with the history and character of the country which gives it birth." Westerners and others who shared the prejudices to which he appealed gladly accepted this old-hand all-knowing manner as the gold coin of authentic knowledge.

4

Of course Hall's authority on the West was unalloyed counterfeit. Primary among Western subjects were true Native Americans, of whom he had no firsthand knowledge apart from a conversation with Black Hawk and a few other transient encounters. As we have seen, he had the curious notion that the way to come by reliable data on the tribes was to ask Cass, Schoolcraft, and other officials in the Indian service. But by their own definition white Westerners knew about Indians and Indian-hating in ways Easterners could not, so Hall held forth at length "On the Intercourse of the American People with the Indians" in several 1831 issues of the *Illinois Monthly Magazine*. That series of articles then became part I of the 1834–35 *Sketches* and, notwithstanding his claim that he took nothing from books, was expanded for *Indian Tribes of North America.*

In the long essay on the North American Indians that concluded volume III (83–345), Hall put the question directly: "Why is it then that they are savages? Why have they not ascended in the great scale of civil subordination? Why are they ferocious, ignorant, and brutal, while we their neighbours are civilised and polished?" Well, they had always been "a restless wandering people," easily tampered with and incited by the British to barbarous attacks on Americans. But the latter were not without responsibility for the current imbecile condition of their red wards, for their Indian policy had been wrong from the beginning. The United States had negotiated treaties with the tribes as though they were free peoples of sovereign nations, and that was absurd. In his most emphatic judicial manner, Hall ruled that "to make a *nation* there must be *government*"; the "systematic anarchy" of the tribes mocked that essential precondition:

> But between *a government* and *no government* there is but one line. There is a clear distinction between a state and a mere collection of individuals: the latter, whatever may be their separate personal rights, cannot have collectively any political existence; and any nation, within whose limits or upon whose borders they may happen to be, has a clear right to extend its authority over them, having regard always to the rights of other nations. It is necessary, for the common advantage and security of mankind, that all men should belong to some government; and those who neglect to organise themselves into regular civil communities, must expect that existing governments will impose their laws upon them. [ITNA, III, 227]

This relatively sophisticated warrant for imperialism simply set aside rights solemnly protected by treaties, since as wandering hordes Indians had no collective rights and could not establish ownership of lands they merely hunted over. It was a matter of power, anyway, not rights, and the quick exercise of power to bring the remaining Indians under "the strong arm of Government."*

* Hall was a strong-arm enthusiast. Like McKenney, he wanted to expatriate blacks, remove Indians—he held the Cherokees and other tribes were better off on the far side of the Mississippi (ITNA, III, 324)—and subject the errant to total institutions. He too shared the "*monomanie*" of the new prison system, though his model was not the congregate system of Auburn but the totally separate or isolated system favored by the Pennsylvania Quakers. In *Letters* he noted admiringly that Western Penitentiary outside Pittsburgh had been established "to discipline the body": "The stupendous building intended for this purpose is nearly completed, and will form a splendid and commodious edifice" (p. 38). Hall was merely more detailed than McKenney in his proposals for confinement of Indians in comparable military camps.

Hall's plan looked toward the future. He proposed to stop the ceaseless warfare and wandering of Indians by having the military collect them in villages where they would be taught to understand the importance of private property. There the red man should be fed well, for "an Indian, like a wolf, is always hungry, and of course always ferocious. In order to tame him, the pressure of hunger must be removed." On the other hand, he should not be pampered, for the bridle had to be placed firmly on his head, as with the wild ass: "in the contest between the man and the brute, between intellect and instinct, the latter must submit. So it is between the civilised and savage man." If the Indian were broken to harness, he would not have to stay in these "villages" forever: "We would deprive him of his natural liberty only so long as should be necessary to bring about that lucid interval, in which he would become sensible of his true condition, and apprised of the means held out for his redemption." Luckily the hordes remaining had recently been sent wandering across the Mississippi, so all that remained was a mopping-up operation and discipline for the survivors leading up to that lucid interval.

All the romance of the "Intercourse of the American People with the Indians" obviously flowered not in the latter, "this deluded race" as Hall called them (ITNA, III, 339), but in the "unconquerable race" of Boones who took possession of the country. In *Letters from the West* the judge recalled his intense interest as a boy "in gazing at the brawny limbs and sun-burnt features of a Kentuckian, as he passed through the streets of Philadelphia. . . . I thought I could see in that man, one of the progenitors of an unconquerable race; his face presented the traces of a spirit quick to resent—he had the will to dare, and the power to execute" (p. 5). Hall imagined plunging like this Boone into *"the boundless contiguity of shade,"* boldly cutting the tie to society:

> The mariner, when he looks abroad upon the vast interminable waste, may feel a depressing, yet awful and sublime sense of danger and solitude; but he has the consolation of knowing that if the solitude of the ocean be hopeless, its dangers are few and easily surmounted: they exist rather in idea than in reality. Boon [*sic*] and his companions could have no such animating reflections. [pp. 252–53]

Captain Ahab might have found these reflections droll, but then I hardly need to point out that we have been cut adrift on Defoe's interminable waste and not on Melville's.

In his letter "The Missouri Trapper" (pp. 292–305) the judge followed "this hardy race" of woodsmen across the Mississippi where he found Defoe's dream come true:

> That which the novelist deemed barely possible, and which has always been considered as marvellously incredible by a large portion of his readers, is now daily and hourly reduced to practice in our western forests. Here may be found many a Crusoe, clad in skins, and contentedly keeping "bachelor's hall" in the wild woods, unblessed by the smile of beauty, uncheered by the voice of humanity—without even a "man Friday" for company, and ignorant of the busy world, its cares, its pleasures, or its comforts.*

The veteran trapper Hugh Glass was actually in more desperate plight than Robinson Crusoe on one occasion after Indians had treacherously stolen his rifle and sought his blood. But Glass made do with only the knife, flint, and steel in his shot pouch, to the admiration of Hall, who quoted the intrepid adventurer: " 'These little fixens,' added he, 'make a man feel right *peart,* when he is three or four hundred miles *from any body* or *any place*—all alone among the *painters* [i.e., panthers] and *wild varments.*' " Wild varments and not people inhabited this wilderness island Hall shared with Glass. Never more than tawny predicates of white subjects like Boone, Glass, and all his other conquering heroes, the judge's Indians practically invited being put out of their misery.

Ever ready to oblige was "The Indian Hater," that archetypal *Western Souvenir . . . for 1829.* The hater was Samuel Monson, and his was the by now familiar story: Like everybody else he had merely despised and feared Indians until one day or rather one night when "wife—children—mother—all perished here in the flames before my eyes." Because of this unbearable personal tragedy "the man of sorrows" shot red men down with unerring aim from that night on "and so long as I have strength to whet my knife on a stone, or ram a ball into my rifle, I shall continue to slay the savages." This lifelong dedication to his calling made sense to contemporaries, for the most part, and certainly seemed worthy of emulation to all the killers who followed, from Paulding's Timothy Weasel to Bird's Nick and Hall's own later and still more celebrated Colonel John Moredock. But there was a problem.

With Eros banished from his pages, Hall had turned his genteel prose toward Thanatos and dwelt in loving detail on arsons, mur-

* Elsewhere Hall referred to "the unpeopled wilderness" and did not neglect McKenney's favorite castaway: In the early days a fort near the great bend of the Ohio "was garrisoned by a small party of soldiers, commanded by a captain, who was almost as much insulated from the rest of the world as Alexander Selkirk in his island of Juan Fernandez"—*Wilderness and the War Path,* pp. 2, 34–35.

ders, and other villainies. One of his letters was "The Harpes" (pp. 265–82); his only novel, *The Harpe's Head* (1833), also dealt with the notorious brothers who spread death and terror throughout the Green River country of Kentucky. Strange to say, the Harpes had

> a savage thirst for blood—a deep rooted malignity against human nature, could alone be discovered in their actions. They murdered every defenceless being who fell in their way without distinction of age, sex, or colour. In the night they stole secretly to the cabin, slaughtered its inhabitants, and burned their dwelling—while the farmer who left his house by day, returned to witness the dying agonies of his wife and children, and the conflagration of his possessions. [p. 269] *

But that was precisely what Samuel Monson had suffered, was it not? Then why did no farmer become a white-hater and elicit the judge's sympathies for his murderous sorrows? Surely because killing wild varments was one thing and killing white folks another. It was unthinkable that Hall should confer the respectability of his judicious understanding or put a "civilized" frame around killers who hunted whites as though they were wild beasts. But as for the killers who hunted reds as though they were, we may consider the Indian-hater as the judge with his robes off.

5

Through the American heartland flowed the Mississippi. One April Fool's Day in St. Louis the "favorite" steamer *Fidèle*—perhaps that listed by Hall as the "destroyed" *Fidelity* in his *Statistics of the West* (1837)—rocked gently in the current, ready for the downriver voyage to New Orleans. Near the captain's office a reward was posted "for the capture of a mysterious imposter, supposed to have recently arrived from the East; quite an original genius in his vocation, as would appear, though wherein his originality consisted was not clearly given; but what purported to be a careful description of his person followed." Now, Hall was a

* A man named Stegal, according to Hall, eventually revenged the foul murder of his wife and the burning of his home by cutting off the head of the big Harpe, Micajah, and carrying the "bloody trophy" to the nearest magistrate—hence the title of his novel. Hall did not say why Stegal stopped there. The small Harpe, Wiley, made his way to Natchez, where he joined a gang headed by a man named Meason or Measan, who also had a price on his head and so on . . .

recent arrival from the East of unclear originality—but view of the purported description was cut off by the surrounding crowd of pickpockets and peddlers. Among them was one who "hawked, in the thick of the throng, the lives of Measan, the bandit of Ohio, Murrel, the pirate of the Mississippi, and the brothers Harpe, the Thugs of the Green River country, in Kentucky. . . ." Now that *had* to be the judge or someone peddling the judge's books, and maybe he was the mysterious impostor! But could he be recognized off the bench?

My expectant surmises aside, so began Melville's surrealistic novel *The Confidence-Man*. Soon appeared another hauntingly familiar face, that of a gray-coated philanthropist who solicited "contributions to a Widow and Orphan Asylum recently founded among the Seminoles." Could Colonel McKenney have put down his inspired idea of balloons to root Seminoles out of their swamps only to take up this grim joke of an asylum for the survivors? But positive identifications were next to impossible in the murky twilight of the boat, with all its "vicissitudes of light and shade."

More certainly than the philanthropist did the Indian-hater come aboard after the *Fidèle* tied up at the "swampy and squalid domain" of Cairo. Or, rather, he came aboard in the person of bilious Charles Arnold Noble, who had strikingly similar florid cordiality and had heard the history of "the late Colonel John Moredock, Indian-hater of Illinois," again and again

> from my father's friend, James Hall, the judge, you know. In every company being called upon to give this history, which none could better do, the judge at last fell into a style so methodic, you would have thought he spoke less to mere auditors than to an invisible amanuensis; seemed talking for the press; very impressive way with him indeed. And I, having an equally impressible memory, think that, upon a pinch, I can render you the judge upon the colonel almost word for word. [p. 161]

In fact, James Hall, the judge, you know, sat right there, and through the ignoble Westerner parroted his Indian-hating formula for Frank Goodman, the cosmopolitan confidence man who was neither frank nor so good. In this perfect setting, just around the bend of the Great River, came the pivotal chapter, "The Metaphysics of Indian-Hating."

By closely following Moredock's history in the *Sketches* but not quite rendering it word for word, Melville tightened his grip on the mysterious impostor. In the novelist's rough hands the judge first launched, as was his wont, into a flatulent disquisition on the unbroken solitude of inland Crusoes (or in this instance of hairy Orsons, an allusion to the medieval romance of a wild man of the woods who had been suckled by a bear). The lure of a world without men and its underlying misanthropy bobbed to the surface out of the Boone legend: "The sight of smoke ten miles off is provocation to one more remove from man, one step deeper into nature." But every step away from the world of man necessarily brought the backwoodsman closer to the world of merciless savages—at this point Melville had all but given away the secret identification of Indians with natural objects. Anyway, as a youth, he had Hall say, the backwoodsman learned from the old forest chroniclers of "Indian lying, Indian theft, Indian double-dealing, Indian fraud and perfidy, Indian want of conscience, Indian blood-thirstiness, Indian diabolism. . . ." To be sure, Charlie Noble later broke in to say, the backwoodsman probably never used those exact words, "you see, but the judge found him expression for his meaning." Just so—Melville called it "sinning by deputy."* Just as, in reverse, the judge in defining Indians very nearly unmasked himself in his own "solemn enough voice, 'Circling wiles and bloody lusts.' "

Through the Chinese-box device of having the bilious Westerner quote the Western judge's lines on yet another Westerner, Melville emphasized that they all, Hall and his neighbors, shared a doctrinal hate for Native Americans. To his due share of this collective abhorrence Colonel Moredock had merely added his private passion to avenge the murder of his family. All gave the killer their sympathy and understanding, not unmixed with admiration —at one of his customary dramatic pauses the judge always insisted the entire company smoke cigars in the colonel's memory —even though as a prime hater Moredock "never let pass an opportunity for quenching an Indian" and though "Terror" became his epitaph. In *Sketches* Hall had warned against assuming that Moredock "was unsocial, ferocious, or by nature cruel. On the contrary, he was a man of warm feelings, and excellent disposition. At home he was like other men. . . . He was cheerful, convivial, and hospitable; and no man in the territory was more generally known, or more universally respected." Aboard the *Fidèle* Melville made the judge put it this way: Moredock had a loving heart, as "nearly all Indian-haters have at bottom," and was convivial, hospitable, benevolent, and "with nobody, Indians

* This quite useful concept has many possible applications to the haters and philanthropists we have been discussing. Even William Gilmore Simms indulged in sinning by deputy through Gabriel Harrison sinning through the bloodhound Dugdale.

excepted, otherwise than courteous in a manly fashion." That exception rather helped his standing in the eyes of other *men*, namely his fellow citizens.

In *Sketches* Hall had noted that Moredock was so universally respected he became a "member of the legislative council of the territory of Illinois, and at the formation of the state government, was spoken of as a candidate for the office of governor, but refused to permit his name to be used." The real-life Moredock was apparently a very singular white man indeed or at least a cut above those who used the chase and slaughter of Black Hawk's band as springboards to political preferment. But witness Melville's reworking of Hall's lines so that aboard the *Fidèle* the judge was moved to add that the reason for the colonel's strange diffidence "was not wholly unsurmised" by those who knew him best:

> He felt there would be an impropriety in the Governor of Illinois stealing out now and then, during a recess of the legislative bodies, for a few days' shooting at human beings, within the limits of his paternal chief-magistracy. If the governorship offered large honors, from Moredock it demanded larger sacrifices. [p. 176]

And there the judge stood exposed, his robes down around his ankles.

Through the American heartland steamed the *Fidèle*, with its cargo of cant and charlatanry, its celebration of killers as pathfinders of progress and as the cutting edge of Christian "civilization," its screaming-eagle rhetoric about the uniformly just and benevolent intentions of the United States, its political paradox "of a people practically oppressed by a magnanimous nation, entertaining towards them the kindest sympathies." The now-you-see-it and now-you-don't truth under the judge's rapidly moving shells was simple: However tawny, real human beings were daily and hourly quenched—raped and shot, looted and uprooted, banished and confined. To palm that truth was the judge's original vocation and in that sense he *was* Melville's "Confidence-Man." But the country was full of recent arrivals from the East, mysterious impostors pretending to be natives and denying real natives their humanity. Their refusal to accept Indians as persons was enshrined by the drafters of declarations and constitutions on parchment, by sculptors in stone, by painters on canvas, by folklorists in legends, and by writers in print. As a lone confidence man, shifty James Hall would have mattered little. But as a representative peddler of the national moral swindle he mattered to Melville and should matter to us. In that sense, through the judge

Burning the Cheyenne Village near Fort Larned, Kansas, from *Harper's Weekly,* June 8, 1867. From Block Island on to the Plains and beyond, Anglo-American arsonists left a trail of burned villages. This engraving shows them setting out (April 19, 1867) to burn about 300 lodges and all the possessions of 1,000 to 1,500 Cheyennes and Sioux. Located on the Pawnee fork of the Arkansas River, the village was on Indian hunting ground supposedly secured by solemn treaties.

and his Indian-hating all the other confidence men were hard at work destroying confidence and betraying men.

Melville knew his Hall. By the 1850s the collective moral confusion was so dumbfoundingly pervasive that Melville's surrealism was in truth the harshest realism, just the means for ripping off false fronts and exposing sham and deception, floating identities, the true patriotism of empty rhetoric. "The Metaphysics of Indian-Hating" gave the reader the best single overview to date of the terrain of racial hatred. It was an acute progress report on the state of the animosity after only two and a half centuries of growth.

In *The Confidence-Man* the judge had given the Indian-hater a prominent place among those who carved out a continental empire: "Though held in a sort a barbarian, the backwoodsman would seem to America what Alexander was to Asia—captain in the vanguard of conquering civilization. Whatever the nation's growing opulence or power, does it not lackey his heels?" The parallel was ominous—why should the hairy conqueror stop short

The Seventh U.S. Cavalry Charging into Black Kettle's Village at Daylight, from *Harper's Weekly,* December 19, 1868. In the post-Appomattox decades the quenching of Indians accelerated and left the prairie "crimsoned," as a military historian has aptly put it. This engraving shows General George A. Custer leading a surprise attack (November 27, 1868) on Cheyennes camped along the Washita River. Custer and his men impartially slaughtered Native American men, women, and children, along with their horses and dogs.

at the Pacific, as Hall once suggested he might, and not go on like Alexander to Asia? As the author of *Moby Dick* well knew, such men as James Kirke Paulding and Charles Wilkes already had come to regard the Western Sea as an extension of the American West. Meville had the unsettling gift of seeing that when the metaphysics of Indian-hating hit salt water it more clearly became the metaphysics of empire-building, with the woodsman-become-mariner out there on the farthest wave, in Melville's words, riding "upon the advance as the Polynesian upon the comb of the surf."

PART FOUR

Civilizers and Conquerors

As a matter of practical policy the annihilation of the Pequots can be condemned only by those who read history so incorrectly as to suppose that savages, whose business is to torture and slay, can always be dealt with according to the methods in use between civilized peoples. A mighty nation, like the United States, is in honour bound to treat the red man with scrupulous justice and refrain from cruelty in punishing his delinquencies. But if the founders of Connecticut, in confronting a danger which threatened their very existence, struck with savage fierceness, we cannot blame them. The world is so made that it is only in that way that the higher races have been able to preserve themselves and carry on their progressive work.

—JOHN FISKE,
The Beginnings of New England, 1889

CHAPTER XVI

The American Rhythm: Mary Austin

> All this time there was an American race singing in tune with the beloved environment, to the measures of life-sustaining gestures. . . . The ritualistic refrain of many of their prayer dances is not, Lord have Mercy on us, but Work with us, work with us!
>
> —MARY AUSTIN,
> *The American Rhythm*, 1923

WHITE AMERICANS such as Mary Austin's family did not sing in tune with the beloved environment but pulled up stakes and moved on so frequently their lives took on a nomadic rhythm. Mostly they moved westward, and in 1879 visitor Robert Louis Stevenson moved with them on an "emigrant train." Later his valuable *Across the Plains* (1892) provided insights into the prejudices the pioneers carried to the Pacific.

Following the woman he loved across ocean and continent, Stevenson had also followed an earlier dream: "For many years America was to me a sort of promised land," he wrote, asking readers to imagine a young Scot "who shall have grown up in an old and rigid circle, following bygone fashions and taught to distrust his own fresh instincts, and who now suddenly hears of a family of cousins, all about his own age, who keep house for themselves and live far from restraint and tradition." These new-found relations indeed liked to think of themselves as the special guardians of freedom and would have been still more charmed by his embrace of one of their favorite sayings: *Westward the course of empire takes its way*. Heavy support for this famous 1726 prophecy of Bishop George Berkeley came from George Bancroft's *History of the United States* (1834–74), the ten imperial volumes of which Stevenson had somehow stowed in his luggage. And even the late James Hall would have appreciated his avowal that Ohio "had early been a favourite home of my imagination."

Yet down the tracks came his corrosive observations on "despised races," a section that would have convinced the judge that Stevenson was merely another "hired agent" in the guise of a traveler. "My fellow Caucasians" on the train, Stevenson found to his disgust, hated their companions in the special "Chinese car" as so many "hideous vermin, and affected a kind of choking in the throat when they beheld them." The Chinese were, he was told, dirty, stupid, cruel, and thievish—epithets that persuaded him the whites had never really looked at them, listened to them, or thought about them. But their "judgments," as Stevenson called them, were "typical of the feeling in all Western America" and had their counterparts in attitudes toward the Indians, "over whose own hereditary continent we had been steaming all these days." The few pathetic families gathered at the way stations moved him to pity, "but my fellow-passengers danced and jested round them with a truly Cockney baseness. I was ashamed for the thing we call civilisation." If oppression drives a wise man mad, he wondered what "should be raging in the hearts of these poor tribes, who have been driven back and back, step after step, their promised reservations torn from them one after another as the States extended westward, until at length they are shut up into these hideous mountain deserts of the centre—and even there find themselves invaded, insulted, and hunted out." Their oppression extended past the extortion of Indian agents and the bad faith of all, down to "the ridicule of such poor beings as were here with me upon the train."

Westward the course of empire had deepened and spread the metaphysics of Indian-hating. In California Stevenson might have added to his list of despised races the "nigger" Afro-Americans, the "greaser" Mexicans, the "Kanaka" South Seas natives, the "raghead" Asian Indians, and of course the native "Digger" Indians. As these epithets suggested, the hatreds had some specificity but also interrelationships and common targets: nonwhites. Hatred of the merciless savage was chronologically prior in North America and, I believe, the paradigm for the others. As the Honorable Frank M. Pixley, who represented San Francisco before the joint congressional committee investigating Chinese immigration, testified in the centennial year 1876: God had assigned Africa to the blacks, America to the reds, Asia to the yellows, Europe and later, somehow, America to the whites, and that meant "the Yellow races are to be confined to what the Almighty originally gave them and as they are not a favored people, they are not to be permitted to steal from us what we have robbed the American savage of." With such a gentle stir, the pigments of Indian-hating shaded off into coolie-hating, the Chinese Exclusion Act (1882), and the "Yellow Peril" hysteria at the turn of the century. In 1877, two years before Stevenson's arrival, San Francisco's Chinatown was repeatedly lighted up by burnings and lootings. As it was, after his return from Monterey in December 1879, he could walk out to the Sand-lot to hear a "vulgar fellow," probably Dennis Kearney, "roaring for arms and butchery."

Just before crossing the Plains Stevenson had remembered a childhood story about Custaloga, "an Indian brave who, in the last chapter, obligingly washed the paint off his face and became Sir Reginald Somebody-or-other, a trick I never forgave him. . . . It offended verisimilitude, like the pretended anxiety of Robinson Crusoe and others to escape from uninhabited islands." Pruned from the published version, probably by one of the author's flock of protectors, was the following revealing sentence: "Just you put me on an uninhabited island, I thought, and then we'll see!" In flight himself from those who had taught him to distrust his own fresh instincts—and more specifically estranged from his father, as Crusoe was said to have been—Stevenson had sought his promised land in the West. Instead he found white cousins treating dark-skinned peoples as despised Fridays. The freedom from restraint and tradition, racist tradition in this instance, was not to be found in Monterey, in San Francisco, or even up on Mount Helena above the Napa Valley, where he and Fanny Osbourne had a brief, idyllic honeymoon. As it happened, he had to recross continent and ocean to create his own imaginary good place, and then it turned out to be Defoe's island decked out with California scenery and a fabulous lot of hidden money, *Treasure Island* (1883).

For apprentice writers Stevenson left behind the magic of his transient presence. In 1888, shortly after her arrival in San Francisco, young Midwesterner Mary Austin daringly drank red wine, torn between wish and apprehension, as she put it, "in the trail of places once frequented by one Robert Louis Stevenson, to whom, Mary gathered, she would be introduced if they met him, which they never did, though Mary forgets whether this was mere accident, or because he was no longer to be found there." Probably the Bohemian legend was no longer there, for that was the year Stevenson merely passed through San Francisco on his way to the South Seas, where we shall encounter him again in a different context. Meanwhile, to follow his apprehensive admirer out to the Golden Gate we must first retrace our steps to the Midwest.

Mary Hunter Austin (1868–1934) was a pioneer and the daughter of pioneers. On her mother's side she traced her descent back to her great-grandmother Polly McAdams, wife of Jarrot Dugger, the son of Pierre Daguerre, who had come over from France with Lafayette to serve freedom. Polly was the daughter of William McAdams, who had served in the Revolution with the North Carolina Continentals and afterward married Mary Hendricks and migrated to Tennessee. At the turn of the century the McAdamses tried to settle in the Illinois country north of St. Louis, but that, Mary Austin noted in her memoirs, "was while the Black Hawk Hunting-Ground was still infested by Indians incited by the French." One day the women and children were rounded up by Indians and held overnight at Spring Cove. At dawn came the prearranged signal of a single rifle shot, at which the women threw themselves flat, covering the bodies of their children, while their rescuers sent a full volley into the Indians. Nevertheless, the McAdamses were pushed back into Tennessee, where Polly became the childhood friend of Sara Childress, who afterward married James K. Polk of Manifest Destiny fame. Polly McAdams herself married Jarrot Dugger and in the 1820s immigrated to Illinois permanently, settling finally in Carlinville, by which time the Hendrickses, McAdamses, Duggers, and other clans had also moved into the new state. Two of her great-granduncles, John and Wesley Dugger, joined in the chase and slaughter of Black Hawk and his people. Abraham Lincoln was a friend of the family; Uncle Sam Dugger helped carry Lincoln to his grave in 1865; and another Dugger served on the commission for his monument—"there is in possession of the Duggers," Mary Austin recalled, "a letter of Lincoln's that the historiographers know nothing about, which has the sanctity of a holy relic."

Mary Austin's family saga read as if it came straight out of the patriotic pages of the 1920s *Saturday Evening Post;* instead it formed the first part of *Earth Horizon*, the autobiography that was published two years before her death in 1934. There she reveled in the women who were so critically important in bringing "civilization" to the West: her mother, Susanna Savilla Graham, who had married the former Yorkshireman George Hunter in 1861 and borne him Mary and five other children; her grandmother Hannah Dugger, who had married the druggist Milo Graham; and most of all her great-grandmother Polly McAdams Dugger, the pioneer matriarch, who had passed her experiences on to the others and been the childhood friend of Mrs. James K. Polk, the "First Lady in the Land. . . . And this was how American women were trained to win the land from savage hordes and to walk discreetly beside their husbands in the highest offices to which they could be called, women and ladies alike."

Those "savage hordes" seem incongruous in the pages of an authority on Indian dance drama. Yet when Mary Hunter Austin continued the westward trek of her grandmothers by taking up a homestead in the arid lands of Southern California, she took with her a full set of their ideas and attitudes. She believed and never stopped believing that the continent had to be won from Indians so nature could be "subdued." She never shed the notion that the fratricide of the Civil War was "romantic" and indeed even willed herself into believing she had seen it in person: "What pageantry, what pomp, what attitudes!" To the end of her life she celebrated men like her father, who had fought for Union and Lincoln—in Captain Hunter's case, his celebrated service had earned him a malarial infection that led to his death in 1878. Always she celebrated the middle-class world of the Midwest and celebrated her ancestors, English, Scotch-Irish, French, and Dutch—all of whom in a sense marched behind her across the Plains. When she was a young woman she learned that Uncle Jeff Dugger, "in his tracing back of the Dugger lineage, had discovered that there was an Indian strain in the particular branch of the Virginia family into which the first Daguerre had married, but said little about it, since in those days, Indians were still 'varmints.' " No doubt the strain went back to Pocahontas—by 1932 one could proudly join the legions who claimed descent from that Indian princess or even acknowledge a Cherokee grandmother somewhere in the safely distant past.

"Wholesome" was a key word in Mary Austin's memoirs. She liked "wholesome young people, immensely worth knowing" but not demanding physically. As a girl she had not cared to play "post-office," since "she thought if she wanted to kiss a boy she'd just kiss him." She rarely had the urge, as she also made clear. A strict Methodist, she did not protest her church's prohibitions against dancing and the theater, since neither meant much to her and her classmates at the Presbyterian Blackburn College in Carlinville and at the State Normal School in nearby Bloomington: "Dancing, in Mary's youth in the Mississippi Valley, had reduced itself to a merely sexual function. . . . For Mary, whose sex perceptions developed late, dancing as she saw it in the eighteen-eighties revealed nothing of its relation to her own interior problem." Five decades later, to those who asked if she was going to tell "everything" in her autobiography, she replied, "Everything that matters": "To satisfy the modern expectation in that direction, Mary would have to tell a great deal more than matters, since to the things Mary cared most about, sex never mattered much." On the contrary, her anxious defense of "a degree of pre-marital restraint" as one of the possible "conditions of success," her bitter attacks on "our intelligentsia" for their Freudianism and their imputation of sexual hypocrisy to the Victorian era, and her curious insistence in *The American Rhythm* (1923) that even for humans "the sex urge is seasonal," all these views and others indicated that sex mattered to her much more than she let herself realize.

As a child she remembered sitting by the door in the summer twilights listening to her mother Susie singing the good old favorites, "Nicodemus the Slave," "Tenting Tonight," and a version of "Yankee Doodle" in which angels poured " 'lasses down on this nigger's head—shoo-fly!" Still, despite this and despite her admission Indians were still considered "varmints" in those days, she maintained that she and the other Hunter children "grew up in a liberal and enlightened atmosphere, where race prejudice did not exist." For proof she offered Mose Drakeford, Carlinville's sole "colored man," who cut wood in the Bottoms, a happy-go-lucky darky who put a straw hat on the head of his mule and "was of the opinion that the 'muel' could keep the flies off his 'hind laigs hisself.' . . . But to Mary he was a voice in the dark, inextricably bound up with the mucky smell of the Bottoms and the malarial vapors that rose out of them after the sun went down." Also she could point out that Grandpa Graham had entertained in his own house two of the Fisk Jubilee Singers when they were in town, "black as stovepiping they were." But the dark looks of blacks hit Mary Austin hard years later when she was living on Riverside Drive in New York and had wandered off in the direction of the section then known as San Juan Hill, where she came upon

> rather shabby houses with a singular effect of darkness, coming out of the doorways and windows, a light of darkness; so strange that I stopped on a street corner to look. And suddenly I saw three black men crossing the street toward me, black, and walking with the jungle stride. They were so black and so freely walking that I was frightened. I looked about, and saw first one and then another black man, and then black women, going up and down the street, walking and disappearing, coming out black as they came toward me, and losing the blackness. [*Earth Horizon*, p. 346]

From then on the black and carnal flesh that advanced with its jungle stride obliged by walking out of its sensuality into reassuring whiteness, as though in early rehearsal for Ralph Ellison's *Invisible Man* (1952). She prided herself on being able *not* to see the blackness of James Weldon Johnson, W. E. B. Du Bois, and other Afro-Americans and thus could blissfully "forget they are black." After one trip to Harlem she did not go back, "popular as it became to do so. I did not wish to see them black, and I feared with much looking, it might come upon me."

So how could someone with such capacity for self-deception ever have gained a feeling for the American rhythm? Part of the answer went back to a summer morning in her fifth year of life when she had stood under a walnut tree in an orchard near her home and suddenly experienced an overwhelming sense of oneness with the universe. Thoroughly conventional otherwise, she was set apart from family and neighbors by this mystical experience and the accompanying discovery of her subconscious, a deeper, intuitive self she called "I-Mary." When she and her mother moved to California in 1888, this second self went along and helped her survive the first hard years of homesteading in the dry lands at the edge of the San Joaquin Valley, the stints of schoolteaching there and in the Owens Valley, the failure of her relationship with Stafford Wallace Austin.

At the time of their marriage in 1891 Austin was a vineyardist near Mountain View Dairy. He came from a still farther West, the Hawaiian Islands, where his father was circuit judge and his family had established themselves as pioneer missionaries. As Mary Austin described him, he was a lackadaisical man, always vainly waiting for the big chance:

> Brought up on a huge carelessly kept plantation, in assumptions of social superiority such as accrue to Nordics living among brown peoples, he had at sixteen removed to the mainland where he spent the next twelve or fifteen years living in boarding-houses and going to school on an allowance from home. He had never heard of such things as budgeting the family income, of competence achieved by cumulative small sacrifices and savings. Now that he did hear of it, he thought it all rather cheap and piffling. [*Earth Horizon*, p. 241]

With only one side of their story before us, we must tread warily here in full awareness of her depreciatory feelings about sex. Still, Wallace Austin seems to have lacked sympathy for his wife's growing rebellion against the status of women and for her dogged determination to become a person and a writer in her own right. Added stress to their relationship came in the discovery that their daughter Ruth, born in 1892, was a congenital idiot who eventually had to be placed in a sanitarium. And at one point Mary Austin seemed to make biology responsible for her husband's failure to contribute to their common life. In the Owens Valley men could always believe something would turn up and would spend their lives in waiting on the releasing discovery, she wrote, while

> in fact it was timeless space that held them. With the women it was not so; they felt, as they hung there suspended between hopes that refused to eventuate, life slipping away from them. For women have "times"; the short recurrent rhythm of well-being, the not-to-be-evaded times of birth, the climaxes of their racial function, the effacing hand of Time across their charm, which points inescapably the periods within which experience is available.

This was part explanation of their incompatibility, part explanation of her own sensitivity to recurrent rhythms in their larger environment, but no explanation at all of why other women seemed oblivious to the same phenomena. The marriage was, in any event, a mismatch from the beginning. Mary Austin put it aside after she went to Carmel in 1904. Wallace Austin secured a divorce ten years later.

2

Like Seyavi, the Paiute basket weaver in *The Land of Little Rain* (1903), Mary Austin found that with mother wit she could live in the desert without a man. For sixteen years, from her arrival in 1888 to her departure for Carmel, her real love affair was with the land and its aboriginal life. "I lapped up Indians," she noted in *American Rhythm,* "as a part of the novelist's tormented and unremitting search for adequate concepts of life and society, and throve upon them." She came to know Yokuts, Paiutes, Washoes, Utes, Soshones, and members of other tribes. When she was ill a Paiute woman suckled her daughter and a couple of years later brought the child dried meadowlarks' tongues in a hopeless effort to make her speech nimble and quick. In the towns between the high Sierras Mary Austin had been shocked by a white man kicking a red woman to death for resisting his sexual advances, after which the Indian husband had wept "in the broken measures of remediless despair. 'My wife . . . all the same one dog.' " She was an angry witness at Lone Pine of the local pastime of "squaw chasing" and horrified by the suicides of two young girls after they had been "captured and detained for the better part of a night by a gang of youths." * Observations at the Indian school nearby made her "a fierce and untiring opponent of the colossal stupidities, the mean and cruel injustices, of our Indian bureau." What she had seen and learned persuaded her in the winter of 1896–97 that she was no longer a Methodist nor indeed a Christian in any orthodox sense. When a Paiute medicine man explained to her that prayer was an inner act, she seemed to have known always that it was just that, an inner reaching out to something that was "outwardly expressed in bodily acts, in words, in music, rhythm, color."

Mary Austin saw how Indians lived with land upon which others would starve and "how the land itself instructed them." She began to study their art and drew from their basket weaving "their naked craft, the subtle sympathies of twig and root and bark." She entered into their lives and danced with them to "the beating of the medicine drum; the pound of feet in the medicine dance. You give way to it through rhythmic utterance." Indians taught her how to write, she said, and the proof was on virtually every page of *Land of Little Rain*. A close observer of cactus, creosote, and other desert growth, she wrote of tree yuccas:

> Tormented, thin forests of it stalk drearily in the high mesas, particularly in that triangular slip that fans out eastward from the meeting of the Sierras and coastwise hills where the first swings across the southern end of the San Joaquin Valley. The yucca bristles with bayonet-pointed leaves, dull green, growing shaggy with age, tipped with panicles of fetid, greenish bloom. After death, which is slow, the ghostly network of its woody skeleton, with hardly power to rot, makes the moonlight fearful. [p. 11]

Out of such torment came beautifully fit prose for both flora and fauna:

> The coyote is your true water-witch, one who snuffs and paws, snuffs and paws again at the smallest spot of moisture-scented earth until he has freed the blind water from the soil. Many water-holes are no more than this detected by the lean hobo of the hills in localities where not even an Indian would look for it. [pp. 27–28]

* In WWII U.S. officials located one of their ten concentration camps near Lone Pine and called it Manzanar. In that camp on December 6, 1942, military police fired submachine guns and rifles into an unarmed crowd of Japanese Americans, killing two, wounding ten, and coincidentally demonstrating the carry-over of racist attitudes from Mary Austin's time in the Owens Valley—for the killings and woundings at Manzanar, see Roger Daniels, *Concentration Camps USA: Japanese Americans and World War II* (New York: Holt, Rinehart and Winston, 1971), pp. 107–8.

And finally the unity of the whole in a sentence with its own inner necessities:

> Twice a year, in the time of white butterflies and again when young quail ran neck and neck in the chaparral, Seyavi cut willows for basketry by the creek where it wound toward the river against the sun and sucking winds. [pp. 169–70]

Quite naturally the Paiute woman tightened the unity by weaving patterns of the plumed quail into her golden russet cooking bowls.

Years later at the Tesque pueblo Mary Austin witnessed Eagle dancers "catch up the rhythm of the wind through the tips of their wing-spread plumes and weave it into the pattern of their ancient dance, to the great appreciation of the native audience." By a wry coincidence, she was instructed in the meaning of a nuance she had misconstrued by Ovington Colbert, a descendant of the same family as McKenney's Chickasaw ward: "the subtle wavering of the movement of the Squaw Dance, which I had supposed to be due to the alternate relaxation and tension of interest, was really responsively attuned to the wind along the sagebrush" (*Rhythm*, p. 30). Colonel McKenney never would have listened to this Colbert nor have looked without fascinated loathing at such exercises in devil worship, no more than had officials from John Endicott to those currently in the Bureau of Indian Affairs. But Mary Austin did and made her great discovery. Indian dances were prayers in motion. She even brought herself to accept the bronze bodies of the dancers as instruments of their rhythmic reverence and attacked the ludicrous prudery of "a people who would undertake to insist that the Corn Dance should be danced in pajamas, lest Deity, to whom the dance is made, should not be able to endure the sight of the bronzed thighs and shoulders he has given to the least of his Americans" (p. 61).

Mary Austin had had to cross the Plains and live close to arid land before she found out that all this while, in the rich black lands she had left behind and in fact all across the continent, there had been "an American race singing in tune with the beloved environment." Before 1900 she came up with that insight and then wrote it into such works of enduring merit as *The Land of Little Rain*, *The Basket Woman* (1904), and *The Flock* (1906).

In her long essay *The American Rhythm*, published in 1923 and as an enlarged edition in 1930, she pursued the implications of her discovery for art and history. Each one of us, women and men, represent "an orchestration of rhythms" (p. 5) set basically by the blood and the breath: "What is the familiar trochee but the *lub-*

dub, *lub*-dub of the heart, what the hurrying of the syllable in the iambus but the inhibition of the blood by the smaller vessels?" That was the biological foundation of the dance and drama, of all art, the inner ebb and flow of the individual merged through movement with those of the group: "If we look for the resolution of intricacies of rhythm called classic, we find it in the dance, and if we go back in the history of the dance we find the pattern by which men and women, friends and foes, welded themselves into societies and became reconciled to the Allness" (p. 9). Naturally in North America particular rhythms grew out of its particular surroundings. On this turtle continent, as Indians called it, authentic poetry would thus not have Old World forms but new ones nourished into being by native soils. Walt Whitman had moved in that direction, she granted, but the singing and dancing of Native Americans had already shaped the basic mold out of recurrent patterns of passionate communion with their Mother Earth.

In a word, she perceived that the Indians *lived* that oneness with the universe she had experienced fleetingly as a child under a walnut tree. "A primitive state of mind is, as nearly as I can make out, a state of acute, happy awareness. Streams of impressions of perennial freshness flow across the threshold of sense, distinct, unconfused, delicately registering, *unselected*" (p. 28), she wrote in words that anticipated Paul Radin's observation that native people live "in a blaze of reality." She perceived that the Indian's state of mind was not simple and perpetually infantile but "one of deeply imbricated complexity." In touch with his body, he used it to celebrate his alikeness, or better his identity, with that which gave him life and well-being. Through body movements, voice, drum, flute, rattle, color, the like, he (or she) joined others in circles of rising consciousness to a level of "uncoerced obedience" to tribal ideals (p. 36). These rhythmic ideals carried on a conversation with creation. "Europeanly derived" men and women, as she called Anglo-Americans, cut off from the mysteries of their bodies and thus cut off from their hearts and the free play of emotions, looked merely through their intellects at these dances and fearfully saw the devil at work. They had cut themselves off from the holy conversation.

What Mary Austin had to reach for haltingly and grasp imperfectly Native Americans had known and felt all this time. In *Lame Deer: Seeker of Visions* (1972), the late respected Sioux medicine man imagined the passing of a freezing winter in the misty beginnings of things and the thankfulness of men and women for the return of the life-giving sun: "I can imagine one of them on a sudden impulse getting up to dance for the sun, using his body like a prayer, and all the others joining him one by one" (p. 198). For fifty long years the Bureau of Indian Affairs had jailed his people if they danced the Sun Dance, Lame Deer pointed out, even though "all our dances have their beginnings in our religion. They started out as spiritual gatherings. They were sacred" (p. 243). In *Land of the Spotted Eagle* (1933), Sioux Chief Standing Bear equated dancing and praying in almost the same words: "And the Indian wants to dance! It is his way of expressing devotion, of communing with unseen power, and in keeping his tribal identity. . . . All the joys and exaltations of life, all his gratefulness and thankfulness, all his acknowledgments of the mysterious power that guided life, and all his aspirations for a better life, culminated in one great dance—the Sun Dance" (p. 257). With no comprehension of such deep and infinite graces the Anglo-American lacked all understanding of the continent he trampled under foot:

> The white man does not understand the Indian for the reason that he does not understand America. He is too far removed from its formative processes. The roots of the tree of his life have not yet grasped the rock and soil. The white man is still troubled with primitive fears; he still has in his consciousness the perils of this frontier continent, some of its fastnesses not yet having yielded to his questing footsteps and inquiring eyes. . . . But in the Indian the spirit of the land is still vested; it will be until other men are able to divine and meet its rhythm. Men must be born and reborn to belong. Their bodies must be formed of the dust of their forefathers' bones. [p. 248]

Mary Austin's major achievement was to divine how irrevocably Indians had made the land their own through dancing prayers—"their own" in the Indian sense of ownership, of course. Her divination lifted her to the distinguished company of Thomas Morton, George Catlin, and the few others who sensed that the "frontier" continent they had invaded possessed its own cherished rhythms.

3

I have remarked that Mary Austin never liberated herself completely from the religious and secular prohibitions and repressions of her youth. To the end she shared with her friend Theodore Roosevelt certain Manifest Destiny assumptions about how the West had to be won from the "savage hordes" that "infested" it.

All her life she shared with contemporaries certain social Darwinian notions about "primitives" and "Dawn Man." Her deep-set conquistador attitude toward the land came out in the very essay that probed for its rhythms: "Thus if we go back far enough into the origin of simple poetic rhythms, we find the gesture by which in the Days-of-the-New the earth was conquered" (p. 9). As Dudley Wynn has pointed out, her warm feelings for native peoples were betrayed by her pathetic celebration of the *Saturday Evening Post* as America's "great folk voice" and of big businessmen as "the high-priesthood of man's economic conquest of the earth." She was capable of true fatuousness, as in her observation in the *Yale Review* of June 1930 that the current attitude of the Department of Interior "is enlightened and sympathetic towards all Indian Art" (XIX, 742), an absurdity that no doubt owed much to the fact that her friend Herbert Hoover was in the White House.

Yet in *Land of Little Rain, American Rhythm,* and a few other perceptive works Mary Austin gave us intimations of a more precious significance than Frederick Jackson Turner's "Significance of the Frontier in American History" (1893). Like other pioneers she homesteaded at what remained of "the hither edge of free land" or at what Turner still more revealingly called "the meeting point between savagery and civilization." Like the others, what she brought in her head to that imaginary point was more important than the geographic fact, though Turner maintained otherwise: "This perennial rebirth, this fluidity of American life, this expansion westward with its new opportunities, its continuous touch with the simplicity of primitive society, furnish the forces dominating American character." For Turner, thus did recurrent new opportunities for acquisition and destruction shape American character. But unlike the historian and his pioneers, Mary Austin took the trouble to enter Indian lives, found therein the reverse of "primitive" simplicity, gave way to their rhythmic utterances, and thereby experienced a true rebirth in the spirit of the land. With her rare openness to visionary experience she brought to the West more than conventionally blinding assumptions and attitudes. In some measure she had readied herself for the kind of experiencing Lame Deer sought:

> I wanted to feel, smell, hear and see, but not with my eyes and my mind only. I wanted to see with *cante ista*—the eye of the heart. This eye had its own way of looking at things. [p. 39]

Even seeing with the eye of her heart half-opened took Mary Austin more than half a continent away from Carlinville, Illinois, and the pioneer world of her grandmothers.

CHAPTER XVII

The Manifest Destiny of John Fiske

> Of course our whole national history has been one of expansion. . . . That the barbarians recede or are conquered, with the attendant fact that peace follows their retrogression or conquest, is due solely to the power of the mighty civilized races which have not lost the fighting instinct, and which by their expansion are gradually bringing peace into the red wastes where the barbarian peoples of the world hold sway.
>
> —THEODORE ROOSEVELT, *The Strenuous Life*, 1901

OUT OF Granger, Wyoming, ran a branch of the Union Pacific called the Oregon Short Line. No *lub*-dub, *lub*-dub of heartthrobs—as Mary Austin heard—but "a-co-she-lunk-she-lunk, a-co-she-lunk-she-lunk, a-co-she-lunk-she-lunk, gliding along, gliding swiftly along" was the very different rhythm of steel wheels on steel rails that John Fiske heard and tried to reproduce for his wife in a letter of June 3, 1887. Outside his Pullman car was such a desert, a "frightful desert. Not a tree, not a blade of grass; mountains rearing their heads on every side, wild and savage mountains parched with thirst." A howling wilderness, a hideous waste, this land whispered nothing to the alienated sojourner: "A land of utter desolation, a land where no man could live! It struck me as being like the moon, yes, these terrible mountains, casting their sharp black shadows across the blazing sunshine are the very mountains I have seen through the telescope in the moon!" Along the Bear River the scenery changed, and in Pocatello, Idaho:

> for the first time I saw wild Indians. At the station I saw a noble savage, with his squaw and two small sons taking nourishment out

> of a swillbox! A few "braves" came capering around on their small horses armed with bows and arrows, and scowled upon us. Anything in human shape so nasty, villainous, and vile must be seen, in order to be believed. You wouldn't suppose such hideous and nauseous brutes could be. [*Letters* (L), p. 539]

Probably these Native Americans were Soshone-Bannocks, though Fiske derisively left them behind to caper unidentified. Shortly the porter, an "amiable Sambo," * called him to see the great falls of the Snake River. The American historian did not remember having heard of the river before but did know Lewis and Clark had "*discovered* this country and won it for the United States." The sight made him resolve to put "poetry" into his account of their expedition when he came to it in his history: "I *feel* it all now; and that alone would be worth the trip."

Five years later Fiske returned to the "New England of the Pacific," as he called Oregon, and on May 11, 1892, delivered the centennial oration "Discovery of the Columbia River." According to him, on that day in 1792 Captain Robert Gray of Boston "discovered the Columbia River. Thereby hangs a tale of empire, which I shall tell" (L, p. 603). Everyone was in Astoria, including representatives of Oregon, Washington, and Idaho, when Fiske rode down to the bar in a ship that symbolized the river: "Lots of steamers came out and greeted us, also a couple of canoes full of the most diabolical painted Indians ever seen outside of the bottomless pit! Such horrible creatures I never saw. The company on our ship was delightful: I was introduced to many old grey-haired pioneers, several who came with Dr. [Marcus] Whitman in 1843" (L, p. 605). In the Astoria Opera House he orated on "this latest chapter in the discovery and occupation of our continent" and was pleased to see the pioneers, "those dear old white heads scattered about in the audience listening to me. It was a great day. . . . We had surpassing mutton for dinner; and in the evening there was a fine display of fireworks on the river." And next morning at breakfast the huge, jovial historian smacked his lips over freshly caught broiled salmon, one of the edible delights of continental conquest.

* A decade earlier Fiske had written his wife from Baltimore that it was a city of "good beer, mammoth oysters and elegant niggers." There to celebrate the first anniversary of Johns Hopkins University, Fiske was one of the guests of honor along with his old teachers James Russell Lowell and Francis ("Stubby") Child. Following one of his lectures, Fiske spent the day "smoking and chinwagging" with Lowell until "Child came in and we talked about Jews" (L, pp. 360, 361). Fiske was glad to say of his summer retreat in Petersham, Massachusetts, "neither Irishman nor negro ever sets foot upon that secluded soil" (L, p. 88). He was not always so restrained in his references to Afro-Americans. On his next train trip to Oregon, for instance, he mentioned the porter as "the nigger [who] arranged our breakfast" (L, p. 602).

Since John Fiske came from what had been Pequot country, he could also participate enthusiastically in a cluster of centennials that went even farther back in time. Thus in that great Columbian year of 1892 he was also named city of Boston orator for "the four hundredth anniversary of the discovery of America" (L, p. 613). Still more fittingly he was invited back to the Connecticut Valley to deliver "The Story of a New England Town" on October 10, 1900, the two-hundred-and-fiftieth anniversary of the founding of Middletown (L, pp. 690–91).

Fiske's story, published that December in the *Atlantic Monthly*, demonstrated his mastery of the filiopietistic style. While telling the celebrants what they most wanted to hear, he made his remarks seem a new revelation of hoary truths. "Our Puritan forefathers" had carved the town out of a wilderness with "Indians everywhere" and with dangerous "dark and unknown recesses." By 1900 he could even count on indulgent smiles when he referred to topics that had "nourished" his youthful years: "predestination and original sin and Webster's Seventh of March speech."

2

Born in Hartford, John Fiske (1842–1901) had grown up in Middletown, where a succession of grandfathers on his mother's side had been town clerks from 1722 to 1847. These maternal ancestors always figured more importantly in his life than his father, Edmund Brewster Green, who had gone West at the time of the Gold Rush and had died a couple of years later in Panama. Even after his mother's second marriage to a wealthy New York lawyer, her son stayed on in the Fisk home with his grandmothers; in 1855 he legally assumed the name of his great-grandfather John Fisk and later added an *e* at Harvard. Surrounded by doting elderly people who happily fell into the habit of looking on him as a prodigy, he much preferred their company to that of "the rough town boys" who used to catch him on the street, as he later explained to his fiancée, Abby Morgan Brooks, and then used "to kick and pound me and tear my clothes, and throw me down in the mud and deface my books, and make life as hideous for me as ever they could" (L, pp. 79–84).

So did the champion of Charles Darwin's "natural selection" and Herbert Spencer's "survival of the fittest" remember his youth. A "model boy," his best biographer points out, he regularly attended Middletown's North Congregational Church, formally joined the congregation in 1856 following a conversion experience, and thought of spending the rest of his life working out God's providences in history. But by then such orthodox Calvinism had "a general flavor of mild decay," as Oliver Wendell Holmes satirically observed. This was only a trifle premature, for the Calvinist denominations were in truth seventeenth-century holdovers, relatively unchanged in their brittle intolerance and in their joyless celebration of human depravity, on the defensive against secular, Unitarian, and Transcendentalist attacks, and on the verge indeed of falling to pieces into those frozen Puritan sects the next century called "fundamentalist." In this state of decay their theology proved incapable of meeting Fiske's demands for certainty. Before long he was reading Alexander von Humboldt and Auguste Comte and calling himself a "positivist," to the dismay of his grandmother, minister, and fellow townspeople. On the eve of his departure for Cambridge in 1860 he wrote his mother of his intense desire "to be freed from the atmosphere of ignorance and religious intolerance which surrounds me" (L, pp. 32–33). At Harvard he read Herbert Spencer and became a disciple, gained notoriety as a religious radical, and laid the foundations for his reputation as a scientific and philosophical exponent of evolution. Through essays and books with such overarching titles as *Outlines of Cosmic Philosophy* (1874), *The Destiny of Man* (1884), and *The Idea of God* (1885), he brought the entire universe under his scientific laws of evolution.

In reality his contribution to evolutionary theory was slight. While he encouraged others to think of him as a scientist, he made no claims to experiments or original observations of his own and contributed to Darwinism only the ancient idea that the long and dependent infancy of the child helped develop human intelligence and the family as a social unit. But even in passing on this suggestion in 1871, Fiske distanced himself from any real relatedness to other forms of animal life: "In the lecture on 'Moral Progress,' " he wrote Darwin, "I have endeavored to show that the transition from Animality (or bestiality, stripping the word from its bad connotations) to humanity must have been mainly determined by the prolongation of infancy or immaturity" (L, p. 207).

Now, Darwin's message on one level had quite simply reminded readers of what many had always feared: they were *part* of the natural process. They were reminded, that is, of what Native Americans had always gloried in. As Chief Standing Bear once observed, "the Indian and the white man sense things differently because the white man has put distance between himself and nature; and assuming a lofty place in the scheme of order of things has lost for him both reverence and understanding." Living, growing things were fellow creatures in a land to which the Indian *belonged:* "He once grew as naturally as the wild sunflowers; he belongs just as the buffalo belonged." But Fiske took himself outside this web of relationships in the very act of discussing prolonged infancy and "man's descent from an apelike ancestor." Both animality and bestiality always had bad connotations for the American evolutionist. It was the mean roughness of those town boys who threw him down in the mud and—a quantum leap deeper—the blackness of those "elegant niggers" in Baltimore and the red savagery of those "hideous and nauseous brutes" in Pocatello. It howled a reminder in his ears of his own "lower" nature rooted in sexual drives from which he was in such hurried "transition." It was sensuality he feared. "The Korân is continually accused of being sensual," he wrote Abby Morgan Brooks, relieved to find that, "on the contrary it is as free from sensuality as the New Testament; and far more so than the Old" (L, pp. 124–25). "Evolution" had lifted him high above a sensual nature that was still the howling wilderness of his Puritan Forefathers.

On the Origin of Species (1859) had a subtitle that seemed to legitimate racism: *The Preservation of Favoured Races in the Struggle for Life*. Moreover, in *The Descent of Man* (1871) Darwin prophesied that "at some future period, not very distant as measured by centuries, the civilised races of man will almost certainly exterminate, and replace, the savage races throughout the world." John Fiske used precisely these social Darwinian terms to laud annihilation of Indians as part of the process whereby "higher races have been able to preserve themselves and carry on their progressive work" (see epigraph, p. 217). For him to speak of *races* higher and lower, and otherwise to ground his argument in current biological theory stamped him a man of his day. Still, in the 1630s Philip Vincent had invoked biology when he contended the English were the world's master colonizers because of their unparalleled fecundity (see p. 50). The contrast between "savagery" and "civilization" was as old as that Roger Williams had drawn between "the Barbarous" and his "countrymen" (p. 53); in Darwin's case it came directly from the ethnological stage-theorists of the eighteenth century. Fiske's social Darwinian racism, that is to say, was old wine in new bottles, a vintage that in his case dated from the origin of the species Anglo-American. Even the words he used to describe those "horrible creatures" at the mouth of the Columbia might have come to the lips of Endicott or Underhill or Mason two and a half centuries earlier: "the most diabolical painted Indians ever seen outside of the bottomless pit!"

Far from being an intransigent rebel against the "doughty Puritans," as Fiske liked to call his ancestors, he was at most a mutation, or better a mutative model child, of their struggle to rationalize the world. John Mason had seen a special providence, God's hand, in making their red enemies at the Pequot fort "as a fiery Oven" (p. 43); John Fiske saw the slaughter as merely part of a general providence that programmed the rise of the higher races and the advance of progressive government over the universe. Mason's God had been the angry Jehovah who had intervened to laugh His enemies to scorn; Fiske's God had come down out of the heavens permanently for steady work as the Great Evolver, as the *immanent* force of His evolving creation. In the oft-quoted quip of Henry Ward Beecher, the renowned preacher, "design by wholesale is grander than design by retail." When Fiske turned to the past, he pictured that wholesale design extending providentially out of the Puritan mission to the whole triumphant sweep of American history.

3

Fiske became the centennial historian par excellence. The one-hundredth anniversary of the United States roused patriotic interest in old glories and brought him an invitation to deliver six lectures at Boston's Old South Church, hallowed meeting place of 1776. Denied the appointment he coveted on Harvard's faculty, he resigned as assistant librarian there to accept the offer, put aside temporarily his philosophical popularizations of evolutionary doctrine, and instead enlisted Darwin and Spencer to help him explain "America's Place in World History." In the spring of 1879 his lectures were a heart-warming success, with the church "*packed full* of the very cream of Boston" (L, p. 378), especially for the last lecture, on the future of Anglo-Saxons. When he visited London a few months later, that final lecture raised the roof: "When I began to speak of the future of the English race in Africa, I became aware of an immense *silence*, a kind of *breathlessness*, all over the room," he wrote his wife (L, pp. 401–2). "All at once, when I came to round the parallel of the English career in America and Africa, there came up one stupendous SHOUT, not a common demonstration of approval, but a deafening SHOUT of exultation. Don't you wish you had been there, dearest? It would have been the proudest moment of your life!" These howls of approbation encouraged him to revise the lecture slightly, retitle it "Manifest Destiny," and make it the centerpiece of the series he brought to London's Royal Institution in May 1880. The lecture was published in *Harper's* in 1885 and in Fiske's *American Political Ideas Viewed from the Standpoint of Universal History* (API) the same year.

"The Centennial has started it," he exulted, "and I have started in at the right time" (L, p. 378). "Manifest Destiny" was for John Fiske what "The Significance of the Frontier in American History" was to be for Frederick Jackson Turner: the foundation of his reputation as a major historian. As Milton Berman noted in his biography, "this essay remained the basis of Fiske's historical writing to the end of his career and was an important factor in his great popularity." By September 1884 he had delivered the lecture some fifty times, three times in London, seven in Boston, four in New York, and at least once in a host of other cities (API, p. 7). In Milwaukee, for instance, he took the Opera House one Sunday evening "and gave 'Manifest Destiny' at 25 cents a head. . . . I had the Governor and several members of the Legislature to hear me and they were profuse in their expressions of delight" (L, p. 460). By the special request of His Excellency R. B. Hayes, president of the United States, William M. Evarts, secretary of state, General William Tecumseh Sherman, astronomer Simon Newcomb, historian George Bancroft, and others, he took the series to Washington, where he was presented to the cabinet and lionized by members and supporters of the administration. Among those who feted him was John Hay, whom we have met in other contexts—Hay had earlier attended Fiske's lectures on evolution and during this Washington visit had him to breakfast twice. "The fact is," Fiske told his wife, "I have got all the brains of Washington to hear me, and they are delighted" (L, p. 432).

Wheresoever he went he evoked delight with his glad tidings. "*Westward the course of empire takes its way*," Bishop Berkeley had written in 1726:

> The first four acts already past,
> A fifth shall close the drama with the day;
> Time's noblest offspring is the last.

In his enormously popular *Our Country* (1885) the Reverend Josiah Strong of the American Home Missionary Society took up "The Anglo-Saxon and the World's Future" and shyly named himself and his countrymen "Time's noblest offspring" (p. 216). But Fiske had already over and over told somewhat different audiences that was their destiny: they could justly derive patriotic pride from belonging "to one of the dominant races of the world" (API, p. 95).*

In *Wonder-Working Providence* (1653) Edward Johnson had declared "the Lord Christ intends to achieve greater matters by this little handful than the world is aware of." In the ten-volume *History of the United States* (1834–74) that Robert Louis Stevenson so laboriously carted across the Plains, George Bancroft had extended this wonderful Providence from little Massachusetts Bay to North America and made the Calvinist deity into a Unitarian who spoke the language of Jacksonian Democracy. In "Manifest Destiny" Fiske made this selfsame Providence still more wonderful by making it global, a Darwinian deity who spoke English with a slight Teutonic accent.† To the concepts of Darwin and Spencer, Fiske added those of the Anglo-Teutonists: English historians Sir Henry Maine, William Stubbs, and the bumptious Edward Augustus Freeman (who is perhaps best remembered for an observation that came out of his visit to the United States in 1881: "This would be a grand land," he wrote a friend from New Haven, "if only every Irishman would kill a negro, and be hanged for it"). Fiske drew on Freeman to show how Teutonic tribes had carried the germ of democracy from the forests of Germany to Britain in the fifth century; there in the chosen home of freedom, away from upheavals on the Continent, it had grown into the English Parliament; then Puritans had carried the idea across the Atlantic for its fullest expression to date, the Constitution of the Founding Fathers that came out of their New England town meeting and the written fundamental laws of settlers in the Connecticut Valley.

To this explanation of the origin of representative and local self-government Fiske joined Spencer's notion of progress from

* In *Our Country* (New York: Baker & Taylor, 1885—rev. ed., 1891), Strong disingenuously noted: "It is only just to say that the substance of this chapter was given to the public as a lecture some three years before the appearance of Prof. Fiske's Manifest Destiny, in *Harper's Magazine*, for March, 1885, which contains some of the same ideas" (p. 208n). But Fiske had given "Manifest Destiny" to the public as a lecture some *five* years before it was published, and internal evidence suggests that priority and dependence were the other way round. But who preceded and drew on whom really does not matter. Both men were swept along into lucrative careers by currents of racist expansionism they articulated in influential lectures, essays, and books. Strong was active in the West and Midwest and more directly reached rural evangelical audiences. Despite his westward treks to the Pacific slope, Fiske was more active in the Midwest and East, especially in New England; he had more direct access to Unitarians, Congregationalists, Presbyterians, Episcopalians, and others in middle- and upper-middle-class urban audiences. Together they reached an incredible number of people with essentially the same thoughts and feelings. (By 1891, 167,000 copies of Strong's little book were in print.)

† Fiske later used the quotation from Johnson's *Wonder-Working Providence* as the epigraph for *The Beginnings of New England* (1889).

simplicity to complexity, militancy to industrialism, and in his own words, from "primeval chaos" to the "millennium" (API, pp. 97–100). The two great branches of the English race would reach the millennium when

> every land on the earth's surface that is not already the seat of an old civilization shall become English in its language, in its political habits and traditions, and to a predominant extent in the blood of its people. The day is at hand when four-fifths of the human race will trace its pedigree to English forefathers, as four-fifths of the white people in the United States trace their pedigree to-day. [API, p. 135]

Thus would the race consummate the Manifest Destiny begun on the Atlantic seaboard by carrying it around the world to usher in that era of universal peace Spencer posited as the goal of evolution. Still, Fiske's conclusion, for all its cheery images, had ominous undertones: "Our survey began with pictures of horrid slaughter and desolation: it ends with the picture of a world covered with cheerful homesteads, blessed with a sabbath of perpetual peace" (API, p. 144).

Whether Asia with its "half-civilized populations" was to be spared this onslaught was not entirely clear, but Africa "will not much longer be left in control of tawny lions and long-eared elephants and negro fetich-worshippers"; Australia had already planted thriving states in what had been "a country sparsely populated by a race of irredeemable savages hardly above the level of brutes"; in New Zealand the English were multiplying faster than almost anywhere else "and there are in the Pacific Ocean many rich and fertile spots where we shall very soon see the same things going on" (API, pp. 118, 133–35). With a Darwinian emphasis that directly echoed Philip Vincent's fertility contentions of the 1630s, Fiske counted on English-speaking whites to outbreed the opposition and, through this mechanism of natural selection, survive and multiply where others would decline and vanish. Still, how were nonwhites, the great majority of the world's peoples, to vanish rapidly enough to suit their replacements? This awkward complication moved Fiske to invoke one "of the seeming paradoxes" of the whitening of the world, namely that "the possibility of peace can be guaranteed only through war" (API, p. 101). And he was comforted by his population extrapolations and by the irresistible military predominance presently enjoyed by the offspring of those old Teutonic tribes:

> So far as the relations of civilization with barbarism are concerned to-day, the only serious question is by what process of modification the barbarous races are to maintain their foothold upon the earth at all. While once such people threatened the very continuance of civilization, they now exist only on sufferance. [API, p. 107]

Just so did Indian-hating become cosmic. Fiske's gratifying fantasy underwrote genocide in what shortly became a global empire.

In 1898 the United States booted Spain out of the Caribbean, took Puerto Rico, grabbed the Philippines, annexed the Hawaiian Islands. In 1899 it helped partition the Samoan Islands and through John Hay addressed an "Open Door" note to the rest of the world. In 1900 it had more than ten million native peoples as "subjects" and had spread "the race," in line with Fiske's prediction, "from the rising to the setting sun" (API, p. 135). Precisely because contemporaries shared his fantasy before they heard it and had already started acting it out, he seemed to them a great man, a philosopher, a scientist, and undeniably the age's most popular lecturer and writer on American history. Though he never revised his confident prophesies in the direction of Henry Adams's somber fear that "the dark races are gaining on us," he did consent to become president of the Immigration Restriction League in 1894 in an effort to keep Manifest Destiny pure at its source. But his was essentially a sunny view of white America's prospects, and that made him the preeminent centennial historian and therefore the proper person to speak to Middletown on its two-hundred-and-fiftieth anniversary.

In his 1900 "Story of a New England Town" Fiske reminded his listeners of a lonely place behind an old tavern about eight miles north of Middletown. If you go there and climb up a steep grade through an orchard, you suddenly come upon an entrancing scene overlooking the beautiful Connecticut: "Turning toward the north, you see, gleaming like a star upon the horizon, the gilded dome of the Capitol at Hartford, and you are at once reminded that this is sacred ground"—sacred because it was there the seeds of federalism were sown in the first written constitution, which became a pattern for those that led to "the development of our nation" (API, p. 151). And just so was Calvin naturalized—or nationalized—on sacred ground where John Mason had hunted and annihilated the Pequots: "The blow which our forefathers struck was surely Cromwellian in its effectiveness. To use the frontiersman's cynical phrase, it made 'many good Indians.' "

After all these bloody years and miles, the Puritans' model child could still stand there gloating and sinning by deputy.

In 1901 John Fiske died on the Fourth of July. But a month before this timely end, he triumphantly queried a correspondent: "I wonder if you are aware of the fact that on May 3 the United States established civil government in the Philippines?" (L, p. 701).

CHAPTER XVIII

Outcast of the Islands: Henry Adams

> Henry Adams's journey to the South Seas is one more ray of light on the fact that the world is fast growing too small for its worldlings. But let him not think that even there he can get out of the crowd—I have many friends in those parts. Please, if he hasn't started, give him my love & tell him that if he wants savage islands he can't do better than come back to the British.
>
> —HENRY JAMES
> to John Hay,
> June 23, 1890

BORN IN 1838, Henry Adams finally sought his desert island in 1890. Shortly he would "be a pirate in the South Seas," he explained to an English friend. "In thus imitating Robert Louis Stevenson I am inspired by no wish for fame or future literary or political notoriety, or even by motives of health, but merely by a longing to try something new and different. Civilisation becomes an intolerable bore at moments" (L, I, 403).* With his unappeasable appetite for irony, Adams almost audibly savored the fact that as he wrote he awaited the last page proofs of a magnum opus that was an imposing monument to "civilization."

Numbed by the suicide of Marian Hooper Adams in 1885, he had since shut himself up in their Washington home on Lafayette Square to finish the *History of the United States during the Administrations of Jefferson and Madison* (1889–91). Therein he had drawn on Auguste Comte's reworking of eighteenth-century stage-of-society theory to help him demonstrate the certain triumph of the centralized national state over kinship or tribal societies. His was first and foremost the study of "the evolution of a race," the Amer-

* For an explanation of the abbreviated references, see p. 510.

ican race, and thus of American character and ideals, for which Jefferson had the centrality we have considered. But now that the historian had finished his massive labors, they bored him. He had come to doubt the inevitability of what he had so painstakingly traced and even to question whether his basic term *race* had any identifiable referent. As he put it later in the *Education*, "race ruled conditions; conditions hardly affected race; and yet no one could tell the patient tourist what race was, or how it should be known." About all the patient reader of his great work could say definitely was that throughout his nine volumes the American race did *not* include reds or blacks or browns, and that these customary exclusions only said what the particular race he had in mind was not. Manifestly the exclusions did not provide a precise definition of race or even the "clue" he sought, without which "history was a nursery tale" (E, pp. 411, 412). For Henry Adams, perhaps our finest historian, serious practice of his profession was over, while his "education" had long since ended: "Life was complete in 1890," he said in a famous sentence (E, p. 316); "the rest mattered so little!"

Before he left for the South Seas, he did his best to persuade John Hay, his next-door neighbor and closest friend, to come along, holding out as bait that Hay's daughter might have "her pirate isle" on which they and the artist John La Farge would establish an ideal republic (Cater, p. 192). In the midst of building a summer home on Lake Sunapee in New Hampshire, Hay had to beg off, though he clearly would have loved to play Alexander Selkirk with them. Their departure sounded "the last ringing of the bell. I now feel I shall never go west, and thence east. I shall never see California nor the Isles of the Sea" (WRT, II, 80). Sorry to leave Hay behind, Adams pulled La Farge away from his family and studio, and in mid-August the middle-aged "common travellers in gaiters" set off across the continent.*

Since Adams detested mountains and execrated the sea, as he was to confess only half-playfully, he was as alienated from his surroundings as John Fiske had been from his Wyoming moonscape. On the train he took instructions in watercolors from La Farge, became interested in light and color, and spent one day trying to dabble sage green for the sagebrush; but this abstract pastime hardly rooted him in what flashed by their windows as they jolted along. In San Francisco they boarded a ship that also

* Surely significant is the failure of some accounts to make clear and of others even to mention that their famous expedition was in fact undertaken by a threesome. La Farge brought along his servant Awoki, whom he called "my little Japanese attendant" (Rem, p. 279).

carried a troupe of cowboys and Indians bound for Wild West shows in Australia. In Hawaii they were entertained by Judge Sanford Dole, the future president of the islands, and amused by King Kalakaua. Finally they reached Samoa in early October and pulled up off the island of Tutuila, where they transferred to the thirty-foot cutter that would take them to Apia sixty miles away.

2

"A few minutes and the steamer was far away," recorded La Farge in his travel diary, "and we saw the boats of the savages make a red fringe of men on the waves that outlined the horizon —a new and strange sensation, a realizing of the old pictures in books of travel and the child traditions of Robinson Crusoe." A few hours later he made absolutely clear they imagined themselves Defoe's hero. Rain squalls forced them into a little harbor at Anua where "palms fringed the shore with shade. A blue-green sea ran into a thin line of breakers—like one of the places we have always read of in 'Robinson Crusoe' and similar travellers." A "splendid naked savage" carried them ashore, and residents of the village nearby, once they had learned they were not missionaries, treated them to their first experience of the *siva*, the legendary dance of Samoans. "So those lovers of form, the Greeks, must have looked," rhapsodized La Farge, "anointed and crowned with garlands, and the so-called dance that we saw might not have been misplaced far back in some classical antiquity." So beautiful and extraordinary was it, "no motion of a western dancer but would seem stiff beside such an ownership of the body." The young men were no less splendid, leading this American son of a Napoleonic officer to "the great wonder that no one had told me of a rustic Greece still alive somewhere and still to be looked at" (Rem, pp. 68–87). The greater wonder, missed even by his companion the aficionado of irony, was that no one had told these Crusoes that some thousands of miles astern, on the continent they had left behind, the original American race had been dancing all this while in rhythmic unity with the beloved land.

Despite the tropical setting, it was as if the clock had whirred back wildly to the seventeenth century and to the first Henry Adams startled by naked bodies in the Massachusetts Bay Colony. Samoa, said the last Henry's best biographer, "swept him off his Puritan feet" (Samuels, p. 11). Built up over the generations from that first Henry to John and John Quincy and through his own father, Charles Francis, the historian's layers of repression seemingly came unstuck all at once. With uncharacteristic breathlessness he wrote Hay that La Farge had been "knocked out" by the *siva:* "Greece was nowhere. I imagine he never approached such an artistic sensation before. For my own part, I gasped with the effect of color, form and motion, and leave description to the fellow that thinks he can do it. To me the effect was that of a dozen Rembrandts intensified into the most glowing beauty of life and motion" (Cater, pp. 198–99). To Elizabeth Cameron, the Pennsylvania senator's wife with whom he was infatuated, he essayed a description of the scene as the girls, naked to the waist, filed into the light:

> The mysterious depths of darkness behind, against which the skins and dresses of the dancers mingled rather than contrasted; the sense of remoteness and of genuineness in the stage management; the conviction that at last the kingdom of old-gold was ours, and that we were as good Polynesiacs as our neighbors—the whole scene and association gave so much freshness to our fancy that no future experience, short of being eaten, will ever make us feel so new again. La Farge's spectacles quivered with emotion and gasped for sheer inability to note everything at once. To me the dominant idea was that the girls, with their dripping grasses and leaves, and their glistening breasts and arms, had actually come out of the sea a few steps away. [L, I, 418–19]

So quickly, apparently, were the languors of "civilization" thrown over for gasps of pleasure at the return of the forbidden body, all those glistening breasts and arms and dripping forms.

In a letter to Hay, Adams enclosed photographs of girls he had snapped with his newfangled Kodak, though he warned that "you must supply for yourself the color, the movement, the play of muscle and feature, and the whole tropical atmosphere, which photographers kill as dead as their own chemicals" (Cater, pp. 202–3). Even so, Hay pored over the images and flayed himself for missing the originals: "I hang over your photographs and contemplate your old-gold girls, and interrogate the universe, asking if there ever was such a fool as I—who shall never *à grand jamais* enter that Paradise" (Samuels, p. 13). More tantalizing still were the *pai-pai* strippers with "beautiful thighs" Adams put before his friend in another letter. Made more taboo by the missionaries than the *siva*, he reported, the *pai-pai* was danced in the dead of night by two or three of the best and most shapely women. In the course of graceful movements their *siapas* or waistcloths worked loose:

> In the *pai-pai* the women let their *lava-lavas,* as they are called, or *siapas,* seem about to fall. The dancer pretends to tighten it, but only opens it so as to show a little more thigh, and fastens it again so low as to show a little more hip. Always turning about and moving with the chorus, she repeats this process again and again, showing more legs and hips every time, until the *siapa* barely hangs on her, and would fall except that she holds it. At last it falls; she turns once or twice more, in full view; then snatches up the *siapa* and runs away. [Cater, p. 216]

These rhythmic legs and hips, his friend's gasps of delight, the setting of a native house, "lighted by the ruddy flames of a palm-leaf fire in the centre, and filled, except where the dancing is done, by old-gold men and women applauding, laughing, smoking, and smelling of cocoa-nut oil"—all were calculated to turn Hay palm-green with envy.

Race and sex were inextricably entangled in these glowing letters from those first few days in Samoa when Adams and La Farge felt "so new again." Better, for that little while the old-gold dancers symbolized sex. Adams then explicitly considered, half-jestingly and half-longingly, an old-gold liaison; that is to say, he half-seriously contemplated a return to what La Farge called "ownership of the body." But such a warm paradise regained became unthinkable when the layers of repression were firmly in place again. Once the Puritan in Adams regained domination, he held there had been nothing there to be attracted to in the first place. Checking any prurient longings his peep show might have aroused in Hay, he declared the *pai-pai* he had just described was not what it seemed: "They are not in the least voluptuous; they have no longings and very brief passions; they live a matter-of-fact existence that would scare a New England spinster" (Cater, p. 219). That proverbially cool old maid was a caldron of passions, in fine, compared with the harmless old-gold girl. Not representing sex, the latter represented no threat: "She has no passions, though she is good natured enough, and might perhaps elope with a handsome young fellow who made a long siege of her" (Cater, p. 217). Adams was safe. He was no longer young and most unlikely to make a long siege of her.

In a letter to his close friend Clarence King, the explorer, Adams bragged that none of the maidens had "inspired me with improper desires" and that his age was not solely responsible: "Fifty-three years are a decided check to sexual passion, but I do not think the years are alone to blame. Probably I should have behaved differently thirty years ago; yet as I look back at the long list of dusky beauties I have met, I cannot pick out one who seems to me likely, even thirty years ago, to have held me much more than five minutes in her arms." That the beauties were dusky doubtless had much to do with his rigid self-control—the more dusky the more unpalatable, for then they shaded "the negro too closely" (L, I, 418, 438; cf. Cater, p. 207). Ernest Samuels, who quoted Adams's long letter to King, noted that it next made a "quixotic and revealing shift, as if following a subterranean and divergent course. 'The moral of this is whatever you please. To my mind the moral is that sex is altogether a mistake, and that no reversion to healthier conditions than ours, can remove the radical evils inherent in the division of the sexes' " (p. 16).* The tone of this defiant pronouncement suggested the moral was not altogether pleasing.

Denying the living human bodies of the "superb men and women" any more reality than passionless "Greek-fauns," Adams simultaneously denigrated their characters and culture. For a historian of his stature to generalize so wildly from a foundation of almost total ignorance was in itself significant. He had readied himself for Samoa hardly beyond reading novels and travelers' narratives—largely because he had not known what to expect, his senses were awakened by those first encounters with "civilisation without trousers." Once there he made no attempt to "meddle with the language. It would be hard enough anyway," he explained to a correspondent, "for every word has different meanings according to its accent, and the field for mistakes is somewhat too wide in a society which is perfectly unreserved in its habits" (Cater, p. 212). So through the filter of an interpreter he desultorily asked about old customs, families, and religion, "chiefly for want of something to talk about during the interminable visits of native chiefs." Even at this remove he discovered that most of what

* From Tahiti Adams wrote Hay no less revealingly about racial and sexual proprieties: "Here, above most places respectable people remain decent, if for no other reason because they are bored by vulgar vice. La Farge and I did not come here to live with mulattoes; one can do that just as easily at home. We take no pleasure in associating with dusky prostitutes whose single idea of enjoyment is to get drunk." Unknown to Adams or to the rest of the "Five of Hearts," as Adams and Hay called their circle of intimate friends, King had two years previously acted out his fantasy of the "primal woman" by taking a black in New York as his lover. Now, irritated by Adams's repudiation of native women, King perceptively pointed out to Hay that the tourists had missed the inner reality of native life. To his mind Adams was "a mere cerebral ganglion vis-à-vis . . . one of the initial centres of human heat." As if he were a Sioux holy man, King wrote that the historian's main trouble was that he saw only with the head and not with the head *and* the heart. In his biography of Hay, Tyler Dennett observed that when King's double life was revealed after his death, "his name was never again mentioned over the 'Five of Hearts' teacups" (TD, p. 157). So much for all the hearty banter about dusky maidens. (For King's common-law marriage with Ada Todd, see Thurman Wilkins, *Clarence King* [New York: Macmillan, 1958], pp. 317–22.)

interested him was secret, became convinced "that the Samoans have an entire intellectual world of their own, and never admit outsiders into it," but cared "too little about these matters to make any searching inquiry, so they may keep their secrets for anything I shall do" (L, I, 441–42). Yet two months after his arrival he wrote Hay: "What I don't know about Samoa is hardly worth the bite of a mosquito" (Cater, p. 224). Still earlier he had written Elizabeth Cameron he was "getting to know my Polynesian too well, and to feel the opéra bouffe side of him." In fifty-three days he and La Farge had "exhausted the islands" (L, I, 449).

In truth Henry Adams had no more respect for the Samoans' culture than Great-grandfather John had had for the Shawnees', Grandfather John Quincy for the Seminoles', or Brother Charles Francis, Jr., for that of the Indian neighbors of Thomas Morton of Merry Mount. In Samoa the natives were "alike each other as two casts from the same mould" (Cater, p. 225). Incapable of mature politics, their concern over rivalries among the imperial powers was "all the play of these brown-skinned children, who are bored for want of excitement, and are quite capable of getting up a fight about nothing" (L, I, 456). They had "the virtues of healthy children" (Cater, p. 220). In fact, "the natives, like all orientals, are children, and have the charms of childhood as well as the faults of the small boy" (Cater, p. 200; cf. Rem, p. 212). Permanently childish, they of course could not think: "As La Farge says, they have no thoughts" (Cater, p. 219). Now, these Fridays had counterparts extending all the way back to the Massachusetts Bay, all indistinguishable from one another, all childish, all mindless. Over the miles and years, all the brown-skinned children were seen as creatures of the viewers' ignorance, prejudice, and fears, and not as the living native peoples of the land.

Accordingly, when the two amiable middle-aged tourists had shipped out of San Francisco, what they brought in their heads and hearts, their "education," shaped how they saw and what they felt about the native peoples of the South Seas. Their intellectual and emotional baggage had predisposed them to see those peoples as infantile, and that seeing had larger implications. La Farge remembered one of his friend's history lessons that gave him "the sense of looking at the world in a little nutshell":

> The Pacific is our natural property. Our great coast borders it for a quarter of the world. We must either give up Hawaii, which will inevitably then go over to England, or take it willingly, if we need to keep the passage open to eastern Asia, the future battleground of commerce. [Rem, p. 153]

The contention that "our great coast" made the Pacific "our natural property" could as easily have been switched to the East and applied to the Atlantic, no doubt to the consternation of Britain and all the "civilized" powers. But it made certain sense if you had been brought up to believe the Pacific an empty space, the watery extension of the once-virgin continent, with the isles of the uttermost seas uninhabited, save for childlike counterparts of the once merciless savages. And so with scarcely a ripple had Adams's continental imperialism flowed down the Pacific slope and out onto stepping-stones for insular imperialism.

3

"Here is an outcast," Adams characterized himself, "who fled from his own country to escape the interminable bore of its nickel-plated politics and politicians; yet when he seeks refuge in an inaccessible island of the South Seas, ten thousand miles from an Irishman, he finds politics running around like roosters without heads" (L, I, 455–56). But just in case, he had prudently bethought himself to bring along to his desert island formidable letters of introduction, and these promptly moved Howard M. Sewall, the young U.S. consul, to show him deference and to take him and La Farge into his own household. No less gratifying was the response of Samoans when they learned that the U.S. man-of-war *John Adams*, familiar throughout the archipelago for its role in recent imperial maneuvers, had been named after Adams's great-grandfather—with mock deprecation he admitted, "I am a great *alí*—nobleman—because all the natives knew the frigate 'Adams' " (L, I, 421).

Then all the more rankling was his reception one October day when he and La Farge climbed up above Apia to "Vailima" where Robert Louis and Fanny Osbourne Stevenson were carving a home out of the mountain side. "There Stevenson and his wife were perched—like queer birds—mighty queer ones too," he wrote Hay:

> He seems never to rest, but perches like a parrot on every available projection, jumping from one to another, and talking incessantly. The parrot was very dirty and ill-clothed as we saw him, being perhaps caught unawares, and the female was in rather worse trim than the male. I was not prepared for so much eccentricity in this particular, and could see no obvious excuse for it. Stevenson has

> bought, I am told, four hundred acres of land at ten dollars an acre, and is about to begin building. As his land is largely mountain, and wholly impenetrable forest, I think that two hundred acres would have been enough, and the balance might have profitably been invested in soap. Apart from this lapse, he was extremely amusing and agreeable. He had evidently not the faintest associations with my name, but he knew all about La Farge and became at once very chummy with him. [Cater, pp. 201–2]

Hay's reply made clear he grasped what had at first graveled his friend: "Now I will have to tell you,—perhaps a dozen fellows have done so—of Stevenson's account of your visit to him. Your account of that historical meeting is a gem of description. I have it by heart. His is no less perfect and characteristic. He writes to N.B.:—'Two Americans called on me yesterday. One an artist named La Farge, said he knew you. The name of the other I do not recall.' Bear up under this, like a man, in the interest of science! It completes the portrait of the shabby parrot" (WRT, II, 86).

No historic meeting ever started off more awkwardly, though Stevenson seemed courteously unaware of Adams's intense dislike: "We have had enlightened society," the novelist wrote Henry James. "La Farge the painter, and your friend Henry Adams: a great privilege—would it might endure" (L, I, 451n). Adams himself wrote Elizabeth Cameron that "we have seen much of Stevenson these last few days, and I must say no more in ridicule, for he has been extremely obliging, and given me very valuable letters of introduction to Tahiti and the Marquesas" (L, I, 446). Yet the ridicule swiftly streamed back full strength: "All through him, the education shows. His early associates were all second-rate; he never seems by any chance to have come in contact with first-rate people, either men, women or [*sic*] artists. He does not know the difference between people, and mixes them up in a fashion as grotesque as if they were characters in his New Arabian Nights" (L, I, 452). So had Stevenson been unpardonably unmindful of the illustrious family name Adams, known even by the natives, and deserved dismissal for his second-rate associates—these second-raters in fact included George Meredith, Fanny Sitwell, John Singer Sargent, Henry James, and his fellow members of the London Athenaeum Club. Adams knew the Stevensons had just broken ground for "Vailima," destined to become perhaps the most elegant home in all Samoa, but preferred to dwell on their temporary quarters in a "two-story Irish shanty," beside which "a native house is a palace." Squalor was the only word to describe their household, "squalor like a railroad navvy's board hut."

Adams could see the Stevensons were clearing the jungle and burning the undergrowth but took them to task for not doing it in white duck and gloves—their dirty hands and soiled clothing the day he and La Farge dropped in, unannounced and unexpected, also bore traces from the sooty old cooking stove they had been trying to set up. But Adams's correspondents had drilled into them that the coarse, squawking parrot named Stevenson plumed himself in squalor and enjoyed being untidy and unwashed. Rather, he absolutely relished dirt: "Stevenson absolutely loves dirty vessels and suffocating cabins filled with mildew and cockroaches; he has gone off to Sydney chiefly, I think, to get some more seadirt on, the land-dirt having become monotonous" (Cater, p. 238).

"Still," observed J. C. Furnas, "it is late in the day to feel the old urge to kick Henry Adams for a snob"—*Voyage to Windward* (1951), p. 371. With less restraint, James Pope Hennessy, another Stevenson biographer, regretted "the sadly provincial judgment of Henry Adams"—*Robert Louis Stevenson* (1974), p. 274. And in his impressive biography of the American historian, Ernest Samuels acknowledged the "violent, disturbing, disquieting" impact of Stevenson on Adams and in explanation offered the latter's "prickly individualism" and tautening prejudices, resentment in having to follow in the more famous man's footsteps to Polynesia, and the "neurotic fastidiousness" that surfaced when he met the Bohemian legend face to face (Samuels, pp. 28–32). All these explanations help but still do not quite account for what was after all an extraordinary outpouring of venom.

While Adams was playing outcast and toying with an old-gold liaison, he met a man who to his horror seemed a real outcast of the islands. To his mind the internationally famous writer had insanely thrown himself off the top rung of society and out of history into native nothingness. By going native, as they say, Stevenson had acted out what Adams had only dared fantasize for a few fleeting days. To damn what he longingly detested, Adams then rummaged through his national heritage and came up with time-tested means: Stevenson was a squaw man. His wife "wore the usual missionary nightgown which was no cleaner than her husband's shirt and drawers," he wrote Elizabeth Cameron, "but she omitted the stockings. Her complexion and eyes were dark and strong, like a half-breed Mexican" (L, I, 425). Yet this was just a bit out of focus, so he later pictured dark Fanny Stevenson dressed "in a reddish cotton nightgown" and seated on a piazza chatting "apparently as well and stalwart as any other Apache squaw" (L, I, 446). Still more revealingly he wrote Hay that Stevenson "goes through fatigues, deprivations and squalor enough to kill a dozen robust Samoan chiefs, not to mention his wife, who is a wild Apache" (Cater, p. 202). To complete this plunge into barbarism and past, the unwashed Fanny and her dirt-obsessed husband had "a mode of existence here . . . far less human than that of the natives" (L, I, 452). Never did that prime Indian-hater Robert Montgomery Bird reveal more of himself in his denigration of white renegades than did Adams of himself in his lines on Tusitala, as the Samoans affectionately called the Scottish "teller of tales." As Adams wrote Elizabeth Cameron after the writer's death: "So I am reading Robert Louis Stevenson's *Letters* which make me crawl with creepy horror, as he did alive" (L, II, 269).

This true great-grandson of John Adams was perceptive in his creepy feeling, for all his slanderous misstatements, about one thing. Stevenson did threaten the very way in which Adams experienced the world. In Samoa the writer had grown healthier, more at peace with himself, more mature. At first the "indecent" dances bothered him, but as he edged further and further from his hereditary Calvinism he developed a passionate fondness for Samoans, whom he did not idealize as he had the Tahitians, and came to accept the way they lived in their bodies. Indeed they taught him not to think sex "altogether a mistake," as Adams put it, but rather the reverse, as he himself observed a few weeks before his death:

> If I had to begin again . . . I believe I should try to honour Sex more religiously. The worst of our education is that Christianity does not recognize and hallow Sex. . . . a terrible hiatus in our modern religions [is] that they cannot see and make venerable that which they ought to see first and hallow most. Well, it is so; I cannot be wiser than my generation. [Furnas, p. 423]

By edging closer and closer to what Furnas called a "wilfully organic intimacy with life," Stevenson had all but stopped his flight from the wild man within. His important, unfinished, posthumous *Weir of Hermiston* (1896) showed that Stevenson had not only found his island but finally strength to face the shadow self that had driven him so far—that inner Caliban he had called Mr. Hyde in one of his most famous and most misunderstood stories.

And that was precisely why he so disturbed Adams. Stevenson was learning a lesson his own education had ruled out, and "nothing attracted him less," Adams later declared, "than the idea of beginning a new education" (E, pp. 316–17). Shortly he wrote Elizabeth Cameron from Papeete: "Like Robinson Crusoe and

Herman Melville, I have been able to turn my mind to nothing except escape from my island." In double-quick time he and La Farge had "exhausted Tahiti which is a true Crusoe island" (L, I, 483). In their scramble to get off the island they did not mark the arrival of a dark genius named Gauguin, whom La Farge would one day half-defensively call "that crazy Frenchman." Unlike these American Crusoes, Paul Gauguin was to find himself in the South Seas and to realize his art in canvases ablaze with the colors of a voluptuous nature under a tropical sun. For him native life turned out to have inexhaustible appeal, and none of his figures was more beautiful or more mysterious than his Tahitian Eve. Plainly a woman and not the babbling child of Adams and La Farge, she was "very subtle, very wise in spite of her naïveté"; she was "an Eve after the Fall," Gauguin granted, "but still able to walk naked without shame and retaining all the animal beauty she enjoyed on the first day." Going the other way, La Farge returned to "civilization" with his paintings—a few of his canvases hinted at Gauguin's opulent innovations but most turned native Eves into Adams's pallid New England spinsters or, more exactly, into his own genteelly draped evasions of superb bodies, as in his sweetly ethereal "Fayaway Sails Her Boat" (Rem, facing p. 68). Adams returned with letters to be posted, an unresolved quest, and a restless aimlessness that made him a tourist through his own life, with every place soon exhausted and no place to stop running.

CHAPTER XIX

The Open Door of John Hay

> For the first time in his life, he felt a sense of possible purpose working itself out in history. . . . As he sat at Hay's table, listening to any member of the British Cabinet, for all were alike now, discuss the Philippines as the question of balance of power in the East, he could see that the family work of a hundred and fifty years fell at once into the grand perspective of true empire-building, which Hay's work set off with artistic skill.
>
> —HENRY ADAMS,
> *The Education*, 1918

UNLIKE Henry Adams, John Hay came from the West—or from what still passed for the West when he grew up in Indiana and Illinois. With pardonable exaggeration his biographers have acclaimed him a "son of the American frontier" (TD, p. 187) or "the son of a frontier doctor, [born] in a small dwelling on the edge of the Western wilderness" (WRT, I, 2).* From Scottish stock, Hay's wandering grandfathers were like Mary Austin's on her mother's side, the McAdamses and the Duggers, and unlike the first Adamses who settled on Thomas Morton's old place and stayed put. *Great-grandfather* Adam Hay located at the foot of the Shenandoah Valley in Berkeley County, Virginia, and there, as William Roscoe Thayer safely surmised, "probably did his share of Indian fighting." *Grandfather* John Hay followed in the footsteps of Daniel Boone and all the other heroes, real and fictional, to the Dark and Bloody Ground, where he settled at Lexington and raised a family; in 1831 he moved to Springfield, Illinois, where he became a friend of Abraham Lincoln, probably knew the Dugger family, and may even have met the illustrious Judge James Hall. *Father* Charles Hay finished his medical training in Lexington, moved to Salem, Indiana, commenced practice there as a country doctor in 1829, married Helen Leonard, a recent ar-

* For an explanation of the abbreviated references, see p. 511.

rival from Massachusetts, and in 1838 named their third child after his own father.

Born six years after the so-called Black Hawk War had cleared the region of Indians, John Hay was a son of the frontier only in the loosest sense. "His grandfathers had belonged to the first line of advance," Tyler Dennett more aptly observed, "his fathers and uncles to the replacement troops," but he was himself from beginning to end a town boy. Still, he did come from the generations of assault and replacement troops, and that origin later helped him write with sympathetic understanding of Lincoln's similar background in "that Anglo-Saxon lust of land which seems inseparable from the race" and which impelled their ancestors to "people" Kentucky and therewith produce such generic figures as Lincoln's Indian-hating Uncle Mordecai (see p. 123).

Across the Mississippi there were still lands to subdue and Indians to subjugate, of course, but John Hay turned his back on them and reversed his grandfathers' line of march by going the other way to make his mark. Through the help of his family and in particular of Uncle Milton Hay, a respected Springfield lawyer, he went to Brown University in 1855. In Providence the frail, bookish youth found a refinement lacking in the Midwest; in the Baptist college he was perhaps best remembered as a rhetorician and poet, with the first four lines of his Class Poem ('58) suggestive of his sensibility:

> Oft let the poet leave his toil and care
> To greet the spirits of the sky and air;
> Let him go forth to learn of love and truth,
> From Nature smiling in eternal youth. [TD, p. 22]

The graduate went back to Warsaw, Illinois, where Doctor Charles Hay had relocated, as to a kind of exile in "a dreary waste of heartless materialism." Unmoved by the great events of the day and especially unsympathetic to the abolitionists, the young poetaster found himself, *faute de mieux*, reading law in Uncle Milton's office and meeting members of the Lincoln circle in Springfield. He thus was on hand to help his friend John George Nicolay with the president-elect's ballooning correspondence. When Lincoln went to Washington to be inaugurated, Hay was invited to go too, to assist Nicolay and to step into his own destiny in the East.

Ever afterward he was known as Colonel Hay, Lincoln's secretary, though he was never really the latter, for Nicolay was, and never really a colonel, for his military title came out of no service in the field or on the parade ground. Lincoln had him commissioned so he could serve as his unofficial secretary and court jester, for he liked the playful wit Hay exhibited and cultivated in the White House. Though the colonel was later to claim Lincoln as the greatest figure since Christ, that conviction was slow in forming. They got on well, but during those Civil War years Hay became the protégé of William H. Seward. Even before Lincoln's assassination Hay had used his connection with the secretary of state to land an appointment as secretary of the legation at Paris, a position he filled for about eighteen months. Still drawing on Seward's good will, he next served (1867–68) as chargé d'affaires in Vienna. Then, even with his patron out of the cabinet, he landed another "fat office," as he put it, as secretary of the legation at Madrid (1869–70). For four of the five years between 1865 and 1870, he served the United States abroad. These were his real student days, Tyler Dennett observed, for on the Continent he matured and made up, at least in some measure, for the intellectual awakening that did not take place at Brown (p. 21). William Roscoe Thayer stated flatly that Hay returned from Paris what he had always wanted to be, "a cosmopolite" (I, 244).

No one could deny Hay's surface polish, world-weary cynicism, and charm as a conversationalist, but there is no evidence he experienced an intellectual awakening anywhere or ever ceased being a provincial. Beyond doubt he was conformist to the core, a dedicated courter of power and respectability, and hence a steadfastly conventional white Protestant American. One of his diary entries, for instance, recorded his impressions of the ghetto in Vienna. From the Judenplatz rolled

> an endless tide of Polish Jews. . . . These squalid veins and arteries of impoverished and degenerate blood are very fascinating to me. I have never seen a decent person in these alleys or on those slippery stairs. But everywhere stooping, dirty figures in long patched and oily black gabardines. . . . A battered soft felt hat crowns the oblique, indolent, crafty face, and, what is most offensive of all, a pair of greasy curls dangle in front of the pendulous ears. This coquetry of hideousness is most nauseous. The old Puritan who wrote in Barebones' time on the "Unloveliness of love locks" could here have either found full confirmation of his criticism or turned with disgust from his theme. [WRT, I, 292–93]

So ingrained a Yankee was Hay, old John Endicott could have counted on him to help shear the Lord of May's lovelock and have found in him an eager, all-around recruit for his war against "the wearing of long haire after the manner of Ruffians and barbarous Indians" (see p. 26).*

Hay fancied himself a Christopher Newman, Henry James's self-made tycoon in *The American* (1877), a fundamental self-misidentification that revealed more about his longings than his circumstances. He was always someone else's man, a lifelong secretary, first as Lincoln's private helpmate and then as Seward's protégé. From 1870 to 1875 he was Whitelaw Reid's editorial assistant on the *New York Tribune*. In 1874, at the age of thirty-seven, he married Clara Stone of Cleveland, and the following summer he was at work in the office of Amasa Stone, bridge-builder, railroad magnate, Rockefeller associate, Western Union Telegraph Company director, and generous father-in-law. Stone gave Hay and his daughter financial security and a new mansion on Cleveland's Euclid Avenue. In 1879 Hay became Rutherford B. Hayes's assistant secretary of state, in which position we met him when he entertained John Fiske so handsomely during the latter's command appearance in Washington. It was during these years too that Hay and Adams and Clarence King, along with Clara Hay and Marian Adams, established their Five of Hearts circle. In 1881 Hay returned to Cleveland to work on the Lincoln volumes he had undertaken with Nicolay, his joint guardian of the Great Emancipator's tradition. In 1883 Amasa Stone committed suicide, leaving Hay a made millionaire. Model robber baron Jay Gould made him a director of the Western Union to take his father-in-law's place.

In 1886 the Hays escaped Cleveland to move into their new Washington home designed by Henry H. Richardson. It faced Lafayette Square and adjoined the house of Henry Adams, also by the same architect, at 1603 H Street. Thenceforth the friendship of Hay and Adams became still more intimate—when apart, the salutations of their letters suggested this closeness: "My Own and Onliest," for example, and "My Beloved" (WRT, II, 99, 100).† And by way of adoption as much an "improvised European" as the "*type bourgeois-bostonien*" Adams discussed in a since famous letter to Henry James (L, II, 414), Hay made numerous trips abroad

* By noteworthy coincidence, down one of those arteries of "degenerate blood" lived an eleven-year-old "Polish Jew" named Sigmund Freud, who would grow up to shed light one day on the inner demons of such individuals as Hay.

† Such salutations led Tyler Dennett daringly to detect "a note of almost effeminate dependence" in their relationship. He did not pursue the matter, no doubt sensing that his observation led to very deep waters indeed. Some day, and not here, obviously, those depths should be explored. At all events, Dennett's formulation bristled with ambiguities of its own. Even in the United States in the late nineteenth century, was *dependence* especially or peculiarly feminine? Or neither?

during this period, ostensibly for his health but frequently for solace from "*civilisation américaine.*" Were it not for family cares, he wrote Adams in 1887, "I would gladly wander on anywhere a mule or canoe could bear us and search till death for the garden of Eden or the fountain of eternal wit, or any other thing we were sure not to find." But Clarence King helped him find a desert island of twelve hundred acres deep in the woods on Lake Sunapee in New Hampshire. In 1890, work on his hideaway there kept him from joining Adams and La Farge, as we have seen, on their jaunt to the real Crusoe islands of the South Seas.

2

That same year appeared the ten volumes of Nicolay's and Hay's *Lincoln*—they had been running as a serial in the *Century* since 1886—and they added luster to Hay's reputation as the poet laureate of the Republican party, the foundation for which had been laid two decades earlier. Then Hay had burst upon the literary scene with a series of periodical articles, *Tribune* editorials and poems, a book of verse, another of essays, followed by, a few years later, a novel.

One of his more interesting articles appeared in the *Atlantic Monthly* of December 1869 (XXIV, 669–78). Titled "The Mormon Prophet's Tragedy," he might better have called it "The Mormon Prophet's Just Deserts": Joseph Smith and his band of "liars and hypocrites" had settled in Southern Illinois, where they practiced polygamy, led "lewd lives" of "open and cynical licentiousness," and kept all of Hancock County—in which the Hay family had also resettled—"in a state of unwholesome excitement" for four years. Finally in 1844, local initiative being what it was, the good citizens formed a militia regiment to carry out the public will: "Resolved, that we will forthwith proceed to Nauvoo and exterminate the city and its inhabitants." Along with a number of leading men who had rallied "at the tap of a drum," Hay's father went along as regimental surgeon, though he dropped out when Governor Thomas Ford ordered the irregulars to disband. Many ignored the order, however, went on to Carthage where Smith was imprisoned, and shot him down in the yard outside the jail. A jury understandably acquitted the mob leaders, for they were "good citizens, educated and irreproachable, who still live to enjoy the respect and esteem of all who know them"—and the writer was one of those who knew them: "In the mind of any anti-Mormon there was nothing more criminal in the shooting of Smith than in the slaying of a wolf or panther." The denigration and dehumanization of those who were different, the pathological reaction to other sexual mores, the ominous will to exterminate—everything was there save color, and that line apparently could be passed in times of acute crisis, of "unwholesome excitement." It was as though the drum were tapping for the Black Hawk War a dozen years earlier. Hay's essay echoed James Hall's tortured justifications of that slaughter unmistakably, if unintentionally, as when he applied the judge's Indian-hating formula: Hay placed the "civilized" frame of his understanding around an act that would otherwise have appeared out-and-out murder.

Pike County Ballads and Other Pieces (1871) laid the foundation for Hay's early fame. James Hall had pioneered in introducing the vernacular of the Old Northwest into his sketches and tales, as when he had the trapper Hugh Glass "feel right *peart*," though the italics underlined the judge's half-apology for daring to bring such language into literature (see p. 209); John Neal and James Russell Lowell in the East, and Bret Harte in the West had experimented with dialect; and shortly Mark Twain would bring that of the Mississippi Valley to fruition in *The Adventures of Huckleberry Finn* (1883). But Hay's slight pieces swept the country when they appeared.

In "Banty Tim," the most intriguing of the dialect poems, Hay seemingly took a strong stand against the racism of the Mississippi Valley, pervasive everywhere and strong in his own hometown of Warsaw, formerly called Spunky Point. As the ballad began, Sergeant Tilmon Joy, accompanied by a black "boy," returned after Appomattox only to be confronted by "The White Man's Committee of Spunky Point, Illinois":

> I reckon I git your drift, gents,—
> You 'low the boy sha'n't stay;
> This is a white man's country;
> You're Dimocrats, you say;
> And whereas, and seein', and wherefore,
> The times bein' all out o' j'int,
> The nigger has got to mosey
> From the limits o' Spunky P'int!

Understanding all this, the sergeant still told the White Man's Committee to go to hell. Why? Because he had been wounded one ungodly day at Vicksburg and left lying on the battlefield with Rebel bullets whizzing all around:

Till along toward dusk I seen a thing
  I couldn't believe for a spell:
That nigger—that Tim—was a crawlin' to me
  Through that fire-proof, gilt-edged hell!

That "boy" Tim had been enough of a man to shoulder the sergeant and take him behind the lines, "His black hide riddled with balls." That was the reason "ef one of you tetches the boy,/He'll wrastle his hash to-night in hell,/Or my name's not Tilmon Joy!" Tim might stay because he had saved the white sergeant that day and refused to leave him, just as William Gilmore Simms's adhesive black Hector saved the life of his master in *The Yemassee* and refused to leave him (see p. 141). And just as there had been room for black Hectors in the bottom caste of South Carolina, so room had to be made at the bottom for black Bantys in Illinois. But once there Tim remained a "nigger" to all concerned, to Tilmon Joy and to his creator, for Hay's use of the epithet in his diary and letters made certain that in this respect at least the sergeant's language was Hay's own (e.g., WRT, I, 161; KJC, p. 48). In fine, Hay merely countered the exclusionist racism of the White Man's Committee with his own paternalistic or accommodationist racism. Just so long as blacks came a crawlin', they could come.*

In 1876 Hay was enraged by the defeat and death of General George Armstrong Custer and his 7th Cavalrymen in the Battle of the Little Bighorn. At the end of the decade he commemorated the event in "Miles Keogh's Horse," a bitter poem about the mount Comanche, sole survivor of "that scorpion ring," and the pathos of his being made to parade riderless with the regiment (CPW, pp. 77–80). Or so the order was issued and it lifted a bit the gloom "That shrouds our army's name,/When all foul beasts are free to rend/And tear its honest fame. . . ." Asking that it be printed anonymously, Hay explained to Whitelaw Reid that it was "a well-intentioned poem, calculated to make people kill Indians. I think H. H. ought to be a Ute prisoner for a week" (January 27, 1880; KJC, p. 232n). The initials referred to Helen Hunt Jackson, whose *Century of Dishonor* would come out in 1881; the playful suggestion was that this Indian-lover be thrown to the Utes, who were allegedly in the habit of raping white women.

At the height of the great railroad strikes of 1877, Colonel Hay

* In 1879, while raising funds for black migrants under attack by white settlers in Missouri, Kansas, and Nebraska, Hay argued for this accommodation in revealing terms. They might be allowed in without danger, he said reassuringly, for blacks were generally "the most domesticated race in the world, a people as devoted to their old plantations as so many cats" (KJC, p. 84).

wrote Amasa Stone of their gloomy prospects: "The very devil seems to have entered into the lower classes of working men, and there are plenty of scroundrels to encourage them to all lengths" (WRT, II, 5). The "scoundrels" in Hay's letter were the industrial counterparts of those frontier seducers or exciters of the natives, and this was the old "outside agitator" refrain: Like Indians before they had been molested or tampered with, workers were content and childishly incapable of self-assertion. The very devil had to stir them into the disorders that menaced men of property: men like Jay Gould, who was currently raiding railroad treasuries and fighting labor unions; Hay's father-in-law, who had left him to tend the shop; and, thanks to Stone, himself. "So an attack on Property," observed William Roscoe Thayer in familiar words, "becomes an attack on Civilization. . . . Those riots of 1877 burnt deep into Colonel Hay's heart" (II, 8, 6). The consequence was *The Bread-winners* (1883), his perfervid, albeit anonymous defense of "civilization."

As the novel opened "a gentleman" sat in front of the library fireplace in one of the great houses on Algonquin Avenue in Buffland—read Euclid Avenue along which millionaires lived in Cleveland. The figure was Mr. Arthur Farnham, who was about Hay's age and who actually looked pretty much as the colonel saw himself. Farnham's face was "wholesome"—that key word again —and like his hands, "of one delicately bred." To be sure, his creator had not been that delicately bred, so this was not made an absolute prerequisite for gentility: Like Hay, Farnham was "one of those fortunate natures, who, however born, are always bred well, and come by prescription to most of the good things the world can give." This daydream-self sat in front of the dying embers in the grate and mused on his childhood, the early death of his parents, studies at St. Cyr in France, return for service in the Union cavalry, and then Indian-fighting with the regulars in "the wide desolation of the West," where he met the sister-in-law of a brother captain, "a tall, languid, ill-nourished girl of mature years."

Farnham had married this wondrous creature because "she was the only unmarried white woman within a hundred miles, and the mercury ranged from zero to −20° all winter." Still, for some reason even the temperature did not bring about their union until that spring when "he seemed to have lost the sense of there being any other women in the world" (p. 51). He placed his hero in the Black Hills about the time Custer was there but prudently removed him from Sioux country just in time but not before the "death of his wife, on a forced march one day, when the air was glittering with alkali, and the fierce sun seemed to wither the dismal plain like the vengeance of heaven." That scene bore some resemblance to the moonscape John Fiske was to describe (see p. 232), and surely Hay was no less alienated from the land. Nevertheless, after his hero's resignation from the army, he still professed to look "longingly back upon the life he had left, until his nose inhaled again the scent of the sage-brush and his eyes smarted with alkali dust. He regretted the desolate prairies, the wide reaches of barrenness accursed of the Creator, the wild chaos of the mountain cañons, the horror of the Bad Lands, the tingling cold of winter in the Black Hills" (p. 8). It had been, in short, an eminently regrettable, howling wilderness.

Once Farnham had returned to Buffland to "take charge of the great estate of which he was the only heir," Hay flicked on the Light Show and had him fall in love, really this time, with the daughter of Mrs. Belding, the widow who lived nearby on Algonquin Avenue. Young Alice Belding was the avatar of Anglo-Saxon good looks and refined virtues. The "tawny gold" of her hair (p. 42), its "yellow lustre" (p. 136), made her describably beautiful:

> Alice rose from the piano, flushing a pink as sweet and delicate as that of the roses in her belt. She came forward a few paces and then stopped, bent slightly toward him, with folded hands. In her long, white, clinging drapery, with her gold hair making the dim room bright, with her red lips parted in a tender but solemn smile, with something like a halo about her of youth and purity and ardor, she was a sight so beautiful that Arthur Farnham as he gazed up at her felt his heart grow heavy with an aching consciousness of her perfection that seemed to remove her forever from his reach. [pp. 318–19]

Naturally, she merely seemed beyond his reach and just as naturally Farnham himself was forever beyond the reach of Maud Matchin, who had set her cap at him and whose "dark hair was too luxuriant to be amenable to the imperfect discipline to which it had been accustomed" (p. 10). Maud had "thick masses of blue-black curls" (p. 114) and had foolishly sought to parlay a high-school education into a marriage above her station.

Yet this lively Light Show burned out when Hay overloaded the circuit. In the first place, he used colors to establish degree of *sexuality* and to sort women into the customary categories: fair versus dark, sexless pink innocence versus sensual blue-black sinfulness, delicate Alice versus lusty Maud. Second, his colors indicated *class,* Alice's elevated origins on millionaires' row and Maud's earthy origins in the home of Saul Matchin, "a carpenter of a rare sort . . . a good workman, sober, industrious, and unambitious. He was contented with his daily work and wage" (p. 19), in sharp contrast with his ambitious daughter, who should have gone quietly into domestic service and been content to marry Sam Sleeny, her father's helper, their boarder, and her suitor. But at exactly this point Hay crossed his class-struggle wires, for Sleeny was so quintessentially Anglo-Saxon as to be, physically at least, the male counterpart of Alice. Sleeny was a fine figure of a man, with a tall, powerful build, "frank blue eyes" and "yellow hair and beard" (p. 303). Lest someone mistake him for an aristocrat and propose a match with Alice, Hay repeatedly emphasized that Sleeny was a dumb worker, "not quick either of thought or speech" (pp. 66, 99), but this hardly uncrossed his lines. Still, Hay pressed on, for he had a third use for colors that made Sleeny's blondness a necessary contrast with the Celtic complexion of his villain. In *The Bread-winners* colors also indicated *caste*.

One glance at Andrew Jackson Offitt sufficed to persuade a fair-minded witness he was not one of those fortunate natures to whom good things come by prescription. A dark "apostle of labor" who wrote for the *Irish Harp*, he was one of those "tonguey vagrants and convicts" who wandered "from place to place, haranguing the workmen, preaching what they called socialism, but what was merely riot and plunder" (p. 215). In Buffland he had organized the secret Brotherhood of Bread-winners, a band of troublemakers and shirks, charlatans and surly brutes (p. 82). One glance at their organizer showed him to be a "scroundrel" after what the rich and well-born had. And when he treacherously attacked and wounded Farnham with a hammer during the strike, he showed not "a particle of regret or remorse" as he counted his loot:

> Money is, of course, precious and acceptable to all men except idiots. But, if it means much to the good and virtuous, how infinitely more it means to the thoroughly depraved—the instant gratification of every savage and hungry devil of a passion which their vile natures harbor. [p. 271]

The yellow-haired Sleeny, who had been temporarily corrupted by this sinister agitator, returned to his senses and in a rage killed him the way a backwoodsman would kill a panther or a snake or an Indian. The jury returned a verdict of "justifiable homicide" and, when the judge made them reconsider, came back to acquit Sleeny on the grounds of "emotional insanity." Everyone knew, however, that Sleeny had been perfectly sane and had performed a public service, so Hay celebrated the jury's verdict more fulsomely than he had that in the Joseph Smith case: It was "thoughtful and considerate," and to "the lasting honor and glory of our system of trial by jury" (p. 307).

All the formulas of the Indian-hating tales and novels were present in *The Bread-winners,* but Hay still could not sin by deputy and get off scot-free. Though there was much speculation about the identity of the anonymous author and praise for his cleverness, Hay's novel seemed to some critics at the time what it was, a vicious attack on labor unions and the Irish (TD, pp. 111–15). The time-worn techniques of Indian-hating could not be applied successfully to a strike novel set in the Cleveland of 1877, in part because within the larger society it was simply much more difficult to *de*humanize a white agitator than a red prophet. Try as he might, Hay could not persuasively make Offitt into a true merciless savage or all Irishmen into "white niggers," with the appropriate devilish passions. When he said Offitt was a tiger, a panther, a snake, "a human beast of prey," his lines merely read as an exercise in reactionary ethnocentrism. Think how much easier his task would have been had his villain been one of Simms's Yemassees, one of Bird's Shawnees, or one of the Seminoles the real Andrew Jackson chased through the Florida swamps. Then his villain would have been truly *outside* white society, as the Celts, despite the Anglo-Saxons, were not; at most, had any of the above Indians been "assimilated," then his red villain would have been identifiably, like black Banty Tim, in one of the bottom castes of a racially stratified society.

Hay's Light Show flared out because what had worked more or less for novelists across the color line would not work within it. It was not enough for him to have Offitt looking like Hay's description of a resident of the Viennese Jewish ghetto, "dark-skinned and unwholesome looking," with an "oleaginous" expression, a mustache "dyed black and profusely oiled," and a face "surmounted by a low and shining forehead covered by reeking black hair, worn rather long," and wearing a black hat and "threadbare clothes, shiny and unctuous" (pp. 74–75, 86, 166; cf. 303). To turn the tables on his creator, all this dirty, greasy walking delegate had to do was wash off some of the grime, shave off his mustache or dye it blond, cut and perhaps dye his hair also, bathe thoroughly, put on a new suit in the current Knickerbocker Club fashion, and run for mayor of Buffland—as he might have done, had Hay not killed him off. The conventions of Indian-hating could not be so directly or so clumsily applied to labor-baiting; for while workers could be called, as they were, "brutes" and "savages," these were simply not the same epithets readers accepted when applied to Banty Tim or to the Indians Mordecai Lincoln stalked so remorselessly. Class and caste were related but *distinct* phenomena, and the disaster of Hay's *Bread-winners* illustrated this home truth.

3

John Hay's novel was an extension of his politics by other means; it was in fact a tract for the times that contained all the essentials of his creed. As he summed that up for a Cleveland audience in 1880, American voters had the duty of living up to standards set by the preeminent race of men from whom they were descended. These forefathers had "believed in order, decency, sobriety; in reverence for all things reverend, for religion and for law" (TD, p. 103). Now himself a great property holder, he stood for sound money and a high tariff. Anonymously a celebrant of murder in his fiction, publicly he was above all else one of the "law-and-order men" who had rallied "to the party which is unquestionably the law-and-order party" (WRT, II, 4). That party was the GOP; it claimed his unswerving allegiance and could count on the support of "the majority of the better sort," as opposed to the Democrats who had deliberately drawn in "the worst elements" of the population (KJC, p. 54). Foremost among the better sort, in his eyes, was James G. Blaine of Maine, the corrupt "plumed knight" he idolized and repeatedly supported for the presidency (TD, pp. 125–30).

For political assets Hay could offer the Republicans his association with the sanctified Lincoln, his reputation as a realist in letters, his talent for invective, his party regularity and unchallengeable patriotism, and, not least, his fortune inherited from Amasa Stone. He could afford to play at making presidents and backed a winner when he joined Mark Hanna and others in paying off William McKinley's debts and in booming the major as "the Advance Agent of Prosperity." Years later Henry Adams noted that Hay had won on this third gamble, after backing Hayes and James A. Garfield and laying out money on John Sherman and Blaine along the way: "I would give six-pence to know how much Hay paid for McKinley. His politics must have cost!" (L, II, 480). They did—about $10,000 for McKinley—but defeat of William Jennings Bryan would have been cheap at twice that price. The Democratic party had been captured by "unclean spirits," Hay declared in his widely circulated pamphlet, "The Platform of Anarchy"; the very devil seemed to have entered into the dangerous classes and their agitators, such men as Bryan, John Peter Altgeld, Tom Johnson, Eugene Victor Debs, Tom Watson, et al. It was 1877 all over again, with revolution imminent—Hay actually wrote Whitelaw Reid that the uprising was "the revolt of Caliban" (KJC, pp. 60–62).

After Caliban had been driven back to the farm and the factory, McKinley paid off by naming Hay minister to the Court of St. James's, the prize he had been after all along. Moreover, he had the good fortune to arrive at his post in 1897, the year of Queen Victoria's Diamond Jubilee and, for all the imperial pomp and circumstance, the year her ministers searched more anxiously than ever for allies to protect the British Empire against the rise of Germany. The latter, said Henry Adams, had effected "what Adamses had tried for two hundred [years] in vain—frightened England into America's arms." That was what he meant in the epigraph that heads this chapter.

Actually the sometimes estranged cousins fell into each other's arms out of mutual need. Great Britain needed a friendly hand almost everywhere, while the United States needed a free hand in the Caribbean and the Philippines. Ten days after McKinley asked Congress for a war resolution against Spain (April 11, 1898), Hay called the relationship he was working to restore "A Partnership in Beneficence." In this important address he said that

> All of us who think cannot but see that there is a sanction like that of religion which binds us to a sort of partnership in the beneficent work of the world. Whether we will it or not, we are associated in that work by the very nature of things, and no man and no group of men can prevent it. We are bound by a tie which we did not forge and which we cannot break; we are the joint ministers of the same sacred mission of liberty and progress, charged with duties which we cannot evade by the imposition of irresistible hands. [Add, pp. 78–79]

To the Royal Society he exulted over the "bonds of union among the two great branches of our race" and affirmed "the object of my mission here is to do what I can to draw close the bonds that bind together the two Anglo-Saxon peoples" (Add, pp. 84, 85).

Hay's reference to "irresistible hands" was John Fiske's Manifest Destiny revived; Fiske had almost two decades earlier held that "the two great branches of the English race have the common mission of establishing throughout the larger part of the earth a higher civilization and more permanent political order than any that has gone before" (see p. 240). Hay had attended a series of Fiske lectures, you may recall, and then had acclaimed "Manifest Destiny" when the historian presented it to the Hayes administration; all of which tempts one to conclude that finally here is a clearly etched illustration of how ideas can ascend from the individual to the policy-making level.* But as in the case of Fiske and Josiah Strong, the question of who drew on whom is largely beside the point. No man and no group of men, as Hay would say, were responsible for the imperial Anglo-Saxonism that by the end of the century covered Washington and London like a ground fog or, better, filled the air like smog. In the quotations just cited Hay not only served up Fiske's leftovers, he anticipated Rudyard Kipling's "The White Man's Burden" (1899) and directly moved Joseph Chamberlain, the British colonial secretary, to rhapsodize in Birmingham over "an Anglo-Saxon alliance": "What is our next duty?" Chamberlain asked. "It is to establish and to maintain bonds of permanent amity with our kinsmen across the Atlantic. There is a powerful and generous nation. They speak our language. They are bred of our race" (WRT, II, 169n). All in the family, then, were these bonds across the Atlantic, and in Hay's words, they had a near religious sanction. *Racism* bound kinspeople not only across space but across time, over the centuries binding together Puritan New Israelites with *fin de siècle* "new" imperialists.

With a sense of elation over the glorious prospects, Hay wrote McKinley from London: "The greatest destiny the world ever knew is ours" (February 20, 1898). Two months later commenced the conflict Thomas Jefferson had predicted in June 1823, when he recommended that the United States not go to war immediately over Cuba, since "the first war on other accounts will give it to us" (see p. 114). Seventy-five years of watchful waiting represented exemplary patience and decorum, though in 1898 the United States did not go to war on other accounts but over Cuba itself, at least ostensibly. Moreover, it went in handicapped by a self-denying ordinance, an amendment of the war declaration by

* In his able recent study *John Hay* (1975), Kenton J. Clymer concluded that "Hay does not appear to have been directly influenced by Darwin, Herbert Spencer, or their popularizers. He was personally acquainted with the popular Social Darwinist, John Fiske, but in virtually none of his writings or private letters does Hay refer to Darwin, Spencer, or evolution" (p. 67). But Fiske believed Hay to be directly influenced by him, and Hay's stress on a racial "lust of land" and other evidence in *Lincoln*, along with that above, bear him out—Fiske's early relationship to Hay was more that of an instructor than of a mere acquaintance. Fiske and George Bancroft were Hay's historians before he became the best friend of Henry Adams, a better historian than either.

Senator Henry M. Teller of Colorado that disavowed any intent to annex or control the island—it took another amendment, that of Senator Orville H. Platt of Connecticut (by rider to the Army Appropriations Act of 1901 and by a treaty of 1903), to make the "Pearl of the Antilles" politically a protectorate and economically a dependent. Cuba was thus "pacified" in less time than it took General Jackson to invade Florida and take the Spanish forts eight decades earlier: On July 26, 1898, Spain requested peace terms, and the following day Hay gave the conflict its memorable name. "It has been a splendid little war," he wrote Theodore Roosevelt; "begun with the highest motives, carried on with magnificent intelligence and spirit, favored by that Fortune which loves the brave" (WRT, II, 337).

Now, these "highest motives" for intervening in Spanish Cuba resembled those John Quincy Adams had advanced to justify the invasion of Spanish Florida (see pp. 109–11). Placed side by side, Adams's White Paper of 1818 and McKinney's War Message of 1898 have haunting similarities. As with the "derelict province" of Florida, Spanish officials had allowed revolutionary Cuba to become a colony of "barbarities, bloodshed, starvation, and horrible miseries." In both colonies there had been the same "wanton destruction of property," the same chaotic conditions that invoked the "necessities of self-defence" for the United States; or, as McKinley put it, "the present condition of affairs in Cuba is a constant menace to our peace." As in 1818, this "intolerable condition of affairs which is at our doors" * cried out for relief and was no longer endurable. So did McKinley call for "enforced pacification" in the same terms as Adams and, like his predecessor, spoke and acted "in the name of humanity, in the name of civilization, [and] in behalf of endangered American interests," all of which came down to the same goal of protecting, in Hay's words, the "sacred mission of liberty and progress" the United States represented. In 1818 Adams had undertaken to demonstrate to Spain that the United States could "as little compound with impotence as with perfidy"; in 1898 McKinley observed that the Spanish government had been unable even to assure the safety of the battleship *Maine* "in the harbor of Havana on a mission of peace and rightfully there." Such impotence had therefore moved him at long last "to secure in the island the establishment of a stable government, capable of maintaining order and observing

* Before he left the White House, Grover Cleveland had implied (December 7, 1896) that the United States might have to intervene in Cuba since it seemingly was in the very doorway: "It lies so near to us," he declared, "as to be hardly separated from our territory," a view of an island neighbor to the south likely to prove fatal had he tried to swim over from Key West.

its international obligations, insuring peace and tranquillity and the security of its citizens as well as our own."

Since McKinley never forgot them for a moment, we should remember those "endangered American interests" in this mix of "highest motives," along with all the others: the "perilous unrest among our own citizens" provoked by the Cuban revolution; the interest of such navalists as Alfred Thayer Mahan in making the Caribbean a U.S. lake; the economic pressures backed up by the prolonged depression of 1893; the growing conviction within influential circles that factories and farms produced more than Americans could consume; the end of the continental land frontier and the drive for a new frontier of overseas markets; and, though it was notably absent during the Indian wars just concluded with the massacre at Wounded Knee, humanitarian distress over Spanish counterinsurgency tactics. All these and other pressures for forcible intervention may be given their due and still not provide more reason for assuming McKinley did not mean what he said in 1898, did not in his heart of hearts believe in *order* and *stable government*, than there had been for disbelieving Adams in 1818. U.S. statesmen have been traditionally loath to countenance impotence.

An incident in the history of American expansion south and west, the Splendid Little War was declared, waged, and concluded while Hay was still in London. Then, in the late summer of 1898 he was called back to Washington to become secretary of state, the position he was to occupy under McKinley and under Theodore Roosevelt until his own death in 1905. By that time his reputation was established as a great statesman, whose Open Door Notes ranked with Washington's Farewell Message, the Monroe Doctrine, and other foundation stones of U.S. foreign policy. And since Hay conducted State Department affairs as the disciple of John Quincy Adams and of William H. Seward (TD, p. 263), his attempts to implement the imperial vision of these formidable predecessors demonstrated further the seamless continuity of fundamental attitudes and assumptions.

Beyond doubt Secretary of State Adams would have found President McKinley's War Message perfectly understandable, I have suggested, and his reasons for intervention immediately acceptable. McKinley's refusal to extend recognition to the Cuban insurgents Adams would have taken for granted, for he had no more respect for Latin Americans than John Hay, who called them "dagoes" (TD, p. 264). Not only was North America the proper dominion of the United States, Adams had held, but all the West Indies were "natural appendages" and Cuba was like a "ripening apple," bound to fall one day—he would only have been surprised that it had taken most of the century for the stem to snap. Adams had been very careful, in turning down Britain's overtures and in persuading Monroe against any self-denying ordinance, to have the latter's doctrine leave the United States free to grab former and present possessions of Spain in "this hemisphere," the boundaries of which were as elastic as those of settlers on Indian land. Moreover, directly anticipating John Hay's Open Door, Adams held in 1842, when he was no longer in the executive branch, that China's exclusion policy was immoral because it violated the Christian command to "love thy neighbor," blocked trade, and was therefore, "an enormous outrage upon the rights of human nature, and upon the first principles of the rights of nations." *

Secretary of State Seward, Hay's patron and other master, was himself a disciple of John Quincy Adams, whom he venerated above all men as teacher and friend. To update and systematize Adams's imperial vision, Seward took the long view, for at some point he must have asked himself the really antecedent question: What was Columbus doing in 1492 if not bumping into outlying islands of the land mass that blocked his passage to India? That question was implied, however Seward had formulated it, by his contention that even the "discovery" of North America and the subsequent "organization of society and government" thereon, momentous as these events had been, were still "conditional, preliminary, and ancillary" to Columbus's true goal and that of European expansionists since. Asia, fountainhead of all "civilization," had been the real destination all along. So were the last four hundred years only the visible tip of an expansionism that went back three millennia, with empire making "its way constantly westward, and . . . it must continue to move on westward until the tides of the renewed and the decaying civilizations of the world meet on the shores of the Pacific Ocean."

With this firm hold on history's motive power, Seward worked out a tight itinerary for his compatriots. From a continental base of power that would surely include Canada and Latin America, the United States had to advance along two routes: south and west through the Caribbean, via a chain of island bases, to the isthmus and thence, via canal, into and over the Pacific; more directly west

* Quoted in Richard W. Van Alstyne's *The Rising American Empire* (Chicago: Quadrangle, 1965), pp. 171–72. In the same context Van Alstyne also quoted John K. Fairbank's *Trade and Diplomacy on the China Coast* (1953). It seems that trading and not governing was the policy of Britain in China since the Opium War of 1844: "This was the spirit of the Open Door, a British doctrine long before John Hay voiced it."

to California, via transcontinental railroad, and thence across by ship to Hawaii, to farther Pacific way stations, and finally to Asia for a triumphant completion of the voyage Columbus started. So citizens of the empire could exploit that illimitable market once they arrived, Seward vigorously supported the Open Door, "this American policy," according to Walter LaFeber, that "was nothing new, dating back to the most-favored-nation clauses in the first American–Chinese treaty in 1844."

Seward did all he could. He helped in the construction of the transcontinental railroad by providing for the importation of Chinese contract laborers (Burlingame Treaty, 1868), those "Celestials" Robert Louis Stevenson saw on his way to San Francisco. He gave his backing to abortive U.S. schemes to build an isthmian canal. In 1867 he purchased Alaska, with its Aleutian finger pointing toward the target. He tried to buy the Danish Virgin Islands and the Spanish islands of Cuba and Puerto Rico, tried to annex Haiti or Santo Domingo or both, and advocated annexation of Hawaii. But as he himself said, "no new national policy deliberately undertaken upon considerations of *future advantages* ever finds universal favor when first announced" (my italics). He knew as well that an industrialized, steam-propelled empire moved more rapidly than one that was sail-driven and horse-drawn and knew what he was proposing shot ahead of where the country was. For all these advantages to be realized, the United States still had to lick its Civil War wounds, complete the basic industrial plant, mop up resistance of the Plains Indians against final continental consolidation, and have a political and economic climate warm enough for the plunge into overseas colonies. Pending that time, Seward left his countrymen a buoy to mark their course: Midway Island, over one thousand nautical miles west of Hawaii, was acquired by the United States in 1867.

Secretary of State Hay returned from London in 1898 to a windfall of ripe apples blown down by the Splendid Little War. Let us follow his efforts to pick them up along Seward's two routes to the Orient, departing from the continental collossus:

North America. The discovery of gold in the Klondike intensified the dispute over Alaska's southeastern boundary. Hay had long supported bringing Canada into the Union and in the early 1890s had joined the Continental Union League with Theodore Roosevelt, Andrew Carnegie, and other annexationists; but that other continental colonizing power continued to elude their grasp, as it had that of all their compatriots since John Adams and Thomas Jefferson. Hence Hay had to content himself with preserving what he could of Seward's purchase; he worked out a *modus vivendi* for the boundary on October 20, 1899—it took Theodore Roosevelt's "big stick" threat of military force and the favorable findings of a commission to satisfy U.S. claims in 1903.

Virgin Islands. Hay tried to buy the islands but haggled over Denmark's price of $5,000,000, offered instead $4,250,000, as though he were dickering for a piece of property with Amasa Stone's money, and had the disappointment of seeing the deal fall through. During the course of the negotiations he demonstrated he had not yet learned the finer points of "the large policy"—when the United States purchased the islands in 1916, amid exaggerated alarms over German threats to the area, the price had gone up to $25,000,000.

Cuba and Puerto Rico. The islands had already fallen into U.S. hands by the time Hay returned to Washington.

Isthmian Canal. Hay abrogated British canal claims under the old Clayton–Bulwer Treaty (1850) by the Hay–Pauncefote Treaty of 1901. Working under the increasingly heavy hand of Theodore Roosevelt, he helped bully the Colombians into parting with their route by the Hay–Herrán Convention of 1903, when that fell through became for the one and only time in his life a "revolutionist" by backing the staged revolt in Panama, and promptly clinched unilateral U.S. control of the "liberated" zone through the Hay–Bunau-Varilla Treaty of 1903.

Hawaii. While still in London Hay had cabled the State Department in support of "prompt annexation [of] Hawaii before war closes as otherwise Germany might seek to complicate the question with Samoa or Philippine Islands" (KJC, p. 124). Hawaii was annexed July 7, 1898, before Hay's return, and was destined to be ruled by a missionary–sugar oligarchy for the next half-century.

Samoa. Hay embroidered Seward's master plan through partition of the country with Germany (December 1899), an agreement that extended U.S. sovereignty over several of the islands, including Tutuila and its harbor Pago Pago, destined to become a strategic naval base.

Philippine Islands. Under McKinley's direction but with his own full concurrence, Hay sent his famous wire (October 26, 1898) to the peace commissioners in Paris: "The cession must be of the whole archipelago or none." He responded angrily to the Filipino nationalists, holding them absolutely unable to run their own affairs and advocating ruthless suppression of the movement led by Emilio Aguinaldo: In June 1899 he cabled Jacob Schurman, chairman of McKinley's First Philippine Commission, that there was "no excuse for further resistance by the Filipinos and if it continues the President will send all the force necessary to suppress insurrection and establish the authority of the United States in the Islands" (KJC, p. 137).

China. With the Philippines holding the key to the markets of Asia, Hay seemingly became the age's master locksmith when

UNCLE SAM CATCHES THE RIPE FRUIT.

he coupled their acquisition with his famous Open Door Notes of September 6, 1899, and his Circular of July 3, 1900. The notes and circular undertook to ensure equal trading opportunities even within the various spheres of influence and the "territorial and administrative integrity of China." Hay pronounced the evasive, noncommittal, even negative responses he received "final and definitive" approval of his principles, and the American public, if not the European colonial powers and China, believed him.

Now, this stark outline of Hay's half-dozen years as secretary of state shows at a glance the massive continuity of his diplomacy with that of the past. *Nothing* he did or attempted was new if we except such embroidery as his acquisition of part of Samoa. On the diplomatic front, therefore, the "new" empire of 1900 deserved to be known as the Adams–Seward–Hay empire, with the efforts of other secretaries of state, including Henry Clay, Daniel Webster, William L. Marcy, Hamilton Fish, William M. Evarts, and Hay's old idol Blaine, thrown in for good measure. All that was new at the turn of the century were *the means* and politically favorable conditions for realizing Seward's "*future advantages.*"

The Adams–Seward–Hay imperial vision had one constant we have traced thus far and that was unforgettably expressed by Thomas Hart Benton in the middle of the Manifest Destiny 1840s. Speaking on the Oregon question in 1846, Benton put the issue for all nonwhite races as "civilization or extinction"; judging Indians inferior to Orientals, the senator maintained that the former had been destroyed so whites could get at the latter and, through conquest, commerce, and intermarriage, sting them into high "civilization." In like manner Seward had granted Orientals no more than "decaying civilizations" and in effect denied native peoples in between Occident and Orient any "civilization" at all. Seward's empire-builders still moved as inexorably through open

◀ *Uncle Sam Catches the Ripe Fruit,* from the *San Francisco Call,* January 29, 1893. Throughout the century, expansionists extended John Quincy Adams's "ripening apple" theory from Cuba to a variety of other countries without doing visible damage to the pleasing inevitability it posited. In the 1850s the *Alta California* described the Hawaiian Islands as "luscious fruit" about to drop in the American lap; four decades later—in a view given concurrent visual form in this cartoon—the U.S. minister to Hawaii pronounced it finally "fully ripe, and this is the golden hour for the United States to pluck it." A succession of golden hours seemingly struck when Hay joined McKinley's administration, though Canada and Mexico remained unplucked.

PORTO RICO
CUBA
PHILIPPINES
LADRONES
ARMY AND NAVY

space, across vacant seas and "unpeopled" lands, as had the conquerors of his continental base. And like the good disciple he was, Hay became livid over the rejection in Bogotá of his convention with Herrán: the Colombians, he said, were "greedy little anthropoids" (TD, p. 376). In his eyes the Filipinos bore no closer resemblance to men and women—Hay assuredly accepted Henry Adams's rejection of them as "the usual worthless Malay type" (L, II, 215). Hay harbored a special hate for Aguinaldo and, as though the Filipino general were a red prophet being seduced by sinister British agents, wrote Henry Cabot Lodge from London that the Germans had been courting Spain "& now they are trying to put the Devil into the head of Aguinaldo" (KJC, p. 127). And after all Hay had done for them with his Open Door Notes, even the Chinese were ungrateful wretches: "We have done the Chinks a great service," he wrote an adviser in 1903, "which they don't seem inclined to recognize" (KJC, p. 156). Hardly more than the Colombians did they resemble persons and deserve to be treated as such. Had China proposed the Open Door for the United States, Hay would have thought that truly preposterous—the protectionist secretary of state did not even consider extending his famous principles to the newly acquired possessions of Puerto Rico and the Philippines.

In the grand perspective, Hay was a product of family work going back even farther than his friend Adams said. U.S. empire-building went back beyond John Adams to the first Henry Adams and the earliest planters. That Hay was a product of this past made him in some ways more important for understanding the deeper currents of U.S. imperialism. With empty space not in front of the advancing empire but in himself, his career had a pathos that calls to mind Thoreau's comment about individuals who came to the grave only to find they had never lived—an ideologue of white nationalism, Hay had so little inner life he could properly call his own, even his anxieties and fears were stock by-products of his time and place. In his speeches all the patterns in the old national comforter stood out. Jefferson's Unavoidable Destiny, John

---

◀ *Uncle Sam's New-Caught Anthropoids*, from the *Literary Digest*, August 20, 1898. Published originally in the *Philadelphia Inquirer*, this cartoon had a legend supplied by an admiring John Bull: "It's really most extraordinary what training will do. Why, only the other day I thought that man unable to support himself." This new imperial relationship of Uncle Sam and John Bull illustrated what Hay liked to call "A Partnership in Beneficence."

Quincy Adams's Apparent Destiny, James K. Polk's Manifest Destiny, and Seward's Higher-Law Destiny became in Hay's hands "a cosmic tendency," a modification that suggested a dash of John Fiske. In a 1904 speech at the Louisiana Purchase Exposition in St. Louis, Hay invoked "that Providence which watched over our infancy as a people" and which, under whatever name, added the vast territory of Louisiana to the Union. It had all been determined or, if you please, predestined, for the secretary held that "no man, no party can fight with any chance against a cosmic tendency; no cleverness, no popularity, avails against the spirit of the age" (Add, pp. 248, 250). And it was left to Hay to declare, while commemorating "Fifty Years of the Republican Party," that the United States had advanced a "general plan of opening a field of enterprise in those distant regions where the Far West becomes the Far East" (Add, p. 284).

In these addresses the themes and attitudes of his youth in Indiana and Illinois, his years in the White House, his tours of duty abroad, even echoes of the Viennese ghetto, all came out time and again, as in his Louisiana centennial Light Show in celebration of the press for its "daily victories of truth over error, of light over darkness" (Add, p. 256). Hay's domestic and international assumptions and attitudes were two sides of the same coin, in part because what had formerly been considered "domestic" Indian-hating had recently been internationalized. In 1904, a year before his death, aware that he was about to step off that stage favored by Providence, Colonel Hay bid young Republicans

> as the children of Israel encamping by the sea were bidden, to Go Forward; we whose hands can no longer hold the flaming torch pass it on to you that its clear light may show the truth to the ages that are to come. [Add, p. 301]

CHAPTER XX

Insular Expert: Professor Worcester

> Too much kindness is very likely to spoil him, and he thinks more of a master who applies the rattan vigorously when it is deserved, than of one who does not. On the other hand, he is quick to resent injustice. . . . With all their amiable qualities it is not to be denied that at present the civilized natives are utterly unfit for self-government.
>
> —DEAN C. WORCESTER, *The Philippine Islands*, 1898

ONE MAY DAY in those distant regions where the Far West becomes the Far East, Commodore George Dewey sank the Spanish fleet in Manila Bay. On receiving word of this glorious victory, President William McKinley supposedly said he "could not have told where those darned islands were within 2,000 miles." At this point the history lecturer usually pauses to quote McKinley's later confession to a group of Methodist ministers: "The truth is I didn't want the Philippines and when they came to us as a gift from the gods, I did not know what to do with them." The president sought counsel from Republicans and Democrats to little avail, walked the floor of the White House night after night, "and I am not ashamed to tell you, gentlemen, that I went down on my knees and prayed Almighty God for light and guidance more than one night." Thereupon the supplicant was enlightened by the Almighty's unequivocal response "that there was nothing left for us to do but to take them all, and to educate the Filipinos, and uplift and civilize and Christianize them." And so, according to the lively presentations of countless lecturers, began "the great aberration."

The more commonplace truth is that Providence hardly figured at all in the U.S. bid for overseas empire. As McKinley's most recent biographer pointed out, the president was inclined from the first to keep the Philippines. Navy Department plans to attack Manila in case of war with Spain were decades old. *Before* February 25, 1898, McKinley had ordered Dewey to destroy the Spanish fleet in the Far East at the outbreak of hostilities, and that meant Theodore Roosevelt's unauthorized cable to that end and of that date merely implemented standing administration policy. *Before* word of Dewey's spectacular victory had been confirmed in Washington, McKinley had decided to send a supporting expedition of troops, and that meant the U.S. presence there was not a gift from the gods or even from the "large policy" expansionists clustered around Roosevelt, Henry Cabot Lodge, and Alfred Thayer Mahan. McKinley knew where his expedition was going and even without prayers why it was going there.

Less "weak" than Spanish Minister de Lôme suggested in his infamous pilfered cable, McKinley tenaciously and shrewdly nurtured the crowd's taste for annexation while bidding for its admiration. Whatever misgivings he may have had about its support evaporated in October 1898 on his off-year election tour of the Midwest and West. His major appearance was fittingly at the Trans-Mississippi Exposition in Omaha, once a staging area in the westward thrust of the continental empire. There, and also at whistle-stops between Ohio and the Dakotas, his references to Dewey and Manila raised hearty cheers and applause and that old battle cry, "Westward ho!" Thus audibly was the voice of the people, the voice he wanted to hear: "Westward ho!" to those darned islands.

At first the commander in chief had for his guidance only a map torn from a schoolbook, but shortly he could follow the action on a Coast and Geodetic Survey chart based on the exploratory findings of expeditions going back a half century and more to Charles Wilkes's detailed maps of the South Seas (see p. 129). Not as ignorant of geography as he let on, McKinley had much to learn about Pacific islands and was still doing his homework when just the man he had been looking for walked into his office.

In December 1898 Professor Worcester of the University of Michigan stopped by the White House on his way to do research in Europe. As he later put it, he wanted "to communicate to President McKinley certain facts relative to the Philippine situation which seemed to me ought to be brought to his attention" (PP&P, p. 87).* Among citizens of the republic, Worcester's qualifications were very nearly unique. Prior to 1898 he had spent almost four years in the Spanish Philippines collecting zoological specimens;

* For an explanation of the abbreviated references, see p. 515.

his article "Knotty Problems of the Philippines" had just appeared in the October issue of *Century Magazine* (LVI, 873–79); and concurrently Macmillan had published his *Philippine Islands and Their People*, a work so timely it had come to McKinley's attention and had already been reprinted twice. In the *Century* summary of his views, Worcester had strongly opposed withdrawal from the islands: "To do so would stultify ourselves in the eyes of the world." Furthermore, the "civilized natives" could hardly be left alone to work out their own salvation: "their utter unfitness for self-government at the present time is self-evident."

This "utter unfitness" of the Filipinos, a phrase that echoed down through the debates of the years to come, had already become the cornerstone of McKinley's policy for the islands. The president thought them no more fit to manage their own affairs than the Cubans, thought it no more "wise or prudent" to recognize their republic than he had the Cuban, and as with the Puerto Ricans, no doubt was relieved not to be handicapped in his dealings with them by a Teller Amendment (the fateful pledge the United States would leave Cuba to the Cubans). Currently in the process of repudiating the alliance between Admiral Dewey's forces and General Aguinaldo's revolutionary Filipinos, McKinley obviously liked hearing the latter were so unfit. He must have been impressed by the authoritative soundness of his visitor's views and by his invaluable firsthand information, for at the close of their discussion he surprised Worcester by asking him to become his personal representative in the Philippines.

By the time the professor had canceled his European trip and agreed to go, McKinley had decided to send a commission headed by Jacob Gould Schurman, the president of Cornell University; accordingly, in January 1899 the president made Worcester a member of this First Philippine Commission. In the spring of 1900 he appointed him to the Second Philippine Commission, headed by William Howard Taft. The only man to serve on both the Schurman and the Taft commissions, Worcester was a commissioner and secretary of the interior of the Philippines from 1900 to 1913. When Taft later became president he reportedly pronounced him "the most valuable man we have on the Philippine Commission." Joseph Ralston Hayden, another Michigan professor and Worcester's biographer, asserted his hero was "the only American who has achieved a secure and important place in history solely as a colonial administrator and statesman" (PP&P, p. v). Worcester's place in history now seems a little less secure than it did in 1929, and even then Hayden's assessment grievously neglected the claims of Colonel Thomas L. McKenney and other colonial administrators within the continental empire. Nevertheless, the leadership of the man he cast as "the foremost apostle of the United States to a backward Malay nation in the Far Eastern tropics" (PP&P, p. 5) continues to merit careful scrutiny.

2

Like John Hay, Dean Conant Worcester (1866–1924) was the son of a country doctor but came by his New England heritage directly and not by adoption. The Reverend William Worcester, founder of his family in America, had been part of the Great Puritan Migration to the Massachusetts Bay in the 1630s. In the 1770s Captain Noah Worcester, his great-grandfather, had been chairman of the Committee of Correspondence in Hollis, New Hampshire. The Reverend Samuel Austin Worcester, his uncle, was the missionary to the Cherokees of *Worcester* v. *Georgia* fame (1832; 6 Peters, 515–96). Doctor Ezra Carter Worcester, his father, practiced medicine in Thetford, Vermont.

The young Worcester grew up in Thetford, attended its academy, and then went to high school in Newton, Massachusetts. To save money he did not go to Dartmouth College, only twelve miles from his home, but went out to the University of Michigan, where he received his A.B. degree in 1889, became an instructor in animal morphology in 1893, and an assistant professor of zoology in 1895. Notwithstanding this junior faculty status, then, Worcester was still a graduate student in his early thirties when he met McKinley.*

Meanwhile Worcester had been in the right place at the right time, for his entire career hinged on his having been in the Far East ahead of the moving empire. In 1887–88 he had spent his junior year abroad in a collecting expedition to the Philippines organized by Chairman Joseph B. Steere of the Michigan Zoology Department. He returned to Ann Arbor to complete his senior year but after graduation found academic life decidedly dull: "Little as I suspected it at the time, the tropics had fixed their strangely firm grip on me during that fateful first trip to the Far East" (PP&P, p. 85). With Doctor Frank S. Bourns, with whom he had made the first trip, he set out to raise $10,000 for two more years in the

* Worcester never earned a doctorate, though he did receive an honorary Doctor of Science degree from his grateful alma mater in 1916. On his appointment to the faculty in 1893 he had married Nanon Fay Leas of Waterloo, Indiana, and the young couple had established a home in Ann Arbor, where they had two children, a girl and a boy.

tropics and somehow, for unexplained reasons, managed to persuade Louis Menage, a wealthy resident of Minneapolis, to put up the money. And since in 1887–88 they had been harassed by Spanish officials who suspected them of espionage, stirring up the natives, or prospecting for gold, Worcester and Bourns welcomed the good offices of President James Burrill Angell of the University of Michigan. Angell helped them work through the State Department for proper credentials and permission from the Spanish government for their tour of 1890–93.

At the time, Worcester maintained, he and Bourns had given no thought to the possibility that the data and photographs they were collecting might be of use to their government or of interest to the public. But "the rapid march of events" caught up with him and his companion. On hearing of Dewey's victory the professor requested leave from his university, sat down with the letters he had written and with materials he had shipped home, and quickly threw together *The Philippine Islands and Their People: A Record of Personal Observation and Experience.* Designed to capture the once-in-a-lifetime market, it was a scissors-and-paste production that credited a single source, Englishman John Foreman's *The Philippine Islands* (1892), for its "historical facts" and then served them up in generous, multipage helpings (e.g., pp. 450–54). A mix of travel book exploitation of strange and remote places and of pseudoscientific commentary on the no less strange and exotic peoples living therein, it closely resembled Thomas L. McKenney's much earlier blend in his *Tour to the Lakes* (1827). Reprinted five times within the year of publication, Worcester's book turned out to be more financially rewarding than the colonel's sketches, and no less revealing.

An early and unwitting revelation was why the tropics had him in their strangely firm grip. With positively Boone-like enthusiasm, Worcester had plunged out of the world of people. The climate of the tropics was "especially severe in its effect on white women and children," he wrote. "It is very doubtful, in my judgment, if many successive generations of European or American children could be reared there" (Phil, p. 67), and by American children he assuredly did not mean those with red, black, brown, or yellow skins (see also Knot, 875). He was drawn, therefore, to a world outside the family shelter and social relationships of his own species or kind and one in which his own kind most probably could not survive. Save for a few Spaniards and other Europeans in Manila and along their route, Worcester and Bourns traversed space occupied only by natives—the former might have appropriately subtitled his work *Isolated in the Islands.* The reader followed the two lone hunters as they traveled through the provinces shooting birds and reptiles and mammals, and collecting data on other mammals that lived in tribes. They coursed through the green jungles of these Crusoe islands in the service of science, career, future economic advantage, and present psychological gratification, that "strangely firm grip."

"Evening found us alone among the savages," reads a typical sentence (Phil, p. 265). On another island, Worcester was pleased to note, no native was "anything like my size" (p. 282). A hulking six-footer, Worcester did not mind being looked up to. Apart from the pygmies, Filipinos averaged about five feet six inches in height and about 125 pounds in weight, rough statistics that place in perspective the difficulty even a strong native experienced in carrying the white hunter across streams and through the surf. Worcester took to riding on native backs as a matter of course but did not enjoy being dropped.

On their way to Siquijor,

> to reach the boat, which was some distance offshore, we had each to ride out astride a man's shoulders. Bourns got on very well, but I was fifty pounds heavier than the fellow who tried to carry me, and the strong swell made him sway and totter in a most alarming manner. Just as I thought myself safe, he stubbed his toe and pitched forward, shooting me over his head. I barely caught the edge of the boat, while the man retained his hold on my ankles, and recovered himself so that I hung suspended. My frantic efforts to get on board without a ducking finally ended in my landing a kick on the nose of my bearer which knocked him into the water, where of course I joined him, to the intense delight of a large crowd of spectators. [Phil, pp. 275–76]

Less understanding than McKenney after a comparable ducking (see p. 183n), Worcester was not soothed by the natives' laughter: "I scrambled aboard in a very poor humour. . . ."

Normally Worcester prided himself on his mastery of situations and natives, possessively calling the bearers of their gear "my Tagalog carriers," "my Mangyans," or "our coolies" (pp. 392, 393, 433). Their porters had plenty to do, for their total kill was heavy even after it had been skinned—all the birds, big and little, crocodiles, deer on Culion, and rare wild buffalo on Mindoro called tamaraus. By accepting Spain's royal order in support of their second expedition, Worcester and Bourns had accepted its sovereignty over the Philippines and interpreted their official sanction to mean they could treat the Filipinos' country as an open hunting preserve over which they could wander at will. But in return for their game and their help, Worcester prided himself on his generosity toward tribesmen: "I had made it a principle to give each savage just twice what he asked for his services, and had left ardent admirers along my whole line of march" (pp. 395–96). A bit less than extravagantly openhanded, Worcester paid off with empty sardine and butter cans, brass rings, bits of copper wire, cheap mirrors, pinches of tobacco, or sometimes even money. In Siquijor he and Bourns "found plenty of men who were glad to serve us for five cents a day" (pp. 285–87, 395). Sometimes he simply conscripted their services (e.g., p. 424), as the Spanish had for centuries; but one way or another he made natives get him where he wanted to go, especially to the wild buffalo on Mindoro, where he stood over his first kill of a tamarau with something of the feeling Buffalo Bill Cody must have had at the outset of his career: No one who has not actually followed their trails days and weeks in vain, Worcester exulted, "can fully realize my feelings when my first bull lay dead at my feet" (p. 404).

Everywhere the white hunter identified with his hosts' attempts to bring "order" to the country. Indeed he was scornful of the Spaniards for having brought so few of the "eighty distinct tribes," as he reckoned their number (p. 57), under effective control and shyly implied white Americans could have done much better in less time.*

No less than Thomas L. McKenney, Worcester was contemptuous of native cultures. To his mind the Filipinos had no literature worth mentioning and dialects almost barren of words of generalization. Like the colonel he enjoyed exposing medicine men as frauds whose charms and visions were shams, a "lot of mummery" (pp. 429–32). "Child's tales" were believed "not only by natives but even by intelligent mestizos," he held, in an opinion implying that admixtures of white blood raised intelligence (p. 272; cf. PP&P, p. 673). Still, he was himself not above exploit-

* Later he explicitly drew a sexual distinction between Spaniards and Anglo-Americans as conquerors. When the former arrived in the Philippines (1565) they had made commendable progress "in subduing" the tribes, he wrote, but then had slowed down and "seemed to have lost much of their virility" (PP&P, pp. 423–24). On the other hand, his countrymen had grown more virile in subduing all the tribes across a vast continent over the same centuries and were more than potent to consummate what the Spanish colonizers had only begun. An avid reader of Rudyard Kipling, Worcester also turned to the poetry of the frail little champion of empire for his criteria of "manhood" and of virile dominance over native peoples (PP&P, pp. 506–08). Not to *subdue* was to *stultify* oneself, one of his key words: Premature withdrawal from the islands "would be to stultify ourselves in the eyes of the world" (Knot, 873; cf. PP&P, p. 264). To Worcester stultification was effeminate ineffectiveness and foolishness, a viewpoint that calls to mind the words of a president decades later who rejected precipitate withdrawal from another distant land to avoid having the United States of America, "the world's most powerful nation," appear "a pitiful, helpless giant"—Richard M. Nixon, "Cambodia: A Difficult Decision," April 31, 1970, *Vital Speeches of the Day*, XXXVI (1969–70), 450–52.

ing natives with Prospero-like magic in "hypnotizing" chickens and in sleight-of-hand tricks (p. 268). Credulous, irresponsible, and dirty, the native was naturally indolent, as indolent as McKenney's merciless savage: "Indolent he surely is, but whether hopelessly so is another question"—Worcester suggested rather harshly that native laziness "might be remedied by increasing their necessities" (pp. 478–79). And like the colonel's Indians, the Filipinos were children—Worcester found "much truth in the statement of a priest who said of them, 'in many things they are big children who must be treated like little ones' " (p. 484).

And this was the best knowledge of the Philippines a citizen could bring to the White House. As for specific proposals, Worcester recommended employing natives in the lower echelons of the colonial regime and taxing the islands sufficiently to keep the machinery of government going. Following the Spanish practice of referring to Filipino nationalists as "*tulisanes*, or professional bandits," Worcester declared that "a vigorous policy in dealing with them would have a very wholesome effect." So far as possible in operations against them "native troops should be used" and "a convenient way to dispose of a part of the insurgent forces would be to retain them in service under white officers." But every effort should be expended to win their confidence, "for if any considerable number of them should take to the mountains, they could cause much trouble before they were run down" (Knot, 878).

Three centuries of Indian-hating had helped prepare McKinley to accept all this as authoritative and to adopt Worcester as his insular expert. Not only did the president offer to make him his personal representative but after their interview immediately cabled the professor's recommendations to Ewell S. Otis, the major general in Manila who had undertaken to make the Filipinos "good Indians."

3

"Not so very long ago," observed Jean-Paul Sartre in a memorable sentence, "the earth numbered two thousand million inhabitants: five hundred million men, and one thousand five hundred million natives." To deal with natives in "our new island possessions" eleven thousand miles away, the men in Washington naturally extrapolated from their accumulated experience in dealing with natives on the continent.

Even the administrative initialisms of the Bureau of Indian Affairs and the Bureau of Insular Affairs matched. In 1899 the Schurman Commission left for the islands technically under the authority of Secretary Hay, instructed by him and ordered to report back to him at the State Department. But by 1900 Filipino resistance made the War Department the responsible executive arm. The Taft Commission went out instructed by Secretary of War Elihu Root and ordered to report back to him. In 1901 McKinley planned to put Hay back in charge, partly for appearance's sake, but the president's assassination and continued hostilities left the Philippines in Root's hands. As in the case of the Indians until 1849, therefore, the Filipinos fell under the authority of the War Department. The Division of Insular Affairs therein, organized under executive authority in December 1898, became the Bureau of Insular Affairs (BIA) under the Organic Act (Sec. 87) of 1902. Somehow historians have failed to note the fundamental identity of the first and second BIA in the War Department. The BIA of which McKenney was godfather and first head and the BIA under which Worcester served were both bureaucracies designed to subdue and control natives, not men.

To deal with restless natives, McKinley just as naturally turned to the time-tested instrument that had fought more than one thousand engagements from Appomattox to Wounded Knee before crushing armed resistance of the Sioux and other tribes. Originally his commander of the Philippine Expeditionary Force had been Major General Wesley Merritt, hero of the Civil War and of Indian-fighting under Custer, but after the United States occupied Manila on August 13, 1898, Merritt had been ordered to Paris to participate in the peace conference. He was succeeded by General Otis, another veteran of the Civil and Indian wars, who had 15,000 troops under his command, led in part by those who had learned their trade fighting Plains Indians; in December Brigadier General Marcus Miller, yet another old Indian-fighter, reached Manila with an additional 2,500 men. Of these and the many other career officers who turned from fighting North American natives to Far Eastern, Major General Henry W. Lawton, who arrived in March 1899, had perhaps the most impressive record: in 1886 he had added to his Civil War Medal of Honor the feat of recapturing the great Apache Geronimo and in 1898 of becoming the hero of El Caney in Cuba.

The big stick in the form of McKinley's expeditionary force was as suited to winning the confidence—hearts and minds—of natives in the archipelago as Custer's 7th Cavalry had been in winning that of Crazy Horse at the Little Bighorn. Moreover,

McKinley's administration treated the Filipinos under General Emilio Aguinaldo as though they were a band of Sitting Bull's Sioux or Geronimo's Apaches. The native forces, once considered useful, were pushed back from their positions around Manila, none too gently, refused a formal hearing in Washington, and shut out of the peace conference in Paris. Aware of the parallels, Rounseville Wildman, U.S. consul in Hong Kong, warned Hay in Washington "that the insurgent government of the Philippine Islands cannot be dealt with as though they were North American Indians, willing to be removed from one reservation to another at the whim of their masters."

Had they known more U.S. history the Filipinos would have been more cautious. Notwithstanding Dewey's denials, Aguinaldo had almost certainly been double-crossed, led by the commodore and other official representatives to cooperate with the United States by their pledges, always unwritten, that they had arrived to help him liberate his country. The build-up of troops and the treatment of his Malolos government understandably increased the Filipino leader's bitter suspicions, but these were assuaged somewhat by those pledges of honorable men and by a careful reading of the U.S. Constitution, which persuaded him it contained no authority for colonies and therefore none for colonialism. Still, he and his army were aware of precedents on the continent, though they acted on this awareness tardily. In a secret cable to John Hay, Jacob Gould Schurman later reported them " 'exceedingly skeptical of our intentions' and [they had] cited the sad case of the American Indians as evidence" (KJC, p. 137). A little less behindhand was the warning an Aguinaldo spokesman passed on to General Marcus Miller in January 1899: To protect the integrity of their republic, he pledged, *"we will withdraw to the mountains and repeat the North American Indian warfare. You must not forget that."* In *The American Occupation of the Philippines* Judge James H. Blount, who fought Filipinos without knowing anything of this background, wryly quoted the spokesman and added simply, "later they did" (1912; pp. 162–63).

"Line up, fellows, the niggers are in here all through these yards!"—so rang out the cry of a Nebraska sentry on the night of February 4, 1899. He had just shot one native or maybe two and touched off what the textbooks cynically call "the Philippine insurrection." Next day an emissary from Aguinaldo requested a truce, but Otis refused, saying the fighting had to go on "to the grim end." It did. By July 4, 1902, when Roosevelt prematurely declared it over, 126, 468 U.S. soldiers had fought and over 4,000 had died in the islands. Almost five times that number of natives had been killed in battles, to which body count one must add another 250,000 civilian deaths from the attendant fighting, malnutrition, and disease. The Philippine–American War was neither splendid nor little.

McKinley's carrot in the form of the Schurman Commission arrived in Manila one month to the day after the war had commenced, a little late to conciliate the natives and keep them from taking to the hills. It was, in any event, an investigatory body and not a peace commission, as its composition made abundantly clear: Dewey, now an admiral, appropriately received the assignment of looking into the naval and maritime needs of the United States in its new possessions; Otis had already undertaken their pacification; along with Schurman and Worcester, the only other "civilian" commissioner was Colonel Charles Denby, former minister to China and an avowed imperialist, who saw the Philippines as stepping-stones to that market.

In a sense President Schurman was the only unknown quantity on the commission. Originally he had opposed annexation on the grounds that the needs of empire—trade, bases, the like—might be met without a costly war of subjugation. Now he never wavered in his insistence the Filipinos had to accept the imperial sovereignty of the United States and later showed plainly in his *Philippine Affairs: A Retrospect and Outlook* that he largely shared Worcester's views of the natives and was by no means averse to the use of force against them: "with a superabundance of force you impress the natives with the plenitude of your power" (1902; p. 109). Yet within weeks of his arrival he had become convinced the war was "truly an inglorious contest"; in June he precipitated a crisis within the commission by proposing direct "negotiations with Aguinaldo to adjust U.S. sovereignty and responsibility with reasonable aspirations [of the] Filipinos, governing and garrisoning mainly through themselves" (PP&P, p. 17). But that would have afforded a measure of recognition to the "insurgents" and was not the "grim end" Otis had in mind; Denby and Worcester also refused to sign Schurman's proposed cablegram, the latter because "it would have stultified us" (PP&P, p. 264). Thereupon Schurman sent off the cable over his own name and received from John Hay the response previously noted: There was "no excuse for further resistance by the Filipinos." McKinley requested the opinions of the other commissioners, and Dewey having sailed for home, Otis, Denby, and Worcester wired back: "Undersigned recommend prosecution of war until Insurgents submit" (PP&P,

p. 18). Forestalled, Schurman shortly thereafter returned home to find few of his compatriots concerned about the reasonable aspirations of natives.

Now, all these details bear recitation, for Schurman's proposal just conceivably offered an opportunity to end the slaughter. The Filipinos were demonstrably ready to negotiate, and a strong stand by his own commission would have been politically embarrassing for McKinley. At the very least, had Denby and Worcester backed Schurman, the men in Washington would have had a much more difficult time with the pretense that only a handful of natives opposed their benevolent assimilation. But Worcester was no less committed to the "grim end" solution than Otis, had in fact joined forces with the general and Denby, and had engineered Schurman's defeat.

From the first Worcester's position as a commissioner had been strengthened by his presumed expert knowledge of the natives and by his working relationship with Major Frank S. Bourns, his old traveling companion, who was currently directing Otis's spy system and liaison with wealthy conservative Filipinos, the so-called *illustrados*, many of whom had supported Spain against their compatriots. Aware that the U.S. colonial government could only rise from the ashes of the Spanish regime through enlisting the same class of collaborators, Worcester and Bourns dangled positions in the new order before them and lined them up to testify before the commission.

A week or so before the blow-up over Schurman's proposal, one of Worcester's letters provided a peek behind the scenes:

> I will tell you that strange as it may seem, the two people that are running the political, and influencing the war end of this thing are Frank [Bourns] and myself. We pull absolutely together. The correctness of Frank's prognostications has been demonstrated time and again. . . . We fill [Otis] full of good ideas, and he communicates them to others as though they were his own. I can't help liking the old gentleman in many ways and he is in many ways a great man. I hope we shall now pull together and do a lot of *good*. [May 26, 1899; PP&P, p. 15n]

Even stripped of their braggadocio, these sentences yield a pretty accurate description of what went on behind the commission's formal deliberations and of how Schurman was outflanked by the young professor.

At the same time Worcester took the war back home. In early June he received a clipping from the *Springfield Republican* "containing all manner of lies about the way things are done out here —women and children killed and what not!" With General Otis's blessing and at government expense he cabled a long rebuttal to the *Chicago Times-Herald* that earned him the accolade of McKinley and Hay. The secretary of state expressed his "profound appreciation of your admirable dispatch," as Hay called Worcester's whitewash: "It is the best thing that has been published from anybody since the trouble out there began. It has been reprinted all over the United States, and has done good everywhere it has been read. We are all delighted with it" (PP&P, p. 16). Before long Worcester was to contend that never had a war been "more humanely conducted" (Rep, I, 184), admit the "water cure" had been applied only sometimes and then "in mild form," and characterize the Batangas concentration camps as "a humane and satisfactory solution of the existing difficulties," indeed a positive boon to the native inmates: "Many of the occupants of his [General J. F. Bell's] reconcentration camps received their first lessons in hygienic living. Many of them were reluctant to leave the camps and return to their homes when normal conditions again prevailed" (PP&P, pp. 215, 222, 224–25).

Meanwhile, accompanied by Colonel Denby, Worcester returned to Washington in September 1899 and worked there until the following March on the Schurman report (Rep). Though in the end all the commissioners, even Schurman, signed the document, Worcester rightly claimed credit, as their sole "ethnologist," for "the scientific portion of the report." It was, though not quite in the way his biographer meant, "a monument to the genius of Worcester" (PP&P, p. 21).

4

Imagine for a moment, before we turn to the report itself, its demands on Worcester's genius. To meet the needs of empire he had to reduce to a geographical expression a people nearly three times as numerous as the American colonists on the eve of their own Revolution. He had to deny the existence, as a people, of more than seven million Filipinos occupying a land area larger than that of New England and New York combined. According to the *Philippine Census of 1903*, their ethnic diversity was less than that of the U.S. population: Worcester had to deny this homogeneity and make them instead into what Theodore Roosevelt was to call a "jumble of savage tribes." Knowing next to nothing about their history, save from his reading of John Foreman and other secondary sources, he had to make authoritative statements about their characteristics and to dismiss as "divers rebellions" their many revolts against Spain's alien power, beginning with that of Lapu Lapu against Magellan in 1521 and culminating with that of the militant nationalists of the late 1800s—a topic illuminated recently by Usha Mahajani in her fine *Philippine Nationalism* (1971). Worcester had to deny this revolutionary nationalism, deny the existence of their unfinished revolution of 1896–98, and deny the fact that it had carried on over into their "insurrection" against successor masters who "owned" their country by virtue of the Treaty of Paris. He had to denigrate the leaders of their revolution as bandits or at best as unscrupulous, self-seeking Tagalog politicians and, in the face of all the evidence to the contrary, assert they had ordered an attack on Otis and his troops without provocation or cause. Denying that the Filipinos wanted independence, he had to make them into little brown natives, in short, too "primitive" to have "civilized" yearnings to be free. The magnitude of his task would have staggered the imagination of one less dedicated.

The careful reader will find all these attempted denials and denigrations in the Schurman report and will discover that the young professor's heroic task very quickly landed him in a seemingly insoluble problem of nomenclature: What should he call these new natives? Of course, "natives" itself served to differentiate them from "men," from "men of various nationalities living in and doing business in the islands" (I, 2), but by itself hardly set them off from all the other nonwhites in the world. In a proclamation of April 4, 1899, the commission had blundered into calling Filipinos "the Philippine people" (I, 4) but that was to utter precisely what should not have been said aloud. The Spanish called them "Indios," thereby perpetuating the old error that Magellan had discovered India—for Worcester to have drawn on that usage by translating the term into "Indians" would have compounded Columbus's misdiscovery with Magellan's. So the awkward best he could do was "the inhabitants of the Philippine Islands, at present collectively known as 'Filipinos' " (I, 11), though without quotation marks the name suggested they had the common characteristics of a people. That, Worcester adamantly insisted, they were not.

In "Part II.—The Native Peoples of the Philippines," the centerpiece of the report (I, 11–16), he undertook to show why they were not: "That the Filipinos do not constitute 'a nation' or 'a people,' will appear from the perusal of the following table, which gives the names of the various tribes so far as known, the regions they respectively inhabit, and where practicable, an estimate of the individuals composing each." The hand-list of the tribes of the Philippines that followed resembled his later *Hand-List of the Birds of the Philippines* (1906). It bore the marks of his training in zoology and botany at the University of Michigan; in its pretensions to scientific taxonomy it derived basically from the Linnaean classification of plants and animals in the tenth edition of *Systema naturae* (1758).

Worcester's table was a triptych, with each compartment containing a "sharply distinct" race, namely, Negrito, Indonesian, or Malayan. The Negritos he ranked lowest physically and mentally: they were "weaklings of low stature" and "in the matter of intelligence they stand at or near the bottom of the human series." Highest physically were the Indonesians of Mindanao: they were "physically superior not only to the Negritos, but to the more numerous Malayan peoples as well. . . . The color of their skin is quite light. Many of them are very clever and intelligent." Intermediate were the Malayans, smaller and as a rule beardless, with brown skin "distinctly darker than that of the Indonesians, although very much lighter than that of the Negritos." Their "civilized tribes" contained "the great majority of the inhabitants of the Philippines." Although "ignorant and illiterate," somehow they possessed "a considerable degree of civilization, and with the exception of the Mohammedan Moros, are Christianized." With seven or eight million natives thus classified by size, color, and intelligence, Worcester divided them into eighty-four tribes, stuffed each into its appropriate racial compartment—twenty-one in the Negrito, sixteen in the Indonesian, and forty-seven in the Malayan—and then as it were stood triumphantly back from what he had wrought. With this schematic before him, how could anyone regard such a potpourri of tribes as "a people"?

Worcester was a graduate student in zoology, not anthropology. At Michigan he did little or no formal work in the latter and we have already sampled his views of Filipinos on his two collecting expeditions. As a commissioner in Manila he secured some old Jesuit manuscripts on "Indios" but his ransacking of these for data hardly made him an ethnologist. Moreover, the quality of his understanding of tribal peoples was fully reflected in his denigration of the Negritos in the report and over the ensuing decades. Their size and color provoked his unbounded contempt, as in his depiction of them as "little, wooly headed, black, dwarf savages"

(PP&P, p. 532). To his mind the stupidity of these despicable runts correlated with their color and could hardly be exaggerated, though "it would be going too far to say that their moral sense has been blunted. It is probably nearer the truth to say that they never had any" (PP&P, p. 530). It is rather nearer the truth to say that these were people whose understanding of flora and fauna put his own to shame and whose thirst for knowledge elicits the admiration of modern observers. Claude Lévi-Strauss approvingly quoted one such observer:

> The Negrito is an intrinsic part of his environment, and what is still more important, continually studies his surroundings. . . . Most Negrito men can with ease enumerate the specific or descriptive names of at least four hundred and fifty plants, seventy-five birds, most of the snakes, fish, insects, and animals, and of even twenty species of ants. . . . and the botanical knowledge of the *mananambal*, the "medicine men and women," who use plants constantly in their practice, is truly astounding. [*The Savage Mind* (1966), pp. 4–5]

Yet these were the people Worcester put at the bottom and later removed from the "human series" altogether. In the November 1913 issue of the *National Geographic* he ranked them "not far above the anthropoid apes" and waggishly anticipated their imminent extinction: "*they are a link which is not missing but soon will be!* In my opinion, they are absolutely incapable of civilization" (XXIV, 1251, 1228).

Had Worcester been a bona fide expert in the arcana of "civilization," his hand-list would have included not eighty-four but about thirty distinct groups, including the Negritos and the Igorots of northern Luzon, whom he properly considered tribal. But approximately ninety percent of the Filipinos would not have appeared in his table at all, for the tribal origins of most of the Christian lowlanders, such as the Tagalogs of central Luzon, had long since ceased to shape their lives decisively. To Worcester, however, they were and always remained "the Malayan savages of the sixteenth century, plus what Spain had taught them, plus what they have so recently learned from us," as if they had been incapable of teaching themselves anything ever, before or after the Spanish conquest (PP&P, p. 687). Hence in the report he classified them as "civilized tribes" and "civilized natives" (I, 15, 159), absurdities of nomenclature he fought for in the face of mounting criticism (PP&P, p. 424).

Yet as Worcester and his contemporaries used the term, a primary meaning of *civilization* was the suppression or destruction of kinship societies. For him to speak, therefore, of a *civilized* tribe or native was for him to call forth an oxymoron, like a *brilliant* moron, a *white* savage, a *wild* man, an *urbane* native. Worcester's oxymorons were the transparent masks of his determination to reduce Filipinos, all of them, to the tribal, nonperson status of natives. In 1913 he still endorsed the proposition that "no Malay nation has ever emerged from the hordes of that race," nor did the Filipinos yet "constitute 'a people' in the sense in which that word is understood in the United States. They are not comparable in any way with the American people or the English people. They cannot be reached as a whole, and they do not respond as a whole" (PP&P, pp. 672–73). A "great dark mass" instead of a people in the Anglo-American sense, they could not have had a revolution against Spain, could not desire independence, and could not take care of themselves. Not men but natives, their "habitats" were as eminently seizable as had been those of natives in Louisiana or California.

From this point of view, the United States had as much right to be in the Philippines as it had to be in Massachusetts and Louisiana and California. In every instance, from William Bradford's Plymouth on into the West, it had moved into "unpeopled" territory, space occupied only by natives. And as John Hay impatiently explained to an anti-annexationist in September 1899: "I cannot for the life of me see any contradiction between desiring liberty and peace here and desiring to establish them in the Philippines" (KJC, p. 139).

Yet Aguinaldo was also right: the U.S. Constitution contained no explicit authorization of colonies. As Aguinaldo, Apolinario Mabini, and their comrades also knew, American citizens from Jefferson to Hay prided themselves on their devotion to the doctrine of the consent of the governed and on how that made their country fundamentally different from the imperial European powers—after all, they had gone to war in the first place with the declared intention of liberating the poor Cubans from Spain's colonial misrule. Not the first nor the last distant revolutionary to be misled by their anticolonial rhetoric, Aguinaldo failed to look behind it to the realities of their westward expansion. Among those realities was a triumph of semantics: the United States never had and would never have *colonies*, only *territories* (or at most "dependencies"). Unhappily for Aguinaldo and his people, they had to experience a lesson Tecumseh, Geronimo, and other Native

Americans had already learned the hard way. A territory did not have to be called a colony for it to have colonized peoples within its borders.*

But "it may be asked," admitted the Schurman Commission, "What's in a name? In this case certainly much" (I, 106). Simply the name of the British "crown colony" made it an embarrassing model, with its "government imposed upon the people from without" and thus with the fatal defect of being "inimical to the habit of self-government" (I, 105). With arresting effrontery, moreover, Worcester and the other commissioners protested "on behalf of all the Filipinos . . . against the suggestion of calling the archipelago a colony." In the native vocabulary no other word was so universally condemned, "so surcharged with wrongs, disasters, and sufferings"—this was dangerously close to an open admission the new natives universally detested the practice, as well as the word, and therefore desired independence. But the commissioners moved on quickly to U.S. territorial governments for more appropriate precedents. More directly relevant than the Northwest Ordinance of 1787 was the act of 1804 Jefferson had drafted to govern Louisiana. The first territory acquired beyond the original limits of the United States, readers were reminded, Louisiana was "all that country west of the Mississippi from the Gulf of Mexico to the Lake of the Woods, extending indefinitely westward." The commissioners merely proposed to extend Jefferson's scheme directly and naturally to the Philippines—in a sense it had been headed that way all along. And in Louisiana, they reported with unconcealed gratification, the author of the Declaration of Independence had not hesitated to suspend its principles and adapt his frame of government "to the condition of the natives" (I, 109).†

* Cf. Frederick Jackson Turner in the *International Monthly* for December 1901: "Our colonial system did not begin with the Spanish War; the United States had had a colonial history and policy from the beginning of the Republic; but they have been hidden under the phraseology of 'interstate migration' and 'territorial organization.' " This past has not been hidden from Native American scholars—see Robert K. Thomas, "Colonialism: Classic and Internal," *New University Thought*, IV (Winter 1966–67), 37–53. Nevertheless, denials of this colonial context continue apace on down to the December 1978 convention of the American Historical Association, as evinced at one of the sessions by Roxanne Dunbar Ortiz in her unpublished "Wounded Knee 1890 to Wounded Knee 1973: A Study in United States Colonialism."

† Always a slippery term, "natives" was especially ambiguous in this context. Jefferson imposed his government on unconsenting white adults of French and Spanish extraction, whom he characterized as being "as yet as incapable of self-government as children" (I, 108). Indians, the true natives, were left out of his scheme altogether or were rather the objects over which he extended U.S. sovereignty—as we have seen, never did Jefferson consider Native Americans in that territory or any other as his "new fellow-citizens." But the commissioners were less concerned with exactitude than with precedents that could be made to stretch over space and time. Furthermore, they were surely right that Jefferson had suspended the principles of the Declaration of Independence whenever natives blocked his path.

With a flourish of impartial assessment and only after "careful consideration and study," the commission held the Filipinos to be presently as utterly unfit for self-government as Worcester had pronounced them to be before he returned to Manila in 1899. For the foreseeable future, therefore, they had to submit to the benevolent tutelage of "a small body of American officials of great ability and integrity, and of much patience and tact in dealing with other races" (I, 98). Could the writer possibly have had himself in mind? To rule natives on tropical islands Worcester was always more than willing to forgo the tedium of teaching undergraduates in Ann Arbor.

5

On January 9, 1900, Senator George Frisbie Hoar, the venerable Republican who had become a leading anti-annexationist, read into the *Congressional Record* the letter of a general officer who had recently returned from the islands to declare in exasperation: "We had been taught (the devil only knows why) that the Filipinos were savages no better than our Indians" (XXXIII, 714). Only the devil fully knew why, no doubt, but the teaching could have been traced to its primary source. The "eighty distinct tribes" in Worcester's patchwork *Philippine Islands and Their People* grew to eighty-four in his "scientific portion" of the Schurman report and then ripened in debates and documents into the ethnologically sanctioned figure.

Almost concurrently with the complaint of Hoar's correspondent, Senator Henry Cabot Lodge released Senate Document 171 (56th Cong., 1st Sess.), a report from his Committee on the Philippines that accepted Worcester's division of Filipinos into "three sharply distinct races" and reproduced his hand-list of tribes. In the 1902 hearings before Lodge's committee, Brigadier General Robert P. Hughes invoked the professor's name as the racial authority in residence. This expert witness learnedly used the cephalic index as "the best index of racial differences" in support of his claim that "the Tagalo and the Visayan are not the same racially. I have so reported to Mr. Worcester, who is an ethnologist." The general's colloquy with the committee went on:

Senator [Fred Thomas] Dubois: You think the difference is greater than among our American Indians?

General Hughes: I think it is, unless you take the Pueblo Indians of Mexico and the Sioux. There is such a marked difference there

> that it might equal the difference between the true Visayans . . . and the Tagalos.
>
> Senator [Eugene] Hale: Do you think it would be as great as that?
> General Hughes: I rather imagine it would be.
> Senator [William Boyd] Allison: Do they speak the same language?
> General Hughes: They speak an entirely different language.
> Senator Dubois: That is true of our Indians, you know.
> [Senate Document 331, 57th Cong., 1st. Sess., I, 537]

"Our Indians" and "our" new natives seemingly had much in common. And in these hearings William Howard Taft repeatedly adopted Worcester's habit of referring to the Tagalogs and the Visayans as different *tribes* and paid the professor the high compliment of quoting him without attribution: the Filipinos, asserted their brand-new governor-general, were "utterly unfit for self-government."

Meanwhile Worcester's definitions of Philippine realities had become lenses through which home-based members of the executive branch saw Filipinos and shaped official policy accordingly. Following his first briefing by Worcester, the president's instructions to the Schurman Commission reflected the professor's views: "the commissioners [should] exercise due respect for all the ideals, customs, and institutions of the *tribes* which compose the population" (Rep, p. 186; my italics). Later McKinley gratefully repeated Worcester's lie that the United States was opposed by only "a portion of one *tribe* representing the smallest fraction of the entire population of the islands" (October 14, 1899; my italics).

In 1900 Theodore Roosevelt became McKinley's running mate and was assigned by Mark Hanna to hold up the administration's side of the great debate over imperialism. Worcester's scientific findings significantly shaped how this "paramount issue," as William Jennings Bryan and the Democrats called it, came before the electorate.

"So far as I am aware," said Roosevelt in Grand Rapids on September 7, 1900, "not one competent witness who has actually known the facts believes the Filipinos capable of self-government at present. . . . Judge Taft, President Schurman, Professor Worcester, Bishop Potter, and all our army officers are a unit in this point." Accepting the vice-presidential nomination on September 15, 1900, he found it unthinkable that the United States would "abandon the Philippines to their own *tribes*" (my italics). The Philippine revolution he called "the Tagal insurrection," and Aguinaldo and his followers he dubbed "a syndicate of Chinese half-breeds": "To grant self-government to Luzon under Aguinaldo would be like granting self-government to an Apache reservation under some local chief." Drawing directly on the formulations of Worcester and the other commissioners, he pointed out that not the author of the Declaration of Independence nor "any other sane man" had held that the doctrine of the consent of the governed applied "to the Indian tribes in the Louisiana Territory . . . and there was no vote taken even of the white inhabitants, not to speak of the negroes and Indians, as to whether they were willing that their Territory should be annexed." The next major acquisition was Florida, purchased like the Philippines from Spain, and there "the Seminoles, who had not been consulted in the sale, rebelled and waged war exactly as some of the Tagals have rebelled and waged war in the Philippines." And so on down through all the acquisitions and the rebellions that had to be stamped out to the present instant: "The reasoning which justifies our having made war against Sitting Bull also justifies our having checked the outbreaks of Aguinaldo and his followers." To an overwhelmingly approving electorate and later as vice-president and president, Roosevelt depicted Filipino revolutionaries as Tagalog bandits, "chiefs," and "hostiles" of a tribe comparable to the Sioux and the Seminole. In sum, it was *The Winning of the West* (1889–96) all over again for the historian–politician; as he had said in that work, war against merciless savages was "the most ultimately righteous of all wars" (III, 30).

The Schurman Commission had been confident that when peace and prosperity were established in the islands and education made general, "then, in the language of a leading Filipino, his people will, under our guidance, 'become more American than the Americans themselves' " (Rep, p. 184). Mainland historians have since often told the triumphant story of how their compatriots thereupon devoted themselves to the Americanization of the Philippines, and that meant their "civilization" or, in its most recent version, their "modernization," as in Peter W. Stanley's fetchingly titled *A Nation in the Making* (1974). Details aside, this historiography of the victors is revealing in its general neglect of the obvious. As Roosevelt knew when he was nominated, no more than the Seminoles were the Filipinos asked to consent before they were governed. As he also knew, on the current Indian reservations "the army officers and the civilian agent still exercise authority, without asking 'the consent of the governed,' " and that was exactly true of the Philippines. Military rule in the islands was not lifted fully until 1902. Then the "civil government" did not rest on Filipino consent but on U.S. military might, and that was the case after the five-man Taft Commission was enlarged to include three "leading" Filipinos and the case even after the Philippine Assembly was inaugurated in October 1907. Under U.S. occupation the "civil government" was, in the words of James H. Blount, "a *military régime* under a *civil* name."* Thus what Worcester said appreciatively of the period ending July 4, 1901, when Taft was installed as governor-general, in fact applied to the next decade and more: "the Philippines were under military rule, which has one great advantage: its methods usually bring quick results" (PP&P, p. 512).

The professor shared the almost absolute power of Governors Taft (1901–3), Luke Wright (1904–5), James F. Smith (1906–9), and W. Cameron Forbes (1909–13). As a legislator, he helped these potentates pass some 1,800 "laws" to get the new regime under way. As an executive, his portfolio as secretary of the interior included the bureaus of public health, public lands, forestry, mining, agriculture, fisheries, government laboratories, patents and copyrights, and the weather, not to mention his pet Bureau of Non-Christian Tribes. As secretary of their interior he threw himself into combat against the childish Filipinos' "almost unbelievable ignorance and superstition" (PP&P, p. 665). Hard on the heels of pacification came the heroic sanitation campaign of his public health officers "to make Asiatics clean up" (p. 354). Unwilling islanders were forcibly vaccinated, and "lazy, vicious" natives laughed at their peril at his orders to "clean up," for he was quite humorless when he intoned, "sane living . . . means sanitary living" (pp. 482, 412). His direct jurisdiction in education, except for the mountain peoples, extended only to a vigorous program in athletics, its public health side: "Before the American occupation . . . the Filipinos had not learned to play. . . . Baseball not only strengthens the muscles of the players, it sharpens their wits" (p. 408). But after a dozen years the job was nowhere near

* *American Occupation of the Philippines*, p. 374. To show how absolute and literally irresponsible the "civil government" was, Judge Blount related an incident that bears on the theme of continuity we have pursued in these pages. In 1903 Roosevelt wanted to appoint Beekman Winthrop, descendant of the Massachusetts Bay Winthrops and recent graduate of the Harvard Law School, to the Philippine bench. Taft asked his secretary to make out a commission but was told he could not do it—a 1901 "law" of the Philippine Commission (No. 136) had set the minimum age for judges at thirty and Winthrop was too young. "I can't eh," his secretary reported Taft as saying, "send me a stenographer." Taft duly changed the minimum age to twenty-five, had his secretary round up the other commissioners, and passed his new "law" (No. 1024) in a few minutes (pp. 443–45). So smoothly did another Winthrop become a magistrate half a world away, and such was the character of the "civil government" that had so pleased John Fiske when it was established (see p. 242). In 1904, incidentally, Winthrop was appointed governor of Puerto Rico—Oscar M. Alfonso, *Theodore Roosevelt and the Philippines, 1897–1909* (Quezon City: University of the Philippines Press, 1970), p. 38.

done, the Filipinos remained almost as utterly unfit as they had been, and the United States, to his mind, had moved "too fast and too far" in giving them a modicum of home rule (pp. 668, 691). Disgusted by president-elect Wilson's "expression of hope that the frontiers of the United States might soon be contracted" (p. 659), Worcester, when he left the insular government in 1913, combated that stultifying possibility with *The Philippines, Past and Present* (1914), a book that "gained world-wide attention," according to his biographer, became a reprinted standard reference work, and helped shape policy in the islands when the Republicans came back into power in the 1920s.

Readers of this monumental handbook for colonial statesmen had before them all the reasons Filipinos so heartily detested its author. In reality the essentials of his attitude toward them differed not a whit from Taft's and that of the other commissioners, but Worcester's imperious bluntness earned him the special enmity of their charges. They were affronted by the very name of his Bureau of Non-Christian Tribes. Under Spanish rule Christian Filipinos had been made to feel ashamed of their tribal origins and had developed what the anthropologist R. F. Barton called "their colonial attitude of being ashamed of their backward tribes." The name of Worcester's bureau slashed into this lamentable but understandable attitude by building into the structure of the new insular government his view of all Filipinos as tribal and of themselves therefore as "Christian tribes," precisely as they had appeared in his Schurman report hand-list. Keeping their psychic wounds open with a series of articles that advertised their exotic backwardness, he then rubbed in the salt of irony by holding them utterly unfit to control their "savage neighbors" or alternatively to protect them—or both! "If the Filipinos were put in control," he asked rhetorically, "would there rise up among them unselfish men who could check the rapacity of their fellows, and extend to the helpless peoples the protection they now enjoy [PP&P, p. 504; cf. p. 495]?"

In the November 1913 *National Geographic*, Worcester illustrated his article on "The Non-Christian Peoples of the Philippines" with a photograph captioned "Some of the men who have established order among four hundred thousand savages" (XXIV, 1194). They were heroes of a stupendous undertaking, a Winning of the Farther West, for in the Mountain Province of northern Luzon alone there were "more fighting men than ever were to be found among all of the American Indians between the Mississippi and the Rocky Mountains, the area of a generation of 'Indian Wars,' " his biographer pointed out proudly (PP&P, p. 29). Worcester and his crew of tribe-tamers, mostly young American adventurers who like himself had originally come out to quell the "insurrection," went among them counting, photographing, and disciplining "known" tribes and "discovering" others. None of this could have come to pass without the network of a thousand miles of trails and roads they had built through the high country: "I have always considered the opening up of adequate lines of communication an indispensable prerequisite to the control and development of any country," Worcester observed, "and this is especially true of the territory of the wild man. No matter how unruly he may be, he is apt to become good when one can call on him at 2:30 A.M., since that is the hour when devils, *anítos* and *asuáng* are abroad, and he therefore wants to stay peaceably in his own house [p. 443]!" With lines of communication open they could reach unruly natives swiftly and teach them salutary lessons: "at the outset we burned towns if their people engaged in head-hunting" (p. 459). Baseball was a means to reach the bodies of "youthful savages" and boarding schools in Baguio, Bontoc, and other villages reached their minds with lessons in American English—the language of instruction, as in the BIA schools on the mainland—on how to be good (cf. p. 666). Precisely like Thomas L. McKenney's mission boarding schools, these drew children away from the presumably baneful influence of home and family and kept them away forcibly, no matter how homesick they became, with the ever-ready cooperation of the paramilitary constabulary. To round off the parallels with the colonel and the first BIA, Worcester and his men "removed" troublesome tribes to special reservations (pp. 445, 473).

To take up where the Spaniards had run out of virility, Worcester had recruited a staff of "absolutely fearless" governors and lieutenant governors for the several provinces. They became "the law and the prophets, and no appeals are taken from any just decisions which they may make, nor is their authority questioned" (p. 447). He always maintained that the life of any one of them would furnish his beloved Rudyard Kipling with material "for a true story of absorbing interest." Jeff D. Gallman, for instance, "was a dead shot with revolver and carbine . . . absolutely fearless . . . of a kindly cheerful disposition. . . . As the years went by, the Ifugaos came to regard him as but little less than a god" (pp. 454–55). Or Walter F. Hale, like Gallman "a man with chilled-steel nerve," who needed his fearlessness "in the early days in Kalinga where the people, who had been allowed to run wild too long, did not take as kindly to the establishment of government control as had the Bontoc Igorots and the Ifugaos" (p.

456). Or consider yet another indomitable official who wandered "for weeks on end through the trackless forests of Nueva Vizcaya in order to get in touch with Ilongot savages who were a good deal more than 'half devil' with the balance not 'half child' but peculiarly treacherous, vicious and savage man" (p. 506). So did the merciless savage of old now lurk some eleven thousand miles from where he started out, though he reappeared as Kipling's "White Man's Burden" (1899): "Your new-caught, sullen peoples,/ Half devil and half child." Worcester drew on the poet for language capable of doing justice to those who bore "their heavy shares of the white man's burden" in assisting "me in the work carried on under my direction for the non-Christian tribes of the Philippines."

"Me" and "my"—of course Worcester had his chilled-steel self in mind when he paid subordinates such handsome tributes. They were loyal to *him,* "the ruler of all non-Christians," as he once introduced himself in a Kalinga village (pp. 431–32). Apart from the harm he did to peoples and cultures, the professor had enormous significance for what he revealed so directly about the Anglo-American attitude toward native peoples. The truth he made plain was how he and his compatriots reveled in being little less than gods. As the natives' absolute law and supreme prophet, Worcester himself transparently lived out the fantasies of omnipotence and omniscience that had given the tropics "their strangely firm grip" on him in the first place. He was an American Prospero who pretended to be a stern but kind parent, but as with Jefferson and McKenney, woe to the natives who questioned his authority, laughed at him, or ran wild too long. And like the original Prospero, he pretended to have no interests of his own: "I am absolutely without political ambition save an earnest desire to earn the political epitaph, 'He did what he could' " (p. 90).

To be sure, Worcester drew economic as well as psychic income from his career in the colonial service. From the beginning he had had an eye on "the enormous natural resources of the archipelago" (Knot, 879). Once in the insular government he worked to bring in American and British capital and resented the childish inability of Filipinos to appreciate the beneficence of their adult protectors: "We must *wean* them from their present hostility toward legitimate business interests" (pp. 692–93; my italics). After 1913 Worcester plunged into business interests of his own, including cattle, coconuts and coconut products, and interisland transportation, established companies in association with Lever Brothers, Ltd., of London, and J. W. Harriman of New York, and bought himself the *Anne W. Day,* a seventy-five foot yacht he generously

made available to the University of Michigan Archeological Expedition of 1922. On Mactan, the island where Magellan had been killed, he built himself a "beautiful, although unpretentious, home, finished in fine Philippine hardwoods and embowered in stately palms and lovely flowering plants" (pp. 41, 70–74, 468). Worcester's epitaph might have been revised slightly without doing damage to the evidence: "He *got* what he could."

But the *control* and the *development* of the country were two sides of the same coin, and so were the psychic and real incomes he derived from the process. In a sense all his acquisitions were food for the body; he had already feasted his soul in playing god to natives. His claim to fame as a colonial statesman rested on the absolute mastery he had shown in subduing the "fierce" Kalingas, the "warlike" Ifugaos, and all the others. And "no other phase of our national effort in the Philippines," asserted his biographer, "has been so universally recognized as being completely what the American people would wish it to be" (pp. 27–28). That was probably true. When they thought about it, Anglo-Americans did not like the idea of anyone running wild anywhere, especially not for too long. And related to this drive to rationalize the world was the major reason McKinley, Roosevelt, Hay, Root, Taft, and all the other conquerors so quickly snapped up Worcester's hand-list of tribes and his insistence the Filipinos were not "a people." The national history of Anglo-Americans had predisposed them to think that their westward-moving empire, then as always, had moved into "unpeopled" territory.

Baguio, the summer capital of the insular government, was Worcester's most fitting monument (pp. 358–87). In the 1890s he had heard of a place in the highlands of northern Luzon that even had occasional frosts. As a commissioner he explored the region and discovered that the marvel really existed and with a climate, as Taft enthusiastically reported to Root, "not unlike that of the Adirondacks, or of Wyoming in summer." The Benguet Road to Baguio was completed in 1905, the Baguio Country Club opened its doors the same year, with facilities that ultimately included a nine-hole golf course, tennis courts, trap-shooting, a polo field, and other Occidental amenities. In this home away from home,

◀ *A Typical Negrito Man with Secretary Worcester*, from the *National Geographic Magazine*, September 1912. Worcester supplied the legend: "This photograph shows the relative size of the Negritos compared with a 6-foot American." His companion does not appear overawed by "the ruler of all non-Christians."

colonial statesmen and their staffs could recuperate from the arduous burdens of ruling new-caught natives, while their military counterparts could relax at Camp John Hay, "The Mountain Post Beautiful" in the pine woods nearby. Baguio was one of the first of the American enclaves, those virtually sealed-off settlements of compatriots abroad that would shortly include the "Zonians" in Panama and one day the "advisers" in Indochina. And the enclave that Worcester pioneered had an importance that transcended its climate. As he said of the annual "Teachers' Camp" at Baguio, it was imperative that U.S. nationals periodically get together with their own kind:

> Americans who spend too many years in out-of-the-way municipalities of the Philippines without coming in contact with their kind are apt to lose their sense of perspective, and there is danger that they will grow careless, or even slovenly, in their habits.

CHAPTER XXI

The Strenuous Life Abroad: "Marked Severities" in the Philippines

I preach to you, then, my countrymen, that our country calls not for the life of ease but for the life of strenuous endeavor.

—THEODORE ROOSEVELT,
"The Strenuous Life," 1899

It is destiny that the world shall be rescued from its natural wilderness and from savage men.

—ALBERT JEREMIAH BEVERIDGE,
"The Star of Empire," 1900

If one starts to play the role of God one cannot avoid playing the role of the Devil as well.

—K. M. ABENHEIMER,
"Shakespeare's 'Tempest,' " 1946

"WESTWARD the Star of Empire Takes its Way"—not everyone was as pleased as young Senator Beveridge by this exultant misquotation of Bishop Berkeley in what became a Republican party campaign document. Decades ago Fred Harvey Harrington ably demonstrated that extension of the empire to the Philippines was opposed by "a strangely assorted group of citizens." Southern Democrats thought it would further exacerbate the "race question" in that era of lynchings and race riots. Northern Democrats, including followers of both Bryan and Cleveland, opposed annexation as a violation of the principles of the Declaration of Independence. A few regular Republicans, such as Senator George Frisbie Hoar and Speaker Thomas B. Reed, broke from party traces on the issue. Independent Republicans—or mugwumps—such as Carl Schurz, Moorfield Storey, and Charles

Francis Adams, Jr., became leaders of the opposition. Outside the usual party structures, other protesters included such reformers as single-taxers and welfare workers, and such academics as William James, William Graham Sumner, and even the wavering Jacob Gould Schurman. The protest was financed by an occasional capitalist angel, such as Andrew Carnegie, and supported by such men of letters as Mark Twain and William Dean Howells. From such odd sources came the movement that established the Anti-Imperialist League in late 1898, denounced administration foreign policy from the platform and in pamphlets, and then quickly subsided—with such notable exceptions as Moorfield Storey and Erving Winslow—following the rout of 1900.

In ways that are just beginning to be understood, historians are indebted to those who participated in this great debate over national destiny. And no one can come to grips with what was in fact at stake, I suggest, without first understanding what the debate was not about. It was not over expansion versus no expansion. Those in the protest movement, David Healy has shown, "betrayed expansionist leanings almost to a man." Bryan, for instance, wanted coaling stations in Cuba and the Philippines, and wanted to annex Puerto Rico if the inhabitants would consent. Richard Olney, Cleveland's former secretary of state, wanted a fortified naval base in the Philippines; like John Hay, Carl Schurz and Andrew Carnegie wanted to annex Canada; and Senator Hoar voted to annex Hawaii and make Cuba a protectorate. Insofar as expansion was the essence of imperialism, then, to call such men anti-imperialists was deceptive.

With Southern senators such as Benjamin R. Tillman of South Carolina and John W. Daniel of Virginia on one side and Northern senators such as Lodge of Massachusetts and Beveridge of Indiana on the other, the debate surely was not over racism. Senator Daniel feared taking up what he called a "mess of Asiatic pottage," while Senator Beveridge favored taking up this mess of "inferior races" as the white man's burden. In a noteworthy essay Christopher Lasch, who quoted Daniel, noted that both sides "accepted the inequality of man—or, to be more precise, of races—as an established fact of life." For the one side racism acted as a deterrent to grasping the spoils of war and for the other it acted as a stimulant. In operational terms, the former failed because they were the "outs," fewer in number, and never within reach of the levers of power.

What then was the debate all about? It was about whether the U.S. empire should be hemispheric or global, and secondarily about the nature of the Constitution: did that document follow the flag? On this less important issue, Senator Francis G. Newlands of Nevada put the issue succinctly on February 20, 1900: "The difference between the imperialists and the anti-imperialists . . . is that the imperialists wish to expand our territory and to contract our Constitution. The anti-imperialists are opposed to any expansion of territory which, as a matter of necessity, arising from the ignorance and inferiority of the people occupying it, makes free constitutional government impracticable or undesirable" (*Congressional Record*, XXXIII, 1996).

On the more fundamental issue of the magnitude of the empire, the opponents of its extension labored at a certain disadvantage politically. Since when had smaller been better in the United States? On the legal and intellectual levels, however, anti-annexationists argued strenuously that to extend U.S. sovereignty over the Philippines upended all the most cherished traditions of the Union.

That was precisely the contention of Thomas Morton's old editor, Charles Francis Adams, Jr.—former president of the Union Pacific Railroad, Henry's brother, and himself a historian with accomplishments recognized in part by his position as head of the Massachusetts Historical Society. In the Christmas season of 1898 Adams entered the controversy with *"Imperialism" and "The Tracks of Our Forefathers,"* a most remarkable address delivered fittingly in Lexington. Adams began by invoking the first Yuletide of the forefathers at Plymouth and voiced anew their children-of-Israel refrain: "Thus, once a year, like the Israelites of old, we, as a people, may take our bearings and verify our course, as we plunge out of the infinite past into the unknowable future." Adams lashed out at the imperialists who had so lost their bearings as to cross over into overseas conquests and away from a "cardinal principle in our policy as a race." Now that he was not defending forefathers from the charges of Thomas Morton, Adams was commendably free from cant about that cardinal principle. Always the Anglo-American had made himself "a terror to poor barbarous people":

> From the earliest days at Wessagusset and in the Pequot war, down to the very last election held in North Carolina [with its outrages against blacks],—from 1623 to 1898,—the knife and the shotgun have been far more potent and active instruments in his dealings with inferior races than the code of liberty or the output of the Bible Society. [p. 16]

It had been "a process of extermination," admitted Adams, but for that very reason "the salvation of the race. It has saved the

Anglo-Saxon stock from being a nation of half-breeds,—miscegenates, to coin a word expressive of an idea." Now that the imperialists had taken their feet from these tracks of the forefathers, Adams added prophetically, their intense racial antipathy toward "the Asiatic" afforded small encouragement for the assumption of "new obligations." But at this point his theme sprang a leak and settled low in the Pacific just off the Golden Gate. Why stop the process of extermination there? Had not the forefathers risked becoming "half-breeds" when they crossed the wilds of the Atlantic in the first place? Why should their sons now be unmanned—that is, become half-native—simply because they were crossing another body of salt water to deal with another lot of natives? In reality, as his subsequent vacillations suggested, Adams had not the slightest assurance the original Magistrate Winthrop would not, had he been alive, have made tracks out to join the Philippine Magistrate Winthrop.*

Yet this farther expansion was not "by natural growth in thinly settled contiguous territory, acquired by purchase for the express purpose of ultimate statehood," defensively asserted Erving Winslow, Adams's neighbor, who was secretary of the New England Anti-Imperialist League and with Gamaliel Bradford, another descendant of the founders of Plymouth Colony, a mainstay of the organization. Expansion that far, Winslow insisted, "cannot be confounded with, or made analogous to foreign territory conquered by war and wrested by force from a weak enemy." He might have been talking about the Black Hills in the Dakotas. Implicit in his argument, and explicit in Adams's, was complete sanction of all the fearful years and blood-soaked miles of the continental conquest.

And that was also the inescapable implication of what Senator Hoar read into the record (see p. 297): had the Filipinos been "savages no better than our Indians," then they would have deserved what they were getting. In reality these new-caught natives differed from "our Indians" in a still more critical aspect for Hoar: they were farther away. On July 5, 1898, in the debate over Hawaii, he supported annexation of those islands but made clear he would not go beyond to "dominion over archipelagoes in distant seas." But why was he willing to go as far as Hawaii? Because there "the people" were closer and "willing and capable . . . to share with us our freedom, our self-government, our equality, our

* Two years later Adams brought himself to vote for McKinley and Roosevelt—see Robert L. Beisner, "Charles Francis Adams and the Election of 1900" in his *Twelve against Empire: The Anti-Imperialists, 1898–1900* (New York: McGraw-Hill, 1971), pp. 107–32.

education, and the transcendent sweets of civil and religious liberty." Who were those lucky "people"? They were the Yankee missionaries and planters who had made these islands fit for annexation, unlike those more distant islands looming up over the debaters. But what if the Hawaiian natives did not want to be annexed? That mattered not at all to Hoar, no more than not consulting "the Indians in Texas or in California or in New Mexico or in Alaska when those territories were taken into the Union." As with the Indians then, so with the Hawaiians now: "it would be as reasonable to take the vote of children in an orphan asylum or an idiot school" (*Congressional Record*, XXXI, 6661–63). Geography and peoples had thus become hopelessly entangled in the senator's mind after he had stretched the imaginary line of the Monroe Doctrine far out into the Pacific (but still not as far as McKinley wanted him to stretch it). A hemispheric imperialist, Hoar believed all governments derived their just powers from the consent of the governed *beyond* his taut imaginary line: in the Philippines but not in Hawaii, not in Cuba under the Platt Amendment, for which he later voted, not in Puerto Rico, about which he made no fuss, and for that matter not in the Indian Territory, denied all transcendent sweets of self-determination and soon to become perforce part of a state (1907). In his twistings and turnings, Hoar himself illustrated the truth of Senator Orville H. Platt's apt observation during their debate. The fundamental principle of the Declaration, said Platt on December 19, 1898, had always depended on "the consent of *some* of the governed" (*Congressional Record*, XXXII, 297; my italics).

I do not deny that it took courage for Hoar and the other anti-annexationists to stand in front of the onrushing empire. To call for a halt to the U.S. empire somewhere was something. Yet their moral position was undercut by their willing acceptance of past conquests, and their arguments were fatally weakened by their own expansionism and racism.* From where they stood, they

* In July 1899 the New England Anti-Imperialist League established a black auxiliary—Daniel Boone Schirmer, *Republic or Empire: American Resistance to the Philippine War* (Cambridge, Mass.: Schenkman, 1972), p. 172. Blacks were thus segregated even within this key organization. Willard B. Gatewood, Jr., mentioned this fact and pointed out also that "for most black Americans the color of the Filipino was important," with many, despite their traditional ties to the Republican party, expressing opposition to McKinley's foreign policy. In the summer of 1899, for instance, Bishop Alexander Walters of the AME Zion Church observed that he did "not think that America is prepared to carry on expansion at this time, especially if it be among the dark races of the earth. The white man of America is impregnated with color phobia"—*Black Americans and the White Man's Burden, 1898–1903* (Urbana: University of Illinois Press, 1975), pp. 200, 204, 209. The segregation of blacks in the Anti-Imperialist League underlined the bishop's point. My reference to this racism obviously does not apply to the Afro-American anti-annexationists, who were pretty much ignored at the time and since.

could not possibly answer the flat statement of Whitelaw Reid, Hay's old superior on the *New York Tribune* and currently one of McKinley's peace commissioners: "The American people is in lawful possession of the Philippines with the assent of all Christendom," said Reid, "with a title as indisputable as its title to California." Nor could they counter Albert Beveridge when he cried in "The March of the Flag" (1898):

> Distance and ocean are no arguments. The fact that all the territory our fathers bought and seized is contiguous is no argument. . . . Cuba not contiguous! Porto Rico not contiguous! Hawaii and the Philippines not contiguous! The oceans make them contiguous. And our navy will make them contiguous.

Even this refreshing frankness about territory the fathers seized made opponents of the enlarged empire squirm.

Many or most of those in what was loosely called the anti-imperialist movement were really liberal imperialists who could not bring themselves to recognize they lamented the passing of a country that never was. The administration had changed his beloved Monroe Doctrine "of eternal righteousness," Hoar complained, into a doctrine "of brutal selfishness." In another anticipation of "the great aberration" thesis, Cleveland muttered that the United States was abandoning all the "old landmarks." And Speaker Reed agreed they were rapidly leaving behind "the foundation principles of our government." Nonsense! thundered the most bellicose expansionist of them all in his campaign speeches of 1900: "We are making no new departure. We are not taking a single step which in any way affects our institutions or traditional policies." And except for the salt water and miles, Theodore Roosevelt was right. Another historian in politics, he quickly punctured inflated notions of just who had traditionally consented to be governed and easily established that Jefferson had applied the principles of his Declaration very selectively indeed. Roosevelt capped his lesson on the realities of the past with a truth liberal imperialists wanted to put behind them: "The history of the nation is in large part the history of the nation's expansion."

"Of course," Roosevelt never tired of pointing out, "the presence of troops in the Philippines during the Tagal insurrection has no more to do with militarism and imperialism than had their presence in the Dakotas, Minnesota, and Wyoming during the many years which elapsed before the final outbreaks of the Sioux were definitely put down." The conquest of the Philippines perhaps had no more but surely no less to do with militarism and imperialism than had the conquest of the Dakotas. As if in mockery of their contention that there was no analogy, Hoar and other administration opponents protested the conduct of the insular war and unwittingly wound up providing valuable documentation of its identity with the traditional strategy and tactics used to crush native resistance on the mainland.

2

Not all the censorship of the Otises nor all the whitewash of the Worcesters could keep the secret from seeping out: the Philippine–American War was waged against the peasant population of the islands. Seventy thousand U.S. conquerors were in the field by the fall of 1900; but even earlier Robert Collins of the Associated Press had cabled that "there has been, according to Otis himself and the personal knowledge of everyone here, a perfect orgy of looting and wanton destruction of property" (MS, p. 11).* And as soldiers will, the invaders chatted about their work in letters to the folks back on the mainland.

Black Americans who had helped take up the white man's burden were in much the same anomalous position they had been during the Plains Indian wars. Willard B. Gatewood's discussion of their letters home shows they were distressed by the color line that had been immediately established in Manila and by the epithet "niggers," which white soldiers "almost without exception" applied to Filipinos. Yet they joined whites in calling them "goo-goos"—as the invaders said, "all goo-goos look alike to me." They too took lovers they called "squaws." A black lieutenant of the 25th Infantry wrote his wife that he had occasionally subjected natives to the water torture. Captain W. H. Jackson of the 49th Infantry admitted his men identified racially with the Filipinos but grimly noted "all enemies of the U.S. government look alike to us, hence we go on with the killing." On the other hand, David Fagen was merely the most famous of the dozen or so black deserters who joined Aguinaldo's forces.† For those who did not

* For an explanation of the abbreviated references, see p. 521.

† General Frederick Funston, who had brought off the capture of Aguinaldo through ruse, forgery, and torture, afterward put a $600 price on David Fagen's head and passed word the deserter was "entitled to the same treatment as a mad dog." In late November 1901 he had the great satisfaction of learning that Fagen's head had been delivered—literally, in a wicker basket—to the U.S. post at Bongabong (Gatewood, *Black Americans and the White Man's Burden*, pp. 288–89). Most white deserters did not join the revolutionists, but no doubt a few somehow made it across the color line. On February 10, 1902, for instance, U.S. officers

make this break, they were between "the 'Devil and the deep sea,' " in the opinion of Sergeant John W. Galloway, who felt "an affinity of complexion" with Filipinos and like other black soldiers saw an integral connection between racism and the imperialism that had carried them out to the archipelago. For such reasons the four black regiments there established relatively cordial relationships with their hosts, rarely figured in the ballooning reports of atrocities, and should be kept in mind, therefore, as exceptions to the general pattern of Anglo-American attitudes and practices.

Letters home from white soldiers frequently supported the view of those blacks who held they were fighting "a race war." On March 20, 1899, for instance, A. A. Barnes of the 3rd Artillery wrote his brother they had burned the town of Titatia the night before in retaliation for the murder of a volunteer: "About one thousand men, women, and children were reported killed. I am probably growing hard-hearted," he acknowledged, "for I am in my glory when I can sight my gun on some dark skin and pull the trigger" (MS, p. 25). Later, an officer who had served in the Philippines wrote reporter Henry Loomis Nelson:

> There is no use mincing words. . . . If we decide to stay, we must bury all qualms and scruples about Weylerian cruelty, the consent of the governed, etc., and stay. We exterminated the American Indians, and I guess most of us are proud of it, or, at least, believe the end justified the means; and we must have no scruples about exterminating this other race standing in the way of progress and enlightenment, if it is necessary. [MS, p. 99]

Besides, extermination of nonwhites brought those traditional fringe benefits the conquerors gloried in.*

By 1901 the civilizers and conquerors in Washington and Manila were sick and tired of "insurrectos" who would not cry quits.

recorded a firefight on Samar that killed the leader of a band: "The leader proved to be a *white renegade*, on whose person was found a commission as a second lieutenant from [General Vincente] Lukban. The name on the commission is John Winfrey" (SD 331, II, 1584; my italics). A *white* Filipino was an oxymoron, in his way more unnerving than a black and obviously no improvement over "a white Injun," than whom, as we heard Robert Montgomery Bird say, "there's nothing more despicable." All the white renegades, from Thomas Morton to John Winfrey, were freaks of nature, *white* savages who had, as John Quincy Adams said of Arbuthnot and Ambrister, disowned "their own natures" (see pp. 163, 110).

* Of these benefits Clarence Clowe, a soldier from Seattle, wrote Senator Hoar on June 10, 1900: "Nor can it be said that there is any general repulsion on the part of the enlisted men to taking part in these doings. I regret to have to say that, on the contrary, the majority of soldiers take a keen delight in them, and rush with joy to the making of this latest development of a Roman holiday" (MS, p. 78). Or this latest development of a Puritan holiday—remember John Winthrop's account of how refreshed the Saints had been after their slaughter of Pequots at Fairfield swamp (see p. 57n).

Before he was relieved in July, Major General Arthur MacArthur, Otis's successor and another veteran of the Civil and Indian wars, reported to the adjutant general that he had used "very drastic methods" against them. MacArthur was replaced by Major General Adna R. Chaffee, who had come up through the ranks fighting Comanches, Cheyennes, Kiowas, and Apaches and had commanded the China Expedition—Colonel Roosevelt had served under him in Cuba, admired the general, and had recently celebrated him in a campaign speech for leading the "drumming guns" of the 6th Cavalry against "the Chinese Boxer, his hands red with the blood of women and children." With Chaffee relocated in the islands and Roosevelt in the White House following McKinley's assassination, the strenuous life descended upon the Philippines in its full weight. On November 4 the proadministration *Manila Times* reported that Brigadier General Jacob H. Smith had been in Samar for ten days and "had already ordered all natives to present themselves in certain of the coast towns, saying that those who were found outside [after a specified date] would be shot and no questions asked." On November 11 the *Philadelphia Ledger*, another proadministration newspaper, printed an officer's letter: "Our men have been relentless, have killed to exterminate men, women, and children, prisoners and captives, active insurgents and suspected people, from lads of ten up, an idea prevailing that the Filipino was little better than a dog." These grim accounts spurred the scattered antiwar forces into one last effort that led to a congressional investigation.

On January 13, 1902, Senator Hoar introduced a resolution calling for the appointment of a blue-ribbon committee to inquire into the conduct of the war. Henry Cabot Lodge headed off this dangerous proposal by agreeing to hold hearings before his Committee on the Philippines. They began in February and finished at the end of June. Lodge suppressed much of the most damaging evidence, allowed his lieutenant, Beveridge, to interrupt and badger "hostile" witnesses and the Democratic minority, barred the public, and allowed only a few pet reporters access to their findings. Perhaps the best commentary on his and the country's attitude toward the Filipinos as a people was his refusal to allow a single victim to come before the committee—efforts to bring Aguinaldo from the islands to testify collapsed in the face of solid Republican opposition (SD 331, II, 1949). All this notwithstanding, the fugitive truth slipped by Lodge to make surprise appearances, and the three thousand pages of the hearings became a little-known treasure trove for those interested in the national character of the conquerors.

In his careful battle plans for the hearings, Lodge led off and concluded with his big guns: the current governor of the islands and the hero who had taken the United States there in the first place. But even this framework creaked.

As Lodge's last witness, silver-haired George Dewey demonstrated conclusively that he had been brilliant only in his white ducks (SD 331, III, 2926–84). Denying any deal with the nationalists, the admiral admitted that he had ordered Aguinaldo be brought back to Manila aboard the U.S. gunboat *McCulloch* and that he had told him to "go ashore and start your army." Under cross-examination Dewey rather ungraciously claimed he had not believed the sincerity of Aguinaldo's proclamations (May 1898) of alliance with the U.S. liberators, since the general was a liar after loot: "I think he was there for gain—for money—that independence had never up to that time entered his head. He was there for loot and money." But why then had he assisted a plunderer to organize an army? "You know the old saying," craftily replied the admiral, "that all things are fair in war." Toward the conclusion of the hearings Beveridge asked him, as a signer of the Schurman report, whether he still accepted its conclusions, including the Worcester line, "the Filipinos are not a nation, but a variegated assemblage of different tribes and peoples, and their loyalty is still of the tribal type." That had been his opinion, replied Admiral Dewey, "and it is still my opinion."

William Howard Taft had been more impressive. He spoke authoritatively of "tribes," their "utter unfitness," and like Worcester twitted Filipinos for their utter inability to control the Moslem Moros. But could the United States govern them? "It is possible," replied Taft, "for us to govern them as we govern the Indian tribes" (I, 329). He held the Filipinos had nobody to blame but themselves for a war "which remains a crime against civilization." Was it then a crime to fight for independence? Taft responded with a variant of John Endicott's point-of-the-sword argument: "It is a crime because it is subjecting their own people, in whose interests they profess to be carrying on the war, to the greatest privation and suffering" (I, 79). On the water cure he reported "some rather amusing instances of Filipinos who came in and said they would not say anything until they were tortured; that they must have an excuse for saying what they proposed to say" (I, 75). And it was his deliberate judgment "that never was a war conducted, whether against inferior races or not, in which there were more compassion and more restraint" (I, 77–78). Alas, the genial governor-general of the Philippines was not being candid. As he delivered this authoritative judgment, he had in his hands an official report from Major Cornelius Gardener in which that "civil" governor of Tabayas Province instanced "the extensive burning of barrios [hamlets] in trying to lay waste the country so that the insurgents can not occupy it, the torturing of natives by the so-called 'water cure' and other methods in order to obtain information, the harsh treatment of natives generally. . . ." *

General MacArthur, the administration's other principal witness, had to be an embarrassment to Lodge and his chief in the White House. In his official report of October 1, 1900, MacArthur had stated frankly that the success of Aguinaldo's guerrilla tactics depended "upon almost complete unity of action of the entire native population" and had discounted terror as the sole explanation, pointing in addition to the archipelago's "ethnological homogeneity" (I, 135–36). Now he repeated that "at one time the people were pretty nearly a unit in opposition" (II, 1943), an admission that should have put to rest for all time the cant of Taft and other administration spokesmen about "the Tagal insurrection." Moreover, MacArthur eschewed Dewey's character assassination. Aguinaldo, he repeated, "was the incarnation of the feelings of the Filipinos" (II, 1926). Yet for all this directness, MacArthur could hardly come right out and say that the absurd body count of five dead Filipinos for every one wounded, a reverse of the usual ratio, indicated ruthless slaughter. The efficiency of the U.S. soldiers accounted for it, he maintained lamely, "a consequence of their training in fire discipline and target practice, and so forth" (II, 1927). And so forth—in his eagerness to help the general over a rough spot, Senator Charles Henry Dietrich, whose fortune came from mining in the Black Hills, may have hurt a little, though his effort revealed continuity of attitude: "Is it not true," rhetorically asked the Nebraska Republican, "that the Filipino wounded, much like the Indians, made every effort to get away, and did not make any report of their wounds?" (II, 1928).

MacArthur was an unusual witness, especially for a military man. With staggering sophistication he called the archipelago not a U.S. colony nor even a new territory but "a tuitionary annex" (II, 1919). He subtly related overproduction economic arguments for its acquisition to psychological motives, and pronounced him-

* For the Gardener report, see Senate Document 331, II, 881–85. Taft transmitted it to Secretary of War Root on February 7, 1902, three days after his testimony—Oscar M. Alfonso, *Theodore Roosevelt and the Philippines, 1887–1909* (Quezon City: University of the Philippines Press, 1970), p. 98. Gardener's reference to "other methods" of torture no doubt included the less common "rope cure" that consisted of tying the victim's neck and torso together and then twisting the rope with a stick. Asked if it worked, one man who saw it administered replied: "A wooden Indian would make a speech if you gave him the rope cure" (MS, p. 62).

self more interested in the latter, the ages-old "expansion to the West," with present currents having swept "this magnificent Aryan people across the Pacific—that is to say, back almost to the cradle of its race" (II, 868).

This Sewardian view was restated in theological terms by another administration witness, James M. Thoburn, Bishop of the Methodist Church for India and Malaysia: "Well, we are there, and I think we ought to do our duty the best we can—if you will allow me to speak as a missionary, I would say—in the field God has put us. We did not seek it" (III, 2672). An old Asia hand of forty-three years' experience, the bishop knew the Filipinos as "treacherous in their character" (III, 2694), in fact "very much, in many respects, like our American Indians, it strikes me. They have no cohesion whatever among themselves" (III, 2669). But as a citizen, did he believe Filipinos should be subjected to the U.S. government without their consent and by force? Yes, the bishop did: "We have acted on the theory for a hundred years with regard to the American Indians, that no matter what they wish or what government they desire we will hold them by force" (III, 2705). And the doctrine historically applied to the Indians was all the more necessary for the Filipinos who "have fallen to us by what we call the fortunes of war." The reader will be pardoned a feeling of *déjà vue*.

A glimpse into what it meant to cultivate Bishop Thoburn's field in which "God has put us" emerged from the fascinating colloquy of Beveridge and yet another administration witness. Colonel Arthur Lockwood Wagner was more accomplished than most, but otherwise his career was illustrative of the many military professionals who left off fighting Plains Indians to take up fighting tropical islanders. Born in Ottawa, Illinois, in 1853, Wagner graduated from West Point in 1875, immediately went into the field against Indians, subsequently taught military science, became chief of the War Department's public relations division, served in Cuba and Puerto Rico, and had just returned to Washington from two and a half years in the Philippines, where he had been adjutant general of Luzon since late 1901. By virtue of this experience and his reputation as the author of well-known works on military history and tactics, Lockwood was one of Lodge's most important expert witnesses. Questioned as to when it was within the rules of war to burn down entire towns and villages, the colonel replied that an officer had to use his judgment and had to "be satisfied that the conduct of the entire community has been such as to deserve it before he would undertake to destroy all the houses in a barrio" (III, 2857–58). There followed Wagner's colloquy with the senator who became a practicing historian:

> Senator Beveridge: When a town or barrio has been notoriously known as a rendezvous, place of departure and return of ladrones [i.e., guerrillas or "bandits"], what then would be a justifiable course to pursue?
> Colonel Wagner: If the town were notoriously a nest of ladrones, if it was impossible to get the rest of the people to yield them up, it would be justifiable and proper to destroy the town, even though we destroyed the property of some innocent people. The Almighty destroyed Sodom, notwithstanding the fact that there were a few just people in that community—less than ten.
> Senator Beveridge: How strange; I was thinking of that instance of Sodom and Gomorrah.

Still, manifestly, did the Almighty thus "Judge among the Heathen," as John Mason had said after making the Pequot fort "as a fiery Oven," and thus had He done the same to Seminole villages, though Andrew Jackson had spelled the biblical names "Sodom and Gomorrow" (see pp. 43, 105).*

3

John Fiske's Manifest Destiny in the West had materialized as General MacArthur's "magnificent and mighty destiny" in the East (II, 870). The underside of that destiny was exposed by a number of witnesses rounded up by the antiwar movement, including, by one of history's little quirks, John Fiske's son-in-law.

Former First Lieutenant Grover Flint of the 35th Infantry had served in Luzon from November 1899 to April 1901 and described himself as presently "independent to a certain extent. I am now writing a biography of my father-in-law, the late John Fisk[e]" (II, 1774). In 1896 he had witnessed Spanish soldiers under General Weyler burning Cuban villages and collecting people in camps, and then in 1900 witnessed U.S. soldiers doing the same thing, burning Philippine villages of fifty or sixty houses where the men, women, and children were apparently engaged in peaceful pursuits: "I think the idea was at the time that the burning of these

* Cf. Secretary of the Navy Paul Nitze in the *New York Times* of August 15, 1965: "Where neither United States nor Vietnamese forces can maintain continuous occupancy, it is necessary to destroy those facilities."

villages would drive the people to the woods or to the towns; a policy of concentration, I think" (II, 1784). He had also seen the seemingly routine torturing of natives, ostensibly to discover weapons and information. One night in May 1900, for instance, he casually watched the water cure being administered to some thirty persons by Macabebe scouts under the command of a U.S. sergeant, "a white man." The next morning he witnessed it applied "to something over twenty men," with members of his own company volunteering to help out. Since he thought a commissioned officer ought to be present, he asked Major Geary, his commander, to be allowed to stay. "All right, if you want to," Geary had indulgently replied, while taking up his own more comfortable position under some trees fifty or sixty yards away. A senator asked Flint to describe the water cure and he obliged:

> A man is thrown down on his back and three or four men sit or stand on his arms and legs and hold him down, and either a gun barrel or a rifle barrel or a carbine barrel or a stick as big as a belaying pin. . . . is simply thrust into his jaws and his jaws are thrust back, and, if possible, a wooden log or stone is put under. . . . his neck, so he can be held firmly.

Senator Julius Caesar Burrows, Republican from Michigan, interrupted to ask, "His jaws are forced open, you say? How do you mean, crosswise?"

> The Witness. Yes, sir, as a gag. In the case of very old men I have seen their teeth fall out—I mean when it was done a little roughly. He is simply held down, and then water is poured onto his face, down his throat and nose from a jar, and that is kept up until the man gives some sign of giving in or becomes unconscious, and when he becomes unconscious he is simply rolled aside and he is allowed to come to. . . . Well, I know that in a great many cases, in almost every case, the men have been a little roughly handled; they were rolled aside rudely, so that water was expelled. A man suffers tremendously; there is no doubt about that. His suffering must be that of a man who is drowning, but who can not drown. [II, 1767]

This was a far cry from the mild, amusing exercise Worcester and Taft discussed so nonchalantly. But given such suffering, why had Flint stopped the torture only in a few cases when it appeared the victim might die? The replies of the Harvard-educated veteran both underlined the banality of evil and said something about the national character: he was only a soldier following orders; while in the service, so far as he could remember, he had never criticized a superior officer; his superior Major Geary was nearby; for him to suggest to Geary it be stopped would have been "improper," just "as it would be improper for your servant to suggest things to you"; and besides, "at the time I did not disagree or disapprove of the process whatever" (II, 1776–80).

Very few of the invaders ever did disapprove of the process whatever and especially not after the war had dragged on into its third year, with guerrillas still active in Batangas Province of southern Luzon and on Samar in the Visayas. The hard truth was that U.S forces had not found islands full of the welcoming natives their leaders constantly invoked, the good Fridays who would adopt them as the protective American fathers who had driven away the mean Spanish fathers. Instead they had seemingly come up against a nation of Calibans who repaid their benevolence with treachery, betrayal, and skulking attacks—General Chaffee called them "gorillas" who hid in the brush. And then there was their "barbarous cruelty common among uncivilized races" that the War Department harped on in its reports (SD 205, I, 2). "I have been in Indian campaigns where it took over 100 soldiers to capture each Indian," reported Brigadier General J. Franklin Bell in 1901, "but the problem here is more difficult on account of the inbred treachery of these people, their great number, and the impossibility of recognizing the actively bad from the only passively so." Or as Major General Lloyd Wheaton reported from Luzon on June 30, 1901: "Unexampled patience was exercised throughout the department in the treatment of these savages, habitually violating all the laws of war as known to civilized nations . . ." (SD 205, I, 50). The "unexampled patience" we have already sampled was palpably wearing thin.

It snapped with what Secretary of War Root called "the base treachery, revolting cruelty, and the conditions of serious danger" of "the massacre" at Balangiga in Samar in September 1901:

> There the natives had been treated with kindness and confidence, liberty and self-government had been given to them. Captain Connell, the American commander, was of the same faith and had been worshipping in the same church with them. With all the assurance of friendship our men were seated at their meal unarmed among an apparently peaceful and friendly community, when they were set upon from behind and butchered and their bodies when found by their comrades the next day had been mutilated and treated with indescribable indignities. [DIL, VII, 189–90]

Now, this was the most spectacular of the Filipino atrocities: seventy-four men of Company C of the 9th Infantry were cut down at breakfast, with less than thirty surviving, of whom twenty-two were wounded; native losses were estimated at from fifty to one hundred and fifty killed. One of the survivors was William J. Gibbs, who received a commendation and promotion to corporal for his bravery after the attack. Currently a piano tuner in Springfield, Massachusetts, he appeared before the Lodge committee on May 9 and 10, 1902. Since the attack set off the still more grim final phase of the war that had just wound down and since in a way it summarized thousands of pages of testimony on the U.S. presence in the Philippines, let us follow along with this trustworthy witness's account of the affair at Balangiga (SD 331, III, 2284–2310).

In July 1901, shortly after they had arrived in the village of some two thousand souls, Gibbs related,

> the officer, Captain Connell, wanted to have things cleaned up around the town, and he went to work and issued a proclamation to have all the natives appear the next morning and clean out the town. The natives appeared to be somewhat reluctant in regard to that. They turned out, but they did not work very hard, and then the next morning they refused to come at all. So he went to work and sent the men, each man to a shack, and forced them to come out, and had a guard placed over the men while they were working in the hot sun. . . . They put the 90 natives in the two Sibley tents, which only held about 16 soldiers; they could not lie down; they had to stand up [all night]. There was not room enough to lie down. They stayed there for two or three days. In the morning Captain Connell would line the natives up and would issue them bolos for the purpose of cutting down the underbrush. . . . The weather was damp, and of course it was the rainy season at that time and it was very unpleasant all right for the natives, and they started to complain about it. They even wanted a little matting to put on the inside of the tent, to keep them from the dampness, from the ground, but he would not allow that at all.

After three or four nights of this the villagers were allowed to go home and come back mornings to work under an armed guard of twelve men.

In the following weeks the involuntary workers had to feed themselves: "We didn't give them anything." And while they were "cleaning up the town," Captain Connell "sent men out from the company to destroy all the rice and fish and everything in the line of food that they possibly could. He thought they were taking them to the insurrectos in the mountains."

But why had Captain Thomas Connell, a West Pointer and veteran of the China Expedition, made the villagers into a forced labor crew? To give the place, as Gibbs put it,

> a semblance of civilization. That is what they were supposed to do. They were supposed to cut down all the trees and underbrush and everything around there that would be an obstruction to seeing things at night on post.

Some two months after this "civilizing" campaign commenced, another seventy-five natives, as it turned out General Vincente Lukban's picked men, obligingly came down from the mountains to help, were issued bolos, and confined nights to the two tents where they also were too numerous to lie down.

A lighter note was struck when Beveridge asked the witness whether it was not a fact that "these people" habitually rested "squatting on their haunches talking to each other for hours at a time?" "Yes, they do," Gibbs acknowledged but in the questioning that followed refused the senator's bait: "No, they do not sleep that way at night."

After about a week of squatting nights on the damp ground in the Sibley tents the mountain men joined the villagers in launching the attack of September 28, pitting surprise and their bolos against three armed guards and the other soldiers, whose rifles were stacked near at hand. As it was, by Gibbs's estimate their own losses were three times greater than those of the soldiers. Beyond doubt the enraged Balangigans chopped up the bodies of their tormentors and thereby indeed committed atrocities. And that was Gibbs's full story of the "massacre" and of the "kindness and confidence, liberty and self-government" Captain Connell had bestowed upon villagers who lived by the fish, coconuts, and rice he had destroyed. Also beyond doubt, I submit, would have been the loud cheers on the mainland had U.S. prisoners of war surprised their armed captors, killed many with their own tools, and escaped from their slave labor camp.

And that is not to say that the American Prosperos did not really feel betrayed by "these savages." From Washington came words like Root's and strenuous injunctions to General Chaffee to put an end to the "insurrection" forthwith. Chaffee in turn passed on word to subordinates that their forces "must be pushed with the utmost vigor." On September 30, 1901, for instance, he wrote General Hughes, the expert on the Visayans' cephalic index, that they had the "duty to suspicion every male inhabitant in these islands" and "while I do not urge inhuman treatment of any person . . . it is necessary that we be stern and inflexible." Chaffee announced that he was reassigning the subdistrict of Samar to Brigadier General Jacob H. Smith, who, "as I am told, is an energetic officer, and I hope he will prove so in command of that brigade." He hardly needed to remind Hughes, who was also another former Indian-fighter, that "it is to our interest to disarm these people and to keep them disarmed, and *any means to that end is advisable*. It will probably cost us a hundred lives to get back the guns lost at Balangiga" (SD 331, II, 1591–93). The phrase I have italicized telegraphed the genocide that followed.*

Brigadier General J. Franklin Bell took charge of scouring Batangas Province in southern Luzon. His orders, reproduced in the hearings, made plain that the demand for "utmost vigor" issued at the top had been passed on down to his subordinates. On December 13, 1901, Bell reminded station commanders that for the past three years the brigade had exercised "forbearance and generosity" in the vain hope "the people might be conciliated and become reconciled to and convinced of the benevolent purposes of the Government." To describe what had happened instead, the former cavalry lieutenant drew on language he may well have used in his campaigns against Native Americans: there had been "skulking operations" and apparently pacific people had "treacherously risen" against the army. Flip back the calendar a quarter of a century and he might have been describing the predicament of Custer's 7th Cavalry in the Dakotas: "We consequently find ourselves operating in a thoroughly occupied terrain against the entire population . . ." (II, 1612–14). The Batangans were more numerous than the Sioux and did not wear war paint to help Bell identify "the actively bad from the only passively so," but otherwise the parallels were close, as a Christmas Eve circular indicated. He proceeded on the assumption

> that, with very few exceptions, practically the entire population has been hostile to us at heart. In order to combat such a population, it is necessary to make the state of war as insupportable as possible, and there is no more efficacious way of accomplishing this than by keeping the minds of the people in such a state of anxiety and apprehension that living under such conditions will soon become unbearable. [II, 1628]

* I use the word *genocide* here—and used it earlier as descriptive of John Endicott's orders before he sailed to Block Island in 1636 (p. 34)—as it was defined by the UN General Assembly in 1946: "Genocide is a denial of the right of existence of entire human groups, as homicide is a denial of the right to live of individual human beings. . . . Many instances of such crimes of genocide have occurred when racial, religious, political and other groups have been destroyed, entirely or in part"—*Journal of the United Nations*, No. 58, Supp. A—A/P, LV, 476.

And as in the Indian wars, the Prospero-like pose of the kind and forbearing parent could shift suddenly into utter contempt and a wrath that had the entire population as its target. "These people need a thrashing to teach them some good common sense," Bell wrote General Lloyd Wheaton on December 27. On New Year's Day he would lead 2,500 men into the field "for the purpose of thoroughly searching each ravine, valley, and mountain peak for insurgents and for food, expecting to destroy everything I find outside of towns, all able-bodied men will be killed or captured. Old men, women, and children will be sent to towns" (II, 1691). Like Worcester and Bourns a decade earlier, Bell and his men did thereafter march through the countryside shooting everything that moved. Under the battle cry of "Remember Balangiga" they shot buffalo—not tamaraus but the indispensable carabaos of the peasants—and other animals, burned homes, destroyed rice and other crops, herded survivors into reservationlike military zones, and forced them into clean-up campaigns and into becoming beasts of burdens the white men would not pick up. But unlike Worcester's collecting expeditions, this time fleeing natives were not excepted from the shooting—at least not those of them who were "able-bodied," a tricky term.

On Samar the energetic General Jacob H. ("Hell-Roaring") Smith issued Christmas Eve orders so similar to Bell's they betrayed their common origin in Chaffee's "utmost vigor." Of course Smith put much of his own notorious Circular #6 in his own words: it was imperative "to adopt a policy that will create in the minds of all the people a burning desire for the War to cease; a desire or longing so intense, so personal . . . and so real that it will impel them to devote themselves in earnest to bringing about a state of real peace" (II, 1571–73). "Burn and kill the natives" soldiers appreciatively called his campaign (III, 2314). "The fear of God" was another name they gave it, James H. Blount recalled—as the former commander of a company of scouts, Blount had reason to know that "the American soldier in officially sanctioned wrath is a thing so ugly and dangerous that it would take a Kipling to describe him." The ugly truth was that the benevolent assimilator had become a Sodom-and-Gomorrah God who played the very devil with natives who got in his way.

Yet another old Indian-fighter, Smith adopted tactics that had worked against Geronimo's Apaches when they were captured, imprisoned, and finally herded into a reservation. He began by ordering all natives, under pain of death, out of the interior. Those who streamed down to the coast were immediately thrown into concentration camps that were not the experiments in hygienic living Worcester described. On January 26, 1902, a member of General Chaffee's party of visitors at the camp in Catbalogan, where Smith had his headquarters, described for the *Army and Navy Journal* what he saw (MS, pp. 90–91). The inmates had a "pinched, hungry look"; the great majority "were in rags" and suffering from diseases that made their death rate "from two to four per day." All of them "had a more or less cowed appearance," the correspondent noted: "They are a miserable-looking lot of little brown rats, and were utterly spiritless. . . . Nothing seems to break their apathy. One to whom I spoke kindly and questioned as to his treatment changed for a moment into a gleam of intelligence, but quickly resumed his original stare." *

Natives still in the back country were shot or in some instances captured, and in any event had their villages burned. On November 23, 1901, for instance, Marine Major Littleton W. T. Waller reported from Basey that the second of his three columns, "in accordance with my orders, destroyed all villages and houses, burning in all 165"; he commended subordinates for taking their men on a "heroic" march where "no white troops" had ever been before; as soon as the men were fit to move again, he would strike a trail that "will lead us out to the east coast near Hernani. First, however, I wish to work southward a little, destroying all houses and crops and, if possible, get the rifles from Balangiga." General Smith endorsed his report, urging another brevet for the marine who had proved himself "an officer of exceptional merit and carries out my wishes and instructions loyally and gallantly"; General Chaffee added his endorsement, noting that "the work of the marines stationed in southern Samar has been gratifying, and they have accomplished much good in the short time they have been there" (SD 331, II, 1603–6).

Shortly after Christmas Waller led fifty inadequately provisioned marines and some thirty *cargadores,* or "coolies," as Worcester would have called them, out of Lanang on the east coast in a disastrous march across the island to Basey. Only the major and a small advance party made it, while the remainder of his column ran out of rations and turned back in a retreat that became a rout through the vine-entangled jungle, with ten marines left behind to die and mark their line of flight. Those who straggled out of the mountains on January 18, 1902, came down seething

* The apathy of these "little brown rats" resembled that of the "Muselmänner" or walking corpses Bruno Bettelheim encountered in Dachau and Buchenwald. Catbalogan was manifestly an "extreme situation" for the inmates, comparable in its psychological impact to the Nazi camps—for the latter, see Bettelheim's essay on schizophrenia in the *American Journal of Orthopsychiatry,* XXVI (1956), 507–18, and his *Informed Heart* (Glencoe, Ill.: Free Press, 1960).

with resentment against their "sullen and unfriendly" natives, who had fared much better than they and had, it was later established, done about everything but carry the white men physically from coast to coast—along with their own gear, the natives had carried the marines' packs and even, toward the end, their rifles. But the distraught leathernecks suspected them of "treachery," of "plotting" against them, of not finding for them the wild food and firewood only they knew how to find. When they had arrived on Samar the preceding October, Waller had charged them to "place no confidence in the natives, and punish treachery immediately with death." So when the carriers voluntarily returned to Lanang, their erstwhile white allies immediately clapped them in irons and shipped them around the island to Basey.

There, on January 20, Waller ordered the mass execution of eleven Filipinos, mostly their former *cargadores*, with marines "clamoring" for turns on the firing squads; two days later Waller informed his superiors it had been necessary "to expend eleven prisoners." When word of this "expending" reached Washington, with the hearings before Lodge's committee getting under way accompanied by a rising chorus of cruelty allegations, Root ordered Chaffee to have Waller charged with murder. At his trial (March 20–April 12, 1902) Waller admitted the killings but contended they differed not at all from the summary executions of prisoners by other expeditions in which he had served, that against the Chinese Boxers in 1900 and that against the Arab cavalry at Alexandria in 1882; he also insisted he had acted well within General Smith's orders. Presided over by Brigadier General William H. Bisbee, "a bald old former Indian fighter," the court-martial thereupon found him not guilty of murder; Chaffee approved the verdict, citing as well Waller's "mental anguish" and physical condition, factors the accused had not pleaded in his own defense. The decision meant, rightly observed Moorfield Storey, "that such killing of allies was not only no murder, but no crime" (MS, pp. 33–35). But it also forced Roosevelt to sign an order on April 21, 1902, for the court-martial of General Smith. Thus, by what was virtually a fluke under the circumstances, did the full truth, or at least most of it, come out about "the wishes and instructions" Major Waller had carried out so gallantly.

Unlike General Bell, Smith had indiscreetly specified what he meant by "able-bodied" or, in his case, by "capable of bearing arms." At his trial

> it was shown that General Smith gave Major Waller the following oral instructions: "I want no prisoners. I wish you to kill and burn;

> the more you kill and burn the better you will please me." He also declared that "the interior of Samar must be made a howling wilderness," and that he wanted all persons killed who were capable of bearing arms *and in actual hostilities against the United States;* and, in reply to an inquiry by Major Waller for an age limit, designated the age of ten years. [DIL, VII, 187; my italics] *

For issuing this order to Waller and to Waller alone, seemingly, were one to have any faith in these proceedings, General Smith was tried not for war crimes, not even for murder, but simply for *conduct to the prejudice of good order and military discipline*. He was found guilty "and sentenced to be admonished by the reviewing authority." In partial explanation of its whimsical decision, the court declared what was manifestly untrue, namely that the accused had not meant everything he said, "notwithstanding that a desperate struggle was being conducted with a cruel and savage foe."

President Roosevelt duly admonished Smith and, since the general was sixty-two years old, retired him from the active list. In approving the findings and sentence, Roosevelt also invoked the merciless savage theme: "I am well aware of the danger and great difficulty of the task our Army has had in the Philippine Islands, and of the well-nigh intolerable provocations it has received from the cruelty, treachery, and total disregard of the rules and customs of civilized warfare on the part of its foes" (DIL, VII, 188). A few days after ordering the Smith court-martial, Daniel Boone Schirmer has remarked, the president made his own attitude quite clear by sending a letter of congratulations to General Bell in Batangas for his scorched-earth campaign that had killed, according to the earlier estimate of the secretary of the province, one-third of the population through shootings, starvation, and war-induced disease. But Bell had commendably said nothing about ten-year-olds. Roosevelt did not order Chaffee tried for his *"any means"* orders; nor his secretary of war for countenancing, if not initiat-

* Moorfield Storey pointed out that Root interpolated the phrase I have italicized and that it was at variance with the words of Smith's own defense counsel, who said simply that the defendant "wanted everybody killed capable of bearing arms, and that he did specify all over ten years of age, as the Samar boys of that age were equally as dangerous as their elders" (MS, p. 33). Undoubtedly Root fiddled with the record as part of his ongoing efforts to keep the home folks from becoming too disturbed over what was happening to distant natives—as it was, their relative unconcern over the growing body of atrocity evidence called into question the depth of their humanitarian concern in 1898 over comparable Spanish tactics in Cuba. But in the context of the extermination in Samar the secretary's interpolation did not mean much. With very few exceptions indeed, every native on the island, as in Batangas, was by official definition "in actual hostilities against the United States."

ing, such orders; nor, needless to add, himself for bringing the strenuous life to the Philippines in such full measure.

Secretary Root's explanation of the wrist-slapping of General Smith, now more scapegoat than hell-roarer, fleshed out the administration's position. Root argued that the general's oral instructions "were not taken literally and were not followed," a contention that was contradicted by Waller's report from the field (quoted on p. 326) and by evidence disclosed at his trial. Whether children under ten and women were or were not killed was never established by any official inquiry and certainly not by the proceedings against Smith. But Root went on to justify "the actual conduct of military operations in Samar" by the "history and conditions of the warfare with the cruel and treacherous savages who inhabited the island" and cited two sustaining "precedents of the highest authority": General George Washington's orders to General John Sullivan in 1779 for the campaign against the Six Nations and the "severity" General William Tecumseh Sherman proposed against the Sioux after "the Fort Phil Kearny massacre in 1866" (DIL, VII, 189–90).

Now, in the process of ransacking War Department records for authorizations of terror, Root had unwittingly disclosed two important and related truths. The first was that the national past contained authorizations of terror and could easily be made to share the guilt of current killings, hurtings, and burnings. The second was that throughout this past, justifications of traditional forms of violence—the "liturgies of inflicting death," in historian Lynn White's words—had remained relatively fixed. Root's high precedents were instances in point.

In 1866 the 27th Infantry had constructed a series of posts in the Dakotas, of which Fort Phil Kearny was one. That was Sioux country, and Sioux warriors soon had the posts virtually cut off from one another and from their supplies. On December 21 Captain William J. Fetterman led a detachment of some eighty men and officers out to relieve a party of besieged wood-haulers but went out under strict orders not to go beyond nearby Lodge Pole Creek. Eager to establish his reputation as an Indian-fighter, Fetterman chased some seemingly panic-stricken warriors across the creek and found himself in a neatly sprung trap. On hearing of the loss of the entire detachment, Sherman wrote General Ulysses S. Grant on December 28: "We must act with vindictive earnestness against the Sioux, even to their extermination, men, women and children. Nothing else will reach the root of this case." That was "the Fort Phil Kearny massacre," and that was the "severity" Sherman ad-

Burial of the Dead . . . at Wounded Knee, New Year's Day, 1891. (Photograph by G. E. Trager, Chadron, Nebraska, courtesy of the Nebraska State Historical Society.)

vocated. As at Balangiga, the brilliant military victory of the foe was quickly labeled a "massacre" by Sherman, and Root was still repeating the label after the turn of the century.* Sherman's advocacy of "vindictive earnestness" was an almost ritualistic expression of the national will to exterminate, and his strategy of driving a continental railroad through Indian country to destroy their buffalo and starve them out in fact waged war against all the Sioux—men, women, and children—by destroying their food base. Hence, in his destruction of animals and crops in Batangas, General Bell was in the same tradition as Generals Sherman and Phil Sheridan and, in fact, George Washington.

* Even General Smith admitted that the actions at Balangiga *"were according to the rules of war,"* save for the mutilation of the dead (MS, p. 37; Smith's italics).

Dated May 31, 1779, and in the handwriting of Alexander Hamilton, Washington's instructions to Major General John Sullivan have been correctly described as "peremptory." Sullivan was to proceed against the hostile tribes of Iroquois, "to lay waste all the settlements around . . . that the country may not be merely *overrun* but *destroyed*." While Bell spoke of *thrashing* the treacherous Batangans, Washington spoke of *chastising* the treacherous Iroquois:

> But you will not by any means, listen to any overture of peace before the total ruin of their settlements is effected. . . . Our future security will be in their inability to injure us . . . and in the terror with which the severity of the chastizement they receive will inspire them.

Equal to the task, Sullivan said "the Indians shall see that there is malice enough in our hearts to destroy everything that contributes

Filipino Dead at Santa Ana, February 5, 1899. (U.S. Signal Corps No. 111-RB-1037, courtesy of the National Archives.)

to their support." In his *Life of Joseph Brant* (1838) William L. Stone described in detail Sullivan's scouring of the countryside, the "war of extermination [he] waged against the very orchards." At the town of Genesee alone Sullivan's army destroyed over a hundred houses, "mostly large and very elegant," laid waste extensive fields of ripening corn and beans, and with "axe and torch soon transformed the whole of that beautiful region from the character of a garden to a scene of drear and sickening desolation"—save for unimportant detail, that might have been a description of the graveyard into which General Smith had turned Samar. Also by way of anticipation, in the north country, observed Stone, "the Indians were hunted like wild beasts."

In his search for supporting precedents, the secretary of war might as well have plunged right back to the beginning. Captain John Underhill spending his days "in burning and spoiling" Block Island in 1636 (see p. 36) was the lineal forerunner of Major Waller spending his days in burning and spoiling Samar in 1901; just as John Endicott's orders "to put to death the men of Block Island, but to spare the women and children, and to bring them away" (see p. 34), were lineal forerunners of Jacob Smith's orders, albeit no one bethought himself to ask Endicott at what age boys became men.

So for once Root was right. In the nation's past there were high and ample precedents for terror. After all, Town Destroyer was the apt name the Senecas had bestowed upon the Father of the Republic himself, as Cornplanter had told Washington to his face in 1792: "and to this day," the chief had added, "when that name is heard, our women look behind them and turn pale, and our children cling close to the necks of their mothers."

CHAPTER XXII

The Strenuous Life at Home: To and Beyond the Louisiana Purchase Exposition

> On all sides the World's Fair represents the evidences of the white man's accomplishment in the Louisiana Territory. The spirit of development abounds. The noise of progress fills the air.
>
> —EXPOSITION SECRETARY WALTER B. STEVENS, 1904

"OUR PRICELESS principles undergo no change under a tropical sun," President McKinley had maintained and been more right than he could have intended. Carried westward by Fiske's Manifest Destiny, Melville's "metaphysics of Indian-hating" had become the current metaphysics of empire-building under a tropical sun. Almost two years after Roosevelt had declared the "insurrection" officially over, John Hay introduced a slight modification by calling United States imperialism "a cosmic tendency"; but that catchy phrase, as previously remarked, was merely another lethal metaphor for the principles of old. To be sure, much else had changed.

Restless as ever after his return from the South Seas and professing "the religion of World's Fairs," historian Henry Adams tagged along with his friend Hay to the Louisiana Purchase Exposition in St. Louis, where the secretary of state unveiled his invincible tendency for the Press Parliament of the World on May 19, 1904. On their trip west, as he related in the *Education*, Adams was astonished by how new everything seemed (pp. 465–68). From Pittsburgh through Ohio and Indiana "agriculture had made way for steam; tall chimneys reeked with smoke on every horizon, and dirty suburbs filled with scrap-iron, scrap-paper and cinders, formed the setting of every town." So revolutionary were the changes, Adams chaffingly "thought the Secretary of State should have rushed to the platform at every station to ask who were the people; for the American of the prime seemed to be extinct with the Shawnee and the buffalo." Millions of Germans and Slavs, "or whatever their race-names," had overflowed the central valleys to become the new Americans, children of steam and brothers of the dynamo, leaving Hay "as strange to the Mississippi River as though he had not been bred on its shores, and the city of St. Louis had turned its back on the noblest work of nature, leaving it bankrupt between its own banks"—one imagines Melville's *Fidèle* floating mockingly in that pollution as Adams and Hay passed by. And then there was the World's Fair itself, with its white Palaces of Electricity and Machinery and Transportation: "The chaos of education approached a dream. One asked one's self whether this extravagance reflected the past or imaged the future; whether it was a creation of the old American or a promise of the new one."

It was both. With his few well-turned phrases Adams had evoked the transition from the agricultural, small-propertied economy of his and Hay's youth to the industrialized corporate capitalism they could see before them in the reeking chimneys and piles of cinders. Though the ailing secretary of state at his side was a reminder "the American of the prime" was not quite yet extinct, the new immigrants did represent political power, some realized and more potential, since they came from European parent nations. Unlike nonwhites, they could not be barred for any visible reason from entering the Lockean consensus and polity; shortly they would find their way in, however abrasively and over whatever obstacles. Their ethnic politics and especially the ethnic politics of their children would make the Roosevelts and the Hays move over to share power even on the national level. They were truly the new Americans. All this and more imaged the future through the extravagance in St. Louis.

Yet the very scrap iron and scrap paper Adams found so esthetically repellent were creations of the past as well as promises of the more littered future. Those calling cards of the industrial transformation of the United States had their origins in the Puritan and Enlightenment rage for order, paradoxically enough, and in the drive of Adams's forefathers and the others for the total control of people and things; they were what was left over in debris from the ongoing efforts since the seventeenth century to subdue and exploit the continent according to increasingly sophisticated computations of efficiency—like the bankrupt Mississippi, that waste lying there was the by-product of those computations.

All the birthday parties at the turn of the century celebrated this historic drive to rationalize the world. "Expositions," said McKinley, "are the timekeepers of progress." The Centennial Exposition at Philadelphia in 1876 had the giant Corliss engine and technology that was updated by the Columbian World's Fair at Chicago in 1893, the Trans-Mississippi Exposition at Omaha in 1898, the Pan-American Exposition at Buffalo in 1901, where McKinley received his fatal wounds, and other expositions in New Orleans, Atlanta, and Charleston. Of them all, none boasted grounds as spacious (1,240 acres) and buildings as large or as numerous (fifteen "stately palaces" or exhibition halls alone) as what the publishers of *The World's Fair . . . Official Photographic Views* (1904) called the "Universal Exposition in St. Louis, celebrating the one hundredth anniversary of the Purchase of the Louisiana Territory. . . . It assembled the industries, the art and processes of the world. It was a meeting of the peoples of the earth, representing practically all nationalities, tribes and kindreds." Even the consent was universal: "By universal consent it surpassed all previous expositions in extent, picturesqueness, variety of detail, and interest" (WF, preface).*

The "stupendous enterprise" took place at "Forest City," the oxymoronic name of the fairgrounds that called attention to nature vanquished and urbanization triumphant. Walter B. Stevens, said to be "the distinguished and versatile Secretary of the Exposition," proudly observed that a mere three years earlier that quarter of Forest Park in St. Louis had been "in a state of demoralized nature. . . . unsewered, unwatered, unlighted, in every respect unimproved." Now it had been moralized and improved, sewered, watered, lighted, and changed from "*Wilderness to Wonderland.*—Where rioted the jungle of briers and vines and rotting vegetation now stand Palaces of Machinery and Transportation with the roadways, paths, rows of handsome trees, lawns and flower beds surrounding them. Education and Electricity are on terraced and adorned islands, once ground cut into deep ravines and made unsightly by patches of brush" (WF, p. 25). Let there be light—those patches of brush half a world away were what Captain Connell was trying to have removed when he was so treacherously cut down by the Balangigans.

The tradition of rationality that launched campaigns to "make Asiatics clean up" and improved green forests into lighted parking lots—one hundred automobiles were on display in St. Louis to entice twenty million visitors into the coming Motor Age—led

* For an explanation of the abbreviated references, see p. 521.

to a proliferation of professional experts such as Worcester and professional boosters such as Stevens. At the exposition, according to the latter:

> In villages under the auspices of the science of anthropology dwell representative families of North American aborigines, of the giants of Patagonia and of the pygmies from Congo land. And finally among all these strange peoples, civilized, semi-civilized and barbarous, assemble the thinkers of all nations to form a "World's University," the International Congress, "to discuss and set forth unification and mutual relations of the sciences, to harmonize the specialized studies." Could the Exposition be more universal in its human activities, body and mind? [WF, p. 4]

Had the trustees of the World's University been alert, they would have appointed Adams to a chair: its goal of the unification of knowledge matched his lifelong quest for unity and its pretensions to universality matched his attempt to work out "laws" of history and ground his discipline in the physical sciences.

For all his intelligence and learning, Henry Adams failed to grasp how much he was himself part of the problem he found so perplexing. There was not the rupture of continuity between the world of 1838 and the world of 1904 that he posited, because rationalization meant perpetual material change and the change itself provided the sequence he denied. Notwithstanding his claims to the contrary, the child Adams was father of the man, just as the adult was the true son of his forefathers' efforts to reduce the world to orderly processes.* Though an early disciple of Alexis de Tocqueville, Adams had forgotten one of his mentor's key insights in *Democracy* (1840), "Why Great Revolutions Will Become More Rare":

> Two things are astonishing about the United States: the great changeableness of most human behavior and the singular fixity of certain principles. Men are in constant motion; the mind of man seems almost unmoved. . . . Men bestir themselves within certain limits beyond which they hardly ever go. They are forever varying, altering, and restoring secondary matters but take great care not to touch fundamentals. They love change but dread revolutions.

* "Out of a medieval, primitive, crawling infant of 1838," he wrote shortly after the trip to St. Louis, "to find oneself a howling, steaming, exploding, Marconiing, radiumating, automobiling maniac of 1904 exceeds belief"—quoted in Ernest Samuels, *Henry Adams: The Major Phase* (Cambridge: Harvard University Press, 1964), p. 320. In Paris he had made himself *au courant* with the purchase of an eighteen-horsepower Mercedes—the irony of the professorial tourist bouncing in his new machine through the French countryside in search of thirteenth-century unity and sixteenth-century windows did very nearly exceed belief.

In St. Louis the truth of Tocqueville's observation could be seen in the many exhibits of secondary changes. Preoccupied with them, Adams missed the still more striking exhibits of fixed fundamentals, those certain principles, or priceless principles, as McKinley had called them.

2

Exposition statues and murals froze for a moment how citizens of the republic continued to perceive themselves and their past. In commemoration of the Louisiana Purchase as a major stage in the great westward thrust of empire, larger-than-life-size explorers of the territory stood beside the wide promenade leading upward from the Palace of Mines and Metallurgy to the Colonnade of States. "These were the white men," explained Secretary Walter B. Stevens, "who traversed the vast region and first made known its resources and possibilities." Among them, "Daniel Boone is leaning forward as if watching a 'varmint,' his rifle lifted, ready to take instant aim" (WF, p. 66). At the head of the main lagoon near the Louisiana Purchase Monument each of four heroic groups represented "an impressive and distinct chapter in the winning of the West" (WF, p. 275): a pioneer in a blizzard; a cowboy at rest while his horse stood nearby, "the Winchester in its place" beside the saddle; " 'A Step to Civilization' which a progressive Indian is urging his son to take, while in the background an old warrior true to his blanket and wigwam stands in discouraging attitude"; and most eloquent of all, "the barbaric buffalo dance performed by Indians." By Solon Hannibal Borglum, brother of Gutzon Borglum of later Mt. Rushmore fame, the stylized frenzy of the statue brought to temporary rest three centuries of fearful clichés about the barbarous madness of Indian dances. As Secretary Stevens remarked, the group was "composed of three Indians who are going through the figures of one of their wild dances. Viewed from any standpoint the impression made by the group is that of wild frenzy" (WF, p. 277). From sculptor Borglum and from interpreter Stevens there was not the hint of an understanding that the Indians might have been using their bodies as instruments of prayer—the secretary's rigidly disapproving description of "The Buffalo Dance" was worthy of Captain Endicott or Colonel McKenney.

In the rotunda of the Missouri Building, largest of the state exhibits, murals developed the same progressive theme in pendentives illustrating the four stages in the territory's history: *pre-*

historic represented by a saber-toothed tiger; *savage*, by an Indian butchering a deer; *developing*, by the Indian giving way to the white man; and *productive*, by a symbolical Abundance surrounded by cultivated fruits and flowers "and by Machinery, Architecture, Science, Literature and Art," in the order listed by Walter Williams, Missouri superintendent of publications. Decorations in the dome also depicted the struggle "of colonization in the wild country yet to be made habitable" and the ultimate victory "of the progressive civilization of peace." Such eighteenth-century notions of the stages of society confronted the fair visitor inside and out.

Of course the whole Louisiana Purchase Exposition was rather heavy allegory solemnizing the growth of the Anglo-American empire. Highly industrialized Pennsylvania granted recognition to that metaphorical meaning through its Governor Samuel W. Pennypacker, who proclaimed August 20 "Pennsylvania Day" at the fair.

In St. Louis on that one-hundred-and-tenth anniversary of the victory of General Anthony Wayne over the Indians at the Battle of Fallen Timbers, Governor Pennypacker delivered a memorable address in which he drew on the authority of Theodore Roosevelt's *Winning of the West* to establish that Wayne's victory was the most decisive "in all our Indian wars." It was of critical importance, for "if the land was to be secured for civilization it must not only be occupied,—it must be won," *won* from "fierce tribes of savages . . . who waged a treacherous and ruthless warfare which spared neither child, woman nor home." So were the merciless savages of old pressed into service, along with their kindred wild beasts, in the governor's twentieth-century rendition of Genesis:

> The wild beasts that filled the caverns below and clung to the limbs of the trees overhead were driven from their lairs. The savages who lurked in the forests and who endeavored to confront the in-pour of emigrants, with treachery and revenge in their hearts and scalping knives in their hands, were after many a fierce struggle, beaten and destroyed.

Winning the West was all the more important now, "if it shall come about, as . . . seems probable, that the American people are to be one of the dominant nations, imposing their race characteristics, seizing the avenues of trade, and extending their institutions," all of which came down to a neat definition of imperialism and demonstration of its continuity. Pennypacker happily claimed as native sons Daniel Boone and John Lincoln, the great-grandfather of the president; acknowledged the bad with the good in "Simon Girty, the thoroughly hated renegade who took part with the savages"; and went on to point out that Pennsylvania "alone of the thirteen original States had a regiment in the Philippines." Thus had the Wilderness Trace led over the Alleghenies and beyond to Batangas and Samar or, as Pennypacker said, "in the United States of America . . . the terms East and West are more or less uncertain in their designation." The historically conscious governor named the most elusive of the cardinal directions that region lying "to the westward of the Allegheny Mountains," a designation of the West that more or less certainly followed the setting sun over the horizon toward Asia and the current westernmost rim of "settlement."

In plaster and paint and word, not one of the priceless principles was more fixed or more central than the national view of natives as merciless savages, brutes akin to the wild beasts of the caverns below and the tree limbs overhead—or the pens and cages between. After the capture of General Aguinaldo, requests poured into the War Department from showmen who wanted to exhibit him. Philip C. Jessup, Root's biographer, noted a typical offer from an Idaho company that asked for " 'a year's lease' of Aguinaldo, the company agreeing to exhibit him in all the principal cities of the United States and to pay over to the government 75% of the gross receipts!" Fortunately the Filipino revolutionary was spared that ordeal—at General MacArthur's urging, he had taken an oath of loyalty on April 19, 1901, asking his compatriots in a proclamation: "Let the stream of blood cease to flow, let there be an end to tears and desolation" (UM, p. 188). Otherwise he very well might have been part of the Philippine Exhibit at the World's Fair.

In fancy one sees Aguinaldo seated, under guard of course, by the Daniel Boone Bridge over the North Lagoon, conversing with Adams and Hay, two or three of Worcester's Bontoc Igorots, local St. Louis novelist Kate Chopin, Secretary Stevens and Governor Pennypacker, and Geronimo, all of whom try to scale language barriers and worlds in order to set forth the unification and mutual relations of the sciences. Their deliberations would be facilitated by the syncopated accompaniment of Scott Joplin, played slowly: "It is never right to play Ragtime fast. . . ."

In fact one could see Geronimo in St. Louis. Previously shown at the Omaha and Buffalo expositions, by 1904 he was a veteran exhibit and star of all the North American entries of "strange peoples." These included Inuits, or Eskimos, from the Arctic Circle in traditional dress and said to be always good-humored "even in July and August days when it made one feel uncomfortably warm to look at the heavy fur clothing" (WF, p. 122), not to mention their panting sled dogs. Along with Geronimo's Apaches, some fifty other Native American tribes were represented, including "the surviving chiefs of the leading tribes which once divided among them all of Louisiana. . . . Stoical to the end, the old Indian puts on all of the feathers and deer teeth and blankets and is photographed in silent useless protest against the [Indian] school . . . and the white man's road generally. There is something of the plaintive in it" (WF, p. 129). Sioux Chief Tall Crane was photographed outside his tepee with "Mrs. Tall Crane, a heavily-built, square-faced squaw" who surprisingly "shows no sign of hard usage" (WF, p. 138). The tepee of the Tall Cranes, the wickiup of Geronimo, and the shelters of other Native Americans were placed appropriately in the western part of the fairgrounds, just beyond the United States Indian School, on "the reservation of ethnology, occupied by communities of many Indian tribes, not entirely confined to the North American Continent. The Patagonians of South America and the Pygmies of Africa have their homes on the reservation" (WF, p. 260). Geronimo had a booth in the Indian School, between Pueblo potters and Pueblo women grinding corn and making bread, where he made and sold bows and arrows, and sang and danced for customers.

In his autobiography Geronimo related that many people invited him to their homes in St. Louis, "but my keeper always refused." Still a prisoner of war, of course, he went nowhere without guards. His keeper allowed him to see the shows on the pike, ride on the Ferris wheel, and visit "strange people of whom I had never heard." One day he went to see "little brown people . . . that United States troops captured recently on some islands far away from here":

> They did not wear much clothing, and I think that they should not have been allowed to come to the Fair. But they themselves did not seem to know any better. They had some little brass plates, and they tried to play music with these, but I did not think it was music —it was only a rattle. However, they danced to this noise and seemed to think they were giving a fine show.
>
> I do not know how true the report was, but I heard that the President sent them to the Fair so that they could learn some manners, and when they went home teach their people how to dress and how to behave.

So declaimed the great old man after taking that step to "civilization." At the time Geronimo was in his brief Christian phase, desperately wanted to return to his own country in Arizona to die, and treacherously said what whites had given him to understand they wanted him to say. Alas, it was not enough to get him home, but for this one revealing moment the ends of the empire came together, victims face to face.

The "little brown people" were the pagan Igorots Ida McKinley had so earnestly wanted to save and Worcester and the other commissioners had just shipped from the islands for the edification of their compatriots on the mainland. Unhappily, no one seems to have thought to take a snapshot of the Igorots' historic meeting with Geronimo—perhaps the latter's guard got in the way—but the next best was an official photograph showing "Receiving Day at Igorrote Town," the village far off in the corner of the Philippine reservation that "no visitor misses. . . . Even the other curious and strange peoples, from every quarter of the globe, leave their World's Fair homes and go to see the Igorrotes. Every day in the week is receiving day with these good natured savages. The picture was taken on the occasion of the ceremonial call paid to the Igorrotes by one of the other Indian tribes [*sic*]" (WF, p. 165). Like the African forest people and other aborigines from faraway places, the Igorots were put on a reservation and seen as another Indian tribe, new-caught natives comparable to the old, except more curious and better-natured savages. As for actual knowledge and understanding, Secretary Stevens and visitors at the World's Fair knew and understood as much but no more about the Philippine Igorots than had U.S. citizens about the Pacific Coast Clatsops in the days of Lewis and Clark.

The science of anthropology moved to close the gap in their understanding by studying "*The Igorot as an Exhibit*. . . . They were measured and photographed and cross-questioned by scientists and studied by many visitors interested in ethnology. They furnished the live object lessons for lecture courses" and good-naturedly "answered questions, pertinent and impertinent, about themselves without asking return information respecting the white people who were studying them" (WF, p. 149). Another photograph showed white people carrying on their study of Igorots by hanging on the outside of the high stockade of bamboo poles around their village and peering in at them in structures with low partitions so that viewers looked "over and into what is very like a pen":

> A little straw thrown on one side is the bed. If a fire is wanted in the house, it is built upon the ground and the smoke finds the way out. As a rule the cooking is done out of doors by the rudest of

Geronimo at the Louisiana Purchase Exposition, 1904. (Courtesy of the Missouri Historical Society.)

> devices. . . . A pot, braced by a rock to give the space underneath for fire, satisfies the Igorrote housekeeper. Fingers are used to turn the boiling meat. . . . With a pen under cover to sleep in and a pot containing a piece of meat, the Igorrote is content. What to wear gives him no concern. [WF, p. 174]

Like the lilies of the field or the pigs of the pen, Igorots gave no thought to the morrow, seemingly, though in the interests of modesty, Secretary Stevens noted, the women had "been induced to wear more clothes than they do at home" (WF, p. 173).

The extravaganza in St. Louis was really a roundup of all the preceding centuries, from the seventeenth through the nineteenth, and not surprisingly the Philippine Purchase was the center of interest as the most recent instance of the victory of "civilization" over "natural wilderness and . . . savage men." Promoters and visitors alike turned to it as the modern counterpart and natural extension of the Louisiana Purchase. The Philippine Reservation exhibited curiosities from the newest possession, just as the Indian Reservation nearby exhibited relics from the oldest. On both reservations persons were shown off as object lessons of what Anglo-Americans were not and must not become. Natives were exhibited as live objects that, save for their human form, were strictly comparable to the Philippine and North American varieties in the giant cage of the United States Bird Exhibit or in the tanks of the United States Fish Commission—the latter

Receiving Day at Igorrote Town, from *The World's Fair: Comprising the Official Photographic Views*, 1904.

boasted a display not only from the Atlantic and Pacific coasts and the lakes and streams in between but from the Philippines as well. Thus the identification of natives as natural objects was on full display in Forest City, in booths and behind bamboo stockades. There persons appeared as bits of national property, to be measured and photographed and cross-questioned, to be consumed by customers who paid to see them dance and portray "the general barbaric life" or to be transformed by tribe-tamers in the Filipino and Indian schools or to be destroyed if they resisted—after all, old Daniel Boone peered out over the fairgrounds and watched over them all with his fabled rifle at the ready, in Secretary Stevens's words, "as if watching a 'varmint.' "

The great white Palaces of Machinery and Electricity notwithstanding, the most representative exhibit at the St. Louis World's Fair was the ingrained racism of the conquerors.

3

"The civilization and the barbarism of the Archipelago are shown side by side in the Philippine Exposition at the World's Fair," wrote Secretary Stevens. "Igorot braves bow to the rising sun, kill a dog and dance on one side of the frail bamboo stockade, while on the other side, at the same hour, the neatly uniformed Scouts from the civilized tribes stand at attention as the United States flag is raised to American music by a Philippine band" (WF, p. 148). In concept and in side-by-side structure the exhibits confirmed the delusion Worcester had done so much to foster, namely, that the Filipinos were a collection of tribes, some "wild," some "civilized." While the professor was in the islands "still bringing people down from the treetops," as he put it, his Department of the Interior furnished numerous "type-casts" for the exposition, including tree-dwelling Lanao Moros, who perched in their bamboo home high above the court of the Ethnology Building (WF, p. 175). Visitors understandably flocked to see Moro tree-dwellers and Igorot dog-eaters, and understandably remembered these "barbarians" from the archipelago long after their memories had faded of the neatly uniformed Macabebe mercenaries or of the handsome, white-duck-attired Visayans who sang the "Star Spangled Banner" in American English. The association was simple and politically handy: Filipinos were "savages" incapable of self-direction, dependents in dire need of the "civilizing" ministrations and permanent political tutelage of their white masters.

The St. Louis World's Fair helped spread this misrepresentation of Filipinos as tribal peoples into homes across the country and drained away whatever residual sympathy there was for them as a people—Anglo-Americans had not got where they were by cherishing tribes. Once this view of them struck root in the United States, as Tocqueville said of an accepted opinion there generally, "it would seem that no power on earth is strong enough to eradicate it." More than two decades later, in *The Conquest of the Philippines* (1926), Moorfield Storey and Marcial P. Lichauco were still trying to eradicate this particular fixed opinion: "These so-called Philippine tribes have been receiving such notoriety to this day that we might as well pause now and see how much weight should be given to these statements" (p. 173). In effect asking Worcester to come center stage for a bow, they vainly sought to throw off the heavy weight of "scientific" sanction: "The existence of tribes has been decided by ethnologists, who claim to see among the brown-skinned natives certain differences which stamp them into the category of tribesmen." Four years later, toward the end of all their decades of agitation for independence, Philippine nationalist Camilio Osias took up the continual "harping on our alleged 'tribal' differences" in his testimony before the Senate Committee on Territories and Insular Affairs:

> I can answer [this harping] by saying that the lack of such differences can be proven here now in this room . . . I ask you to look at the various Filipinos here and see if there is any more difference among us than there is between an American of the Connecticut Tribe, let us say, and one of the Michigan Tribe, or of the Pacific Coast Tribe, or some other American Tribe. [UM, p. 384]

But the retentionists—that is, those who wanted to hold on to the islands—had never depended on seeing for believing that once a native always a native.

The Louisiana Purchase Exposition had also been an educational experience for the "leading" Filipinos whom Worcester and Frank S. Bourns had lined up to testify before the Schurman Commission and later, under Taft's nurturing and guiding hand, had organized into a political party, the Federalistas. Assimilationists, if one shuns the harsher term "collaborators," these wealthy politicians aspired to a role under the United States comparable to the one their class had enjoyed under Spain, and they did indeed rise to high positions in the new regime, with three of their leaders on the Taft Commission itself. Aside from their immediate objectives of office and patronage, they sought incorporation of the Philippines into the Union and then full statehood. But this end was in a way more farfetched than independence—as Secretary of War Root said, "statehood for Filipinos would add another serious race problem to the one we already have" (UM, p. 330). Then in St. Louis the Federalista members of the honorary board of commissioners discovered for themselves just how idle was their dream of statehood. All too clearly the United States would not welcome as citizens dark-skinned natives from a distant archipelago. Once home even these Americanistas, as they were also called, had to advocate eventual independence.*

Not so accommodating and not given at all to placating petitions were the Moslem Moros of Mindanao and the Sulu Archipelago. Those at the World's Fair proved "disappointingly unprofitable from the showman's point of view" and so fiercely repelled photographers a sign had to be "posted warning the kodak army that they invade at their peril the village"; their chiefs demonstrated total unwillingness "to exhibit themselves to the Americans and . . . to perform in accordance with the wishes of the showmen" (WF, pp. 154, 175). Seemingly the Moros still failed to recognize the realities of their subjugation. Of all the official views of native peoples, perhaps the most fascinating and certainly the most ominous was of some Moro families who had obviously been coerced into lining up to smile at the birdie but who defiantly faced the camera with what Secretary Stevens saw as "rather repellent expressions" on their faces (WF, p. 155). And indeed the fist of a man squatting in the front row, raised menacingly toward the photographer, augured more Moro misbehavior.

Under the command of General Leonard Wood, a retaliatory expedition was mounted against them two years later on the island of Jolo. In the angry words of Mark Twain, "a tribe of Moros, dark-skinned savages, had fortified themselves in the bowl of an

* The truth the Federalistas belatedly discovered about the Lockean polity was critical, as we have noted in several contexts: Apart from Afro-Americans under the largely inoperative Fourteenth Amendment, non-Europeans in ancestry were nonpeoples in the United States. As originally drafted, the U.S. Senate version of the Philippine Independence Act excluded Filipino immigrants completely. In the act's final form (1934), it assigned an annual quota of fifty immigrants and after the withdrawal of sovereignty the islands were to be subject to the same U.S. immigration exclusion as other Asiatic countries—Public Act No. 127, 73rd Cong., Sect. 8 (A) (1). Not until the U.S. immigration law was revised in 1946 could Filipinos be naturalized. During their Commonwealth period (1936–46) they were anomalously citizens of the Philippines and "nationals" of the United States.

extinct crater not many miles from Jolo; and as they were hostiles, and bitter against us because we have been trying for eight years to take their liberties away from them, their presence in that position was a menace." The ensuing slaughter of some six hundred, counting women and children, called forth Twain's most memorable outburst:

> General Wood was present and looking on. His order had been, "Kill *or* capture those savages." Apparently our little army considered that the "or" left them authorized to kill *or* capture according to taste, and that their taste had remained what it has been for eight years, in our army out there—the taste of Christian butchers.

But by 1906 the Philippines had become a bore. Apart from Twain, Moorfield Storey, and a few other veterans of the antiwar movement, no one cared much what happened to Moros on Jolo or to other dark-skinned "hostiles" on other remote islands. Twain put his "Comments on the Killing of 600 Moros" in the posthumous privacy of his *Autobiography* (1924), Theodore Roosevelt cabled Wood his congratulations "upon the brilliant feat of arms," and that was about it.

Of course the conquest also had a sunny side, the Americanization side. From the *Report of the Twenty-ninth Annual Lake Mohonk Conference of Friends of the Indian and Other Dependent Peoples* (1911), Peter W. Stanley quoted the assurances of an American businessman from Manila: "A splendid job is being done, and the influence of it on the Filipino is most marked." The prestigious conveners of Lake Mohonk had glided smoothly from annual conferences on the oldest "Dependent Peoples" to the newest—with stopovers along the way "on the Negro Question" (see p. 495)—and no doubt were pleased to hear these familiar assurances of benevolence and of splendid efforts being undertaken in the field. And in fact colonial statesmen had carried to the islands polo and trap-shooting for themselves and baseball for their new wards, along with soap and toothbrushes for everybody. They built schools to teach "God's language" and combat illiteracy, mounted clean-up campaigns, and administered vaccinations that lowered the death rate and stemmed epidemics of cholera and smallpox coming out of the war years. Unlike Worcester, many of them evinced an agreeable willingness to throw, over the realities of subjugation, the beneficent cloak of their partnership with Filipinos in building an Asian outpost of republican institutions.

From 1913 to 1921 Governor Francis Burton Harrison tried to

The White Man's Burden, from the Admiral Dewey number of *Harper's Weekly*, September 30, 1899. "Now this is the road that the White Men tread," said Rudyard Kipling, "when they go to clean a land." The bar of implement Wilson's expressed desire to contract the frontier by a program of "Filipinization" of the insular government that would lead to independence. But even Harrison was unable to bridge "the growing gulf between the races," as Governor James F. Smith had called it in 1909. "The social status of an American woman who marries a native,—I myself have never heard of but one case," remarked Judge Blount, "is like that of a Pacific coast girl who marries a Jap." A white male who married a Filipina was ostracized by the U.S. "community" in Manila. And for all his good intentions, Harrison himself was by no means free of the racist virus—on his one trip to visit the Igorots, Stanley reported, "he had carried with him a cake of carbolic soap and had washed himself whenever possible after shaking hands with an Igorot."

The racism in Manila reflected and reinforced that in Washington. There too the Filipinos remained, after all those years of tutelage, "utterly unfit for self-government." Warren Gamaliel Harding had to tell them bluntly "the time is not ripe for your independence," and Calvin Coolidge had to reiterate they had not "as yet attained the capability of full self-government" (UM, p. 264). Henry L. Stimson, their governor from March 1928 to early 1929, held that as Malays they "were capable of hopeful progress while under our supervision" but like Worcester believed they had an inborn tendency to slide back into "barbarism" (UM, p. 266). When it was finally promised in 1934, their independence came not because of any upsurge of contrition or any major shift in this estimate of their innate capabilities but because traditionally proindependence Democrats came into power in 1933, the protectionist farm bloc wanted an end to competition from Philippine imports, and the Filipinos themselves had continued to agitate for at least partial U.S. withdrawal.

One day in February 1932 Herbert Hoover and Secretary of State Stimson had a striking conversation. Stimson asked the president if

> he really believed that the United States was not enough of a governmental power and did not have enough of a constitutional

---

soap passing from hand to hand at the lower right will begin the native's enlightenment. It would be risky for him to spurn this generosity—"sane living," said Secretary Worcester, "means sanitary living." Here soap—of course the advertiser's—appears as an actual yardstick of "civilization." By no means a parody of this burden of whitening the world, the advertisement soberly and suitably appeared in a weekly that advertised itself by subtitle as *A Journal of Civilization*.

King of the Caroline Islands, from the *Literary Digest*, September 10, 1898. Forerunners of the modern transnationals, firms such as the Singer Manufacturing Co. and Pears' Soap had their maps in hand and were moving with the rest of the empire-builders into the "blank spaces"—spaces, that is, inhabited only by natives—all over the world. Thanks to this corporate Manifest Destiny, the "King" shown here will shortly shed his grass skirt and don properly stitched white ducks.

> freedom to evolve [a] relationship to another country like the Philippines similar to the relationship of England to the British Commonwealth of Nations.
>
> Hoover: Well, that's the white man's burden.
>
> Stimson: Yes, that's what it comes down to and I believe in assuming it. I believe it would be better for the world and better for us.

But traditional Republican party retentionism had momentarily blinded the hardheaded Stimson to a truth he knew very well: there was more than one way to skin a cat and there were other possible colonial relationships. In spite of the act providing for Philippine independence that passed over Hoover's veto the following January, the westward course of empire was not permanently halted, much less reversed. The white man's burden has never been dropped lightly.

PART FIVE

Children of Light

In seeking to identify the causes of the war and apportion responsibility for its outbreak, one must begin with the fact established by the testimony of all the whites and most Indians that the Pequots were blatantly and persistently provocative and aggressive.

—ALDEN T. VAUGHAN,
"Pequots and Puritans," 1964

CHAPTER XXIII

The Occident Express: From the Bay Colony to Indochina

> In a frontier society, someone had to impose a semblance of justice and order.
>
> —ALDEN T. VAUGHAN, *New England Frontier*, 1965

> Tonight Americans and Asians are dying for a world where each people may choose its own path to change.
>
> This is the principle for which our ancestors fought in the valleys of Pennsylvania. It is the principle for which our sons fight in the jungles of Vietnam. . . . This will be a disorderly planet for a long time.
>
> —LYNDON B. JOHNSON, "A Pattern for Peace in Southeast Asia," April 1965

IN PARTS Two, Three, and Four, the epigraphs quoting Timothy Dwight, George Bancroft, and John Fiske underline the abidingness of mainstream Anglo-American views of national origins —mavericks like Morton and Melville were so exceptional as to prove the rule. The epigraphs also underline and extend a commonplace repeatedly demonstrated by the text. In war the first casualty is indeed truth, but it has an identical twin: conscience. For three hundred years the war against the Indian shaped what became American nationalism by driving underground the ultimately undeniable truth about his humanity and by hardening Anglo-Americans to what they were doing. They proceeded with dispossession and extermination or uprooting in an atmosphere fouled by self-congratulation and by sighing regrets over the Native American's utter unworthiness of their disinterested benevolence. Coming early in this tradition and in clearly etched outline, the Pequot War had a lasting significance far out of proportion to its magnitude. It set the pattern for justifications of rooting natives out of their swamps and of plucking them down from their mountain fastnesses. It represented "a moral precedent," observed N. Scott Momaday, "upon which a tradition of oppression has been based." Then, just as the Kiowa writer put his insight into print, there appeared stunning new confirmation of its truth.

"A prima facie case exists for the claims of Puritan apologists," claimed Alden T. Vaughan in the April 1964 *William and Mary Quarterly*, "and its partial validity is undeniable. Unfortunately, it is not the whole story." More unfortunately still, it was with that part of the story that Vaughan, an assistant professor of history at Columbia University, began in his essay on "Pequots and Puritans": "One must begin with the fact established by the testimony of all the whites and most Indians that the Pequots were blatantly and persistently provocative and aggressive" (PP, 262, 267).*

For a moment let us assume that by a straw vote whites had unanimously supported the proposition Vaughan was pleased to call "Pequot perfidy" (PP, 266). Would that have established its existence? Or would the whites not have been saying more about themselves, as Vaughan was, than about the Pequots? Under the circumstances, it would have meant merely that they subscribed to "the new creede" Thomas Morton indicted, to wit, that "the Salvages are a dangerous people, subtill, secreat and mischeivous" (see p. 19). Morton was an unbeliever. Or consider the more direct testimony of William Wood two years *before* the opening of hostilities:

> The *Pequeants* be a stately warlike people, of whom I never heard any misdemeanour; but that they were just and equall in their dealings; not treacherous either to their Country-men, or *English:* Requiters of courtesies, affable towards the *English*. [*New Englands Prospect* (1634), pp. 52–53]

Vaughan began with a "fact" that was not a fact, and his contention that "all whites" subscribed to the prejudice was untrue. As to whether "most Indians" shared it, who could possibly have taken their depositions and made the professor privy to the re-

* For an explanation of the abbreviated references, see p. 529.

sults? Actually, the Pequots were on the point of forming an alliance with the Narragansetts, and the proposed confederation was defeated only by the prodigious efforts of Roger Williams, who stirred up old enmities and suspicions (see p. 41). Had it been established, then most New England Indians would have been allies of the Pequots.

The Bay Colony, wrote Vaughan, "had been overripe for Endicott's blow at Satan's horde" (PP, 266). And "early on the morning of April 23, 1637, two hundred howling Pequot braves descended on a small group of colonists at work in a meadow near Wethersfield, Connecticut" (PP, 260–61). In his attempt to establish them as bona fide merciless savages, Vaughan unintentionally fell into a largely successful demonstration of Puritan perfidy and murderousness, as in the expedition under John Endicott to kill all the men on Block Island. Furthermore, his attempts to tie the Pequots to the ostensible first victims of the war simply spread confusion. In 1634, he stated flatly, the Pequots "made their first hostile move against the English with the assassination of Captain John Stone of Virginia and eight other Englishmen" (PP, 258), an assertion undercut by the authoritative admission of Captain John Mason that the Indians responsible "were not native Pequots, but had frequent recourse unto them" (see p. 38). Since the evidence nowhere showed a link between the Pequots and the death of the no less quarrelsome John Oldham, Vaughan fell back on the unsupported and almost certainly untrue assertion that "the few surviving assassins [*sic*] sought refuge with the Pequots" (PP, 260), an amusing if dubious instance of guilt by hospitality.

For all the alleged merciless savagery that left blame lying "somewhat more heavily upon the Pequots than the Puritans," Vaughan concluded that "at bottom it was the English assumption of the right to discipline neighboring Indians that led to war in 1637" (PP, 269, 268). Yet when his essay became—"with slight variations"—a chapter of *New England Frontier* the following year, the conclusion had undergone metamorphosis. Now "most of the blame for the war," Vaughan contended, "must fall on the Pequots," and so on to their "blatant and persistent aggression" (NEF, pp. 135–36). Coming out of unaltered or only slightly varied evidence, the conclusions collided in a sign sufficient of the historian's inner unease, confusions, divided feelings about the world of the forefathers and about how it had been won. Momaday's acute assessment of another writer on the Pequots had direct application to Vaughan: "In the same breath he deplores and defends the morality in which his own prejudices have their roots." The second conclusion meant the Columbia professor had given his prejudices their head, but that by no means solved his problem.

The anemia of the analysis in *New England Frontier*, its starts and stops and blusterings, was a dead giveaway that the author was not explaining but explaining away. "By necessity, as well as by inclination," read his preface, "I have concentrated on the acts and attitudes of the Puritans toward the Indians and have not, for the most part, attempted to account for the actions and reactions of the natives." By inclination as well as by necessity, then, the natives were left out of the account, for the most part, even an account subtitled *Puritans and Indians*. By chaining himself to the Saints' explanations of their acts and attitudes and then by remaining on the surface of those, Vaughan became a bond servant of their self-justifications and bound himself over to the writing of a twentieth-century Puritan cover story: "the New England Puritans followed a remarkably humane, considerate, and just policy in their dealings with the Indians" (NEF, p. vii).

No doubt an unintended consequence of this filiopietistic historiography was the reduction of the formidable Saints to a set of Sabbath school lessons with no more depth than the canvas of a John Endicott portrait. Let us content ourselves with three illustrations of what Vaughan said the Puritans were not.

The Puritans were not covetous of Pequot lands: "Although the war had not been fought to wrest land from the Pequots, most of the conquered territory was annexed by the English as the spoils of war" (NEF, p. 152). Land hunger was unquestionably among the motives that led the Puritans to war. Hard evidence from Lion Gardiner showed that leading Saints were plotting at least as early as 1635 to wrest land from the Pequots (see p. 39).

The Puritans were not sadists: After the roastings in the "fiery Oven" at Mystic, "a key element in the anti-Puritan syndrome has been an indictment of the colonists for reveling in the destruction of the Pequots. Rather than reflecting sadistic pleasure, Puritan writings show how deeply they believed in God's direct manipulation of history" (NEF, p. 145n). This superficial psychology was mere obfuscation. Hell had always been a magnificent repository for the Christian lust to punish; the flaming Pequot fort was the realization of that lust in this world. Even if the Puritans dutifully assigned the ultimate pleasure to God, they could still revel in having been His avenging angels: "we were like Men in a Dream," exulted John Mason; "then was our Mouth filled with Laughter, and our Tongues with singing" (see p. 56)—if that was not revelry, it was a pretty fair Puritan substitute.

The Puritans were not racists: "And since most of the tribes of

New England were on the white man's side, the Pequot War cannot accurately be described as an 'Indian war' in the usual sense. This was no racial conflict between white men and red, no clash of disparate cultures or alien civilizations" (NEF, p. 135). The *white* man's side? Then since Custer employed Crow scouts, the Battle of the Little Bighorn was not a racial conflict? Still, the refrain of *New England Frontier* was that the colonists did not see the Indians as a race apart, "did not portray relations with the Indians as determined by any fundamental distinction of race. Winthrop, Bradford, and their contemporaries treated the encounters of Europeans and Indians as though they were determined by all the complexities of human behavior" (NEF, pp. viii, 62n). Precisely the reverse was the case, as we noted in Winthrop's May 1637 warning to Bradford: "looke at the pequents, *and all other Indeans*, as a common enimie" (see p. 52). It was also untrue that to argue the Indians should have acted together as a race "was to be far more race-minded than Puritans or Indians ever were" (NEF, p. 324)—remember Miantonomo's desperate plea (quoted on p. 60): "So we are all Indians . . . and say brother to one another." As though unaware of what his other hand was doing, Vaughan himself quoted Winthrop on the Indians not recognizing political divisions among the English, assuming "that we were all as one" (NEF, p. 159). Vaughan also obligingly outlined special laws that proposed to ban racial intermarriage in Massachusetts (seemingly found to be superfluous); prohibited the sale of guns, horses, and boats to Indians; forbade their settlement within a quarter of a mile of any Connecticut town; enabled the sale of Rhode Island Indians into slavery upon conviction of theft involving over twenty shillings; and going the other way, established stiff punishment for any Connecticut white who dared run away to the Indians (NEF, pp. 185–210 passim). Sheer mystification, then, was his final restatement of all this: "The New England Puritans had treated the Indians not as a race apart, but as fellow sinners in God's great universe" (NEF, p. 338). Never were Indians *fellow* sinners of the Puritans, never politically, never legally, and never religiously. As Neal Salisbury has demonstrated in precise detail, even John Eliot's Christian Indians prayed in racially segregated churches and lived as a lower caste in racially segregated hamlets.

By inclination Vaughan considered the natives objects of Puritan acts and attitudes. Accordingly, his chapter "The Indians of New England" (NEF, pp. 27–63) treated them with poorly concealed contempt as objects, cultureless merciless savages. His characterization of their religion was a marvel of Eurocentrism:

"Indian religion was largely amorphous; it possessed neither scriptures nor universal dogma, and it lacked a priesthood in the European sense." Politically, all the tribes "were essentially 'monarchical'—in the eyes of the Puritans—" and their eyes were Vaughan's as he proceeded to project European hierarchy onto the Indians, finding below their leaders "simply subjects, though below them were often servant-slaves who had been captured in war." As for their materia medica, it hardly existed: "The Indians had no *bona fide* medicine to speak of, and in times of illness depended on the powwow to expel the evil spirits from their suffering bodies"—contrast this dismissal with Thomas Morton's admiring mention of their excellent midwives and skillful physicians and surgeons (see p. 17).*

By seeing Indians through Puritan eyes Vaughan necessarily saw them as an inferior race apart. While he "eschewed 'savage' and 'barbarian' " as descriptive terms (NEF, p. ix), he depicted Native Americans as both in everything but name and made his pages a compendium of racist clichés. At best they were "sons of the New England forest" (NEF, p. 282). Frequently they were naughty sons, however, forcing the Puritans into their "chastisement of the Pequots" (NEF, p. 309). To the colonists, said Vaughan, all natives looked alike, or tended to, much as all "googoos" looked alike to U.S. volunteers in the Philippines: "Indians were difficult for Englishmen to identify, and they could easily be concealed by other Indians. Therefore, the practice of selling Indians into slavery in the West Indies emerged as a drastic but effective method of eliminating chronic troublemakers" (NEF, pp. 206–7). As with natives everywhere, their efforts to organize against their oppressors were manifest "conspiracies": "The General Court of Connecticut wrote to Massachusetts for help in

* Or consider Governor Winthrop's famous remedy for ulcers: "one ounce of crabbe's eyes and four ounces of strong wine vinegar" and that at a time when Peruvian bark, the Indian herbal now known as quinine, helped so many colonists, once they learned of it, expel malaria from their suffering bodies—A. Irving Hallowell, "The Backwash of the Frontier: The Impact of the Indian on American Culture," *The Frontier in Perspective*, ed. Walker D. Wyman and Clifton B. Kroeber (Madison: University of Wisconsin Press, 1957), pp. 229–58. Vaughan admitted that "considering the primitive state of European medicine in the seventeenth century . . . the gap between the red man's medicine and the white man's may not have been very impressive" but could scarcely countenance the fact that in some respects Indian medicine was superior—his discussion of the Indian medicine man or powwow as conjurer or "witch doctor" was a classic instance of how not to understand other peoples. For respectful consideration of Indian materia medica in New England, begin with Gladys Tantaquidgeon, "Mohegan Medicinal Practices . . .," *Forty-third Annual Report of the Bureau of American Ethnology* (Washington: Government Printing Office, 1928), pp. 264–70; for North America as a whole, see Virgil Vogel's impressive *American Indian Medicine* (Norman: University of Oklahoma Press, 1970).

stamping out the conspiracy before it materialized" (NEF, p. 159). Not only did the Pequots "howl" like wolves, they were a pack or "horde": A Saybrook contingent "was attacked by a horde of Pequot warriors" (NEF, p. 130). Vaughan slightly modified the usual "rooting" of natives out of their swamps in his reference to the colonists "ferreting out the escaping Pequots" (NEF, p. 148).

Apart from modernisms, the Reverend William Hubbard would have found Vaughan's language and sentiments laudable; and the reverse, apart from archaisms, was also true. In fact the historian approvingly quoted the Indian-hating Puritan divine as an authority on Pequot bloodthirstiness: after a taste of victory, said Hubbard, they "began to thirst after the Blood of any Foreigners" (NEF, p. 123). Unlike Timothy Dwight, Vaughan did not indulge the sentimental tear over their destruction: the year 1637 "ended with encouraging solutions to New England's problems, both military and religious . . ." (NEF, p. 160).

For Vaughan the revered forefathers represented the *order principle:* they were alternatives to "anarchy outside the settled areas" in New England (PP, 268). In his hands they became disembodied instruments of "civilization": their "migration changed New England from a forest preserve into a frontier of western civilization" (NEF, p. 51), a statement that takes us back to William Bradford's "unpeopled" wilderness or, more strictly, to a sort of game covert or stream. Order was another name for *fate:* "Fate had dealt harshly with the Indians of New England" (NEF, p. 320). And fate was another name for *progress:* Two societies met in colonial New England:

> One was unified, visionary, disciplined, and dynamic. The other was divided, self-satisfied, undisciplined, and static. It would be unreasonable to expect that such societies could live side by side indefinitely with no penetration of the more fragmented and passive by the more consolidated and active. [NEF, p. 323]

Seemingly Vaughan remained unaware of how this amazing active–passive metaphor undercut his verbiage about the forbearing Puritans and the blatantly aggressive Pequots. But his revelation of the "penetration" of the latter by the former brings us to his final attribute: the unloving Puritans were rigid exemplars of the *masculine principle.*

The 1630s and the 1960s met and embraced in *New England Frontier.* While Vaughan extolled the Saints for bringing order to their frontier, President Johnson sent their sons into action against disorder on a frontier that had become planetary. By calling on the future, Vaughan avoided direct reference to this latest frontier in his defense of the Bay Colony:

> It will be interesting to see what sort of justification is devised if American spacemen discover a sparse population of neolithic *Homo sapiens* on a strategic planet. The seemingly egocentric and imperialistic rationalizations of our ancestors may yet enjoy a revival. [NEF, pp. 369–70n]

Say rather that it would be fascinating if the planet turned out to be *strategic*—but "our" ancestors? Say the Kiowa ancestors of N. Scott Momaday? Or a prospective revival of Puritan rationalizations among, say, the handful of Pequot survivors?

2

In and out of the history profession, reviewers acclaimed *New England Frontier:*

> A most important book, since it provides a test case of Puritan morality. . . . In every way a model for historians—and as exciting as any book involving Indians. [*Choice*, II, 904; cf. *Library Journal*, XC, 4072]
>
> Alden Vaughan shows how the Puritans endeavored to treat Indian tribes of New England with decency and concern. [*Christian Science Monitor*, November 18, 1965]
>
> Will remain the standard work in the field. . . . One of the most important contributions of the book is the demonstration that the Puritans did not deal with the Indians on the basis of a superior race subjugating an inferior one or regard all Indians as conforming to a single type. The Indians were seen to be men with as much individual variety as white men. [*Church History*, XXXVI, 228–30]
>
> Mr. Vaughan has exhaustively examined the records and written a book of indispensable value to any student of colonial New England. It throws much new light on the nature of Massachusetts Indian society. . . . The author also vindicates the much misunderstood Puritan character, absolving it of many charges of blind bigotry. [*New York Times Book Review*, November 7, 1965]
>
> The book—at least around the Boston area—has received general, as well as professional, notice. We can be grateful; for the myth of unscrupulous treatment—invented by the white man and imposed on himself—dies hard. [*New England Quarterly*, XXXIX, 422–24]

Obviously Vaughan had met a keenly felt need. Of a dozen or so reviews, only novelist William Eastlake was sharply critical: "Kidding Our Puritan Consciences" (*Nation*, CCII, 22–23). Somewhat skeptical was historian Theodore B. Lewis, Jr., who properly questioned in the *William and Mary Quarterly* (XXIII, 500–2) how much "negative evidence lies hidden in Mr. Vaughan's footnotes" but—short of the kind of analysis to which I have just subjected the reader—could only conclude lamely: "The Puritan record was relatively good, but it was not perfect." The rest of the reviews were raves.

The critical success of *New England Frontier* was assured by the enthusiastic response of the reigning doyen of colonial historians. In *Book Week* of September 19, 1965, Edmund S. Morgan wrote that Vaughan had demonstrated Puritan fair dealing with Indians in a study "heavily buttressed with facts":

> He raises questions that will affect future historical consideration of Indian–white relations throughout the country. . . . In our haste to disavow racial superiority, we fail to recognize how tightly the idea of race has trapped us. By doing what a historian should, by trying to see the situation as men of the time saw it, Mr. Vaughan escapes the trap.

But more than the "idea" of race is a trap. Seeing through Puritan eyes by inclination and sources, Vaughan had seen "the situation as men of the time saw it," and by that Morgan had to mean *men* as distinct from *natives*. Over a century earlier Henry David Thoreau had written a journal entry with contemporary Indian-haters such as Parkman in mind, but his observation nevertheless has continuing application: "It frequently happens that the historian, though he professes more humanity than the trapper, mountain man, or gold-digger, who shoots [an Indian] as a wild beast, really exhibits and practices a similar inhumanity to him, wielding a pen instead of a rifle."

Or reconsider Herman Melville, who saw that some men have fingers like triggers and predicted: "And Indian-hating still exists; and, no doubt, will continue to exist, so long as Indians do." The indisputable achievement of Alden T. Vaughan, his original contribution to knowledge, as it were, was to show the ease with which the Indian-hating of the first half of the seventeenth century could be brought down intact into the second half of the twentieth. The high praise lavished upon his exploit demonstrated that this traditional Anglo-American detestation of natives still existed, nay flourished, however subterranean and genteel in expression. Besides, in Boston, New Haven, and points westward, it was a comfort to be reassured the forefathers were not land-hungry racists and hence not responsible for the terrible spirits in the air across the land.

Not only did *New England Frontier* have a vertical significance for citizens, back through time. It had a more immediate relevance, across space, for a nation at war with natives in Indochina. The parallel was not lost on William C. Kiessel of the Army and Air Force Exchange Service. In the *American Historical Review*, Kiessel gladly granted the book would take "its place as the standard authority on Puritan–Indian relations" but touched on a more current theme in noting what happened when two cultures collided, the one

> unified, visionary, disciplined, and dynamic while the other was divided, self-satisfied, undisciplined, and static. . . . Basing his work on wide-ranging research, yet never pedantic, Vaughan describes how the civilized, Gospel-centered (but not bigoted) Puritan society inexorably expanded and finally controlled the Neolithic world of the Indians. The narrative moves like the advancing frontier it describes, marshaling widely separate facts into a documented story that seldom loses impetus. [LXXI, 1421–22]

Indeed, the story had seldom lost impetus since the Pequot War. In 1965, given U.S. reverses in Vietnam, it was a real comfort to be reminded how inexorably fated had been the advance of the Anglo-American frontier through native worlds. Along with its many other distinguishing characteristics, Vaughan's work was a worshipful tract on the beginnings of the counterinsurgency that had seldom paused and never stopped.

3

Yet the lean young New Frontiersmen who moved into the White House in 1961 have been credited with the "new military strategy" of counterinsurgency (COIN). And it was true that John Fitzgerald Kennedy accepted the Democratic presidential nomination by proclaiming the New Frontier "here whether we seek it or not." It was "here" in the sense of everywhere, for shortly after his inauguration he saw the United States locked in "relentless struggle in every corner of the globe" (GPP, II, 801). And on September 9, 1963, just two months before his assassination, he still saw China as so large, as looming "so high beyond the frontiers, that if South Vietnam went, it would not only give [the Chinese] an improved geographic position for a guerrilla assault on Malaya, but would give them the impression the wave of the future in southeast Asia was China and the Communists" (GPP, II, 828). Throughout his brief administration COIN was Kennedy's primary weapon against all such hostiles beyond the frontiers.

It was also true that Kennedy had a personal passion for COIN. His famous reading of the guerrilla warfare texts of Mao Tse-tung and Che Guevara was one index, and students of his foreign policy list others: the COIN schools set up in Washington for eager deskbourne warriors (Bruce Miroff has pointed out that before going to Third World countries all State Department personnel had to take a special COIN course); the orders, over Pentagon objections, that Army Special Forces wear the elite green berets by which they became known; the strategic-hamlet program in Vietnam; the proposal to name Edward Geary Lansdale, the legendary counterinsurgent, ambassador to Saigon; the staff position of cold warrior Walt Whitman Rostow, modernization enthusiast and author of *The Stages of Economic Growth* (1960)—a blueprint for nudging Third World countries into the "takeoff" stage and thence into "self-sustained growth"; and the appointment of Maxwell Davenport Taylor, the authority on brush-fire wars, first as special military representative of the president and then as chairman of the Joint Chiefs of Staff. Kennedy did all this and more, but that did not make him the architect of modern COIN.

For that union of the covert operations of the Central Intelligence Agency (CIA) and the special operations of the U.S. Army, one has to go back to a seminal document circulated within the Eisenhower administration. Titled "Training under the Mutual Security Program" and dated May 15, 1959, it came primarily from the pen of Brigadier General Richard Giles Stilwell, the gifted young officer who was serving on a special presidential committee and who was said by David Halberstam to be "a former CIA man" —an identification a little like calling a large and ferocious member of the cat family "a former leopard." It incorporated a good deal of material written by other cold warriors and in particular a "Confidential memorandum prepared by Colonel E. G. Lansdale, Office of Special Operations, Department of Defense [DOD]." Its overall analyses and recommendations foreshadowed countless misdeeds in the following decades. Air Force Colonel L. Fletcher Prouty, from 1955 to 1963 the key officer for contacts between the CIA and the DOD on military support of special operations, reproduced it verbatim in his *Secret Team* (1973; pp. 442–79) and declared it "one of the most influential documents of the past quarter-century." The claim is persuasive.

"Training under the Mutual Security Program" posited a world in which there was little mutuality and even less security. It was a battleground, a terrain of hatred, with the Free World pitted against the forces of "Communist subversion." With its associates, the United States *was* the Free World, or interchangeably, the West. Winning the West was, therefore, the defense and extension of "freedom," and that applied, with a touch of geographical madness, to the South, to "Afro-Asia"—as the manpower planners termed those continents—to Latin America, and to the Middle East.

In this southern half of the globe were the "fledgling" or "neutral nations," the "low income" or "less developed countries," in urgent need of military hardware and economic capital. But the document contained only fleeting reference to the threat of "armed dissidents" and the attendant necessity that "appropriate elements of the army should be equipped and trained for unorthodox warfare." Rather the focus was on political and social inputs, the indispensable follow-up "development of requisite institutional frameworks, managerial organizations and individual talents to effectively use the physical resource inputs." Stilwell contemptuously cast aside "the myth of non-interference" as little more than a cloak for "timidity, lack of assurance, sometimes want of moral courage when confronting issues which, admittedly, run close to national nerve centers and traditions." But with "sophisticated handling" the "human resources" of the host country could be developed and its institutions made over in the U.S. image, with elites or "leadership cadres" groomed for the superstructure, and masses conditioned for the infrastructure: "An educational support program," with a price tag of several hundred million dollars, would raise the general level of comprehension "of the subversive forces at play" and be effective in "breaking down the myth of imperialist exploitation, of indicating our interest in individual opportunity and social democracy." Seemingly, social democracy could draw on the officer corps as "a rich source of potential leaders"—with enlivening candor Stilwell accepted the likelihood of military dictatorships and approvingly cited as a model the Westernization of Turkey by Kemal Ataturk: "One must reckon with the possibility—indeed probability—that the Officer Corps, as a unit, may accede to the reins of government as the only alternative to domestic chaos and leftist takeover." Finally, Stilwell called on Colonel Lansdale's testimonial that "the U.S. Army helped build José Rizal into the Philippine national hero;

and did the same with respect to the legendary figures who today furnish inspiration for the armed forces of Vietnam. . . . The record is witness to the tremendous influence exerted by a few dedicated Americans over the policies and points of view of key decision-makers; the value of their efforts, both to the country concerned and the United States, has been inestimable." *Incalculable* might have been more apt. We shall return to these few dedicated Americans later on and attempt an estimate of their role.

Here, in 1959, were all the elements of Kennedy's COIN strategy, including Stilwell and Lansdale, who became stars on the New Frontier team; manipulation of the domestic affairs of other countries from the outside; co-optation and subornation of local elites; and on down to admiring reference to "noteworthy 'civic actions' in the Philippines, Vietnam and Laos."

Yet the document also invited the thoughtful reader back in time and space. The "less developed" countries of "the middle third of the world" were those "backward territories" of the nineteenth century not under the sovereignty of any "civilized" Western power. Indeed, except for lack of woods in some instances, they resembled in every particular Alden Vaughan's seventeenth-century forest preserve, that "unpeopled" wilderness. In backward territories lived backward races or peoples, dark-skinned natives. Stilwell's contempt for their culture, or rather lack of culture, was as ill-concealed as Vaughan's: In Africa and Asia, he wrote, "the political and social revolution has uprooted most of the symbols, beliefs and concepts to which men previously clung. The gap must and will be filled," and if the United States did not fill it—win the hearts and minds of the "middle billion"—then the Soviet Union would. To be sure, members of the "country team" should have knowledge of local traditions and "superstitions," but that was primarily so they could identify "major vulnerabilities in the local structure" and propose "remedies in keeping with native mores."

Stilwell's native was Vaughan's, save for a few centuries and some thousands of miles. The general and the historian shared an unshakable belief in the self-evident superiority of the norms of the West or, as Stilwell put it, of the "balance, stability, confidence and progress within the American society." Both regarded the native as a child or a "savage," while eschewing the word, an empty vessel into which the advanced white Westerner could dump inputs, "our widely heralded social traits," in Stilwell's words, or salvation or Yankee know-how or whatever the redeeming remedy in vogue.

Although the frontier had become global, President Kennedy saw himself, in imagery and substance, standing at Frederick Jackson Turner's "meeting point between savagery and civilization," fending off the forces of chaos and darkness: " 'We in this country in this generation,' " read a sign at the Green Beret base at Nhatrang, Vietnam, " 'are by destiny rather than choice, the watchmen on the walls of world freedom'—JFK." This quotation of their patron saint graced a camp that struck an observer as "not unlike a frontier stockade." Not so far away, all the U.S. Marines living in the Queson Valley, according to the *New York Times* of February 28, 1970, considered "all areas outside their small circular fortresses . . . to be 'Indian Country.' " And on the Vamcotay River near the Cambodian border, according to the *New York Times* of February 17, 1970, U.S. Navy enginemen piloted patrol boats through what they too considered "Indian Country." All were seeing territories "beyond the frontiers," as had their by then dead commander in chief. And while still alive he had made clear that the hostiles lurking there were the traditional merciless savages in a new setting. On December 14, 1961, Kennedy wrote Premier Ngo Dinh Diem of the indignation of the American people that had "mounted as the deliberate savagery of the Communist program of assassination, kidnapping and wanton violence became clear" (GPP, II, 805).

Under the heading "The 'Old Stockade Idea' Works," editor Harold H. Martin of the *Saturday Evening Post* identified Kennedy's strategic hamlets for his readers by quoting a U.S. counterinsurgent: " 'It's the old stockade idea our ancestors used against the Indians,' an American said triumphantly. 'Now the Cong can't just walk in and take what they want' " (November 24, 1962, p. 16). In anticipation of Mao Tse-tung's famous statement that "guerrillas must move among the people as fish swim in the sea," U.S. commanders in Batangas and Samar had also undertaken at the turn of the century, as we have seen, to drain the sea into concentration camps. But the *Post* editor was right in believing the idea older still. As if in confirmation of that fact and of his awareness of the strategic-hamlet system designed by Professor Eugene Staley of the Stanford Research Institute, Alden T. Vaughan wrote that in the 1670s "Puritan military strategy would have been served best by the formation of an outer ring of armed villages, from which patrols could range the woods between the stockades while one or more armies sought a showdown with the enemy" (NEF, p. 315). But the idea was proposed way back then, and shortly all "Praying Indians," as John Eliot's flocks were called, were collected in "specified towns" that were forerunners of strategic hamlets, just as the concentration camp on Deer Island in Boston Harbor, where they wound up, was a forerunner of all the concentration camps to come.

These telling illustrations might be multiplied indefinitely—Denis Warner even witnessed the water cure being administered in South Vietnam to heighten our sense of *déjà vue* (*Reporter*, September 13, 1962). Of course the men playing Cowboys and Indians on this farthest frontier were given to extravagant language. Outside the war room of Admiral Harry D. Felt's headquarters in Honolulu was a notice headed "Injun Fightin' 1759. Counter-Insurgency 1962" and listing what were presumably the standing orders of Rogers's Rangers in the French and Indian War two centuries earlier—one order read sensibly: "Don't sleep beyond dawn. Dawn's when the French and Indians attack."

Or consider the testimony of General Maxwell Taylor before the Senate Foreign Relations Committee on February 17, 1966:

> We have always been able to move in the areas where the security was good enough. But I have often said it is very hard to plant the corn outside the stockade when the Indians are still around. We have to get the Indians farther away in many of the provinces to make good progress.

Figures of speech, certainly, but they did not have playful consequences, and they pointed to an underlying reality: Taylor spent his career in an army that had cut its eyeteeth fighting French and Indians, British and Indians, or just simply Indians. As for "our" Indians, the *Pentagon Papers* (1971) made plain that as ambassador to their country (1964–65), Taylor could call restive South Vietnamese generals to the Embassy and read them the riot act with all the imperiousness of a General Cass dressing down Ojibwa chiefs at Fond du Lac (NYT, pp. 336–37, 379–81). And like Cass, McKenney, and all the others, Taylor was exasperated by his natives and even wondered aloud whether some racial characteristic might not account for their factionalism and lack of national spirit: "Whether this tendency is innate," he said to a working group in Washington, "or a development growing out of the conditions of political suppression under which successive generations have lived is hard to determine. But it is an inescapable fact that there is no national tendency toward team play or mutual loyalty to be found among many of the leaders and political groups within South Vietnam" (GPP, III, 668). Given the innate illogic of racism, the general recognized no incongruity in puzzling a few sentences later at the ability of hostile natives to rebuild their units and make good their losses, considering it "one of the mysteries of this guerrilla war. . . . Not only do the Viet Cong units have the recuperative powers of the phoenix, but they have an amazing ability to maintain morale." Taylor might have been General Philip H. Sheridan describing the Sioux.

After the assassinations of Diem and his brother Ngo Dinh Nhu and of Kennedy himself three weeks later (November 22, 1963), the strategic-hamlet program collapsed. Under President Johnson the war became more orthodox, or "conventional," so to speak, with saturation bombing, napalm, free-fire zones, defoliation, and the rest that cost a million to two million Indochinese lives and some fifty thousand U.S. lives—body counts that made the suppression of the Filipinos seem a limited war and the extermination of the Pequots hardly worth mentioning. But COIN carried on in Operation Phoenix, the CIA program designed for "rooting out the Viet Cong infrastructure." Presumed National Liberation Front (NLF) cadres were summarily "neutralized"—Frances FitzGerald rightly observed that "the program in effect eliminated the cumbersome category of 'civilian'; it gave the GVN [Government of South Vietnam], and initially the American troops as well, license and justification for the arrest, torture, or killing of anyone in the country, whether or not the person was carrying a gun." Phoenix Director William Colby set quotas on the number of civilians to be captured or assassinated each month; after Richard Nixon named him head of the CIA, Colby testified that 20,587 NFL suspects had been killed in the first two and a half years of Phoenix.*

Naturally some cold warriors advocated still more drastic measures along the way. But even hardheaded systems analysts blinked when Air Force General Curtis E. LeMay advocated bombing the North Vietnamese "into the Stone Age." Concurrently Professor Vaughan was going the other way by attacking the Bay Colony sachems for having encouraged tribespeople to resist Puritan missionaries: "There is some evidence," wrote Vaughan, "that a far greater number of them would have thrown off the shackles of the Stone Age if their sachems had not been so reluctant to jeopardize their own power and wealth" (NEF, p.

* Under critical fire later, Colby stuck to his guns: "In view of the world situation and our national policy, we are not spending much of our effort on covert operations. We are keeping our musket and our powder dry"—*The CIA File*, ed. Robert L. Borosage and John Marks (New York: Grossman, 1976), p. 205; on the Phoenix program, see pp. 62, 190; see also James P. Sterba's "The Controversial Operation Phoenix: How It Roots Out Vietcong Suspects," *New York Times*, February 18, 1970.

326). Either way Stone Age natives were in bad odor—LeMay's notorious proposal was to stop swatting flies in frustration and go after the source by hitting the piles of manure. What is not well known is that his recommendation came up in a running dialogue with McGeorge Bundy, special assistant to the president:

> LeMay: We should bomb them into the Stone Age.
> Bundy: Maybe they're already there.

In the summer of 1963 Henry Cabot Lodge, Jr., grandson of the "large policy" senator, took the mandatory COIN course in Washington and then took up his duties as ambassador in Saigon.* By then the Buddhist self-immolations, or "monk barbecue shows" as Madame Nhu jeeringly called them, and continued NLF successes combined to make the Kennedy administration impatient with its client regime. Accordingly, on August 24 Lodge received the since famous cable that was in effect authorization and encouragement of a coup against Diem (GPP, II, 734–35); the ambassador in turn had CIA agents so inform plotters within the Saigon military. Washington officials had second thoughts, the Saigon generals failed to act promptly, and the coup was put off by uncertainties that only came to an end in October with the partial suspension of funds for the Diem government. But during all the vacillations Lodge held firm against any act that would abort the coup, arguing specifics and general conviction:

> My general view is that the U.S. is trying to bring this medieval country into the 20th Century and that we have made considerable progress in military and economic ways but to gain victory we must also bring them into the 20th Century politically and that can only be done by either a thoroughgoing change in the behavior of the present government or by another government. [GPP, II, 790]

* In "the tracks of our forefathers"—in 1953 Kermit ("Kim") Roosevelt, another "large policy" grandson, organized the covert action that overthrew Iranian Premier Mohammed Mossadegh, who had nationalized his country's oil industry. This kept the Shah on his throne—*CIA File*, pp. 20–21. In another operation Roosevelt put strongman Gamal Abdel Nasser in power in Egypt or at least was given credit for doing so by the CIA, in which he was special assistant to the deputy director for plans (i.e., clandestine services) and known as "Mr. Political Action"—Joseph Burkholder Smith, *Portrait of a Cold Warrior* (New York: G. P. Putnam's Sons, 1976), pp. 211, 245. Major Quentin Roosevelt, still another "large policy" grandson, continued family tradition by following the empire to the Farthest West. In WWII this Roosevelt became the Office of Strategic Services representative to Chiang Kai-shek's government—his and Kermit's father was the former head of the United China Relief and a longtime friend of the generalissimo—R. Harris Smith, *OSS: The Secret History of America's First Central Intelligence Agency* (Berkeley: University of California Press, 1972), p. 266. In their penchant for the game of nations and secret intrigue, this generation of Roosevelts marched in their grandfather's footsteps.

What the United States had propped up, the United States could knock down. No evidence has come to light that the CIA was directly involved in the subsequent assassinations of the Ngo brothers, and both Lodge and Kennedy seem to have been surprised by this foreseeable upshot of all their plottings, pressures, and withdrawal of support. By the same token no evidence has come to light that U.S. officials took positive steps to ensure safe conduct out of the country for native allies who had become as expendable as the great Narragansett sachem Miantonomo had become for the colonists, though the Saints were rather more demonstrably involved in that exemplary murder of a former ally (see p. 60; cf. NEF, pp. 161–66).

The task of throwing off "the shackles of the Stone Age" or alternatively of bringing countries out of the Dark Ages into the light of the twentieth century was never a burden to be shouldered by those with no stomach for bloodshed. "Westernization," "Americanization," "modernization," "nation-building," or "progress"—under whatever name the process was always an assault on family structures and on the village. Natives who resisted these assaults on their persons and cultures provoked exterminatory attacks in a continuing tradition of Indian-hating that underscored the interrelationship of COIN and modernization.

Alden T. Vaughan celebrated the beginnings of this process in the Bay Colony and praised the Christian Indian towns in which John Eliot penned up his native flocks in a preindustrial form of urbanization. Another professor, Samuel P. Huntington, celebrated the most recent manifestation of the process in his well-known essay "The Bases of Accommodation" in the July 1968 *Foreign Affairs* (XLVI, 642–56). Head of the department of government at Harvard University, Huntington went to the heart of the matter: "The war in Viet Nam is a war for the control of the population." In rural areas the NLF had effective control of its constituency within the population and could not be dislodged so long as that population stayed put. Ergo, the answer was to drain the sea in which the guerrilla fishes swam or, more directly, to drive the people out of the countryside into urban slums: "The urban slum, which seems so horrible to middle-class Americans, often becomes for the poor peasant a gateway to a new and better way of life." And through the " 'direct application of mechanical and conventional power' . . . on such a massive scale as to produce a massive migration from countryside to city" the poor peasant was being driven to the way of life these latter-day saints had chosen for him; consequently "the Maoist-inspired rural revolution is undercut by the American-sponsored urban revolution":

> In an absent-minded way the United States in Viet Nam may well have stumbled upon the answer to "wars of national liberation." The effective response lies neither in the quest for conventional military victory nor in the esoteric doctrines and gimmicks of counterinsurgency warfare. It is instead forced-draft urbanization and modernization which rapidly brings the country in question out of the phase in which a rural revolutionary movement can hope to generate sufficient strength to come to power.

The lines of the answer had the beauty of simplicity. Through his application of power "on such a massive scale" Huntington was in effect proposing to bomb the Vietnamese *out* of the Stone Age so that "history—drastically and brutally speeded up by the American impact—may pass the Viet Cong by." And aboard this Occident Express the United States would speed the survivors on into a new and better future.

At both ends of the line Professors Vaughan and Huntington used the rhetoric of the behavioral sciences to put the "civilized" frame of their understanding around appalling atrocities of the flesh and of the spirit.

And so did the New Frontier sink into the swamps of Indochina. In the words of Richard J. Walton, "Vietnam is Kennedy's most lasting legacy." And since the New Frontier was an extension of the old, to ask why it bogged down in Vietnam was to extend prior questions: Why were Anglo-Americans in the Bay Colony? In Louisiana? California? The Philippines?

Still, there were those "few dedicated Americans" who led the way there in the first place. *How* the United States established its physical presence in Indochina—and with what consequences—is the remaining item of old–new business.

CHAPTER XXIV

The Ugly American

> *The Ugly American* is an inspiring book and a fighting book; it is not the complete answer to Graham Greene's *The Quiet American*, but it does show that we are beginning to ask ourselves seriously and objectively the questions to which Mr. Greene gave an ugly bias.
>
> —*Catholic World*, April 1959

"SLASHING . . . NOT TO BE IGNORED" read a puff for *The Ugly American* (1958) by William J. Lederer and Eugene Burdick. The review in *Time* (October 6, 1958) from which it came was a bit more balanced: "This book is a slashing, over-simplified, often silly and yet not-to-be-ignored attack on the men and women who have taken up the white man's burden for the U.S. in Southeast Asia." Anyhow, Lederer and Burdick, two young men who were old Asia hands, had teamed up to produce this slashing series of fictional sketches based on "fact." It underwent twenty printings from July through November, became the October Book-of-the-Month Club selection, remained on the bestseller list for seventy-eight weeks, and ultimately sold over three million copies. It became "an affair of state" when Senators John F. Kennedy, Clair Engle, and others presented every member of the U.S. Senate with a copy and when President Eisenhower read it and appointed a committee to investigate the entire aid program. It read like a collection of *Saturday Evening Post* features and fittingly ran as a serial in that traditional voice of middle Anglo-America. It could also pass as a scenario in search of a studio, a search that ended when Universal-International snapped up the motion picture rights. It was, finally, a home-grown answer to the title it echoed, Graham Greene's *The Quiet American* (1955).

Time notwithstanding, *The Ugly American* was not a broadside against the men and women who had taken up the white man's burden in the Far East. It was rather an attack on the bad guys, the legions of official fatheads and timeservers who did not trouble to learn native languages and who by preference lived with other "clean-cut Americans" in their own golden ghettos within the capital cities. Representative of this great majority of second-raters was the old political warhorse Louis Sears, the imaginary U.S. ambassador to Sarkhan, an imaginary country out toward Burma and Thailand. Marking time until he received his promised federal judgeship, Sears had initially brushed off the appointment because "I just don't work well with blacks" and had grudgingly accepted only after being told the Sarkhanese were browns. Once at his post he still saw them as "strange little monkeys" and suspected that "the damned little monkeys always lied." Sears and all the others like him were the envoys of U.S. arrogance, ethnocentrism, overt racism.

But there was a tiny handful of good guys, men like Ambassador Gilbert MacWhite, Sears's successor, who had conscientiously prepared for his position and still discovered he did not know enough about "the Asian personality." Before he was forced to resign, MacWhite learned more, particularly from the few white men he found in all Southeast Asia who "were at all valuable in our struggle against Communism," as he put it in his final letter to the secretary of state: "One of them was a Catholic priest, one an engineer, one an Air Force Colonel, one a Major from Texas, and one a private citizen who manufactures powdered milk. From this tiny handful of effective men I learned some principles." Without such men, he warned, the life-and-death struggle would be lost: "If we cannot get Americans overseas who are trained, self-sacrificing, and dedicated, then we will continue losing in Asia. The Russians will win without firing a shot, and the only choice open to us will be to become the aggressor with thermonuclear weapons." Hence, on its negative side *The Ugly American* was more accurately a slashing attack on *how* the white man's burden had been taken up; on its uplift side it was a series of parables on "How to Handle Natives So the Fiendishly Clever Russians Will Not Win (without Firing a Shot)." On both sides it cast natives as counters in the battle between the huge empires of the West and of the East; the distance between the sides was the short step between overt and covert racism, as three of the more interesting parables demonstrated.

1. *A Catholic priest.* In 1952 Father John X. Finian, S.J., was forty-two years old, a promising Jesuit intellectual, and already the author of two well-received books, one a social interpretation of medieval religious visions, the other on the challenge of Communism. Ordered to Burma, he took up his mission with zeal and

in full knowledge "that Communism was the face of the devil." Following the Irrawaddy River a thousand miles above Rangoon, Finian studied his notes on Burmese culture and prepared for the final leg of his journey into the mountains near the Chinese border. He had determined to eat the food of natives and to learn their language; even with his body racked by dysentery he chewed into the language "like a cold chisel driven into granite" and in four weeks hammered out a command over simple sentences. But his main problem was "to find at least one native Catholic who was courageous," no easy task in Burma or seemingly in all of the Far East. After stealthy manipulation and testing that was painful for the archbishop's jeep driver, Finian found in him his courageous Friday, the native "with whom I can trust God's work" and with whom he would dare plunge into the jungle to confront the devil. Seven additional recruits came into the fold with relative ease, since he "had not required that they be Catholic; only that they be anti-Communist and that they be honest and have courage. 'Come to think of it,' Finian said to himself, 'that was asking quite a bit,' " and of course it was, especially of natives. With his small band the missionary set out to unmask the enemy and that too was no easy task, "for we were only nine against three thousand active Communists." With cunning worthy of the Great Serpent himself, Finian and his followers published a newspaper, *The Communist Farmer*, on a ditto machine he had thought to bring along. Through it and other ingenious means they made their enemies a laughingstock. It was a miraculous feat, truly, but why had Finian found it necessary to come halfway around the world to discredit Burmese Communists? Because from that distance and farther, for that matter, the white missionary could see them as the faces of evil. "The evil of Communism," he wrote in his diary before leaving Burma for Sarkhan, "is that it has masked from native peoples the simple fact that it intends to ruin them." Without the help of this Celtic-American father and others like him, in fine, native peoples were too simple to unmask that simple fact.

2. *A Major from Texas*. Major James Wolchek was a paratrooper who came over from Korea for duty as an observer with the French forces operating out of Hanoi. When he arrived, he struck Major Monet of the Foreign Legionnaires as tough as "whang leather." Wolchek had to admit his nickname was "Tex": "In a way, he thought, it was ironic that he should look so much like the imaginary Texan." It was ironic because his parents had migrated to Fort Worth from Lithuania in 1922, two years before his birth, and they had been short, dark, and small-muscled people. Still, "they

had always dreamed of the American frontier; they found the American magic in Texas," in real estate and in a half-dozen tall, muscular children. Now, just as Tex was about to accompany Monet on a drop into Dien Bien Phu, they received word of that disastrous defeat. In the nightmare that followed, Ambassador MacWhite showed up from Sarkhan and with Tex and Monet discovered too late that the Communists were winning one victory after another because none of the French officers had ever read Mao Tse-tung on guerrilla warfare.* As Tex remarked, "conventional weapons just don't work out here. Neither do conventional tactics." It was sad, especially for the French: "Imagine a nation which produced Napoleon, Foch, and Lyautey," said Monet in a soft voice, "being beaten by so primitive an enemy."

MacWhite, Tex, and Monet were on hand to witness the French evacuation of Hanoi. It was a bitter experience made more interesting for the reader by the fact that the Russians in the book were suave supersubversives such as Ambassador Louis Krupitzyn, MacWhite's adversary in Sarkhan, while their Asian comrades were, well, natives—the contrast very nearly made explicit the covert racism of the text. Consider the authors' account of the Viet Minh padding into Hanoi, some merely wearing breechcloths and many barefooted: they looked as though they were people "who were fighting a war that should have taken place three hundred years before." An officer on a bicycle held up his hand, whereupon the guerrillas disappeared "as fast as the slithering of lizards . . . into doorways and gutters." Tex was left to reflect on his frontier heritage and on his awareness "that around all of Hanoi a huge, silent, and featureless army of men, each of them no more impressive than these, were oozing into the city which they had conquered." The Viet Minh were definitely natives: proper European Reds did not slither or ooze.

3. *An Air Force Colonel.* MacWhite first heard of Colonel Edwin B. Hillandale from President Ramón Magsaysay of the Philippines. Magsaysay had opined "that average Americans, in their natural state, if you will excuse the phrase, are the best ambassadors a country can have" and had cited "The Ragtime Kid—Colonel Hillandale" as a prime illustration: "He can do anything. But I hope you don't steal him from here." Of course MacWhite schemed to do just that, and with his help the colonel stole the

* John F. Kennedy's famed reading of Mao was certainly stimulated if not initiated by Lederer and Burdick's bad book. Ambassador MacWhite recommended to the State Department that no Americans be allowed to serve in his country without first having "read books by Mao Tse-tung, Lenin, Chou En-lai, Marx, Engels, and leading Asian Communists," a recommendation that foreshadowed Kennedy's compulsory COIN courses.

spotlight from Tex Wolchek, Father Finian, Homer Atkins, retired heavy-construction man and the ugly American of the title, and other leading competitors to become the central hero of this tiny troupe of dedicated Americans. He was called the Ragtime Kid because, as his nickname implied, he played jazz and native tunes on his harmonica. He was a knight-errant wandering up hill and down dale out into the barrios and boondocks to "show the idea of America to the people." Though Lederer and Burdick did not say so, the Ragtime Kid was a sort of twentieth-century reincarnation of Johnny Appleseed, warning folks in the backcountry against modern merciless savages and handing out the seeds and saplings of American democracy. In the authors' hands the colonel had mythic pretensions, if not proportions.

"In 1952 Colonel Hillandale was sent to Manila as liaison officer to something or other," wrote Lederer and Burdick in a disingenuous explanation of their hero's official status. In reality the authors' real-life model—to whom we turn in the next chapter—was not in his natural state at all and the something or other was his position as head of the CIA's covert operations in the Philippines. But the earnest authors depicted Hillandale as a happy-go-lucky character who somehow became Magsaysay's unofficial adviser in the presidential election of 1953. In that great campaign a trouble spot developed in a province north of Manila where "the Reds had persuaded the populace that the wretched Americans were rich, bloated snobs, and that anyone who associated with them—as did Magsaysay—couldn't possibly understand the problems and troubles of the Filipino." So Hillandale rode up there on his red motorcycle with "The Ragtime Kid" painted on the gas tank in black, parked on a main street, sat down on the curb, played his harmonica, and pretended to be broke—to destroy forever the canard that Americans were rich, bloated snobs. In the election that followed ninety-five percent of the inhabitants of that province voted for Magsaysay: "Perhaps it wasn't the Ragtime Kid who swung them; but if that's too easy an answer, there is no other." Sadly, there was another answer, and the authors were treating their readers as though they too were inhabitants of the province and too simple to know the truth.

Natives were superstitious. When MacWhite got the Ragtime Kid on loan from Manila for two months, the latter ambled down the streets of the Sarkhanese capital noting with interest all the palmistry and astrology establishments. It so happened that these were items in this modern Prospero's bag of tricks. Shortly at a dinner party the colonel took the prime minister away from the rest of the guests, read his palm, and made a definite hit. The two men came out of the study "arm in arm, and the Prime Minister was gazing up at the Ragtime Kid with obvious awe"—in fact the palmist told the statesman so much about himself he had him "sweating like a Westerner." Unhappily, a stupid chargé d'affaires failed to follow up through official channels the once-in-a-lifetime invitation that Hillandale read the king's hand and cast his horoscope. After this fell through, the angry colonel explained to MacWhite that "the key to Sarkhan—and to several other nations in Southeast Asia—is palmistry and astrology." The key had been placed in his hands and then knocked away before he could open all the doors:

> "You see, sir," said Colonel Hillandale, beginning to get somewhat excited, "the Chinese Communist Armies have been mobilizing near the northern border. I knew that if I could once get to the King, I could tell him that the stars ordered that he send the Royal Sarkhanese Army up north for maneuvers. If this were done the Communists would interpret the move as a clear indication that Sarkhan was definitely pro-American and anti-Communist. It would have been a defeat for the Commies and would have been a great propaganda victory for us throughout all Asia. And, sir, I am positive that the King would have done what the stars ordered him to do."

Why was he so positive? Because the king was a native, and natives were superstitious.

And that this "novel" became "an affair of state," with Eisenhower, Kennedy, and millions of their compatriots taking it seriously, testified to its genuine "Made in America" stamp. It had a full complement of all the familiar assumptions and attitudes. Never for a moment was the superiority of the norms of the West questioned. Natives existed to be manipulated and hoodwinked. Their touching faith in the occult was a beautiful vulnerability for philanthropists who had been properly "briefed" in Washington. U.S. backwoodsmen should go after natives where they were, most of them, in their "primitive" villages and help them see truths too complex for their simple minds. Knowing their language helped these envoys of "progress" get inside their heads and reveal the "truth" about Reds or, more exactly, about the infinitely despicable Asian Commies. Feigned concern for their culture masked the cold warriors' contempt for their "Stone Age" existence. Along with all this, *The Ugly American* picked up a real modern counterinsurgent and helped make him a legend every bit as authentic as General George Armstrong Custer.

CHAPTER XXV

The Secret Agent: Edward Geary Lansdale

> To stay in the game. Desires like that—not just for power, but for action, activity, excitement—become an addiction.
>
> —DANIEL ELLSBERG, *Look*, October 5, 1971

"THE Report the President Wanted Published" duly appeared in the *Saturday Evening Post* of May 20, 1961. The editors introduced the anonymous contribution with a note explaining that "an American Air Force officer," whose name could not be revealed "for professional reasons," had visited a little village in South Vietnam, and his account of the people there had come to Kennedy's attention shortly after he took office. First scanning and then reading it "with absorbed attention," the president found behind the official language therein "a story of human valor and dedication to freedom, a reminder that Communism is *not* the wave of the future," and had thereupon written a memorandum suggesting it would make "an excellent article for a magazine like the *Saturday Evening Post*." Kennedy was right, and the editors understandably obliged, for in outlook and tone the story was a perfect fit. It seemed inspiriting proof that the republican values of middle America flourished in the swamps of Southeast Asia. And happily now it can be told—"an American Officer" was none other than Lederer and Burdick's Ragtime Kid, Colonel Edwin B. Hillandale—who had already appeared in the pages of the *Post*—or in real life, as it were, the recently promoted Colonel Edward G. Lansdale. And actually the story that followed was merely an annex to Lansdale's long and then unpublished report of guerrilla successes in Vietnam, the document that had reached the White House in early February and spurred Kennedy into his multifarious COIN projects (DOD, XI, 1–12).*

* For an explanation of the abbreviated references, see p. 534.

The Village That Refused to Die was the more descriptive title of the later television film based on the annex. Lansdale reported that in 1959 the village of Binh Hung had been established in the mud flats and swamps of the southern tip of Vietnam by Father Hoa, a Chinese Catholic priest, and "his flock of Christian refugees" from Communist tyranny. On his arrival the previous January, Lansdale had been greeted by twelve hundred "settlers" lined up in military formation: "I was startled and deeply touched. Many of them raised their hands in a familiar three-fingered salute. They were former Boy Scouts and it was the only military gesture they knew." It was the old familiar story of settlers establishing a Christian outpost in terrain infested by hostile Reds—when the mass media made their story widely known, the villagers received shipments of CARE "Settler Kits" from Anglo-Americans who had not forgotten their own pioneer days. The Binh Hung settlers had at first fought off the surrounding guerrillas "with Boy Scout staves and knives."* Recently they had acquired more sophisticated weaponry, including two mortars they had calibrated to zero in on all the approaches to the village. The past Christmas Eve Father Hoa had been singing the midnight mass when scouts rushed in warning that a guerrilla attack was imminent:

> Father Hoa paused to give the firing order. It was right on target. The Viet Cong fled with their wounded. A prisoner later said the guerrillas believed the accuracy of the midnight mortar fire was achieved by sorcery.

So were the superstitious heathens routed by the fighting man of God and his flock, another set of stock figures to American readers and viewers. Indeed, despite his Chinese origins, Father Hoa seemed suspiciously like the navy chaplain of the World War II song hit "Praise the Lord and Pass the Ammunition."

Readers and viewers had been secretly subjected to a bit of "black"—that is, secret—propaganda by the nation's leading practitioner of the craft. Lansdale was not even the anonymous "Air Force officer" mentioned by the editors, since his military title was cover for his real position as a senior CIA operative. He covertly did good to his mainland compatriots—the "target" audience—by enlisting their sympathies on behalf of heroic Christians besieged on the other side of the world by merciless enemies

* Though a former Boy Scout myself, I must report with chagrin that I have absolutely no recollection of our staves.

whose favorite tactic was "the ambush." The Binh Hung story, Lansdale had no doubt, "will interest other men who may someday face Communist guerrillas in combat." And who could think of abandoning settlers who had been discussing "What Freedom Means to Me" for the two months prior to his visit? "They were still going strong on the subject when I left," he added. "The light in their eyes when they talked about freedom showed that it was not mere oratory. Freedom is precious to them, a personal thing. . . . Repeatedly they asked me for assurance that the United States would stand firm in its policy in Asia, and particularly in Laos." *Particularly* in Laos, where the CIA had already gone some fair way toward making a mess of the country?

2

Lansdale's *Post* article was a good illustration of his *modus operandi*. Plainly a legend must be approached circumspectly, and that is all the more true if the legend has written a memoir that is misleading from beginning to end. On January 30, 1954, Brigadier General C. H. Bonesteel III wrote a "Memorandum for the Record" in which he identified Lansdale as one of the four CIA representatives present at a meeting of the President's Special Committee on Indochina; seventeen years later that datum was published in the *Pentagon Papers* (GPP, I, 447). But with that finally in the public record Lansdale never mentioned his work for the CIA—also known as the "Company"—in his account of making a president in the Philippines and of propping up a premier in Indochina. It was as if General Billy Mitchell had finished his career pretending not to be an aviator. And Lansdale's publishers did nothing to make readers more witting. In their "About the Author," Harper & Row merely identified him as having been born in Detroit in 1908, as having worked in advertising after graduating from the University of California in Los Angeles, as having served in the Office of Strategic Services (OSS) during WWII, and as having then become "a career officer in the U.S. Air Force, rising to the rank of Major General before his retirement in 1963. . . . General Lansdale now lives near Washington, D.C., with his wife Helen, in an area that has been home to his family for the past three hundred years." On the jacket appeared this blast of flatulence: "Now, for the first time, Lansdale himself sets the record straight."

In the Midst of Wars (IMW) was a twentieth-century version of

The Confidence-Man, save that now the mysterious impostor had been institutionalized and there was no Melville to guide us past mawkish platitudes and specious surfaces to the underlying deception and distraction. Surely Lansdale and/or his ghost kept both eyes open for unwanted revelations, and we may assume he submitted his manuscript to the Company for policy clearance. At CIA headquarters in Langley, Virginia, his associates must have gone over it line by line. What remained was there because it seemed harmless. Hence the problem of lifting the many masks of Lansdale's masquerade becomes ticklish. Still, flagrant falsities can be pinned down by other documents, and there is always the intriguing chance that his Company censors overlooked choice "secrets" that were home truths.

Probably Lansdale was born in Detroit, schooled in Los Angeles, trained in a San Francisco advertising agency, and launched by the OSS, but here the trouble starts: "I served with the OSS for a time, then became an army intelligence officer. The end of World War II found me in the Philippines" (IMW, p. 4). When and why had he left the OSS? Or had he left? Throughout the war General Douglas MacArthur had barred the organization from the South Pacific. In early 1943 General William J. ("Wild Bill") Donovan had dispatched none other than Worcester's protégé and biographer, the now aging Professor Joseph Hayden of the University of Michigan, to Australia with a plan for OSS guerrilla operations in the Philippines, but MacArthur had rejected it and refused to have Donovan's agents under his command. Afterward some OSS officers in Washington hatched a "Penetrate MacArthur" project that supposedly failed. Though unsubstantiated, it still remained conceivable that Wild Bill Donovan had Lansdale switch to army intelligence so he would at least have a man underground—a "mole" in CIA lingo—behind MacArthur's wall. Lansdale had not necessarily "left" the OSS and assuredly would have found the role of a double agent exciting. In any event his service in the OSS, for however long, was formative and an integral part of his career in its successor, the CIA.

For a century and a half the rising American empire had managed without a central agency for the collection of information and the conduct of covert operations against "distant natives." Plainly, the leading reason was that within that growing continental empire natives were not all that distant. Jefferson received word of what was going on among restless tribes through the spies of Harrison and other officers of his administration and from his own spies in the Spanish territories. Later McKenney drew on his Indian agents for comparable reports, and down through the decades the BIA spied on its wards as one of its functions. At the turn of the century Worcester and Bourns helped General Otis improvise and maintain a spy network among the Filipinos, and under the second BIA, insular espionage was systematized. At home the creation of the FBI in 1908, then known as the Bureau of Investigation, subjected reservation Indians to additional investigations and ongoing surveillance. But it was not until World War II that U.S. intelligence operations became the global responsibility of one newly created bureaucratic entity, the Office of Strategic Services.*

The OSS Lansdale joined was the long-delayed institutional expression of what John Hay used to call "a partnership in beneficence" or of what Colonial Secretary Joseph Chamberlain had referred to in the same vein as "an Anglo-Saxon alliance" (see p. 268). The British Empire had a Secret Intelligence Service (SIS or MI-6) that traced its origins back to the days of the Spanish Armada and that had since performed yeoman service in controlling "subject peoples"—that is, "niggers" or "wogs"—in actually distant colonies through "special means." By "special means" MI-6 meant paramilitary strikes, "black" propaganda tactics to divide and conquer, sabotage, assassinations, the whole panoply of dirty tricks. Now with their empire on the wane and their very existence as a kingdom threatened, the jaded British resignedly handed on the torch of the Anglo-Saxon mission to their still fresh and lusty cousins across the Atlantic. With British cooperation from the highest level down, Director Donovan patterned his new organization on MI-6 and had his first operators learn clandestine operations and techniques from old hands in secret British schools in Canada. Despite his Celtic origins, Donovan was himself an Anglophile, brought in others at the highest echelons, and sought recruits with the right mix of "Wall Street orthodoxy" and U.S. nationalism. They streamed in—from the corporate world, the professions, legal firms and banks, the entertainment industries, Ivy League faculties, leading families (such as J. P. Morgan's, the Roosevelts, the Mellons, and the du Ponts), and the advertising agencies (such as J. Walter Thompson's and even the unidentified San Francisco firm for which Lansdale worked). Within the OSS the old-boy network resembled that in MI-6, and the letters were half-seriously said to mean "Oh So Social," according to *The CIA and the Cult of Intelligence* (1974) by Victor Marchetti and John D. Marks:

* The transition from continental to global clandestine services left the FBI responsible, supposedly the national agency exclusively responsible, for watching over citizens and others on the mainland. Of course this division of labor did not work out quite that neatly.

> [As in its predecessor] most of the CIA's top leaders have been white, Anglo-Saxon, Protestant, and graduates of the right Eastern schools. While changing times and ideas have diffused the influence of the Eastern élite throughout the government as a whole, the CIA remains perhaps the last bastion in official Washington of WASP power, or at least the slowest to adopt the principle of equal opportunity. [p. 278] *

As an adopted son of the Golden West, Lansdale had to make up for his lack of snobbish background in the East and to the end was still seeking status as a true son of the American Revolution, as the "About the Author" note suggested: "General Lansdale now lives near Washington . . . in an area that has been home to his family for the past three hundred years." Maybe, but what had happened to the family homestead in unhallowed Detroit, his putative place of birth?

3

The "British Connection" was of direct importance to Lansdale. In one of the more trustworthy sections of his memoir he held forth on "OCCULTISM. The Vietnamese had small need of Filipino ghost stories. Vietnam was so filled with the arcane that I used to advise Americans to read Kipling's 'Kim' and pay heed to the description of young Kimball O'Hara's counterintelligence training in awareness of illusions" (IMW, p. 243n). Like Professor Worcester who had preceded him to the Philippines, Lansdale was a devotee of Kipling's works and found therein a model for himself, his "team," and all the other American imperial agents who had stepped forth to relieve MI-6 of the main weight of the white man's burden.

* For all their preoccupation with "black" propaganda, "black" flights, "black" cargo, and all their other "black" or secret undertakings, American spooks were not fond of blacks personally. Even in the 1950s, according to Joseph B. Smith, there were no black officers in the CIA: "black officers did not appear until a decade later when Africa became a target of attention and someone thought they might fit in over there. In the fifties, blacks worked on the night office cleaning crew, and some drove the CIA buses that shuttled employees around to other agency buildings"—*Portrait of a Cold Warrior* (New York: G. P. Putnam's Sons, 1976), p. 73. In the less snobbish FBI the racism was still more blatant. William C. Sullivan, former head of the Domestic Intelligence Division, said that J. Edgar Hoover told him he would never have a black agent as long as he was director. Sullivan testified before Senator Frank Church's Select Committee on Intelligence that Robert Kennedy had asked Hoover how many black agents he had. Hoover told the new attorney general he did not categorize people by race, creed, or color. Kennedy said that was laudatory but that he still needed to know how many black agents there were. 'Hoover had five black chauffeurs in the Bureau', Sullivan said, 'so he automatically made them special agents' "—quoted by David Wise, *The American Police State: The Government against the People* (New York: Random House, 1976), p. 298.

On the surface *Kim* (1901) was merely the marvelous children's story of fond memory. The orphaned son of a nursemaid in a colonel's family and of a sergeant in an Irish regiment, Kim lived as a native in the home of a half-caste woman, played in the streets of Lahore "burned black as any native," joyously entered into the intrigues of lovers ("Kim could lie like an Oriental"), became known to one and all as "Little Friend of all the World," and then wandered off as the disciple of a Tibetan lama who had once been "a master hand at casting horoscopes." But Kim was destined for better, more mysterious things as "a Sahib and the son of a Sahib." From the beginning he possessed a birth certificate, that precious piece of paper his father had left as his estate, with the prophecy it "would yet make little Kimball a man"—and by that the ex-sergeant (and Kipling) meant a *man* as distinct from a *native*. Once other sahibs knew about it and discovered he was white, "white as you or me," he was sent off to St. Xavier's in Lucknow to learn to write and figure in English and to be made into "a man," or for redundancy's sake, into "a white man." One "Colonel Sahib" who posed as an ethnologist took him under the wing of the Secret Service and had him thoroughly grounded in native superstitions, their "Lamaism, and devil dances, and spells and charms" for warding off "all the powers of darkness." Masters of disguise taught him how to "change thy colour" and dress and language to match those of different castes and religions. This training in awareness of illusions prepared him "to enter into another's soul" to pry loose news of a confederacy of natives "who had no business to confederate," of gun shipments, of suspicious explorers, or "of some near-by man who has done a foolishness against the State." On this deeper level, then, Kim had become the sharpened instrument of his masters' imperial mission and had been fully outfitted to play the Great Game of control of natives, "the Great Game that never ceases day and night, throughout India." Unreflectively Kim threw himself into these endless deceptions with a schoolboy's delight in being the bearer of *"the keys of secret things,"* though his constantly changing identities made him uncertain about the boundaries of his own: " 'I am Kim. I am Kim; and what is Kim?' His soul repeated it again and again." But he never forgot his skin color, even when it was dyed, and the lesson he had learned at St. Xavier's: "St. Xavier's looks down on boys who 'go native altogether.' One must never forget that one is a Sahib, and that some day . . . one will command natives."

"I am Air Force Colonel Lansdale. I am Air Force Colonel Lansdale; and what am I?" one imagines Kim's American counterpart asking again and again. "You should know one thing at the beginning," he confided in his account of helping "our Asian friends in the Philippines and Vietnam cope with wars of rebellion and insurgency":

> You should know one thing at the beginning: I took my American beliefs with me into these Asian struggles, as Tom Paine would have done. Ben Franklin once said, "Where liberty dwells, there is my country." Tom Paine had replied, "Where liberty dwells not, there is my country." Paine's words form a cherished part of my credo. My American beliefs include conviction of the truth in the precept that "men are created equal, that they are endowed by their Creator with certain unalienable Rights" and in the provisions of our Bill of Rights to make that great precept a reality among men. [IMW, pp. ix–x] *

These personas of imperishable freedoms were Lansdale's entrée to the Asian struggles. With his guileless good looks and open, folksy manner, he seemed the very personification of a larger-than-Kim-size "Friend of all the World." Even his crew-cut briskly suggested he had nothing to hide, from the top to the bottom of his lanky six feet of clean-cut Americanism. With this make-up rooted in place, Lansdale played a U.S. version of the Great Game with Kim's boyish "pure delight" in playing it for its own sake. Like his model, Lansdale never forgot he had been born a sahib, though he did not use the term, of course, and preferred to put the matter less bluntly by chiding compatriots "who went too far by going native" (IMW, p. 366). Another bearer of *"the keys of secret things,"* he had access to untold (and untallied) riches and hence had every reason to appreciate what Kipling had said in his own voice: "One advantage of the Secret Service is that it has no worrying audit . . . funds are administered by a few men who do not call for vouchers or present itemised accounts." Finally, Lansdale also seemed orphaned, cut off from family, home, all the customary ties.

By his own account he had a family—a mother and brothers in Los Angeles, and a wife and two sons in Washington—but terrorist threats persuaded him, so he said, that they had best be left behind: "they stayed home while I went to the Communist battlegrounds" (IMW, p. 11). Apart from "a brief taste of family life" now and then over the years, he was the fabled American Child of Light on a risky journey to face merciless savages all by himself, a

* Now, why do you suppose Lansdale left out an essential part of that great precept, to wit, that *all* men are created equal?

lone Boone on the Dark and Bloody Grounds of the world. In the summer of 1950, just as he had torn the walls and plumbing out of an old Washington house he had thought of remodeling himself, he was ordered back to the Philippines. He put down his hammer and left behind his sawhorses and plumbing fixtures, left his wife to cope with the mess and with carpenters and plumbers, left her his share of householder's chores and family responsibilities, and left behind his own paper-shuffling desk job and all the dull daily routine of a Washington bureaucrat. He said he felt guilty for leaving his family in that predicament, but what surfaced in his half-apology was a sense of release and a savoring of delicious expectations: "It was close to playing hooky, to go to war" (IMW, p. 15). Wrapped up in his one set of orders were danger, adventure, a role as a mysterious agent from the West, the heady satisfactions of being the secret proconsul of the world's greatest power, and the irresistibly appealing prospect of dominion over natives—those creatures created by the colonial imagination he shared in full measure. Like Worcester a half century earlier, Lansdale jumped at the chance to abandon kith and kin for tropical Crusoe islands and the challenge of dealing there with yet another lot of native "insurgents."

CHAPTER XXVI

Covert Savior of the Philippines

A good smile is a great passport. Use it!

—EDWARD GEARY LANSDALE,
In the Midst of Wars, 1972

IN THE Philippines the origins of the peasant uprising known as the Huk Rebellion (1946–56) were deep-rooted. U.S. colonial officials had carried on the policy, dating back to Taft and Worcester, of collaborating with and using national and provincial elites. The political power of these *illustrados* had thereby been strengthened, and their dependence on peasant clients and tenants lessened. American colonialism also worked to the economic advantage of mainland investors and wealthy Filipino businessmen, spread cash-crop agriculture, particularly sugar cane and rice in central Luzon, accelerated the breakdown in traditional landlord–tenant relationships, and encouraged the landed gentry to make every customary obligation subordinate to maximization of profits. Colonialism and capitalism thus joined to make peasants increasingly expendable. The roots of the villagers' anger reached back at least to the Philippine–American War; back to the outburst of agrarian unrest of the 1930s, when they organized to make landowners live up to their obligations and curb the abuses of their private armies; and back to the early 1940s, when they banded together in *Hukbo ng Bayan laban sa Hapon* (Hukbalahap), or the People's Anti-Japanese Army, to resist the new set of invaders and their collaborators, many of whom were the selfsame *illustrados* the United States had found so useful. At the war's end Huk veterans were treated with harsh disdain by Americans and their recycled collaborators, refused the back pay given other guerrillas, harassed, and in some instances imprisoned by their "liberators." In 1945 peasant organizations joined others, including the *Partido Komunista ng Pilipinas* (PKP), or the Communist Party of the Philippines, in forming the Democratic Alliance to run or endorse candidates for public office. A PKP proposal to run separate candidates for president and vice-president was decisively rejected by a majority of alliance leaders. In the 1946 campaign that followed, the Democratic Alliance nominated and elected six out of the eight congressmen from central Luzon. *Illustrados* and their U.S. mentors parried this threat to their hegemony by refusing to seat the congressmen-elect and by stepping up their campaign of terror against the peasant movement—"the only good Huk is a dead one" intoned a meeting of central Luzon mayors who missed by just a bit the cadence of the immortal words supposedly uttered by General Sheridan (HUK, p. 190).* Killings and hurtings, including the water cure, and general repression of an intensity hard to exaggerate drove peasants from protest to rebellion.

Now, even if my incomplete outline of the revolt's background is inadvertently inexact or misleading in some particulars, it still suggests the enormous complexity of forces we are just beginning to understand. Unfortunately for the peasants, subtle analysis of tangled evidence was never Lansdale's strong point. In *The Huk Rebellion* (1977) Benedict J. Kerkvliet quoted a secret March 1946 report from the then Major Lansdale, chief of U.S. military intelligence in the islands. The Huks were "Communist-inspired," Lansdale stated as fact. "Like all true disciples of Karl Marx," they believed

> fully in revolution instead of evolution. They have made their boast that once their membership reaches 500,000 their revolution will start. Meanwhile, in the provinces of Pampanga, Nueva Ecija, Tarlac, Bulacan, and Pangasinan, they are establishing or have established a reign of terror. So ironclad is their grip and so feared is their power that the peasants dare not oppose them in many localities. Upon liberation, their members were about 50,000; sources now report some 150,000 tribute-paying members. [The Huks are now organized] into trigger men, castor oil boys, and just big strong . . . ruffians to keep the more meek in line. [p. 147]

Kerkvliet's painstaking research and analysis, on which I have drawn freely, eviscerated Lansdale's allegations. In a country with three-fourths of the people living on the land, the tiny Philippine Communist Party (PKP) had virtually no cadres at the village level and only a slippery foothold among workers in Manila. The PKP relationship with the Hukbalahaps and the later *Hukbong Mapagpalaya ng Bayan*, or People's Liberation Army (HMB or still, confusingly, Huks), was always tenuous and at critical points strained or snapped by disagreements and conflicts. The Huk rebellion was neither "Communist-inspired" nor Communist-controlled,

* For an explanation of the abbreviated references, see p. 534.

let alone started, though Lansdale's misrepresentations became the accepted line of U.S. and Philippine authorities: the normally tame natives of central Luzon had been driven wild by "outside agitators," those ever-useful mainland imports. Indeed in 1946, the very year of Lansdale's report, the PKP came out against the rebellion and urged instead continuation of the "legal struggle." Two years later the PKP did an about-face, judged the time ripe for "armed struggle," and rushed to catch up with and attempt to lead rebels already in the field. But at no time did PKP state Communism become the vision of the peasant movement. With rare exceptions aside, by no stretch of the imagination did the Huks qualify, had they wanted to, as "true disciples of Karl Marx" (HUK, pp. 230, 262–66).

Nevertheless, these were the "Communist-inspired" and "Communist-led" (IMW, p. 2) rebels Lansdale came out to crush. He returned as an air force lieutenant-colonel, but that was cover for his real position as chief of the Manila station's Office of Policy Coordination (OPC), and that was in turn the CIA's cover title for covert psychological warfare (psywar), covert political action, and covert paramilitary operations—or, simply, COIN (CW, p. 103). In the all-out CIA effort that followed, he had a bottomless budget of untallied COIN dollars that supplemented the $67 million the U.S. Congress appropriated between 1951 and 1954 for military assistance to the Philippines (HUK, p. 244). And he came back with the great advantage of having a handpicked Friday as the recently installed Philippine secretary of defense.

Lansdale had met Ramón ("Monching") Magsaysay when the then Liberal party representative came to Washington in 1948 to confer with lawmakers about Filipino veterans' benefits. Impressed by his energy, earnestness, and experience as a guerrilla leader against the Japanese, and by his possibilities generally, Lansdale took him in hand and "brought him together with a number of high-echelon policy formulators" (IMW, pp. 13–14). Also impressed, the formulators put pressure on President Elpidio Quirino to make the obscure congressman his defense chief. Consequently, Secretary of National Defense Magsaysay had been sworn into office a week before the arrival in September 1950 of his American sponsor and mentor.

Also on "my team," as Lansdale called his cohorts, was Colonel Napoleon ("Poling") Valeriano, a close aide in this and subsequent operations. The former chief strategist of the campaign against the Huks, Valeriano had commanded the infamous Philippine constabulary units known as "Nenita" or "skull squadrons" after their skull and crossbones emblem. Under Valeriano the Nenita reportedly made generous use of the water cure, broke bones and cut off heads, cordoned off areas and killed innocent people within, and left bodies floating in rivers (HUK, pp. 160, 196–98, 211). In his *Counter-Guerrilla Operations* (1962)—of which the coauthor was Charles T. R. ("Bo") Bohannan, another member of the Lansdale team—Valeriano himself described the Nenita as a "hunter-killer detachment [organized] to seek out and destroy top leaders of the Huk" (p. 97). Turning these forerunners of the CIA's Phoenix assassination squads loose and dropping bombs, including napalm, on villagers summed up Valeriano's strategy: it consisted in using the heaviest stick at hand to pound the Huks into submission.

Lansdale's basic innovation was to add a carrot to the stick, or in more apt metaphor, to drain away peasant support so guerrilla fishes still unnetted would be left flopping high and dry. Under his mentor's guidance, Monching Magsaysay launched or talked about launching a series of agrarian reforms. Perhaps the most widely publicized was the Economic Development Corps (EDCOR) that Lansdale described in familiar terms:

> Magsaysay would undertake to obtain land that was judged to be in the public domain, ready for homesteading. He then would grubstake a group of settlers on this land, the settlers being retired or nearing-retirement soldiers and their families along with former Huks who were neither indicted nor convicted by civil courts and who desired to be "reeducated in the democratic, peaceful, and productive way of life." [IMW, p. 52]

Save for mention of the Huks and their reindoctrination, this was How the West Was Won all over again, with the grubstaked settlers readying their wagons for the long pull across plains and mountains. In fact less than a thousand families were resettled by EDCOR in Mindanao and less than three hundred of these had been rebels (HUK, p. 239). Obviously the intent was not to solve the tenancy problem but to give the peasants hope that it might be solved and that their government was at last responding to their needs. The intent was to counter the Huk cry, "Land to the landless!" with Madison Avenue means. Ironically, EDCOR and other programs did persuade many Huks to surrender and thereby to disprove Lansdale's thesis that they were Communist-led, for the PKP vainly denounced government programs and promises as empty and deceitful. And in truth, Lansdale's commitment to the principles of homesteading was less than firm.

"Land reform is a gimmick," he confided years later to his

friend Richard Critchfield of the *Washington Star*. "The military war is a gimmick. It's fundamentally a question of forming a political base." Land reform was a psywar gimmick, and Lansdale thought of "psywar in terms of playing a practical joke" (IMW, p. 71). Aficionados of this admittedly "low humor" praised both Lansdale and E. Howard Hunt, another OSS veteran and then the OPC station chief in Mexico City, for having the "black minds" that made them the Company's outstanding operational heroes. In the early 1950s the then novice Joseph B. Smith heard a Lansdale story circulating within the CIA: After ordering "a careful study of the superstitions of the Filipino peasants, their lore, their witch doctors, their taboos and myths," Lansdale had a small aircraft fly over Huk areas broadcasting mysterious Tagalog curses on villagers who dared support the rebels: "He actually succeeded in starving some Huk units into surrender by these means" (CW, p. 95). Lansdale called this his "eye of God" technique, a psywar gimmick he extended to posting baleful staring eyes in likely places—they were printed inside a triangle and copied from what he could remember of Egyptian guardian eyes over the pharaoh tombs (IMW, pp. 74–75). Teammates Valeriano and Bohannan pronounced this play upon peasant superstitions "surprisingly useful" in the overall campaign to "fool 'em" (IMW, pp. 155–56).* Its success was less than surprising, really, for with the eye of God upon them, villagers not unnaturally assumed His Anglo-American surrogates were close behind.

Perhaps you will recall that shortly after the turn of the century Professor Worcester had extolled opening up trails and roads into the Philippine backcountry as the only way to get at the "wild man" at 2:30 A.M. and terrorize him into tameness, "since that is the hour when devils, *anítos* and *asuáng* are abroad, and he therefore wants to stay peaceably in his own house" (see p. 302). Now, or rather in a 1972 interview with journalist Stanley Karnow, Lansdale recounted an *asuáng* story to top all *asuáng* stories:

> One [Lansdale-initiated] psywar operation played on the superstitious dread in the Philippine countryside of the *asuang*, a mythical

* In Vietnam the CIA later introduced a variation of Lansdale's "eye of God" technique. In a 1965 forerunner of its Phoenix Program, veteran CIA agent Patrick J. McGarvey recalled, "some psychological-warfare guy in the CIA back in Washington thought of a way to scare the hell out of the villagers. When we killed a VCI [Viet Cong Infrastructure suspect] in there, they wanted us to spread-eagle the guy, put out his eye, cut a hole in his back, and put his eye in there. The idea was that fear would be a good weapon. The funny thing was, people got squeamish about cutting out his eye, so they got hold of some CBS eyes [the network's logo] and began dropping them all around the body"—quoted by Seymour M. Hersh, *Cover-Up* (New York: Random House, 1972), p. 85.

> vampire. A psywar squad entered an area and planted rumors that an *asuang* lived on where the Communists [*sic*] were based. Two nights later, after giving the rumors time to circulate among Huk sympathizers, the psywar squad laid an ambush for the rebels. When a Huk patrol passed, the ambushers snatched the last man, punctured his neck vampire-fashion with two holes, hung his body until the blood drained out, and put the corpse back on the trail. As superstitious as any other Filipinos, the insurgents fled from the region. [CIA, p. 28]

The action violated all the written and unwritten rules of land warfare, but to appreciate its humor fully one needed to know that such psywar operations were administered by a division of Magsaysay's staff known as the Civil Affairs Office: Lansdale's squad was therefore a "civic action" team making its contribution to social intercourse. Too good to be kept "black," the story was repeated in his memoir. To Lansdale it was plainly not a *pro confesso* account of the killing of a *man* but merely a tale of the vamping of a *native* (IMW, pp. 72–73). In truth it appeared to be another decisive demonstration of perdurable colonizing attitudes that went back to Worcester, to McKenney and Hall, to Jefferson, and beyond.

"Only a dead Huk is a good Huk," said Major General Mariano Castañeda, who also missed the sweep of the aphorism attributed to General Sheridan. Chief of staff of the Philippine armed forces, he used the army more as a club than a stick, mounting ponderous encirclement operations, tightening the areas surrounded into smaller "killing grounds," and then blasting those with artillery, thereby hitting or infuriating noncombatants and missing guerrillas who had long since slipped through his lines. To reform a service that was about as inefficient and corrupt and abusive as the Huks charged, and to remake it into an effective COIN instrument, was the other half of Lansdale's strategy. When Castañeda and his allies resisted Magsaysay's "housecleaning," Lansdale said he went to General Leland S. Hobbs, chief of the Joint United States Military Advisory Group (JUSMAG): "Hobbs agreed that Magsaysay deserved support and pleaded his case with President Quirino and General Castenada. I added my own arguments in less formal meetings with the two" (IMW, p. 43). Lansdale modestly refrained from indicating just how powerful his "arguments" were: Unless Quirino ousted Castañeda, U.S. military aid would be withheld. At the height of the struggle Castañeda supposedly threatened to shoot Lansdale as an "interfering bastard," while a

Magsaysay bodyguard stood protectively over him, "set to shoot it out with Castaneda, wild-West style" (IMW, p. 44). But reason (and dollars) prevailed when President Quirino reluctantly "retired" his good friend the general.

Lansdale's Wild West allusion was not passing literary fancy. As a New World Crusoe and Kim's grown-up cousin, his Great Game wherever he went was Cowboys and Indians, that entertaining institutionalization (for all ages) of national patterns of extermination. Wherever he went, to the Philippines and later to Indochina, peasant guerrillas were of course the Indians, merciless savages all, with their "savage" attacks, their campaign that "continued savagely for months and years," their "ruthless savagery," their "guilt of premeditated savagery," and their terror that "was the old nightmare of savagery with murders and kidnappings" (IMW, pp. 22, 27, 85, 301, 352, 369, 375, et passim). As for the Company cowboy, Lansdale was ever the man with the gun (or guns) who pointed more than his finger; he was the generic Westerner as depicted in, say, John Ford's film *My Darling Clementine* (1946), in which Wyatt Earp anticipated him by shooting holes in the old nightmare of savagery when there was "no law west of Kansas and no God west of Fort Scott." Lansdale had just wandered later to a farther Wild West: After an attack on Magsaysay's home barrio in Zambales, "we all took off after the Huk force. We were more like a posse than a military unit in pursuit of the enemy" (IMW, p. 66). Or the year following: "My memory of those first months of 1951 is a blur of events: Magsaysay placing his paratrooper carbine so he could grab it quickly as we drove into ambush terrain on provincial roads (he was a crack shot, firing from his hip gunslinger style)" (IMW, p. 86).

Lansdale himself tied his COIN operations to the Winning of the West: "Of course, my concept of having soldiers behave as the brothers of the people wasn't original. The U.S. Army has a long history of similar civic action, from the opening of the West and protection of settlers in our country to such endeavors in foreign countries as the founding of public school and public health systems" (IMW, p. 84n). For this history that led back past Worcester's efforts "to make Asiatics clean up" and on to the civic actions of Custer's 7th Cavalry, *Fort Apache* (1949), another Ford film, had direct relevance, for it underscored who "the people" had been all along. They were not the Indians who blocked the opening of the West to "civilization" and who therefore had to be wiped out or driven away by the last reel. Just so was it with the Huks: their weariness and lack of supplies, government reforms and promises that thinned their ranks, reorganized COIN forces that kept them on the run, and the overwhelming military might the United States threw into the struggle, all combined to make their plight increasingly desperate from late 1951 on. Lansdale recorded the customary happy ending with an overall body count (1950–55) of 6,874 Huks killed, 4,702 captured, and 9,458 surrendered (IMW, pp. 50–51).

2

Among those captured in 1952 was William Pomeroy, an American Communist party member who had served as a historian with the U.S. Fifth Air Force in WWII and then established beginning contact with the Huk movement, later returned to Manila as a journalist and as a student at the University of the Philippines, there met and married Celia Mariano, a Filipina, and with her joined the rebels in the Sierra Madre in 1950. Lansdale related that a surprise raid on "Stalin University" in the Sierra Madre sent Pomeroy scurrying for safety, dropping his glasses in flight, and myopically running "smack into a tree, knocking himself out" (IMW, p. 98). After their capture Celia Mariano and William Pomeroy were sentenced to life imprisonment for "rebellion complexed with murder, robbery, arson, and kidnapping"—in *An American Made Tragedy* (1974) Pomeroy contended that the previously unknown crime in the Philippine penal code had been concocted by or within JUSMAG (AMT, p. 81). "After serving ten years of this sentence, both were pardoned, and Pomeroy was deported to the United States," wrote Lansdale, characteristically omitting mention of the refusal of the Philippine government to allow Celia Mariano to accompany her husband or to join him elsewhere.

Pomeroy was predictably regarded within the CIA as a white renegade, another despicable Simon Girty who "took part with the savages," in Governor Pennypacker's words (see p. 339)—at a COIN seminar in Washington he was identified as "a renegade American leader among the Communist Huks" (CW, p. 91). As a zealous Marxist, Pomeroy jumped at the chance to claim PKP leadership over the whole peasant revolt and to write of it as "the Communist-led movement" (AMT, p. 79), thereby adding his misrepresentation to Lansdale's. But Pomeroy also wrote *The Forest* (1963), a remarkably well-written and moving personal record of his part in the guerrilla struggle. It was immeasurably more honest than Lansdale's memoir and distinguished by an unquestionably sincere solidarity with those who were being victimized. That courageous commitment aside, Pomeroy's narrative also revealed more shared assumptions and attitudes with the secret agent than either of them were likely to have thought possible and surely more than they could have comfortably acknowledged. In certain respects, and in them only, the two covert philanthropists were mirror images of each other.

"In my youth," Pomeroy wrote in *The Forest,* "the Indian trail was always a great fascination; now we hurry on forest paths, cross glades, and wade streams, that might have been met by LaSalle, Champlain, and the *coureurs de bois*" (p. 103). From his youth on he had identified not with the Indians who made the paths but with the explorers and frontiersmen who rushed down them to open up the West: one of their guides he thought of as he did "of the old mountain men of the western lands, who guided the wagon trains over the plains" (p. 123). Or at a Huk camp in the Sierra Madre, "smoke plumes up in the rich sunlight from huts that could be log cabins at the rim of the clearing. They might be homes of settlers in the wilderness in the Boone days of Kentucky. And those men going up the trail to [the] forest could be frontiersmen with long rifles out for game" (p. 77).

Like the settlers of that earlier Dark and Bloody Ground, Pomeroy had few good words for the real forest people, those he called "Dumagats" or "the aboriginals, the primitive ones, the original inhabitants of the forest." His prejudice against them matched Professor Worcester's earlier contempt, for the "Dumagats" were "creatures lost in time," "stone age throwbacks," or in short, treacherous merciless savages:

> For a pound of salt or tobacco they will sell out any friend, any ally. They are possessed with the terrible honor of the primitive: molest a woman, insult a man, and they are savage silent ambushers with poisoned darts. . . . With these people we must live, for our own survival. When the Huks first met Dumagats and suffered from camps raided because of their misunderstood betrayals, there were enraged reactions. One FC [Field Command] commander ordered the liquidation of all Dumagats, but one cannot liquidate people who merge like their own *anitos* into rock and tree. So we have organized them, as much as they can be organized, promising land and fertilizer, schools and medical care, in the society to come. [pp. 185–86]

If you cannot liquidate them, organize them.*

These unimproved natives merged with and were inseparable from unimproved nature, the forest that was for Pomeroy not a sacred grove, not an embracing shelter, nor even an indifferent refuge. It was a vine that caught at him in a ravine, a root that tripped him up, a tree malevolently blocking his way, or a thorny creeper he cursed for its "cunning malice," while the whole was a metaphor for the colonial system:

> And then, at times, the forest is all the evil forces that have held back the advance of civilization, and I am man, fighting his way through the dark underbrush of ignorance, intolerance, and misunderstanding, toward an open world of enlightenment, of freedom and of brotherhood. [p. 203]

Rarely has the truism that Marxism is a Judeo-Christian heresy been more vividly demonstrated. Pomeroy was as alienated from the land he presumed to liberate as the Puritans had been from their howling wilderness, John Fiske from his Wyoming moonscape, or Henry Adams from his detested mountains and execrated sea.

Both the secret agent and the guerrilla propagandist had come to the islands to unmask the enemy, to wit, each other. Both considered natives too simple to do their own unmasking without what the latter called "firm centralized guidance": "Now we are tearing away the blindfolds," wrote Pomeroy, "letting them see their own land clearly" (p. 25). Peasants were being led into the future by those with "a scientific knowledge of society and with precise guidance of main forces, allies and reserves, based on the most nearly correct estimate of given situations" (p. 68).

3

Meanwhile, back at the JUSMAG compound the secret agent led the Filipinos into a different future by putting the finishing

* This pervasive contempt for the forest people had serious consequences. An extreme example was later related by Jesus Lava, a leading Huk, to account for their failure to make much progress in the Ilocos area: A rebel leader had "raped a Negrito woman. This closed off all mountain routes, which went through Negrito territory, making travel for our missions very difficult"—HUK, p. 235. Ironically, Magsaysay and Lansdale solicited Negrito help early on, traveled out to one of their hiding places in the mountains of Zambales, and persuaded them to cooperate with the joint U.S.–Philippine operations, "point out the mountain hideouts of the Huks and assist in their extermination"—Carlos P. Romulo and Marvin M. Gray, *The Magsaysay Story* (New York: John Day, 1956), p. 135.

touches to the Lansdale–Magsaysay saga. Word passed up through Company channels that President Quirino's Liberal party government was no better than the Huks charged; it was said to be "corrupt beyond belief or salvation" and likely to topple the Philippine Republic if reelected in 1953 (CW, pp. 106–7). Two years before that election took place, Desmond FitzGerald, deputy chief of the CIA's Far East Division, sent out Gabriel Kaplan, a New York lawyer and Republican politician, to help Lansdale elect Magsaysay president. Kaplan helped with the election of 1951 and put together a network of civic organizations—the chambers of commerce, the Jaycees, Rotary, Lions, League of Women Voters, and veterans groups—in the National Movement for Free Elections (Namfrel) in support of the defense chief's candidacy. In his account of Namfrel, Lansdale said only that it had been formed by "a group of public-spirited citizens" (IMW, pp. 90, 95n). In fact, these public-spirited folk had been rounded up by Kaplan and funded by the CIA station (CW, p. 108). For the 1953 crusade and what the secret agent called "a magnificent chapter" of Philippine history, he was backed by his team; by Namfrel; by Embassy Councillor William Lacey and Admiral Raymond Spruance, the U.S. ambassador; by Admiral Arthur Radford, chairman of the Joint Chiefs of Staff; by Allen Dulles and his CIA boss's brother, Secretary of State John Foster Dulles; and by the Eisenhower administration's "bottomless blank-check tactics" to keep the election on track (ST, p. 107; cf. CIA, p. 27). How many millions of secret U.S. dollars Lansdale could put his hands on was perhaps roughly indicated by his wide-eyed professions of poverty:

> I was surprised to hear my name included in the rumors as having given Magsaysay three million dollars for his campaign. The rumormongers certainly didn't know the tiny budget of government and personal funds on which I was operating. I was close to being flat broke. In the midst of all this political hubbub, I concentrated on writing a plan for safeguarding the integrity of the 1953 election. [IMW, p. 104]

So the nearly flat-broke Ragtime Kid on the subject of "integrity" —he was dissembling, of course, but that was just a bit of disinformation within the larger untruth, that the Filipinos were writing their own "magnificent chapter."

Add to the awesome clout of Lansdale's slush funds the assets of his candidate. Ramón Magsaysay was the son of a carpenter and had been born in the tropical counterpart of a log cabin. He had himself worked with his hands as a mechanic and was the first serious aspirant for high Philippine office not to come from the gentry. Moreover, he was colorful, a man of the people, and his CIA mentors noted delightedly, possessed of "charisma." Lansdale packaged his candidacy as the island equivalent of Indian-fighting Andrew Jackson and of the proverbial struggle from log cabin to White House, save that Huk-fighting Magsaysay had to struggle from barrio to Malacanang Palace.

One problem was troublesome. Magsaysay belonged to the Liberal party, of which Quirino was certain to be the nominee. The Nationalist party almost certainly would nominate Senator José Laurel or Senator Claro M. Recto, both of whom had been leaders of the puppet government under the Japanese. But as Joseph B. Smith remarked, "conning one another is the second favorite activity of con men" (CW, p. 109). The deal was for Magsaysay to change parties like shirts, for Laurel to stand aside, and for Lansdale's man to become the Nationalist nominee. By joining Lansdale's American team, Laurel and Recto put to rest their collaborationist pasts and could hope to control the man they regarded as a stupid soldier: "I thought it amusing to arrange a deal with the American military who spent most of their time unjustly defaming me," Recto told Smith years later. "As for Ramón, he was so dumb I knew I could handle him."*

Inauguration day at the presidential palace, said the secret agent, reminded him

> strongly of the cartoons and stories of Andrew Jackson in the White House. The public went wild. Vast throngs of citizens at the Philippine inauguration actually tore the clothes from Magsaysay's back as he made his way to the stand. After he was sworn in, he threw open the doors of Malacanang and invited everyone to come for a visit. They came by the thousands, the well-clad and the barefoot, jamming into rooms of state. . . . This exuberant scene of democracy triumphant was a fitting way to mark the close of my mission and to return home. [IMW, p. 124]

Or as Magsaysay said more simply, "the people will have their own way." Actually, the Manila CIA station oversaw all his im-

* Joseph B. Smith went to the Philippines in 1957, just after the death of Magsaysay, to become the CIA deputy branch chief and to attempt to become a second General Landslide, as the original had been dubbed after the great triumph of 1953. Recto had broken with Magsaysay in 1955 as a U.S. puppet, whereupon the CIA had denounced Recto as a Communist agent who had infiltrated into the Philippine Senate, and subjected him to assorted dirty tricks. When Smith went through the files, "I found something that absolutely astounded me. I saw a sealed envelope marked 'Recto Campaign.' I opened it and found it filled with condoms, marked 'Courtesy of Claro M. Recto—the People's Friend. The condoms all had holes in them at the place they could least afford to have them" (CW, p. 280).

portant foreign-policy decisions, wrote his speeches, and made him a favorite of the island and the mainland press (CW, pp. 252–54). Lansdale's man, Colonel Napoleon Valeriano, became his senior military aide. Never really "the showcase of democracy in Asia," as the *New York Times* benignly editorialized (September 17, 1953), the Philippine Republic now had become the showcase of Company colonialism.

"It was soon no secret," *Time* breezily acknowledged, "that Ramon Magsaysay was America's boy" (November 23, 1953). Once a native, always a boy? Even a Magsaysay? One of his mentor's postelection anecdotes inadvertently answered that question in the affirmative:

> In keeping with U.S. policy, I had avoided associating with him in the campaign. (Oddly enough, my absence in his campaign didn't prevent gossip that I was present at Magsaysay's side during much of it. People swore that they saw me on the platform with him, including the time when I was in Indochina. Later, Magsaysay confessed to me that he had generated some of this gossip himself, believing that the idea of his closeness to Americans would help in his election. He aided the image by having his aide Manuel Nieto grow a mustache like mine. Nieto, of Spanish ancestry, could pass for an American.) [IMW, pp. 121–22]

Clearly, a dark-skinned native, even a Magsaysay, could never pass for an American, and that was to say a fair-skinned adult. This home truth was one of the choice secrets overlooked by Lansdale's WASP censors, who in turn had naturally assumed "an American" possessed their own bright good looks.

In January 1954 Lansdale returned to Washington the covert savior of the Philippines from "Communism." With the election of Magsaysay he had also given the Filipinos a hero to place beside José Rizal, the model nonmilitant that William Howard Taft had bestowed upon them—or so Lansdale and Taft's other successors thought (see p. 366, cf. CW, p. 283). And with the Magsaysay formula down pat, why should he not take his team of Americans and Filipinos on the road to save other countries and create other national heroes on demand? Turn the page of his memoir and come to his next chapter, "Assignment: Vietnam"!

CHAPTER XXVII

Closing the Circle of Empire: Indochina

> Besides, it is well known that empire has been travelling from east to west. Probably her last and broadest seat will be America. Here the sciences and the arts of civilized life are to receive their highest improvement. . . . Elevated with these prospects, which are not merely the visions of fancy, we cannot but anticipate the period, as not far distant, when the AMERICAN EMPIRE will comprehend millions of souls, west of the Mississippi.
>
> —JEDIDIAH MORSE,
> *The American Geography*, 1789

HIGH ABOVE the South Seas the secret proconsul mused over his mission. Secretary of State John Foster Dulles had told him he was "to help the Vietnamese help themselves." But could he help, with the French certain to insist upon their colonial prerogatives and embittered by the war they had all but lost? Just the preceding April the Viet Minh surrounding Dien Bien Phu had pressed their attacks "savagely home," and on May 7 the fortress had fallen: "The French defeat at Dien Bien Phu had shown that Asians could beat Europeans decisively in battle." Now what would happen to the millions of Vietnamese who had a vital stake in the ongoing Geneva Conference? "Was I going to be able to help them?" wondered Edward Geary Lansdale, according to his recollections, as his plane touched down at Tan Son Nhut, Saigon's airport (IMW, pp. 126–31).*

At precisely that moment on June 1, 1954, the United States of America finally thumped down on the far side of the Pacific and, as General Arthur MacArthur would have said, closed the circle of empire by returning "to the cradle of its race." Just as it had

* For an explanation of the abbreviated references, see p. 534.

arrived in the Philippines from which Lansdale's plane took off, the United States arrived in Indochina to intervene directly in a revolution and to supplant another colonial power. The circle was closed physically by Lansdale and his team, for they were the first American fighting men in Indochina—as distinguished from such military observers as Lederer and Burdick's Tex Wolchek—of the hundreds of thousands to come. Our chapter epigraph from Jedidiah Morse's *Geography* serves both to commemorate this historic moment and to remind us that all these dedicated Americans had in a very real sense been on their way for centuries—Vietnam lay west of the Mississippi, and millions of souls, led off by Lansdale, did their best to make it part of the American empire. As for the more immediate background of their arrival, the secret agent's memoir was masterfully misleading.

Thanks to one of history's good jokes, it was a former Lansdale protégé and teammate named Daniel Ellsberg who provided us with the documents necessary for establishing that immediate background and for pinpointing the imperial agent's falsities about how he got to Saigon and why. The *Pentagon Papers* Ellsberg made available demonstrated that the cold war attitudes and misconceptions usually attributed to later administrations actually had their surface roots in the years when Harry S. Truman was in the White House (GPP, I, 15–55). On May 23, 1961, for instance, Vice-President Lyndon B. Johnson indeed formulated a variant of the "domino theory" when he affirmed to President Kennedy that his recent trip to Southeast Asia had made plain "we must decide whether to help these countries to the best of our ability or throw in the towel in the area and pull back our defenses to San Francisco and [a] 'Fortress America' concept" (DOD, XI, 164). But Johnson's absurd notion that the "loss" of Indochina would lead inexorably to such a drastic rollback of empire had its origins in the 1949 "loss" of China and thus predated the outbreak of the Korean War in June 1950 (GPP, I, 82). It had been adumbrated back in 1945 when the United States had encouraged France to give Vietnam greater self-government, lest the continent be swept by a wildfire of "ideologies contrary to our own or . . . a Pan-Asiatic movement against all Western Powers" (DOD, VIII, 22–25).*

* Melancholy footnotes to this fear of contrary ideologies and native independence movements were the eloquent letters Ho Chi Minh sent Truman in 1945–46 pleading for recognition and for support "of Annamese independence according to [the] Philippines example" (DOD, VIII, 61). No doubt the revolutionary had in mind the postcolonial Philippines and of course had no way of knowing what "independence" had in store for that new republic. But his September 2, 1945, Declaration of Independence of the Democratic Republic of Vietnam showed that well into the 1940s he still admired the United States and drew from its Declaration the paradigmatic principles of what he presumed was the leading anticolonial power in

The extravagant foolishness of the "domino theory" has long diverted attention from its serious origins. It was an updated, internationalized version of the old fear of pan-Indian movements that went back beyond the Pequots and the Narragansetts. Now as then it was the revealing outcrop of deep-seated dread: let one lot of natives slip from under Western control and then other lots would contagiously rise up one after another to throw off that "association of the philanthropic, the pious and the profitable" called colonialism and would therewith send the pieces tumbling down all the way back to the Pacific slope or even—in its earliest incarnations—all the way back to the Atlantic strand. This underlying and historic fear motivated Truman's administration and those of his successors, I am suggesting, along with their more commonplace political and economic motives in both the domestic and the foreign spheres. They feared exactly what Ho Chi Minh avowed in 1946: "the white man is finished in Asia" (GPP, I, 49). In 1954 the day of the white man's departure seemingly came a quantum leap closer at Dien Bien Phu, the world-turned-upside-down victory that Lederer and Burdick discussed with undisguised loathing in their fiction, and Lansdale with unconcealed apprehension. That "unprecedented victory of Asian over European," in the words of the Pentagon analysts, had been lacerating for the French and hardly less so for their U.S. allies, who had looked on in scared disbelief as barefooted and sneaker-shod natives overwhelmed a modern mechanized army they had helped keep in the field with more than $1.5 billion of economic and military assistance (GPP, I, 180, 211).

All too clearly the French had let their natives get out of hand. On January 30, 1954, and thus prior to the debacle at Dien Bien Phu (March 13–May 7), Eisenhower's Special Committee on Indochina heard Admiral Radford of the Joint Chiefs warn "that the U.S. could not afford to let the Viet Minh take the Tonkin Delta. If this were lost, Indochina would be lost and the rest of Southeast Asia would fall." Director Allen Dulles of the Central Intelligence Agency (and *not* his brother) "inquired if an unconventional war-

---

the West—excerpts from the DRV declaration appear in George McTurnan Kahin and John W. Lewis, *The United States in Vietnam* (New York: Dell, 1967), pp. 345–47. Like Aguinaldo and all the Native American rebels who preceded that earlier revolutionary, Ho Chi Minh had to learn the hard way that official American protestations of anticolonialism were the assurances of confidence men (some of whom had admittedly conned themselves). As for this great lost opportunity to support one of the anticolonial movements sweeping the world after WWII, the Pentagon analysts put the question squarely: "If the U.S. choice in Vietnam really came down to either French colonialism or Ho Chi Minh, should Ho be [or have been] automatically excluded? [GPP, I, 51]." Unfortunately, their analysis of why this exclusion was indeed virtually automatic barely scratched the surface of the three and a half centuries of history that had rendered all but a few Anglo-Americans unfit for support of native liberation.

fare officer, specifically Colonel Lansdale," could not be added to the group of liaison officers already assigned to Vietnam, and Radford thought that might be done (GPP, I, 443–47)—it was this meeting that Lansdale attended as a CIA representative (see p. 382). Well before the French defeat of May 7, therefore, Dulles moved to intervene directly in the conflict by sending his expert native-handler to Indochina.

In the intervening four months until Lansdale quietly entered the fray, events and decisions combined to make his brand of unconventional warfare almost all the administration had to fall back on. Proposals to use tactical nuclear weapons were considered and rejected. Conditions for conventional warfare seemed scarcely more promising, as the debriefing of Major General Thomas J. H. Trapnell, Jr., the returning chief of the Military Assistance Advisory Group (MAAG) in Indochina, made clear on May 3:

> The battle of Indochina is an armed revolution which is now in its eighth year. It is a savage conflict fought in a fantastic country in which the battle may be waged one day in waist-deep muddy rice paddies or later in an impenetrable mountainous jungle. The sun saps the vitality of friend and foe alike, but particularly the European soldier. [DOD, IX, 406]

Secretary of the Army Robert T. Stevens, remembered best perhaps as Senator Joseph McCarthy's mild-mannered adversary, also worried about what would happen to *white* GIs in that terrain: "It has a tropical, monsoon climate with pronounced wet and dry seasons and the disease and morale hazards are high for Caucasian troops." As if that were not enough, "the population, when not hostile, is untrustworthy" (DOD, IX, 475). In any event, Eisenhower had already decided on May 7, the day Dien Bien Phu fell, not to send in combat forces to fight alongside the French as partners in "a 'white man's party' to determine the problems of Southeast Asian nations" (GPP, I, 501–2). What the president meant by a "white man's party" was spelled out on May 15 in one of Secretary Dulles's cables to the U.S. Embassy in Paris: France could not equivocate "on completeness of independence" for Vietnam, whatever that meant, "if we are to get [the] Philippines and Thailand to associate themselves. Without them, [the] whole arrangement would collapse because we are not (repeat not) prepared to intervene purely as part of a white Western coalition which is shunned by all Asian states" (DOD, IX, 466–67). But what Eisenhower, Dulles, and the rest of the Special Committee were not prepared to do overtly they had already set out to do covertly. They had made arrangements for a secret "white man's party," no French invited.

"Was I going to be able to help them?" wondered the secret Samaritan; but we need not pause to marvel at such spontaneous philanthropy. A serviceable summary of his actual assignment, one he may have had a hand in drafting, appeared as a statement of the problem by the President's Special Committee:

> To set forth a program of action without resort to overt combat operations by U.S. forces, designed to: (a) secure the military defeat of Communist forces in Indo-China, and (b) establish a western oriented complex in Southeast Asia incorporating Indo-China, Thailand, Burma, Malaya, Indonesia, and the Philippines. [DOD, IX, 333]

Lansdale was to work his Magsaysay magic on the Vietnamese, in fine, so they would further U.S. interests while thinking they were furthering their own. Moreover, he was to exploit his Magsaysay triumph by bringing Filipinos and other Asians into his operation as native alter egos who would camouflage the ultimate objective of a *Western*–oriented complex, and that was to say, as John Foster Dulles linked the terms, *white Western*–oriented and *white Western*–dominated. U.S. racism was built into Lansdale's mission and landed with him and the empire at Tan Son Nhut on June 1, 1954.

2

Privileged behind-the-scenes glimpses of "a 'cold war' combat team" and of the secret agent's actual nurturance of the Vietnamese came to light with the publication of the 21,000-word "Lansdale Team's Report on Covert Saigon Mission in 1954 and 1955" (GPP, I, 573–83). With this and other previously secret documents as anchors, we can face up to Lansdale's memoir and observe how the imperial agent plied his craft.

Lansdale's assignment in Vietnam was to do precisely what the United States pledged solemnly not to do: in its unilateral declaration on the Geneva Accords (signed July 21, 1954) it undertook "to refrain from the threat or the use of force to disturb them" (DOD, IX, 671). Under nominal cover as the U.S. Embassy's assistant air attaché, he hurriedly drew together the nucleus of his team of disturbers—called the Saigon Military Mission (SMM)—before the cease-fire on August 11 put a ceiling on the number of military personnel the United States could legitimately keep in the country. First to come aboard was Major Lucien Conein, Missourian of French–American parentage, veteran of both the French Foreign Legion and the American OSS, and an old Indochina hand who had "gone too native," or so U.S. officials in Saigon came to believe. An able "paramilitary specialist," in Lansdale's more supportive judgment, Conein was "assigned to MAAG for cover purposes," as were other key CIA agents, including William Rosson, Arthur Arundel, Joseph Redick, "an expert in countersubversion and small arms," and Rufus C. Phillips III, whom the Company had recruited in New Haven: "Phillips, who had played football at Yale, was a towheaded six-footer, a giant among the smaller Vietnamese. He stood out like a sore thumb" (IMW, p. 236). Charles T. R. Bohannan, "a former team-mate in Philippine days, was in town"—in his memoir Lansdale recalled Bo turning up unexpectedly, saying "he figured he might be missing some fun and games" (IMW, p. 287). And of course present was the indispensable Colonel Napoleon Valeriano, "who had a fine combat record against the Communist Huks" and who had come over to help organize "the guard battalion for the Presidential Palace."*

In the philanthropy of Lansdale's memoir: "I had split my small team in two. One half, under Major Conein, engaged in refugee work in the North" (IMW, p. 168). In the relative reality of his secret report: "Major Conein was given responsibility for developing paramilitary organizations in the north, to be in position when the Vietminh took over." Conein was "to beat the Geneva deadline" by having these underground groups, trained and supported by the United States, in place "ready for actions required against the enemy" and otherwise through sabotage and terror to disrupt international agreements reached only after eight years of bloody warfare. In late September 1954 they "learned that the largest printing establishment in the north intended to remain in Hanoi and do business with the Vietminh. An attempt was made

* So smoothly had the Huks become simply "Communists," after having first been "Communist-inspired" and then "Communist-led." Apparently Valeriano joined up at considerable personal sacrifice: "When Valeriano was in Vietnam helping Lansdale in the early days of the Diem regime, he carried off to Saigon the wife of a wealthy Filipino businessman. The injured husband immediately put out a contract on Valeriano, and he was never able to set foot in Manila again. It would have meant instant death. Subsequently, Valeriano worked in the Pentagon, trained the Cuban brigade preparing for the Bay of Pigs invasion, and was involved in post–Bay of Pigs activities with which Lansdale was concerned" [CW, p. 105].

by the SMM to destroy the modern presses, but Vietminh security agents already had moved into the plant and frustrated the attempt." The northern team was there when Hanoi was "evacuated" on October 9 and was disturbed by

> the contrast between the silent march of the victorious Vietminh troops in their tennis shoes and the clanking armor of the well-equipped French. . . . The northern team had spent the last days of Hanoi in contaminating the oil supply of the bus company for a gradual wreckage of engines in the buses, in taking the first actions for delayed sabotage of the railroad (which required teamwork with a CIA technical team in Japan who performed their part brilliantly), and in writing detailed notes of potential targets for future paramilitary operations. (U.S. adherence to the Geneva Agreement prevented SMM from carrying out the active sabotage it desired to do against the power plant, water facilities, harbor, and bridge.) [GPP, I, 579]

No doubt such restraint was admirable, especially in "the last days of Hanoi," but why Lansdale believed sabotage of buses and railroads was permitted by the Geneva Accords, while sabotage of power plants and bridges was not, has to remain one of the mysteries of the clandestine mentality. The paramilitary actions of his team flagrantly violated those agreements and the U.S. pledge not to disrupt them, as well as the fundamental international-law rights of the Democratic Republic of Vietnam (DRV) as a sovereign state.

Though merely one of his "cover duties," Conein did in fact engage in refugee work in the north by generating dislocation and deracination. One shrewd stroke was "a black psywar strike in Hanoi: leaflets signed by the Vietminh instructing Tonkinese on how to behave for the Vietminh takeover of the Hanoi region in early October, including items about property, money reform, and a three-day holiday of workers upon takeover. The day following distribution of these leaflets, refugee registration tripled." CIA lies sped the refugees, mostly Catholics, south to the Free World, and by the spring of 1955 their flow clogged the movements of the northern team: "Haiphong was reminiscent of our own pioneer days as it was swamped with people whom it couldn't shelter. Living space and food were at a premium, nervous tension grew. It was a wild time for our northern team." It was wild Dodge City all over again, save this time the frontier was rolling backward, with the hostiles taking over Haiphong on May 16.

3

Meanwhile, back in Saigon "our Chief's reputation from the Philippines had preceded him. Hundreds of Vietnamese acquaintanceships were made quickly." Lansdale put his experience with Asians to use by making big Rufus Phillips director of a subtle psywar strike in the form of

> an almanac for popular sale, particularly in the northern cities and towns we could still reach. Noted Vietnamese astrologers were hired to write predictions about coming disasters to certain Vietminh leaders and undertakings, and to predict unity in the south. The work was . . . based on our concept of the use of astrology for psywar in Southeast Asia. [GPP, I, 582]

The Ragtime Kid held palmistry and astrology to be "the key" to the Far East, and here he was in the flesh, life foreshadowing art.

COIN operations bred such instant experts as the secret proconsul and brought together, in Washington and in the field, men "who had little experience . . . and even less idea of the political situation in the countries involved," wrote L. Fletcher Prouty. "It was a shattering experience to attend some of these meetings and to hear men, some high in the councils of government, not even able to locate some of these countries [on the COIN "watch list"] and to pronounce their names" (ST, p. 136). Since Lansdale had been to Indochina on a brief visit in 1953, he could locate the country; otherwise he was a major case in point. In between his Philippines and Vietnam assignments, he had hoped to "read into the scene," as they say in the CIA, but had not had time (IMW, pp. 127–28). He had thus arrived in Saigon as grossly ignorant of the Vietnamese and their culture as Taft had been of the Filipinos or McKenney of the Ojibwas. Grounded firmly in the national tradition of disdain for natives, Lansdale never extended his fabled old-hand expertise beyond their putative superstitions and psychological vulnerabilities (cf. IMW, pp. 226–27, 243n).

After all, the secret agent had come over to help because they were natives: "there were some practical things that needed doing, and there were no Vietnamese to do them" (IMW, p. 234). To bring them up to snuff, he bombarded them with "a Thomas Paine type series of essays on Vietnamese patriotism" through the good offices of a "publisher known to SMM as The Dragon Lady," an apt comic-strip touch. He set up "a small English class conducted for mistresses of important personages, at their request" and of course so the SMM could get a hold over the generals, politicians, and entrepreneurs who were their lovers.

Apart from those "brief tastes of family life," Lansdale himself seems to have practiced self-imposed chastity, partly to remain acceptable to his celibate premier but mainly to armor himself against enemy and adversary agents. Within the "intelligence community," as he well knew, the body was a prime target of dirty tricks, and its appetites were weaknesses to be mercilessly exploited. Besides, Lansdale was a product of a tradition leading back to John Endicott, the tradition in which energies and impulses were channeled into quite different "fun and games."*

The *overt* Lansdale encouraged the formation of a new organization called the Freedom Company of the Philippines: "I felt that the presence in Vietnam of such people, so visibly dedicated to the principle of man's liberty, would have a heartening effect upon Vietnamese nationalists" (IMW, pp. 213–14). The *covert* Lansdale mentioned the organization in his diary-form 1955 report but was still more explicit in a 1961 memorandum for General Maxwell D. Taylor, "Resources for Unconventional Warfare, S.E. Asia" (GPP, II, 643–49). By then renamed the Eastern Construction Company, the Freedom Company had started in 1954 as

> a non-profit organization, with President Magsaysay as its honorary president. Its charter stated plainly that it was "to serve the cause of freedom." It actually was a mechanism to permit the deployment of Filipino personnel in other Asian countries, for unconventional operations, under cover of a public service organization having a contract with the host government. Philippines Armed Forces and government personnel were "sheep-dipped" and served abroad. Its personnel helped write the Constitution of the Republic of Vietnam, trained Vietnam's Presidential Guard Battalion, and

* In a sense secret agents had updated and bureaucratized all John Endicott's layers of repression. From being a vessel of sins and lusts in the 1630s, the body had become a vessel of welcome weaknesses in the 1950s, though sex remained distinctly dirty. The perforated condoms illustrated the dirty-story sensibility of the spooks, as did their attempt to produce a phony pornographic film of Indonesia's Sukarno and a blond Soviet spy and their other ventures into blue film making. Or take this funny story from Joseph B. Smith's tour of duty in Buenos Aires: Argentine agents ran into difficulties getting "a confession from a stubborn suspect even with the help of their favorite interrogation aid—an electric rod they attached to his testicles to give his memory a jolt. He finally told them, 'Look, I've been an electrician for forty years, I've been shocked at least ten thousand times—a few more just don't bother me' " (CW, pp. 239–40, 368). To catch a glimpse of the systematic attacks on the sex of countless victims—live wires wrapped around penises or inserted into vaginas and attached to nipples with clothespins—in U.S.-sponsored detention cells in Latin America, see A. J. Langguth, *Hidden Terrors* (New York: Pantheon Books, 1978), pp. 164–65, 248–52.

> were instrumental in founding and organizing the Vietnamese Veterans Legion. [GPP, II, 647–48]

Johnny ("Frisco") San Juan, past national commander of the Philippines Veterans Legion and one of Magsaysay's key aides, became president of the mechanism, and most of its cadre, like Valeriano, had been on Lansdale's team in the islands. Bohannan developed "a small Freedom Company training camp in a hidden valley on the Clark AFB [Air Force Base] reservation" in Luzon, where recruits learned the arts of psywar, smuggling, sabotage, commando raids, kidnapping. After their sheep-dipping, these liberty-loving Fridays served abroad, to hide the truth that the secret proconsul was throwing a white man's party.

For the *overt* Lansdale rays of Philippine sunshine broke through the Vietnamese clouds "in the form of a remarkable group of Filipino volunteers who made up the first contingents of an endeavor called Operation Brotherhood. This happy and hard-working crew had only one aim. They wanted to ease the suffering of their fellow Asians" (IMW, pp. 168–69). For the *covert* Lansdale "OB," as he called it, was masterful guile. His pride bubbled over as he reported that former teammate Oscar Arellano, now vice-president for Asia of the International Junior Chambers of Commerce, "stopped by for a visit with our Chief; an idea in this visit later grew into 'Operation Brotherhood,' " another mechanism of

> volunteer medical teams of Free Asians to aid the Free Vietnamese who have few doctors of their own. Washington responded warmly to the idea. President Diem was visited; he issued an appeal to the Free World for help. The Junior International adopted the idea. SMM would monitor the operation quietly in the background. [GPP, I, 575, 576]

In 1961 Lansdale was still more specific about the quiet monitoring of these Free Asians: they were under "a measure of CIA control." OB had been from the beginning part

> of the 1955 *pacification* and refugee program. Initially Filipino teams, later other Asian and European teams, served in OB in Vietnam. Their work was closely coordinated with the Vietnamese Army operations which *cleaned up* Vietminh stay-behinds and started stabilizing rural areas. [GPP, II, 648; my italics]

Operation Brotherhood was run by the Freedom Company, fittingly, and fully funded by the CIA (CW, pp. 179, 251–52). Even in their refugee work the Free Asians found time to scout out potential agents on the other side. But their nonwhite skins were considered still more critically important in "winning over the population" through the pacification program devised by SMM and adopted by Diem. Thus the fabulous Free Asian Brothers eased the suffering of their fellows by their integral participation in Diem's 1955–59 campaign of terror to clean up the countryside of stay-behinds, members of dissident religious sects and political factions, and of just plain suspects (FF, pp. 140–41).

Freedom was indivisible. The *overt* Lansdale had his team carry it up into the Central Highlands to share with the peoples there —the two dozen or so mountain tribes the French called Montagnards: "Diem agreed to reforms in administration and official attitudes in tribal areas. . . . Officers on my staff started sporting bracelets that denoted their adoption into tribes" (IMW, p. 327). Diem's reforms were to deny the Montagnards the autonomy they had enjoyed under the French, to settle some 200,000 refugees on their traditional lands, and later to embark upon their removal to reservations (GPP, I, 312). Lansdale's agents were in fact the first organizers of the Secret Army, as Fred Branfman has called it, that came to include some 45,000 Montagnards, Nung, and other tribal groups in South Vietnam alone. Behind their shield of Filipinos, CIA operatives extended these irregular forces to Thailand, even to Burma among the Shan tribes, and of course to Laos among the Meo and the numerous Lao Theung tribal groupings (cf. CW, p. 252). By 1961 the *covert* Lansdale could report to Maxwell Taylor that

> about 9,000 Meo tribesmen have been equipped for guerrilla operations. . . . Estimates on how many more of these splendid fighting men could be recruited vary, but a realistic figure would be around 4,000 more, although the total manpower pool is larger. . . . Command control of Meo operations is exercised by the Chief CIA Vientiane with the advice of Chief MAAG Laos. [GPP, II, 646]

With this excerpt before us, we can understand why the anonymous "Air Force officer" of "The Report the President Wanted Published" (see p. 382) went out of his way to have his Christian Chinese refugees beg for assurance the United States "would stand firm in its policy in Asia, and particularly in Laos": the secret agent had a vested interest in the mercenary army in that tiny and formerly peaceful country. Natives there and elsewhere in Southeast Asia constituted a manpower pool that CIA operatives could dip up by false promises, threats, force, bribery. Before the decade was over many more splendid fighting men bit the dust in battles that were not theirs and that nonetheless decimated their peoples.

The tribal recruiting operation Lansdale started came straight out of traditional Western attitudes toward natives as objects. His Freedom Company and huge Operation Brotherhood betrayed freedom and denied brotherhood on a scale that would have left Melville's confidence man gasping with admiration.

And the joke was that Asians saw around the Filipinos to the reality of Lansdale's surprise party. "As a Chinese told me in Singapore one time," recalled one of his fellow agents, "the Filipinos 'have brown faces but they wear the same Hawaiian sport shirts the Americans do' " (CW, p. 252).

CHAPTER XXVIII

The Quiet American

> Now and then, Thuc or Nhu would be with Diem when I arrived for a meeting with him, but they would excuse themselves after we exchanged greetings. I saw Nhu's children more often, since they had the run of the palace, and I often stopped for a brief game of cowboys and Indians with them. Once, when I had tumbled dramatically to the floor under their pointed finger "guns," there were gasps behind me. I looked up to see a group of diplomats arriving for a protocol call. They were staring at me deadpan, not a smile among them. The children and I thought the moment hilarious.
>
> —EDWARD GEARY LANSDALE,
> *In the Midst of Wars*, 1972

FREE FILIPINOS. Free Vietnamese. Free Asians. Free World. In the looking-glass world of the secret proconsul, fact and fancy intertwined, pirouetted, and interpenetrated to create endlessly mirrored confusions. In his COIN-resources memorandum for General Maxwell Taylor, for instance, Lansdale noted that "there is also a local veteran's organization and a grass-roots political organization in Laos, both of which are subject to CIA direction and control and are capable of carrying out propaganda, sabotage and harassment operations" (GPP, II, 647).* Had he forgotten that he was not speaking to natives but writing *secrets* for the witting? Or did he really believe there could be *grass-roots* organizations controlled by the CIA? In naive Age of Reason fashion I have distinguished between the *overt* and the *covert* Lansdales by rule of thumb, laid bare what I view as the former's double-dealing, and still have penetrated no more than his false front. What if there were nothing beyond, no real behind, as in the set of a false-front Western? Or better, what if we laboriously lift the many

* For an explanation of the abbreviated references, see p. 534.

masks of the agent's masquerade only to find beneath his last philanthropic phrase an actor who had become his part in the national scenario? What was Kim? What was Lansdale?

The outrider of empire was fond of repeating that he took his American principles with him into Asian struggles:

> Americans too have an ideology. It sees man as a free individual, graced with spiritual values from which come human rights. It is in keeping with what Paul of Tarsus wrote to the Corinthians nearly two thousand years ago: "Where the spirit of the Lord is, there is liberty." When we Americans give of our substance to the people of other countries, we should give as generously of our ideology as we do of our money, our guns, our cereal grains, and our machinery. In sharing our ideology, while making others strong enough to embrace and hold it for their own, the American people strive toward a millennium when the world will be free and wars will be past.

In the midst of wars the secret agent foresaw eternal peace (IMW, pp. 105–6). On the other side of the earth from where the Saints set out, Lansdale was still building New Israel. The undercover representative of a redeemed people, he built veterans' groups and the like in countries that would thereby become strong enough one day to share and hold *American* freedom for their own. So, yes, his Company could foster organizations at the rice roots, since it did not represent the selfish nationalism of the *colons* he despised in French intelligence. Rather, his CIA represented the anticolonialism and selfless internationalism invoked by the very term *Americanization*.

"For most Americans," Claude Julien wrote in *America's Empire* (1971; p. 7), "when the United States intervenes in world affairs it is not to defend their interests or their national ambitions but to serve selflessly an international order." Their long onslaught upon the nature both without and within had created a defactualized universe in which disinformation became the official medium of exchange. Lansdale was no more than the repository and instrument of their hereditary belief that the United States was fundamentally unlike and better than the European imperial powers. He implemented *national* policies that by collective sleight of hand became the abstractions of *international* mission—getting the Vietnamese to further U.S. interests while thinking they were furthering their own was from beginning to end a national swindle. And in my view of the evidence, this Lawrence of the Free World lived out the empty abstractions of his role with impenetrable innocence: Turned on in his most secret memoranda, he still sounded like a recording of the Pledge of Allegiance. As an *American* legend in search of substance, he not surprisingly received the tribute of having characters in three novels based upon him. The best of the three cut through to the heart of the moral problem he posed.

In Graham Greene's *The Quiet American* (1955) the legend ascended to the level of literature in Alden Pyle, the title character who was also "impregnably armoured by his good intentions and his ignorance," also in a clandestine service, also in search of a dependable Friday to help him save the country from Communism. There when Lansdale led the Americans in to supplant the French, Greene had seemingly studied the secret agent intently, for Pyle's principal characteristics matched Lansdale's down to and including Boy Scout mask: Pyle had "an unmistakably young and unused face," a crew-cut, gangly legs, great earnestness, and all the other familiar features and trappings of American "innocence." Narrator Thomas Fowler was his counterpoint, a Greene-like British correspondent, cynical and sufficiently knowledgeable about Asia to know how little he knew, all in all a fit representative of European "experience." When they met, Fowler tried to get Pyle to look at the lovely feminine forms coming up the street toward them in "white silk trousers, the long tight jackets in pink and mauve patterns split up the thigh," but the American preferred his abstractions to the sensual realities of these delicate women:

> He was absorbed already in the dilemmas of Democracy and the responsibilities of the West; he was determined—I learnt that very soon—to do good, not to any individual person but to a country, a continent, a world. Well, he was in his element now with the whole universe to improve. [pp. 16–17]

Pyle already knew all about Asia, how to "save" its peoples, even in spite of themselves, by mobilizing a "Third Force" of indigenous nationalism against the old colonialism and the new Communism; he knew all this because he was a student of the "very profound" York Harding, a roving American columnist who wrote books called *The Advance of Red China, The Challenge to Democracy*, and *The Role of the West*.

The Role of the West: Pyle believed England and France so tainted with colonialism they could not hope to win the confidence of Asians, while America had come in "with clean hands." Fowler wondered about Hawaii, Puerto Rico, and New Mexico (p. 123). Pyle knew Asians "don't want Communism." Fowler thought they wanted enough rice and "don't want our white skins around telling them what they want" (p. 93). Pyle knew the Vietnamese would never hate the Americans as they hated the French. Fowler wondered:

> "Are you sure? Sometimes we have a kind of love for our enemies and sometimes we feel hate for our friends."
> "You talk like a European, Thomas. These people aren't complicated."
> "Is that what you've learned in a few months? You'll be calling them childlike next."
> "Well . . . in a way."
> "Find me an uncomplicated child, Pyle." [p. 174]

With a few well-chosen words Greene ripped the false front from the Pyle philanthropy of winning the hearts and minds of childlike, superstitious natives: winning hearts and minds was the Great American Con.

Fowler admitted that Pyle's motives were good, as were his country's; but the moral problem was that this good man was a killer, responsible for at least fifty deaths before he was stopped —in the novel, not in the flesh. Pyle selected as his Third Force leader a terrorist general named Thé, supplied him with smuggled explosives, and helped him set up bombings to be blamed on the Communists. Winning the East for democracy, Pyle retained his essential innocence even in the presence of the civilian victims, seeing them only as counters in the ideological war: "When he saw a dead body he couldn't even see the wounds" (p. 31). After one of Pyle's bombs went off in a Saigon square during the shopping hour, Fowler lost control and thrust him into a pool of blood or "into the pain," as the correspondent thought of it: "I remember how he turned away and looked at his stained shoe in perplexity and said, 'I must get a shine before I see the Minister' " (p. 61). Fowler had to tell him that he had the Third Force and national democracy all over his right shoe (p. 162). Such innocence, Fowler concluded, "always calls mutely for protection when we would be so much wiser to guard ourselves against it: innocence is like a dumb leper who has lost his bell, wandering the world, meaning no harm" (p. 36). The simile blew the agent's cover, to expose the national mission he secretly represented.

American reviewers of *The Quiet American* could hardly believe "the malice toward the United States that controls the novel"—*Christian Century*, August 1, 1956. "Its two principal characters are so fantastic that one is half inclined to suspect that the whole thing has been devised as an elaborate leg-pulling"—*New York Herald Tribune*, March 11, 1956. "It is hard to know whether the author is presenting a thesis or a burlesque"—*Christian Science Monitor*, March 22, 1956. But just a couple of years later *The Ugly American* came along to prove that Eisenhower, Kennedy, and millions of their compatriots took the Ragtime Kid seriously and gloried in the leprous innocence Greene had not burlesqued but dissected. And of course it was hardly a decade later that Indochina was very nearly swamped by Pyles who streamed in by the hundreds of thousands. Only a novelist of genius could have caught so perceptively the inner meaning of their coming while they were still far beyond the horizon.

Graham Greene was also brilliantly prophetic in a piece he wrote for the *New Republic* the year his novel came out. He pictured Ngo Dinh Diem sitting in Norodom Palace in Saigon, "with his blank brown gaze, incorruptible, obstinate, ill-advised, going to his weekly confession, bolstered up by his belief that God is always on the Catholic side, waiting for a miracle. The name I would write under his portrait is the Patriot Ruined by the West" (CXXXII, 12).

Lansdale could claim a heaping helping of the collective credit for the ill advice and for the ruins.

2

Once, all sorts of people rushed to claim credit for "The Biggest Little Man in Asia," as the *Reader's Digest* called him (February 1956). A Worcester-like professor named Wesley Fishel said he had "discovered" Diem in 1950 during a Far Eastern intelligence assignment—the young political scientist arranged to have the Viet patriot visit the United States, later came out to train his police and proffer advice as head of the Michigan State University team of nation-builders, and became a pioneer in glorifying his "Democratic One-Man Rule," as Fishel called it in a *New Leader* article: "The peoples of Southeast Asia are not, generally speaking, sufficiently sophisticated to understand what we mean by democracy and how they can exercise and protect their own political rights" (XLII, 10–13). Early supporters of Fishel's find included such powerful figures as Francis Cardinal Spellman, Senators Mike Mansfield and John F. Kennedy, Supreme Court Justice William O. Douglas, Leo Cherne and Joseph Buttinger of the International Rescue Committee, and the claque of publicists and political philanthropists who lobbied and propagandized as the American Friends of Vietnam. They made Diem known in Washington ruling circles; in the spring of 1954 they may very well have played a major part, though the exact nature of their role remains unclear, in moving Emperor Bao Dai to name their man his new premier (GPP, I, 296).

Yet in the 1950s South Vietnam was primarily a CIA operation (ST, p. 196; CW, p. 176–78). Just as that truncated country was essentially a creation of the United States, so was Diem essentially a creation of the Company, whatever his support in other quarters. Once the principal CIA man on the scene persuaded himself and succeeded in convincing his boss that Diem was the shadow leader they sought, Allen Dulles shifted the weight of the CIA behind their choice and brought along his brother at State, Admiral Radford and the Joint Chiefs, and the rest of the Eisenhower administration. Lansdale did not "invent" Diem, as has been suggested with pardonable exaggeration, but surely handpicked him as the Indochinese Magsaysay, his second American viceroy in Asia—in Vietnamese, "My-Diem," the American Diem, as the new premier's enemies charged. In his headlong rush to duplicate exploits, however, Lansdale paid no heed to an essential difference between his viceroys: Diem's record suggested he would be a less amenable Friday than Magsaysay.

In his memoir Lansdale maintained he had never even seen Diem before the new premier sped by him in a limousine after his arrival at Tan Son Nhut on June 25, 1954. He thought Diem should have come into the city slowly to make an impression on the people lining the streets and wondered what "further errors of judgment Diem's advisers might be making. Then the thought occurred to me that perhaps he needed help, since he had been away from Vietnam and the Vietnamese people for years" (IMW, p. 157). Plainly a Welcome Wagon full of help was on its way, for the secret proconsul had been in Vietnam for some twenty days by then and was fully prepared to brief the Asian statesman on what had happened during his years of self-imposed exile. Lansdale promptly tracked Diem down in his palace and, with the help of an interpreter, modestly presented him with "A plan of overall governmental action," as he put it in his 1955 report: "It called for fast constructive action and dynamic leadership" (GPP, I, 576). Though his plan for eliminating Diem's political rivals was not put into immediate effect, it laid the foundation on which he built their fabulous relationship.

"To enter into another's soul," as Kim was trained to do, was the secret agent's specialty. But the man he described as looking like "the alert and eldest of the seven dwarfs deciding what to do about Snow White" (IMW, p. 159) was a difficult target, perhaps the most ticklish mark he ever had. Diem (1901–63) came from a mandarin family but seemingly exaggerated its status by simply fabricating high lineage for himself (FF, p. 107). Curiously, he was a Viet nationalist who prided himself on his family's early conversion to Roman Catholicism. In his youth he had had some experience in public administration; in 1933 he was appointed minister of the interior at Bao Dai's Imperial Court in Hué but resigned after a few months in protest against French heavy-handedness. His political retirement, or retreat as they say in the church, began then and extended over the next two decades, through WWII, political upheavals, revolution (GPP, I, 296). Committed nationalist or aloof monk, Confucianist Catholic or Catholic Confucianist, shy or arrogant, shrewd or paranoid—Diem slipped out of the categories Westerners tried to clamp on him. "A man of contrasts," uncharacteristically faltered *Time* in its appreciative profile "Chastity & Stuffed Cabbage":

> His eyes peer out distantly from beneath heavy lids. He is a lonely man, unused to self-expression, who lets others bring up the subject and then blurts, interminably and at random, not always expressively. . . . Monkish and inward-looking . . . he long ago pledged himself to chastity; he is so uncomfortable around women that he has none on his personal staff and he once put a sign outside his office: WOMEN FORBIDDEN. Yet Diem is also indulgent and demonstrative, downs huge breakfasts of such dishes as stuffed cabbage. . . . He may erupt into sudden violence. Considering someone he dislikes, he will sometimes spit across the room and snarl, "dirty type!" [LXV (April 4, 1955), 23]

Later, in a detailed description designed for use by his superiors, Lansdale was more reassuring about his short (five feet four inches) understudy: Diem's "feet seem barely to reach the floor when he is seated. However, he is not defensive about his short stature and is at ease around tall Americans. He has a very positive approach to Westerners, not the least bit concerned about differences such as Asian–Caucasian background[s]. When the Vice President [Johnson] sees him, he will find him as interested in cattle as any Texan and as interested in freedom as Sam Houston" (DOD, XI, 36). These Lone Star analogies revealed more about the Company cowboy than they did about the compact premier. For Lansdale to enter into another's soul was not necessarily for him to grasp what lay hidden there, for by no stretch of the imagination could Diem be sensibly viewed as a sort of Annamese cowpoke.*

Lansdale understood and naturally applauded Diem's fierce anti-Communism, his elitist abhorrence of revolution from the bottom up, his authoritarianism, and his unwillingness to tolerate real difference. He accommodated his personal life to the "puritanism" of the premier's strain of Catholicism, his traditionalism, and his family centralism that became nepotism. Even the vacuities of personalism—Diem's personal philosophy that was an amalgam of papal encyclicals, Marxism, and Confucianism—struck responsive chords in his own sonorous sermons on Paul of Tarsus and Tom Paine, though he did urge the premier toward broader ideological formulations (GPP, I, 300–1). But the secret agent never understood that differences between Asian and Caucasian backgrounds did vitally concern the premier, who saw himself as a man of destiny in the struggle between the East and the West. Lansdale never grasped, in short, that Diem's destiny diverged and ran counter, at critical points, to the American Manifest Destiny the agent was tirelessly trying to foist off upon the world.

One pictures Lansdale in Norodom Palace or "Independence Palace," as it was renamed, playing Cowboys and Indians with the Nhu children, sharing breakfasts of stuffed cabbage, and joining the entire family for supper. Afterward the talk was no doubt of freedom, with Diem holding forth on that favorite topic into the wee hours and with the secret agent in rapt attention, eyes fixed on his mark, bravely trying to keep up with his flow of revelations about the inscrutable Orient, and vainly attempting now and then to interject a few words about the necessity for fast constructive action and dynamic leadership. Such sessions required endurance and courage of sorts, for Diem knew little English and preferred to speak French to Westerners, while Lansdale not only knew no Vietnamese but little or no French and next to nothing about Indochina's history and cultures. But Diem was the key to his whole operation, so he wormed his way through these obstacles into the bosom of the Ngo family and participated intimately in what he liked to call "the 1954–55 birth of their nation" (DOD, XI, 33). It was a sensational performance.

To be sure, Lansdale brought more than his virtuosity to their relationship. The 900,000 refugees his team helped speed to the

* Johnson wisely sidestepped Lansdale's appeals to his native-son vanities. On his trip to Vietnam in 1961 he shot wildly but still closer to the mark when he pronounced Diem "the Winston Churchill of Asia" (FF, p. 96).

south provided Diem with a Catholic constituency (virtually his only constituency) within the predominantly Buddhist population. When General Nguyen Van Hinh, Diem's ambitious army chief of staff, threatened a coup in the fall of 1954, Lansdale bluntly told him "that U.S. support most probably would stop in such an event"; to keep it from happening, at a critical point he inveigled two of the general's key aides to come along on a junket to Manila "to see some of the inner workings of the fight against the Filipino Communists" (GPP, I, 578, 580). Concurrently he helped Diem launch a policy designed to divide and conquer the armed sects: the Cao Dai syncretic religious movement that controlled the countryside northwest of Saigon; the Hoa Hao dissident Buddhist movement that controlled most of the Mekong Delta southwest of Saigon; and the Binh Xuyen secret society under Le Van ("Bay") Vien, the Mafia-like ruler of both the underworld and the police of Saigon–Cholon (GPP, I, 293–95, 303). These were the sects the French depended upon to hold off the Viet Minh in the countryside and in Saigon itself.

Along with his brothers Nhu; Monsignor Thuc, the Catholic archbishop; Luyen, who later served as a diplomat abroad; and even Can at a distance in Hué, Premier Diem adroitly maneuvered the sects into mutual distrust and opposition. Yet his intrigues would never have moved forward had they not been lubricated by the enormous slush fund Lansdale poured into them. With his customary innocence, the agent observed in his memoir that French "soreheads" who had attempted to assassinate his character and then his body had also circulated "convoluted fictions about my bribing Vietnamese with huge sums of money" (IMW, pp. 218, 224, 318). In 1955, however, he had more forthrightly reported that the United States furnished "financial support" for Trinh Minh Thé, the Cao-Dai Third Force terrorist responsible for the street bombings Graham Greene incorporated in his novel (GPP, I, 578). In *The Two Viet-Nams* (1967) Bernard Fall asserted—with Lansdale's denial duly footnoted—that Thé cost $2 million and other sect leaders about $7 million (pp. 245–46). These bribes enabled Diem to isolate the Binh Xuyen.

The crisis with that formidable sect came to a head in the spring of 1955. In the early morning hours of March 28 a paratrooper unit loyal to Diem attacked the Saigon police headquarters. On the night of March 29 the Binh Xuyen struck back, dropping mortar shells on the palace grounds. According to Lansdale, Diem

> was desperately trying to get French and US help to remove the Sureté [police] from the control of the Binh Xuyen. French and US

> reactions to the problem were in the form of advice to proceed slowly, to act with caution. Events would not permit this. [GPP, I, 230]

At least one set of helping hands behind these incredibly tangled events was familiar. Lansdale played a double role as the Company confidant of Diem and as his champion within the U.S. Mission and in Washington, via CIA channels. His second role brought him into sharp conflict with the French forces under General Paul Ely and extended all the way back to Paris, where Prime Minister Edgar Faure finally insisted he be removed from Vietnam (GPP, I, 238). It also put him at odds with General J. Lawton Collins, Eisenhower's ambassador and personal military representative. Ely and Collins understandably opposed civil war in the streets of Saigon and urged conciliation upon Diem. Perhaps even less in favor of that course than the premier, Lansdale implied that Collins had fallen for French horror stories:

> The French had daily fed us the latest French propaganda line (Diem was weak, Diem was bloodthirsty, the VNA [Vietnam Army] had low morale . . . Americans didn't understand the Vietnamese, all whites must encourage only selected Vietnamese loyal to the French because the remainder would turn against all whites in another "night of the long knives" similar to that of 1946). [GPP, I, 231]

Confident of his own talents as a native-handler, Lansdale contemptuously dismissed the French line and assuredly did not counsel Diem to pursue cautious tactics. We may assume that with his backing, Diem rejected the urgings of the ranking French and U.S. authorities. Thereupon, Collins flew back to Washington and, with the approval of Eisenhower, on April 27 persuaded Secretary John Foster Dulles to consider a replacement for Diem (GPP, I, 233).

Just at this point, with even his staunchest Washington supporters looking for a fallback Friday, Diem's chances for survival had become virtually nil. When he learned through his own channels that he was about to be dumped, Diem called his only hope to the palace on April 28, the day following Dulles's decision in Washington. Through his interpreter Lansdale told him "that it looked as [though] the Vietnamese still needed a leader, that Diem was still President, that the US was still supporting him" (GPP, I, 233). With only the support of the secret proconsul and against the advice of everybody else—the French, overt U.S. officials, and his own cabinet—Diem moved against the Binh Xuyen that very day;

the ensuing battles left hundreds of casualties in the streets of Saigon, and over the next nine hours his army units drove Bay Vien's forces back into Cholon (GPP, I, 183, 234). Immediately Lansdale cabled Washington that Diem had asserted his leadership and established control of Saigon; Dulles reversed himself and instructed the U.S. Embassy there to disregard an earlier message that had implemented his decision to dump the weak, maladroit Diem; the American Mission accordingly burned his message and rallied behind the strong, dynamic Diem (FF, p. 107; GPP, I, 234, 303). Long live the vice-king. On his return Collins accused Lansdale of inciting "mutiny," as in a way he had, but the general accommodated himself to the *fait accompli* and joined the rest of the Eisenhower administration in supporting the Lansdale line that Diem was the strong shadow they had been looking for all along: the "Diem-is-the-only-available-leader syndrome," as the Pentagon analysts put it, was the cornerstone of U.S. policy well into the 1960s (GPP, II, 9).

On April 28, the day the battle was joined, Emperor Bao Dai ordered Diem to return to France (GPP, I, 234). Diem defied the summons, broke away from his legal mandate to office, and committed his forces to the combat that by mid-May succeeded in driving the Binh Xuyen into the Rung Sat swamp east of Saigon (GPP, I, 303). Lansdale naturally encouraged this revolution from the top, for it placed Diem more firmly in his hands and hastened the day of French departure from Vietnam (FF, p. 106; cf. IMW, pp. 298–300).

3

Under the Geneva Accords, by July 20, 1955, the Saigon and Hanoi governments had to consult on the elections scheduled for the next year. In his memoir Lansdale gave Diem full credit for refusing to talk with the Viet Minh and for defying the international agreements (IMW, pp. 324–26). The Pentagon analysts also declared that the United States did not "connive with Diem to ignore the elections" (GPP, I, 245). But Secretary of State Dulles publicly expressed the administration's desire to postpone the elections as long as possible and, with his brother, surely connived through their proconsul to keep them from taking place—it was general knowledge among Joseph B. Smith's Company associates at the Singapore station that they were helping "Diem maneuver his way out of holding the general elections to unite the country required by the Final Declaration of Geneva" (CW, pp. 179–80). Lansdale and his superiors had no intention of submitting Diem to an election he was certain to lose—six years later the secret agent was still unwilling to have the plebiscite promised at Geneva unless "a clear majority can be counted upon to vote for freedom" (DOD, XI, 31). Freedom was still named Ngo Dinh Diem.

A clear majority was ensured in the scheme he devised to provide Diem's government with the semblance of a legal foundation. According to his memoir, in the fall of 1955 he told the premier that he had the rare opportunity to become father of his country, just like George Washington. When Diem followed his advice to set up a "nationwide" referendum limited to South Vietnam, Lansdale claimed he cautioned him to be fair and warned against stuffing of the ballot box by his brother Nhu's front organization called the Movement for National Revolution:

> Cheating would be building the future on a false foundation and this would mean that whatever he did next would be short-lived. . . . The most that Diem might do, if he felt that custom and superstition might work against him, would be to add some subliminal insurance by the use of color in the ballots, printing his own in red, the Asian color of happiness. Bao Dai's ballots could be printed in black or blue or green. [IMW, p. 333]

On October 23, the date that became South Vietnam's National Day, Diem became president and father of his country by what the Pentagon analysts called a "dubiously overwhelming vote" (GPP, I, 246). Not content with a respectable margin, brother Nhu efficiently used his network of spies and ballot-box stuffers to dredge up for Diem 98.2 percent of the six million votes cast—in this outpouring of the people's will, Saigon, with 450,000 registered voters, showered 605,025 votes on Diem. Presumably this triumph for freedom owed something also to the psywar stroke of making Bao Dai's ballots green. Lansdale genteelly refrained from explaining that green signified cuckoldry and hence was repellent to superstitious natives.

Seemingly the Magsaysay formula had worked to perfection. Diem's immediate enemies had been bribed into submission, crushed, or put to flight. Along with their Emperor Bao Dai, the French had been outmaneuvered and forced to withdraw. The Geneva Accords had been neatly scuttled. Venality had been bested. The voice of the people had been heard in support of freedom. The secret proconsul had created another president and given another people a national hero. At the end of 1956 Lansdale returned to Washington from "Assignment: Vietnam" with vivid memories, again the covert savior of a country. At least three sets of recollections made unforgettable vignettes.

First, Lansdale remembered that one of Diem's finest moments was when the Binh Xuyen shelled Norodom Palace in March:

> The French colons in the Saigon bars told a story with great glee of how Diem had hidden under his bed quivering with fear. What he actually did was typically different. He went out in his night-shirt into the Palace grounds where some of the Guard Battalion had abandoned their artillery to take cover, and drove them back to their guns with a tongue-lashing while padding around the yard in a pair of old slippers. [DOD, XI, 37]

The imperial agent related the anecdote proudly, as though he were speaking of a spunky boy who had daringly risked catching cold. And paternal possessiveness unmistakably underlay his instruction to the diminutive premier: "I told him to take care of himself and to stop running around outdoors in his pajamas" (IMW, p. 262).

Second, at a critical point in the fighting for Saigon–Cholon, Trinh Minh Thé's troops tried to cross the Canal de Dérivation and were caught in a trap. "Damned mad," Lansdale stormed into the palace to find out why Thé had no artillery support. The stunned Diem issued the necessary orders and asked Lansdale to stay for a talk. Seated by each other on a couch, Diem chided the agent for his strong feelings over a former peasant who would have to take his chances with other military men, some of whom were better educated and more experienced:

> I told Diem to stop this line of talk. It was leading to a point where he might say something that would make me angry all over again.

Nhu came into the room to tell them that Thé had been shot in the head and killed:

> I turned to Diem and told him, "We have lost a true friend." I couldn't trust myself to say more. Diem looked at my face and started crying. Great sobs racked his body. I sat down again and held him in my arms. He asked me brokenly to forgive him for the way he had talked about Trinh minh Thé a little earlier. I told him there really was nothing to forgive, but he must always remember.

> True comrades were rare. He must never turn away from the unselfish ones who served freedom. [IMW, pp. 306–9]

And so came the fade-out, with the weeping Friday cradled in the arms of his secret Robinson.

Actually, Thé was their bought-and-paid-for Cao Dai terrorist who had once perhaps been instructed in the use of plastic charges by the CIA. Thé had a surgical ruthlessness that greatly appealed to Lansdale and no doubt for that very reason made Diem uneasy. His death was either a stroke of good luck that would have made no one in Diem's entourage weep or it was the result of a carefully laid assassination plot. The circumstances of Thé's death were "mysterious": he was shot in the back of the head at point-blank range, with powder burns around the wound. He thus had to be shot by someone "close" to him in both senses of the word, a fact that has led to charges that he was killed on Diem's orders. Of course, Lansdale's account would suggest the contrary, but it was surely constructed with an eye to the "plausible deniability" he relished. To him as well, I suggest, "true friend" Thé may have become an expendable leader of the "Third Force." Lansdale had admittedly regarded him as a possible alternative to Diem and had involved himself deeply in commitments that could easily have become awkward. As he told his friend Robert Shaplen:

> I had felt from the start that we shouldn't put all our eggs in one basket, that it was always important to look for alternatives. As a matter of fact, if worse came to worse and if the whole thing collapsed and the Binh Xuyen or someone else took over, I felt we ought to get ready to fight back from the hills, like guerrillas, and I had even made some preliminary surveys, with the help of Thé and others, about where we would withdraw to . . . I didn't tell all this to Diem, of course. [*The Lost Revolution* (1965), p. 125]

Had he been told, the supposedly weeping Diem might have stirred uneasily in his mentor's arms and reflected bitterly on the pertinence of his observation: "True comrades were rare."

Third, at the height of the slaughter in the streets, Lansdale braked to a stop in front of his office:

> A crowd of French officers had gathered in the street. . . . One of them held a little Vietnamese girl in his arms, her leg bleeding from a shrapnel wound, and was making an impassioned speech to the others. "Look what Lansdale has done," he said. "He makes war on children!" As the crowd glowered at me, I took the girl away

> from him and handed her to Lieutenant Colonel George Melvin, who had come up, so that he could take her to our dispensary a few yards away for treatment. The loss of his human prop didn't deter the speaker, who continued describing my villainy. It seems that I had been directing the fighting against the Binh Xuyen . . . had my hands dripping with the blood of innocent people, and was to blame for "everything." [IMW, p. 293]

The passage read as though it had come straight out of *The Quiet American*. Like Pyle, Lansdale did not really see the bleeding little girl but a human prop in the arms of an ideological adversary. He handled her as a native prop, with all the solicitude he would have shown in handing on a sack of rice. "Let's get out of here," he said to big Rufus Phillips and another teammate. "We have a meeting with General O'Daniel."

CHAPTER XXIX

The New Frontier

> For I stand tonight facing west on what was once the last frontier. From the lands that stretch 3000 miles behind me, the pioneers of old gave up their safety, their comfort and sometimes their lives to build a new world here in the West. . . . [But] the problems are not all solved and the battles are not all won, and we stand today on the edge of a new frontier—the frontier of the 1960s, a frontier of unknown opportunities and paths, a frontier of unfulfilled hopes and threats. . . . For the harsh facts of the matter are that we stand on this frontier at a turning point in history.
>
> —JOHN F. KENNEDY,
> Los Angeles Coliseum, July 1960

FROM 1957 to 1963, according to Lansdale's memoir, he was "stationed at the Pentagon" (IMW, p. 377).* But where in that huge symbol of empire was he stationed and to what end? In 1959 his testimonial to the dedicated Americans who were remaking Southeast Asia—incorporated in General Stilwell's seminal COIN document discussed earlier (see p. 365)—placed him in the Office of Special Operations, a euphemism for the covert operations that drew their basic authority from a 1955 National Security Council directive, NSC 5412/2. Lansdale was thus in the DOD office responsible for coordinating with the CIA and supplying the military muscle necessary to implement plans that came out of the subcommittee of the NSC named after its directive, Special Group 5412/2 or 54/12. On occasion, moreover, he represented the DOD on SG 54/12, but that by no means meant he had left the Company for the orthodox military establishment. The director of Central Intelligence (DCI) could count on his secret agent and on William Putnam Bundy, who came from State in the same dual role, to

* For an explanation of the abbreviated references, see p. 534.

back his proposals and hasten them on their way to the NSC. Who Lansdale was had thus come to baffle nonnatives: "in those crucial early years of Vietnam," asked L. Fletcher Prouty, "did McNamara and Rusk look upon Lansdale and Bill Bundy as Defense and State men under their command and control, or did they recognize them as CIA agents under the direction of the DCI?" (ST, p. 134). What was Lansdale?

One of his 1959 special operations gave him a leading role in *Shadow over Angkor*, a film produced by neutralist Prince Norodom Sihanouk of Cambodia. According to the secret agent, he had met "this plump playboy" only once and that was in 1953; nonetheless, "years later, he made a movie in which he as the hero bested a villainous American spy who was a caricature of me. Evidently he loved fantasy" (IMW, p. 113). The film centered on what Sihanouk called "the Dap Chhuon plot" in which a general by that name spearheaded a CIA attempt to overthrow his government. Lansdale contended he "never met nor had dealings with Dap Chu'on, although I wish I had, since he was beloved by the people of Cambodia" (IMW, p. 183n). Now, how could he possibly have known that?

The agent's denial had only to be reversed for it to point toward pay dirt. In *My War with the CIA* (1973) Sihanouk remarked on the unprecedented number of high U.S. officials who visited the Angkor temples in February 1959, just before he sent troops to arrest Dap Chhuon in his villa nearby: after Admiral Harry Felt and General J. Lawton Collins passed through, "next to sign the visitors' book was none other than Colonel Edward Lansdale, the renowned CIA specialist in cloak-and-dagger operations. It was he who had established Ngo Dinh Diem in power in Saigon, and had helped kick out the French" (pp. 107–8). The prince also claimed that he and his parents narrowly escaped assassination six months later in "the laquer-box attentat"—a bomb so disguised and allegedly mailed from a U.S. military base in South Vietnam exploded in the Royal Palace and killed their chief of protocol when he opened the parcel. Most unfortunately, in *Alleged Assassination Plots Involving Foreign Leaders* Senator Frank Church's Select Committee on Intelligence failed to include these claims (AAP; 1975). The Cambodian leader contended "the CIA was in the forefront (except, when it suited their purposes to remain concealed) of every plot directed against my life and my country's integrity" (pp. 110–11). An inquiry into Shihanouk's allegations by the Senate Intelligence Committee might have revealed what role, if any, Lansdale played in those plots.

Nevertheless, we can be certain the prince loved fantasy less than the secret agent suggested. Sihanouk's allegations find staggering if veiled confirmation in a secret report of the U.S. Operations Coordinating Board dated February 10, 1960:

> [Sihanouk] has emerged with added power and prestige from the abortive coup plots and subsequent activities mounted against him in 1959 by *ostensibly* anti-communist elements. In the process many of these elements were eliminated and the revelation of their real or fancied association with the United States and other free world countries undermined Cambodian confidence in U.S. motives and became an obstacle to the pursuit of our objectives. Moreover, Sihanouk has given further evidence of political astuteness in the domestic arena, has displayed increased alertness to communist subversion, and has shown no inclination to tolerate any challenge to his pre-eminence. Policy guidance, therefore, should be directed conspicuously and specifically at the problem of dealing with Sihanouk, by all odds the major single factor in Cambodia and the principal target of U.S. policy. [DOD, X, 1249; my italics]

Political astuteness and increased alertness to Communist subversion had made Sihanouk an even more inviting target.

A year later Lansdale tried to get him in his sights through a proposal to send "a task force of journalists"—some of whom would undoubtedly have Company affiliation and other assignments—to Cambodia to report "on policies being implemented by Sihanouk and other officials" (DOD, XI, 29). But it was not until almost a decade later that a military coup and an armed intervention finally hit the number-one U.S. Cambodian target—Sihanouk regarded the events leading up to the March 18, 1970, coup that deposed him as "simply a re-vamped version of the 'Dap Chhuon Plot' of early 1959" (p. 55). By then presumably no longer directly involved, Lansdale could still claim credit as a pioneer agent of his downfall, and that is to say the agent was probably well cast in *Shadow over Angkor*. Certainly Sihanouk's response to the many Westerners who thought the film "the product of an overly active imagination" spoke directly to his predicament: "Alas, I did not at that time possess an imagination sufficiently fertile to foresee the grotesque and fantastic schemes which the CIA was dreaming up" (p. 111).

In early January 1961 the recently promoted Brigadier General Lansdale, assistant to the secretary of defense for Special Operations, made a quick (twelve-day) trip to Vietnam and returned with the disquieting news that had such an impact upon incoming

President Kennedy (DOD, XI, 1–12).* Seemingly his Magsaysay formula had all but failed. Just the preceding November an abortive coup had sent bursts of heavy machine-gun fire into Diem's bedroom and made him still more suspicious of associates, including senior U.S. officials, especially Ambassador Elbridge Durbrow. Nhu's power and influence over his brother had grown. The "Viet Cong"—the term was invented by the Diem regime to damn the National Liberation Front (NLF)—had made shocking advances in the countryside. To stop this "downhill and dangerous trend" Lansdale proposed his brand of unconventional warfare, including psywar and the fostering of a loyal grass-roots political opposition. He was critical of Diem but not too critical, for he was figuratively "one of the men who had invented Diem," as David Halberstam noted, "and you do not knock your own invention." More critical of Durbrow and other U.S. officials, he proposed a new team, "a hard core of experienced Americans who know and really like Asia and Asians, dedicated people who are willing to risk their lives for the ideals of freedom." And who was more experienced and more of a dedicated American than Diem's old case officer? Lansdale wanted to get back in the field as secret proconsul, wanted to play hooky again by getting back into the war.

He almost made it. Roger Hilsman, director of intelligence and research at the State Department, reported that Lansdale's presentation made Kennedy think about sending him to Saigon as the next ambassador (GPP, II, 26). And one Sunday morning, according to David Halberstam, the president summoned the secret agent to a special breakfast meeting in the White House, greeted him graciously, and turning toward Dean Rusk, asked: " 'Has the Secretary here mentioned that I wanted you to be ambassador to Vietnam?' Lansdale, caught by surprise, mumbled that it was a great honor and a marvelous opportunity. He was deeply touched, and even more surprised, for it was the first he heard of the idea, and also, as it happened, the last." Had the appointment not been blocked by State and perhaps by Defense, where Lansdale also had enemies, Allen Dulles would also have been touched, for it would have meant that his Company had taken over the Saigon Embassy, lock, stock, and barrel.

There remained the task force of dedicated Americans. On April

* On his flight out to see the Annamese George Washington, Lansdale "carried with him a gift 'from the U.S. Government,' a huge desk with a brassplate across the base reading, 'To Ngo Dinh Diem, The Father of His Country.' The presentation of that gift to Diem by Lansdale marked nearly seven years of close personal and official relationship, all under the sponsorship of the CIA" (ST, p. 60).

12, 1961, Walt Whitman Rostow proposed to Kennedy that they turn to "gearing up the whole Viet-Nam operation" and for openers suggested "the appointment of a full time first-rate back-stop man," or Lansdale by name (GPP, II, 34–35). Working closely with Roswell L. Gilpatric, deputy secretary of defense, Lansdale wrote a long working paper that recommended the president establish a task force on Vietnam; in the drafting he built in a major role for himself: "Fullest use should be made of the existing position of personal confidence and understanding which General Lansdale holds with President Diem and other key Vietnamese" (DOD, XI, 32; GPP, II, 35). On April 27 Gilpatric sent the task force draft to Kennedy. It called for an acceleration of COIN as the primary means of blocking "the Communist 'master plan' to take over all of Southeast Asia" and otherwise bore marks of the secret agent's handiwork, especially in its stress on vigor, enthusiasm, and dynamic leadership. In his covering memorandum, Gilpatric told Kennedy that "Brigadier General E. G. Lansdale, USAF, who has been designated Operations Officer for the Task Force, will proceed to Vietnam immediately after the program receives Presidential approval" (DOD, XI, 42–56). The secret agent made arrangements to be in Saigon on May 5 and had already started rounding up his old CIA teammates, including Joseph P. Redick, his indispensable interpreter of "the uninhibited communications between President Diem and myself" (GPP, II, 38). Alas, this was the high point in Lansdale's Vietnam policy-making. In the bureaucratic infighting that followed, his and Gilpatric's supreme role on the task force was simply excised from the draft by George Ball, then deputy undersecretary of state: "State objected, successfully, to having an Ambassador report to a Task Force chaired by the Deputy Secretary of Defense, and with a second defense [*sic*] official (Lansdale) as executive officer," concluded the Pentagon analysts, laboring under the illusion the secret agent was now merely an air force general (GPP, II, 43–44). Thereafter, despite four requests from Diem, and others from Maxwell Taylor and William Bundy, that Lansdale be sent to Saigon, he did not get out there for duty again until August 1965 (GPP, II, 17, 653; DOD, XI, 422).

2

Anyway, "I had been shunted from Washington work on Vietnamese problems in 1961," the secret agent explained "and had been busy with other duties" (IMW, p. 378). Those other duties were spelled out in the Church committee's *Alleged Assassination Plots:*

> As a result of the Bay of Pigs failure [April 15–17, 1961], President Kennedy distrusted the CIA and believed that someone from outside the Agency was required to oversee major covert action programs. Rather than appoint his brother, Robert Kennedy, to head MONGOOSE, as proposed by [Richard] Goodwin, President Kennedy gave General Edward Lansdale the task of coordinating the CIA's MONGOOSE operations with those of the Departments of State and Defense. Lansdale had developed a reputation in the Philippines and Vietnam for having an ability to deal with revolutionary insurgencies in less developed countries. Kennedy appointed General Taylor Chairman of the Special Group Augmented [SGA]. Robert Kennedy played an active role in the MONGOOSE Operation, a role unrelated to his position as Attorney General. [AAP, p. 140]

Lansdale's secrecy and lies and dirty tricks, supposedly so effective in handling natives in the so-called less developed countries, had come home to roost in the Oval Office. Distrusting the CIA after its abortive Cuban invasion attempt, Kennedy unwittingly appointed a senior CIA operative to oversee CIA dirty tricks against Fidel Castro—if we are to believe the Church committee. Fourteen years later, furthermore, Senator Church and his colleagues still missed the wry truth—at least in print—that Kennedy had appointed a superfox to oversee the foxes: "[John] McCone and [William] Harvey were the principal CIA participants in Operation MONGOOSE. Although [Richard] Helms attended only 7 of the 40 MONGOOSE meetings, he was significantly involved. . . ." And therefore Lansdale, in charge of Operation Mongoose from November 30, 1961, until it was disbanded on October 30, 1962, had been merely an air force general and not a principal CIA participant? What was Lansdale?

"We were hysterical about Castro at the time of the Bay of Pigs and thereafter," testified Robert McNamara in 1975 (AAP, pp. 157–58). Like his predecessors in the 1890s, President Kennedy unhesitantly intervened in the affairs of natives living on an island "so near to us," in Grover Cleveland's words, "as to be hardly separated from our territory." The New Frontiersmen waged a secret war to redeem the derelict province through means that included the economic embargo, acts of economic sabotage, and terroristic commando raids. The CIA directed clandestine operations from the JM WAVE station in Miami that was staffed by over three hundred employees and case officers, who in turn had control over several thousand Cuban agents. In addition there were the CIA mercenaries known as "paramilitaries" or, *mirabile dictu*, "cowboys." Ray Cline, then CIA deputy director for intelligence, later explained how the national game of Cowboys and Indians had become global:

> You need to understand the national consensus of the 1950s and '60s when we believed the world was a tough place filled with actual threats of subversion by other countries. The Russians had cowboys around everywhere, and that meant we had to get ourselves a lot of cowboys if we wanted to play the game. You've got to have cowboys—the only thing is you don't let them make policy. You keep them in the ranch house when you don't have a specific project for them.

Cline was quoted by two contributing editors to *Harper's*, who also reported that William ("Rip") Robertson, one of the top-notch Company cowboys, offered a commando fifty dollars for a Cuban ear; when he was presented with two, he laughed, paid the hundred dollars, and had the team over "to his house for a turkey dinner." Thanksgiving for native ears was in character for Rip Robertson, as the commando described him: "Rip was a patriot, an American patriot. . . . He'd fight anything that came against democracy. He fought with the Company in Korea, in Cuba, and then he went to Vietnam. He never stopped. . . ."*

Under "higher authority"—a circumlocution for President Kennedy—and under the Special Group Augmented (SGA), Lansdale became the ramrod of a secret outfit extending from the highest

* Taylor Branch and George Crile III, "The Kennedy Vendetta," *Harper's*, CCLI (August 1975), 57–58. This single incident would suggest that the going price for native parts had changed relatively little over the centuries. In 1697 Hannah Duston chopped up her sleeping captors (and their children) and returned with her bloody trophies in the hope of claiming the current bounty of ten pounds apiece on Native American scalps—see Leslie Fiedler's discussion of Duston as an exemplary hero in *The Return of the Vanishing American* (New York: Stein and Day, 1968), pp. 98–108.

Of course the price was not always available, and when it was it varied according to location and circumstance. In 1901, you may recall, General Frederick Funston willingly paid out $600 for the head of David Fagen, the Afro-American "mad dog" who had gone over to the side of the Filipinos. In 1962 the representative of the DOD and the Joint Chiefs on the Mongoose working group forwarded for Lansdale's consideration an attempt to systematize the bounties offered for disrupting the Cuban regime. Called "Operation Bounty," the concept was described as a "system of financial rewards, commensurate with position and stature, for killing or delivering alive known Communists," with $5,000 to be paid for an "informer" on up to $100,000 for "government officials." Lansdale testified that he "tabled" the proposal as something that should not be undertaken "seriously" (AAP, pp. 144, 276). Considerations of practicality and not principle led to this reaction, for the secret agent was by no means opposed to assassinations as such.

levels in Washington down to the JM WAVE station in Miami and on to the paramilitary bases scattered throughout the Florida Keys and beyond. The basic objective of Operation Mongoose, as he formulated it, was to have "the people themselves overthrow the Castro regime rather than U.S. engineered efforts from outside Cuba" (AAP, pp. 140–48, 333–37). From outside Cuba he engineered thirty-two "tasks" in January 1962 for the participating U.S. agencies and exhorted one and all to dynamic leadership: "it is our job to put the American genius to work on this project, quickly and effectively. . . . we are in a combat situation—where we have been given full command" in Washington and, if the Mongoose could fly, over the people themselves in Cuba. Shortly he added "Task 33," which called for biological and chemical warfare against the Cuban sugar-crop workers. In February he added to his "Basic Action Plan" a proposal for attacks on the cadre, including key leaders of Castro's government:

> This should be a "Special Target" operation. CIA defector operations are vital here. Gangster elements might provide the best recruitment potential for actions against police—G2 [intelligence] officials. [USSR] Bloc technicians should be added to the list of targets. CW [Chemical Warfare] agents should be fully considered.

Lansdale testified that his proposal to recruit Mafiosi to attack key leaders indeed meant the targeted killings of individuals, or in the word he carefully avoided, *assassinations*. The SGA tabled his "Basic Action Plan" and the record, according to the Church committee, showed that he was instructed to proceed with intelligence collection only. Under his direction, nonetheless, Company cowboys were sent into Cuba, with ten to fifteen agent teams dispatched in June alone. Sabotage actions escalated as well. On August 30, 1962, for instance, the secret agent targeted for the SGA a mine, refineries, and nickel plants, and proposed "encouraging destruction of crops by fire, chemicals, and weeds." Hardly more than the technology had changed since Endicott and Underhill ravaged Block Island.

At an SGA policy meeting on August 10, 1962, someone explicitly raised the topic of the assassination of Castro (AAP, pp. 161–69, 319–23). Though several of the sixteen persons present openly discussed assassination, the minutes of the meeting, kept by CIA officer Thomas Parrot, revealingly contained no reference to the discussion. Subsequently the top officials present were afflicted with failing memories and changed stories about who said what or even if the topic had come up. Lansdale testified he believed that McNamara had brought it up and that he "was usually very brief and terse in his remarks, and it might have been something like, well, look into that." Look into it the secret agent had, for three days later, on August 13, he directed CIA officer William Harvey to develop plans, "including liquidation of leaders." Upon receipt of his memorandum, Harvey excised the offending words, took exception to making assassination a matter of official record, and called Lansdale's office to point out "the inadmissability and stupidity of putting this type of comment in writing in such a document."* McNamara had blundered in calling assassination *assassination*, and Lansdale had blundered still more egregiously in putting his "liquidation of leaders" in writing, thereby undercutting everyone's "plausible deniability." Before the Church committee the secret agent testified that when Castro's assassination surfaced at the meeting, the SGA "consensus was . . . hell no on this and there was a very violent reaction." Well, then, why had he gone ahead three days later with the assassination proposal?

> General Lansdale: . . . I don't recall that thoroughly, I don't remember the reasons why I would.
> Q: Is it your testimony that the August 10 meeting turned down assassinations as a subject to look into, and that you nevertheless asked Mr. Harvey to look into it?
> General Lansdale: I guess it is, yes. The way you put it to me now has me baffled about why I did it. I don't know.

Lansdale's testimony, aptly observed Senator Howard H. Baker, Jr., was "not a model of clarity." The secret agent also had to deny the accuracy of two reporters who had quoted him as saying he was ordered to develop an assassination plan by higher authority —in the *Washington Star-News* he was said to have named Robert Kennedy as the intermediary who brought him his orders.

Yet even in this tangle over whether the Kennedy brothers did or did not directly order the assassination of Castro, members of the Church committee revealed they too considered that ordinary

* Between 1960 and 1965 the CIA concocted at least eight separate plots to assassinate Castro through means that included high-powered rifles, poison cigars, poison pens, poison pills, and deadly bacterial powders (AAP, p. 71). In April 1962, just a few months before Lansdale's indiscreet August 13 memorandum, William Harvey sent four poison pills into Cuba to be administered by a Cuban "asset" established through Mafia lieutenant John Rosselli of the gambling syndicate (AAP, pp. 83–85). Hence Harvey was not opposed to assassinations in principle or practice but only to any open reference to them. In charge of the CIA operation to develop "Executive Action" capability, Harvey reminded himself in notes to use this euphemism and its alternatives, "last resort beyond last resort," "the magic button," etc., and sternly enjoined himself: "never mention word assassination" (AAP, p. 183).

native lives came cheap. They did not investigate the acquisition of ears and other trophies in Cuba nor, as Gary Wills has pointed out, "the large-scale terrorist assassinations in the CIA's Phoenix program. . . . [The] whole emphasis was on plans to kill foreign *leaders*. Other kinds of ambush, terrorism, and 'liquidation' do not seem to count." Indubitably Lansdale had worked out plans for targeted killings, proposed chemical and biological warfare against the Cuban people, and sent in cowboys to blow up bridges, burn down structures, destroy crops, and cut down those hostiles unlucky enough to get in their way. And after all, he was ramrod of a secret outfit whose very name hinted that its targets were rodentlike natives. Lyndon B. Johnson exaggerated not at all when he reported learning shortly after he moved into the White House: "we had been operating a damned Murder, Inc., in the Caribbean" (AAP, p. 180n).

The funny, relatively nonlethal side of the secret agent surfaced in one of his plans that was never implemented. Still drawing on his old-hand knowledge of superstitious natives, Lansdale proposed to spook them by spreading word among Cubans of the imminent Second Coming of Christ and the no less imminent Departure of Anti-Christ Castro. On a certain day there would be manifestations of His Coming "and at that time," in the words of the SGA's secretary, "just over the horizon there would be an American submarine which would surface off of Cuba and send up some star-shells. And this would be the manifestation of the Second Coming and Castro would be overthrown. Well, some wag called this operation . . . Elimination by Illumination" (AAP, p. 142n). Say, rather, another Elimination by the Children of Light.

3

The Philippine Islands. Vietnam. Laos. Cambodia. Cuba. Covert philanthropy under three presidents—Truman, Eisenhower, and Kennedy. And then Vietnam again: On November 2, 1963, on the very day Major General Lansdale "had been retired from the air force," came the "shock" of Diem's assassination. As he explained on the last page of *In the Midst of Wars*, "I had been shunted from Washington work on Vietnamese problems in 1961 and had been busy with other duties. The coup and murders in Saigon seemed incredible." Living his cover to the end, the agent was inexact.

Despite those "other duties" we have just scrutinized, Lansdale had been advising the Joint Chiefs to the end of 1961 on how to go about "getting Diem to play ball" (DOD, XI, 427; XII, 440–41). With the disbanding of his Mongoose program following the Cuban missile crisis of October 1962, he stayed on for a year as McNamara's special COIN assistant, a vantage point from which he no doubt observed Diem's rapid trajectory toward the point of no return. Somewhere along the way he himself despaired of his too-obstinate Friday ever learning how to play ball properly. On September 10, less than two months before the coup, Rufus Phillips, his longtime understudy, flew in from Saigon and brought to a meeting of the National Security Council the bad news that the strategic-hamlet program he was running had broken down and that Diem and Nhu had lost touch with their people and with reality. "In the turning around of Phillips," David Halberstam rightly remarked, "a bench mark had been passed. It was a symbol of Lansdale turning as well; the people who had invented Diem were now leading the assault against him." And it was Lucien Conein, Lansdale's man who had since become a "colonel," who played the major CIA role in the events leading up to the coup and the murders (AAP, pp. 217–23). The senior secret agent may even have known that Conein had reported through the Saigon Embassy to Washington that his old friend Duong Van Minh—"Big" Minh, the general who had long been a virtual auxiliary of the Lansdale team—and the other plotters were considering, among their three alternatives, the assassinations of Diem's brothers Nhu and Can (GPP, II, 766–68). Given his own experience with "Special Target" operations, in any event, Lansdale could hardly have found the coup and the murders truly "incredible" nor have been truly shocked that Diem had been uninvented.

Of course, still other patriots of sorts were waiting in the wings to be ruined by the West. By the time Lansdale finally got back to Saigon in August 1965, the whirligig of coups, demicoups, and countercoups had come to rest with a ten-general directorate nominally headed by Premier Nguyen Cao Ky, who knew the secret agent by reputation as a kingmaker and who greeted him by saying: "Now that you are here, you will have no problems, because I have no intention of becoming a king." Lansdale quickly charmed his way into Ky's confidence, and under his tutelage the young air force general shifted from expressions of admiration for Adolf Hitler's fighting spirit and ironhanded leadership—expressions that matched his own actions—and began to sound more like an Asian Tom Paine, dedicated to an "unending quest for peace," an "attack on hunger, ignorance and disease," "land reform," and all the other laudable ends the secret agent helped embed in the Declaration of Honolulu, the Great Society document issued by Johnson and Ky in February 1966. But after General Nguyen Van Thieu took over the reins of government, he refused to deal with Lansdale on the ground he was too close to Ky. The secret agent, alias Major General Lansdale, USAF (Ret.), thereupon returned to Washington in June 1968. He had backed the wrong Friday.

Really none of his old magic had worked this second time around. He had come out in 1965 with a fresh team of young cold warriors, including Daniel Ellsberg, to breathe life into the moribund COIN programs, but the war had already become massive, with almost 200,000 U.S. troops in the country and more to come, with General William C. Westmoreland readying his ponderous "search and destroy" missions, and with the ongoing "Rolling Thunder" bombings in the north. The Americanization of the war had produced a cancerous spread of U.S. agencies, services, missions, institutes, and programs, and had created a bureaucratic muddle that made Saigon seem Washington Far East—or Farthest West. Aside from the CIA Phoenix program, the human hunt that Rip Robertson and some of the Company cowboys came from the Caribbean to join, the rest of the war was for bombers and shellers and burners of villages, and not for grass-roots cloak-and-dagger specialists.

Under nominal cover as minister counselor to the U.S. Embassy, Lansdale had been pushed to the sidelines by Henry Cabot Lodge and limited to work on "pacification," though that venerable undertaking—the word dated back to 1573, when Spain's Philip II ordered that it be used in place of "conquest"—currently was more elegantly called "rural construction and development." Under whatever name, it was not possible in a countryside fast being destroyed, so the veteran agent stayed pretty much in his villa—talking to cronies from the exciting first days of Diem; to visiting firemen and correspondents, whom he could always fill in on the inside story of the latest disaster; and to the new generation of enthusiasts, who were enthralled by his talk of the Third Force and the Mission of the West (FF, pp. 358–59, 482). It was only later that Daniel Ellsberg found the strength to shake off his "fun-and-games" addiction: " *'Good colonialists,'* " he wrote in *Papers on the War*, "was what we were all trying to be. . . . It was, and is, the wrong aspiration to have" (1972; p. 145). But for Lansdale and the vast majority of their compatriots, it was inconceivable that the American helpers had not been all that different from the French *colons*. After Ellsberg blew everyone's official cover by giving the secret history of Vietnam to the press, Lansdale sadly

Edward G. Lansdale, 1965. Named Ambassador Henry Cabot Lodge's "special assistant" on August 31, Lansdale was identified by the legend of this newsphoto as "a retired Air Force officer" and as "a counter-insurgency expert . . . appointed to head a mission seeking to revive the pacification program in South Viet Nam." (By permission of Wide World Photos.)

pronounced his erstwhile apprentice "independent and brilliant but lacking in security discipline." For the secret outrider of empire, security discipline was his life, and the colonial aspiration his reason for existence.

Just a few days before the NLF's great Tet offensive in 1968, Lansdale sent a memorandum to Ambassador Ellsworth Bunker: "At Tet with the advent of the new year," he advised Lodge's successor,

> it would be useful to keep in mind that certain of the dates in the year ahead will be looked upon by most Vietnamese as inauspicious days. The dates are given below for the guidance of those who wonder at the Vietnamese reluctance to undertake new projects on such days:
>
> February 3, 12, 21
> March 3, 12, 21
> April 2, 11, 20. . . .

And so on into the really inauspicious days for the American Mission after Tet 1968. In his last days of Saigon, so to speak, the ministerial secret agent went out as he came in, still proposing astrology as the key to unlock the minds and hearts of childlike Asian natives.

CHAPTER XXX

The Problem of the West

Our fathers wrung their bread from stocks and stones
And fenced their gardens with the Redman's bones;
Embarking from the Nether Land of Holland,
Pilgrims unhouseled by Geneva's night,
They planted here the Serpent's seeds of light;
And here the pivoting searchlights probe to shock
The riotous glass houses built on rock,
And candles gutter by an empty altar,
And light is where the landless blood of Cain
Is burning, burning the unburied grain.

—ROBERT LOWELL,
"Children of Light," 1944

IN LATE April 1975, just a month shy of twenty-one years after the secret proconsul's arrival at the Tan Son Nhut airport, the United States airlifted its dedicated nation-builders out of Saigon. Taking off from Tan Son Nhut and from such pick-up points as "Dodge City," as GIs called the annex of the defense attaché's office, they left behind the wreckage and craters from some fourteen million tons of bombs and high-explosive shells, some five to six million refugees, a million to two million dead bodies, expenditures of over one hundred and fifty billion dollars, and a countryside reminiscent of John Fiske's imaginary Wyoming moonscape (see p. 232). They left the way they came, though by subtitle and text CIA analyst Frank Snepp suggested the contrary in *Decent Interval: An Insider's Account of Saigon's Indecent End* (DI; 1977). As "in the last days of Hanoi" two decades earlier, the party was over, not the city itself, and it had been an indecent happening from beginning to end.

Just as on the way in, so on the way out the CIA—by then grown to a contingent of three hundred in Saigon alone—frantically cast about for a serviceable man Friday to retrieve the irretrievable. Despite his close working relationships with Nguyen Van Thieu and his staff, Station Chief Thomas Polgar decided this most recent figurehead had outlived his usefulness and was blocking the possibility of negotiating with Hanoi and the National Liberation Front (NLF). Going over the head of Ambassador Graham Anderson Martin, Polgar secretly cabled Washington "that Thieu's days were numbered" and implicitly requested authority to seek his replacement. To Polgar's despair, CIA Director William Colby vetoed his proposed coup d'état on grounds not of principle but of "the adverse political fallout that could be expected at home and abroad if it ever became known the CIA was again in the business of overthrowing governments" (DI, pp. 132, 289–91, 319–21, 433). Waiting in the wings had been the customary cast of military statesmen, including Nguyen Cao Ky, whose cause had the support of Lansdale's old friend Robert Shaplen of the *New Yorker*, and Big Minh (Duong Van Minh), Thieu's chief rival, the Third Force general who had been close to the Lansdale team and had played a leading role in the 1963 coup and the murders of Diem and Nhu. It was like old times, save for the absence of the master scene shifter himself and the Company's apprehensiveness about its image as "a guardian of life."

In the final hours of the U.S. presence in Saigon, as Frank Snepp's grim account made clear, the operative principle became "save the gentlemen in the white skin" (DI, pp. 517–62). At the Hotel Duc, for instance, a hundred or so "locals" or "indigenous employees" were abandoned when their CIA patrons took off by helicopter from the rooftop pad and went whirling off to the fleet. Four blocks away, at the Embassy, thousands of panic-stricken Vietnamese "friendlies," lovers, camp followers, employees, and their families jammed up against the gate and the ten-foot-high wall. Caught in the crush, Keyes Beech, dean of the American press corps, called for CIA help to get inside the compound: "Once we moved into that seething mass we ceased to be correspondents," Beech reported of his group of American, Vietnamese, and Japanese journalists. "We were only men fighting for our lives, scratching, clawing, pushing ever closer to that wall. . . . We were like animals. Now, I thought, I know what it's like to be a Vietnamese. I am one of them. But if I could get over that wall, I would be an American again." Kicking nearby Vietnamese down, U.S. Marines lifted Beech and Bud Merick of *U.S. News & World Report* over the top: "Goddamnit, though!" exclaimed Beech inside the Embassy. "The marines left our Vietnamese and Japanese friends outside the wall. They couldn't, or wouldn't, pull them over the wall." As the last chopper took off from the Embassy roof, desperate "locals" came surging up over the edge of the pad and vainly plunged for the wheels. Left behind hopelessly awaiting evacuation, mostly in the courtyard below, were over four hundred natives, not all of whom were Vietnamese, according to one witness: "There were a few Koreans and some others—Filipinos, I think—in the group." Evidently no Anglo-Americans were among these relics of Lansdale's Operation Brotherhood.* Embassy Administrative Counselor Henry Boudreau checked the parking lot just before his own evacuation and later reported with relief: "Didn't see any white faces out there."

Indeed, the whole party had been a whiteface affair. True, there were other considerations: each president from Truman through Nixon had feared the domestic consequences of "losing" Indochina, as Daniel Ellsberg has argued; each moved to counter the Soviet threat and Chinese expansionism, real or imagined; and each sought to advance national economic interests in the area and to enhance U.S. credibility and prestige. But all these aims were pursued within a historical context for which racism had never been peripheral. With the deepening disaster, policy-makers took a last-ditch stand that reduced other considerations to a single supreme test of the *national will:* "Time and again I was told by men reared in the tradition of Henry L. Stimson that all we needed was the will, and we would then prevail," recalled James C. Thomson, Jr., one of the more reflective policy aides at State in the 1960s. "Implicit in such a view, it seemed to me, was a curious assumption that Asians lacked will, or at least that in a contest between Asian and Anglo-Saxon wills, the non-Asians must prevail." It was a curious assumption, one with racist underpinnings, but still more curious was the collective unawareness, amounting to self-imposed ignorance (shared to some degree by even so insightful a scholar as Thomson), of the roots of that assumption: three and a half centuries of conquest had made more self-evident than questionable the Anglo-American conviction that in any contest with nonwhites, dusky natives would surely lose.

"The United States is a Pacific power," wrote Richard M. Nixon in the October 1967 *Foreign Affairs*. "Europe has been withdrawing the remnants of empire, but the United States, with its coast reaching in an arc from Mexico to the Bering Straits, is one anchor of a vast Pacific community. Both our interests and our ideals propel us westward across the Pacific, not as conquerors but as partners . . ." (XLVI, 112). Propelled by the traditional westward thrust of their empire, Anglo-Americans were trying to take over

* See p. 411. John Stockwell, another CIA veteran of Vietnam, recorded that in the ignominious evacuation, the Company "left 250 of its Filipino employees behind at the mercy of the communists"—*In Search of Enemies: A CIA Story* (New York: W. W. Norton, 1978), p. 185.

from the departing Europeans without appearing as new masters but as partners and protectors. After the 1968 elections the "new" Nixon Doctrine set out to hide its whiteface origins while extending the U.S. imperial role in Asia at cut-rate costs in dollars and lives. But "Vietnamization," as Ambassador Ellsworth Bunker reportedly declared, simply meant changing "the color of the corpses." Far from being new, the doctrine was shopworn, having been on the counter at least since the spring 1954 policy statement of the President's Special Committee on Indochina: to establish a Western-oriented complex in Indochina, "the eventual goal must be the development of homogeneous indigenous units with a native officer corps" (DOD, IX, 334).* Lansdale's Saigon Military Mission had been only the first systematic attempt to reach the goal that still eluded Nixon two decades later. It always expressed, implicitly or explicitly, concern for the color of the corpses and the living bodies, beginning with Eisenhower's determination not to appear openly as a partner in a "white man's party." His successor fully shared that repugnance. "The war in Vietnam," Kennedy told Arthur Schlesinger, Jr., "could be won only so long as it was *their* war. If it were ever converted into a white man's war, we would lose as the French had lost a decade earlier."

But with first the French and then the American involvement, it had never been *their* war, and from that truth flowed the basic dilemma facing Washington policy-makers. They believed every effort had to be exerted to keep the conflict from becoming "a white man's war"; yet their every step further into the conflict made it still more visibly "a white man's war." In October 1961, for instance, William Bundy gave McNamara his candid assessment of their chances: "It *is* really now or never if we are to arrest the gains being made by the Viet Cong," he argued. "An early and hard-hitting operation has a good chance (70% would be my guess) of *arresting* things and giving Diem a chance to do better and clean up [*sic*]." Yet everything depended upon Diem's "very problematical" effectiveness: "The 30% chance is that we would wind up like the French in 1954; white men can't win this kind of fight" (DOD, XI, 312). But subsequent steps, especially the U.S.-sponsored overthrow of Diem two years later, put white men still more visibly into precisely that kind of fight.

The odds had worsened by July 1965 when George Ball proposed to President Johnson that the United States try to find some way out of South Vietnam at minimal cost:

* For an explanation of the abbreviated references, see p. 534.

> No one can assure you that we can beat the Viet Cong or even force them to the conference table on our terms, no matter how many hundred thousand *white, foreign* (U.S.) troops we deploy.
> No one has demonstrated that a white ground force of whatever size can win a guerrilla war—which is at the same time a civil war between Asians—in jungle terrain in the midst of a population that refuses cooperation to the white forces (and the South Vietnamese) and thus provides a great intelligence advantage to the other side. [GPP, IV, 615; Ball's italics]

Insofar as there were differences of opinion within the Johnson administration, Bundy and Ball were on opposite sides of the question of leaving or staying in Vietnam—in 1965 Bundy recommended "holding on" (GPP, IV, 610–15). Yet Bundy and Ball essentially agreed on the crucial issue of the white man's chances in that kind of fight, and both agreed that the United States was fighting "a white man's war."

Now, the logic of those agreements inevitably cut the other way as well. Obviously only those of a certain identifiable complexion could wage "a white man's war." Believing like Eisenhower and Kennedy that their views would remain secret, Bundy and Ball simply did not trouble to hide their historic assumption that the United States of America was *a white man's country*.

Of course that key assumption was normally kept out of sight if not out of mind Hence application of the Magsaysay formula (see p. 401) continued to exert its seductive appeal. In 1965, for instance, John T. McNaughton, who had worked closely with Lansdale as one of his nominal Pentagon superiors, granted that Vietnam was being "lost" but then was heartened by the thought of the secret agent's exemplary exploit: "The situation could change for the better overnight, however. This happened in the Philippines. This is another reason for d—— perseverance" (GPP, III, 683). President Johnson and his successors did persevere in fighting "a white man's war" without appearing to do so, in having Asian proxies, as it were, nail their coonskins on the wall. But in the end the white war managers let even that fig leaf drop as they turned the full force of their frustration and fury against the ungrateful Vietnamese people, North and South.

2

From top Washington officials back in "the world," as the invaders called their homeland, on down to cowboys out in "Indian

Country," the nation-builders tested their Free World teammates and not surprisingly found them unable or unwilling to affirm U.S. patterns of racism. Maybe, speculated Ambassador Maxwell Taylor, their utter lack of any "national tendency toward team play" was "innate" (GPP, III, 668). Maybe, it followed, they were no more fit for self-government than any of the other natives Anglo-Americans had encountered. On May 23, 1966, Taylor's successor also angrily commented on their "internal squabbling":

> It is obviously true that the Vietnamese are not today ready for self-government, and that the French actively tried to unfit them for self-government. One of the implications of the phrase "internal squabbling" is this unfitness. But if we are going to adopt the policy of turning every country that is unfit for self-government over to the communists, there won't be much of the world left.

Simply substitute certain terms, and this literally could have been Senator Henry Cabot Lodge six decades earlier commenting on the Filipinos' "utter unfitness" for self-government. And exactly like his "large policy" grandfather, Ambassador Henry Cabot Lodge was congenitally unprepared to believe natives might know what they wanted or what was best for them: "Some day we may have to decide how much it is worth to deny Viet-Nam to Hanoi and Peking—regardless of what the Vietnamese may think" (GPP, IV, 99–100).

In the award-winning documentary film *Hearts and Minds* (1975), Lodge's military counterpart, General William C. Westmoreland, appeared on the screen declaring authoritatively that "Orientals" placed a lower value on life than "we" do. In *A Soldier Reports* he complained that his filmed remarks "on the Oriental's value of life were used completely out of context, juxtaposed on scenes of Orientals wailing for their dead" (1976; p. 414). But Westmoreland's contempt for "Orientals" pervaded his curious memoir of victory—he even held the South Vietnamese responsible for the final debacle and asserted the United States could still pride itself on never having lost a war. As for "the enemies," they were the dehumanized "wild varments" that had somehow followed the empire across land and sea to this tropical frontier, as the general had made plain in 1967 while explaining why the total number of U.S. forces in the tiny country had to be limited:

> If you crowd in too many termite killers, each using a screwdriver to kill the termites, you risk collapsing the floors or the foundation. In this war we're using screwdrivers to kill termites because it's a

> guerrilla war and we cannot use bigger weapons. We have to get the right balance of termite killers to get rid of the termites without wrecking the house. [FF, p. 460]

Merely the most recent in a long line of innocent pest exterminators, the general mounted massive "search and destroy" operations that eradicated the village culture not of humans called peasants but of enemies called termites. Like Pyle, Westmoreland could not *see*—for all his "body counts"—the dead bodies and the suffering produced by his attitudes and actions. To the end he could not see that "Orientals wailing for their dead" was exactly the right context for his anesthetizing remarks about the value they put on life.

Expressions of fierce resentment at the squabbling, insensitive, backward, dirty, lazy, childish natives were no less characteristic of the middle levels of the U.S. Mission. The Vietnamese peasant was "at the end of the line politically and emotionally," said Embassy Public Affairs Officer John Mecklin, the former *Time-Life* correspondent Lansdale had praised for acting as a conduit for his CIA handouts (GPP, I, 581). In his *Mission in Torment* Mecklin drew on the accumulated racial arrogance of three centuries to depict the illiterate peasant as cut off from "civilization." He had an untrained mind that "atrophies, like the shriveled leg of a polio victim. His vocabulary is limited to a few hundred words. His power of reason, the greatest of nature's gifts to mankind, develops only slightly beyond the level of an American six-year-old, again because it is never trained." And like his friend the secret agent, Mecklin believed this pathetic creature "ridden by superstition, incapable of distinguishing between truth and falsehood, except at the most primitive level" (1965; pp. 74, 76). At what was illogically considered a still more "primitive level," parallels with past natives were drawn still more explicitly. Testifying before a congressional subcommittee in 1971, Hugh Manke, head of the International Voluntary Service, cited the instance of a U.S. adviser in Pleiku who asserted that "the Montagnards have to realize they are expendable" and who "compared the Montagnard problem to the Indian problem and said *we could solve the Montagnard problem like we solved the Indian problem*."

Outside the command structure, at least nominally, the U.S. press corps in Saigon expressed somewhat different attitudes from those directly engaged in fighting, bribing, and manipulating the Vietnamese. This was less true of such old hands as Keyes Beech, for whom the Vietnamese at the last hour were like animals, and Mecklin himself when he still worked for Time Inc.; it was more true of such younger correspondents as David Halberstam of the *New York Times* and Malcolm Browne of the Associated Press. When the accounts of the latter were not sufficiently "positive," Ambassador Frederick E. Nolting enjoined them to "stop looking for the hole in the doughnut," Halberstam recorded, and Admiral Harry Felt abruptly told Browne: "Get on the team." But the critics raised only minor questions and accepted the major premises of the official line that the United States was there to do good to the Vietnamese. Even when they complained against military censorship, they emphasized their indispensable, historic role as loyal auxiliaries. On August 7, 1962, Browne forwarded a list of reporters' complaints to General Paul D. Harkins, along with this reminder: "Like soldiers, we, too, have our traditions, in the light of which this is a strange war. The AP has been covering wars for 100 years, and has had its share of battle casualties. An AP correspondent died with Custer at the Little Big Horn (an ambush remarkably similar to some of the fights here) and there have been many others since." Browne's allusion to the Little Bighorn may not have been exactly encouraging to the U.S. Mission, but it was surely revealing.

That allusion and countless others like it "put the Vietnam War into a definite historical and mythological perspective," as Frances FitzGerald observed in her superb analysis of American Prosperos in *Fire in the Lake:*

> The Americans were once again embarked upon a heroic and (for themselves) almost painless conquest of an inferior race. To the American settlers the defeat of the Indians had seemed not just a nationalist victory, but an achievement made in the name of humanity—the triumph of light over darkness, of good over evil, and of civilization over brutish nature. Quite unconsciously, the American officers and officials used a similar language to describe their war against the NLF. [FF, pp. 491–92]

Since instances abound, consider only a few names of their air and ground operations: "Rolling Thunder"; "Prairie"; "Sam Houston"; "Hickory"; "Daniel Boone"; and even "Crazy Horse," a 1966 operation carried out appropriately by elements of the 1st Cavalry Division (GPP, IV, 280, 292, 412, 447, 453).*

* It was as if Cowboys and Indians were the only game the American invaders knew. Its imagery spread through the Westerns shown in Saigon theaters and in the field. In August 1961 Theodore H. White wrote President Kennedy how discouraging he had found it "to spend a night in a Saigon night-club full of young fellows 20 and 25 dancing and jitterbugging (they are called 'la jeunesse cowboy') while twenty miles away their Communist contemporaries are terrorizing the countryside"—Arthur M. Schlesinger, Jr., *A Thousand Days* (Boston:

Yet where were the targets of such operations with their surefire names? Unobligingly, they "would not stand up and fight" but glided back into their leafy hiding places or into their villages. Spectral enemies, they were at once nowhere and everywhere in their land with its mines and booby traps, rice paddies you sloshed through waist deep, and mountainous jungles you could hardly penetrate; with its heat and dust of the dry season and mud of the wet; with its clouds of flies and mosquitoes and its stench of animal and human dung; and with its dysentery, malaria, and other hazards Army Secretary Stevens had warned would be high for "Caucasian troops." And not only where but *who* was the enemy in all of this? In one of the supposedly unpeopled wildernesses known as "free-fire zones," they were those who were there or at least those who ran when pursued. Once bagged, they were statistics fed into Westmoreland's computer, with their severed ears on occasion tied to the antenna of a troop carrier as trophies and verifications of the body count. Alive in the villages, however, hostile and friendly natives were all one mass of little yellow peasants. The only certain identification came after the fact: "Anything that's dead and isn't white is a VC," joked GIs in Quang Ngai Province.

Quang Ngai was "Indian Country" on the northeast coast of South Vietnam; the village of Son My on the Batangan peninsula was a cluster of hamlets and subhamlets that included My Lai 4. On March 16, 1968, a company from the 11th Brigade of the Americal Division entered My Lai, searched the subhamlet and found three weapons, and destroyed at least 347 old men, women, children, and babies. Then the foot soldiers systematically burned homes and huts with the bodies "lying around like ants." The official report stated that 128 North Vietnamese "enemy soldiers" had been killed in the intense firefight; Westmoreland congratulated the company "for outstanding action." Officers up the chain of command knew at the time that it was slaughter and not a "victory"—they joshed one another about it that evening during

Houghton Mifflin, 1965), p. 544. A few years later Robin Moore described the reaction of a South Vietnamese strike force to the showing of a Western at Nam Luong. Shown against the side of a building through a small projector, the cowboys and Indians appeared tall and thin: "However, the strikers loved the action and identified themselves with it. When the Indians appeared the strikers screamed 'VC,' and when the soldiers or cowboys came to the rescue the Nam Luong irregulars vied with each other in shouting out the number of their own strike-force companies"—*Green Berets* (New York: Crown, 1965), p. 122. And even the Americans' predecessors had played at being "the conquerors of virgin territories," according to Jean Lartéguy's *Yellow Fever* (New York: Dutton, 1965), p. 251. Paul Résengier, the French antagonist of the Quiet American figure in the novel, had the distinction of being called "cowboy"; out in the paddy fields his Mekong commandos had adopted for their own "the song of the American pioneers . . . 'The memory of a gallant pal Scalped by the Indians in the war.' "

the cocktail hour at brigade headquarters. Nonetheless, the officers and men involved, with some help from their friends, kept the massacre secret for the next twenty months. It was revealed only after tenacious effort by an ex-GI named Ronald Ridenhour and after the November 1969 antiwar mobilization in Washington had encouraged the press to relax its self-censorship. In response to the first sensational revelations, the Pentagon appointed Lieutenant General William R. Peers to head up a board of inquiry. When the Peers panel finished its work in March 1970, charges were brought against fourteen officers, including Major General Samuel W. Koster, former commander of the Americal Division and current superintendent of West Point. None was ever convicted of participating in the cover-up. Only First Lieutenant William Laws Calley was ever found guilty of murder at My Lai.

In 1902 Chairman Henry Cabot Lodge of the Senate Committee on the Philippines had headed off calls for the appointment of a blue-ribbon committee by holding closed hearings of his own on atrocities in the islands (see p. 315). In 1969 the response in Congress was identical, except that this time calls for an independent inquiry were bottled up in the other branch. Chairman L. Mendell Rivers of the House Armed Services Committee directed an investigating subcommittee headed by F. Edward Hébert to hold closed hearings on My Lai. On July 15, 1970, Hébert released a heavily censored report that concluded reassuringly:

> What obviously happened at My Lai was wrong. It was contrary to the Geneva Conventions, the Rules of Engagement, and the MACV [Military Assistance Command, Vietnam] Directives. In fact, it was so wrong and so foreign to the normal character and actions of our military forces as to immediately raise a question as to the legal sanity at the time of those men involved.

My Lai was obviously wrong, but so was the subcommittee's conclusion—hence obviously wrong about the current searching and destroying in Indochina and hence obviously wrong about the "normal character and actions" of U.S. military forces. Massacres were at least as American as the timeworn assertions that they were foreign abnormalities.

Over at the Pentagon the Peers panel soon had incontrovertible proof that My Lai was not an isolated incident. On the very day of the butchery there, another company from the same task force entered the sister subhamlet My Khe 4 with one of its machine-gunners "firing his weapon from the hip, cowboy-movie style." In this "other massacre," members of this separate company piled

My Lai Victims, from *Life* Magazine, December 5, 1969. Combat Photographer Ron Haeberle explained this snapshot: "Guys were about to shoot these people. I yelled, 'Hold it,' and shot my picture. As I walked away, I heard M16s open up. From the corner of my eye I saw bodies falling, but I didn't turn to look." (By permission of Ron Haeberle, *Life* Magazine, © 1969 Time Inc.)

up a body count of perhaps a hundred peasants—My Khe was smaller than My Lai—"just flattened that village" by dynamite and fire, and then threw a few handfuls of straw on the corpses. The next morning this company moved on down the Batangan peninsula by the South China Sea, burning every hamlet they came to, killing water buffalo, pigs, chickens, and ducks, and destroying crops. And as one of the My Khe veterans said later, "what we were doing was being done all over." Said another: "We were out there having a good time. It was sort of like being in a shooting gallery." Detailed evidence of this forgotten massacre remained a closely guarded secret until Seymour M. Hersh obtained a set of the forty or so volumes of the Peers report and summarized them in *Cover-Up* (1972), the source of the above data and quotations.

Presumed to be investigating a cover-up of My Lai, the Peers panel had added one of its own by suppressing evidence of another massacre. No one was tried for murder at My Khe. Up and down the chain of command the routine response to allegations of war crimes, in these instances and others, was to elevate to an operative principle the old Army motto, "Cover your ass." But that had also been the tactic if not the language, you will recall, of Secretary of War Elihu Root in handling charges of "marked severities" in the Philippines (see p. 521). Official and military attitudes toward native victims of U.S. forces had changed glacially if at all over the intervening decades. Indeed, in 1902 General Jacob H. Smith had been court-martialed for his order to kill all the males on Samar over ten years of age. In 1971 General Samuel W. Koster had the pleasure of learning that the charges against him had been dropped, though his men observed no such age limitation in the slaughter on the Batangan peninsula and though Army Secretary Stanley Resor admitted "a great deal of information suggesting that a possible tragedy of serious proportions had occurred at My Lai was either known directly to General Koster, or was readily available in the operational logs and other records of the division" (*New York Times*, May 20, 1971). Otherwise, the parallels were exact. Like General Smith, General Koster was let off with a reprimand and forced into early retirement. Native lives still came cheap.

In fact, the cheapness of native lives gave rise to a rule that had real, if unofficial, standing in military justice. For ordering or encouraging his sergeant to shoot an unarmed prisoner, for instance, First Lieutenant James B. Duffy was found guilty of premeditated murder and then had his conviction reduced to involuntary manslaughter by a military court: "The court didn't want to make Duffy suffer that badly to get the Army off the hook," said a young law officer quoted by the *New York Times*, March 31, 1970. "To a lot of us it looks like another example of the M.G.R.—the mere gook rule—being applied." As reported the following day, Duffy's sentence indicated that one Vietnamese peasant cost six months in the stockade and forfeiture of $150 in pay, but not dismissal from the service—Duffy's sentence was virtually identical with that imposed in 1902 on another first lieutenant for precisely the same crime in the Philippines (see p. 522). Though the going price of a "goo-goo" in the islands and of a "slope," "dink," or "gook" in Indochina may not have fluctuated much over the century, the MGR systematized white assessments of natives as less than full human beings. The army's informal MGR was very nearly formalized by its original charge that Calley had murdered at least 102 "Oriental human beings." How many "Oriental human beings" were equal to one real human being?

In his useful *Nuremburg and Vietnam* (1970), General Telford Taylor, a professor of international law and former chief U.S. prosecutor of Nazis at the original war-crime trials, considered the disturbing implications of My Lai and tried to place the massacre in its proper setting: "As in the Philippines 65 years ago," he noted, U.S. forces were thousands of miles from home in dangerous and unfamiliar terrain that "lends itself to clandestine operations in which women and children frequently participate, the hostile and the friendly do not label themselves as such, and individuals of the yellow race are hard for our soldiers to identify" (p. 152). Hard for, say, a Japanese–American leatherneck from Los Angeles?* Natives—whether Jefferson's "merciless Indian Savages" in the Alleghenies two hundred years ago, the volunteer's "goo-goos" in the Philippines sixty-five years ago, or the GI's "gooks," the army's "Oriental human beings," and Taylor's "individuals of the yellow race" in Indochina more recently—looked alike only to those conditioned not to look at them closely or, sadly, to treat them respectfully. Sorely troubled by the onslaught against the Vietnamese people, and to his credit publicly wondering whether Westmoreland and the other war managers should

* See the testimony of Scott Shimabakuro, who served with the Marine Corps in Vietnam, *Congressional Record*, CXVII (April 6, 1971), 2851–52; see also the testimony of Evan Haney, a Native American from Oklahoma currently living on Alcatraz in San Francisco Bay, ibid., 2890. Shimabakuro's sergeant told him he should not marry his Vietnamese girl friend "because she was a gook, which struck me as kind of funny because I was a gook also. But besides this, he said, 'She's not civilized, you know. You'll be so embarrassed with her when you get her to the United States that you'll want to get rid of her because she's a savage. She doesn't know how to live like a human.' "

not also be tried for war crimes, Taylor still revealed that he shared the conditioning that helped produce the terrible phenomena he was pondering.

But were the soldiers at My Lai misfits, as the Pentagon claimed, or berserk, as Hébert's subcommittee suggested? In his trial testimony and in his tape-recorded memoir, devious and self-serving as both were, Lieutenant Calley revealed himself to be hypnotizingly ordinary, a "run-of-the-mill average guy," as he himself put it. At five feet three inches in height, Calley was an inch shorter than the late diminutive Diem but nevertheless still felt like other Westerners in Vietnam: "I felt superior there," he recalled. "I thought, *I'm the big American from across the sea. I'll sock it to these people here.*" But why was he there? Like Lansdale and the others who had taken the empire there in the first place, Calley "really felt, *I belong here.* It may seem ridiculous saying this. Why in the world would a guy just commit himself to South Vietnam? Well, why would a guy commit himself to South Dakota? . . . *I'm an American officer and I belong in South Vietnam.*" And what had he done at My Lai? "We weren't in My lai to kill human beings, really. We were there to kill *ideology* that is carried by—I don't know. Pawns. Blobs. Pieces of flesh, and I wasn't in Mylai to destroy intelligent men. I was there to destroy an intangible idea." Such leprous innocence qualified "Rusty" Calley as a scaled-down Quiet American, the comic-book counterpart of all the Alden Pyles who could not see the wounds they inflicted on human bodies.

The infantrymen in Calley's Charlie Company were also everyday "grunts." Like the other soldiers in the My Khe shooting gallery nearby, they lived out their Indian-hating fantasies with relish. In Joseph Strick's *Interviews with My Lai Veterans* (the 1971 Academy-Award-winning documentary), Vernado Simpson, Jr., a black antipoverty worker in Jackson, Mississippi; James Bergthold, a white part-time truck driver in Niagara Falls, New York; and Gary Garfolo, the unemployed white son of a barber in Stockton, California, reminisced about their former messmates:

> Simpson: They would mutilate the bodies and everything. They would hang 'em . . . or scalp 'em. They enjoyed it, they really enjoyed it. Cut their throats.
>
> Bergthold: They cut ears offa guy, and stuff like this here, without knowing if they were VC or not. If they got an ear they got a VC.
>
> Garfolo: Like scalps, you know, like from Indians. Some people were on an Indian trip over there.

Elsewhere a My Lai veteran equated "wiping the whole place out" with what he called "the Indian idea . . . the only good gook is a dead gook." The Indian idea was in the air of Vietnam. Specialist 4 James Farmer, a soldier from a different battalion, put it this way: "the only good dink is a dead dink." And from within the Marine Corps came this echo: "the troops think that they're all fucking savages."

"We all belong to the unit Lieutenant Calley belonged to," said one of the hundred honorably discharged Vietnam Veterans against the War at their Winter Soldier Investigation in Detroit (January 31–February 2, 1971). When Senator Mark O. Hatfield read their testimony into the *Congressional Record* (CR) of April 6 and 7, 1971, the westward march of atrocities was demonstrated beyond doubt for anyone who would look at the evidence (CR, CXVII, 2825–900, 2903–36). Had they been willing to place that evidence beside the Lodge committee's 1902 *Hearings before the Senate Committee on the Philippine Islands,* Congressman Hébert and his subcommittee would have had compelling proof of "the normal character and actions of our military forces" when mobilized against those seen as natives. As a sole example, consider the testimony of former foot soldier Robert A. Kruch, who had been told by his colonel in mid-1969 that he did not "want anymore of your —— prisoners. I want a body count." Kruch said that after that he and his buddies had been told to, "well, we were told we were in a free fire zone and anybody we saw that was over 12 years of age that we thought was a male, was to be considered the enemy and engaged as such" (CR, CXVII, 2928). If we lower the age of the male targets by a couple of years, Kruch's colonel might well have been hell-roaring General Smith turning Samar into a howling wilderness sixty-seven years earlier (see p. 328).

Need I pause here to acknowledge the obvious—the scarcely comforting truth that such atrocities were not invented by Anglo-Americans? One thinks of the Australians in Tasmania, the Germans at Lidice, the British over Dresden, the Russians in the Ukraine, the French in Algeria and Indochina, and so on, ad nauseam. But the massacres at My Lai and all the forgotten My Khes in Vietnam had a basic continuity with those of Moros on Jolo and of Filipinos on Samar at the turn of the century, and of Native Americans on the mainland earlier—all the Wounded Knees, Sand Creeks, and Bad Axes. That linkage of atrocities over time and space reveals underlying themes and fundamental patterns of the national history that lawmakers, generals, and so many of their compatriots were eager to forget.

"Thus we spent the day," recorded John Underhill in the 1630s, "burning and spoiling the country." In the 1960s the "Zippo" cigarette lighters of servicemen were handy for burning Vietnamese "hootches"—one company of leathernecks made a name for themselves as the "Burning Fifth Marines," with a "zippo inspection" routine and the habit of burning at least half the villages they passed through (CR, CXVII, 2883). Better yet for burning was "a tank-mounted flamethrower, nicknamed 'zippo' by soldiers," shown in an Associated Press Wirephoto of August 8, 1970, as it fired "a stream of napalm at foliage . . . to help clear possible enemy hiding places." An even more awesome spoiler was the air force's seven-and-a-half-ton "command vault" bomb that completely flattened an area the size of two football fields and killed everything within a square mile—people, monkeys, water buffalo, pheasants, everything. And the feared night, always the hiding place and friend of the hunted, was to be turned into day by electronic battlefields with sophisticated gadgetry for detecting enemy movements—the high-flying klieg light of Aerosystems International, Inc., that was planned to provide a "battlefield illumination system," and the air force contract awarded to Westinghouse and the Boeing Company to study the feasibility of a mammoth space mirror to be blasted into permanent synchronous orbit 22,000 miles above Indochina to brighten this Farther West's darkest hours. Even on the level of the foot soldier, there was no comparison between the firepower of Underhill's day and that of a GI with his M-16 capable of firing ten bullets a second: "Terrified and furious teenagers by the tens of thousands have only to twitch their index fingers and what was a quiet village is suddenly a slaughterhouse," observed William Barry Gault in his thoughtful "Remarks on Slaughter," *American Journal of Psychiatry*, CXXVIII (October 1971), 450–54.

Yet these spectacular advances in the technology of destruction merely embedded in modern hardware the traditional will to burn and spoil. With the will to destroy native peoples who stood between them and the West, the Saints and their descendants always found a way. Native Americans such as Navaho Ben Muneta have perceived this continuity and readily remarked on parallels between the destruction of ecosystems on the mainland, in the Philippines, and more recently in Indochina, while Anglo-Americans have not been quick to see the same elementary phenomena. An exception was E. W. Pfeiffer of the University of Montana, a zoologist who was sent to Indochina by the American Association for the Advancement of Science and who returned to tell a session of his colleagues about the havoc wrought by the "command vault'' bomb, the tens of millions of craters created by other bombs, and the meaningful context of all this desolation: ''The present-day U.S. policy of massive defoliation, crop destruction, bombing and plowing [with giant bulldozers] of Indochina can be viewed as a modern counterpart to the extermination of the bison in the American West,'' he perceived. ''This modern program . . . has as destructive an influence on the social fabric of Indochinese life as did the ecocide (destruction of ecology) of the American West upon the American Indian'' (*Washington Post*, December 28, 1971).

A century before the ecocide in Indochina, Kit Carson efficiently used the means at hand to devastate the land of Ben Muneta's Navaho people, ''burning homes, farms, orchards, livestock, food stores'' (*Akwesasne Notes*, Early Summer 1974, p. 26). Carson's Long Knives were forerunners of the Burning Fifth Marines, and so were General Winfield Scott Hancock's cavalrymen, who lacked zippo lighters and flamethrowers but nonetheless made do with flaming torches to burn the Cheyenne village near Fort Larned at about the same time—whether burning hogans or tepees or hootches, all three sets of arsonists had common attitudes and assumptions that made their means relatively inconsequential. Moreover, not all their means had undergone spectacular change. Throughout, index fingers twitched triggers and hands held torches or lighters. In 1967 Herman Kahn of the Hudson Institute recommended substituting German police dogs for Vietnamese soldiers on night patrol (FF, p. 483)—in his memoir Westmoreland even came up with a canine body count of 36 KIA and 153 wounded. But the ancestors of the K-9 Corps were the bloodhounds used in the Second Seminole War (see p. 201), the fierce best friends of frontier whites commemorated in Horatio Greenough's *Rescue Group* (see p. 121), and in fact the mastiffs early English colonizers brought over to terrorize natives—my guess is that none of the scout dogs in Vietnam was the peer of Dugdale, the Indian-killing hound immortalized by William Gilmore Simms's *The Yemassee* (see p. 142).

And after all, ''wiping the whole place out,'' that ''Indian idea'' invoked by the My Lai veteran, had its paradigmatic expression when the technology of death was relatively backward. Yet in less than an hour, with firebrand and powder, musket and sword, the Saints leveled the Pequot fort and piled up a body count at Mystic in 1637 that equalled or exceeded that at My Lai in 1968. As John Underhill said at the time: ''We had sufficient light from the word of God for our proceedings.''

3

Four centuries after Columbus, Frederick Jackson Turner had pronounced the continental frontier "gone, and with its going has closed the first period of American history." Three-quarters of a century later the second period closed, with the westward course of empire rolling back from the Pacific rim: "I saw that we had overreached ourselves," said *Look* foreign editor J. Robert Moskin on his return from a trip to Vietnam. "America's historic westward-driving wave has crested" (November 18, 1969). That wave has carried us along, and with its ebbing we can conveniently haul up here and bring to an end our own long passage to Indochina. But before turning about and facing home again, we had better cast an apprehensive glance over our shoulder. What Turner said of the second period applies with a vengeance to the third period of American history: "He would be a rash prophet who should assert that the expansive character of American life has now entirely ceased."

In the hauntingly evocative lines of his famous paper "The Significance of the Frontier in American History," presented at the Columbian World's Fair at Chicago in 1893, Turner had made that expansive character the key to the origins of American democracy: "The existence of an area of free land, its continuous recession, and the advance of American settlement westward, explain American development." Like James Hall before him, Turner had a single topic, the West; he also spent a lifetime ringing the changes on the same basic theme. In 1896 he returned to his theme with "The Problem of the West" and repeated that expansion of the frontier was the key factor in the growth of the American economy and society: "For nearly three centuries the dominant fact in American life has been expansion," he observed in words Theodore Roosevelt was to adopt as a refrain in the campaign of 1900. Though expansion had been checked at the Pacific coast, Turner wrote, the current call for a vigorous foreign policy, an interoceanic canal, and enhanced sea power, "and for the extension of American influence to outlying islands and adjoining countries, are indications the movement will continue." The historian might have been laying out the course of our own odyssey—remember that Homer's hero also sailed westward—to outlying islands and adjoining countries. Turner surely anticipated the gist of countless expressions of U.S. foreign policy, from John Hay's Open Door Notes to and beyond Richard M. Nixon's 1967 statement on Indochina: "Both our interests and our ideals propel us westward across the Pacific." Despite his geographical determinism, gross exaggeration of the importance of "free" land, anti-intellectualism, and other rough edges that kept his frontier hypothesis from being the master key to American history, Turner saw clearly the direction the empire was headed in and even foresaw the inner identities of the New Frontier and the old long before John F. Kennedy so christened the former.

The enormously influential historian spoke so authoritatively and so prophetically because he was even closer than John Fiske to the heart of collective desires and could speak more directly from the heartland of conventional wisdom built up by three centuries of expansion. While invoking the "perennial rebirth" of American life on the ever-advancing frontier, Turner partook in the perennial revoicing of the national creation myth. In flight from Europe and "from the bondage of the past," his white settlers had conquered a wilderness continent in West after West and in transit had conquered an identity for themselves as a people, with their dominant individualism, restless energy, practicality, and Davy Crockett-like "exuberance which comes with freedom." But now, with the passing of the "free" land that had made all this possible, the problem was to find a substitute somewhere else —in a "remoter West," in education, in science. But this formulation of "the problem of the West" was itself symptomatic. It made dispossession of Native Americans—who were "the rightful owners of the country," said Cooper's Leatherstocking—the defining and enabling experience of the republic. From it flowed ineluctably the inference that the European immigrants had built their economy of possessive individualism and their relatively open polity of Lockean liberalism (popular sovereignty, representative government) on the red man's unfree grave and had fenced their gardens with his bones. That this inference did not bother Turner and his compatriots was why the West had become such a problem in the first place.

The generic frontier was the magic margin of American history. It was, said Turner, "the meeting point between savagery and civilization." Behind that moving "line of most rapid and effective Americanization" were the people. And *who* were "the American people"? They were the immigrants who planted Jamestown on Chesapeake Bay and built New Israel on Massachusetts Bay. They were their descendants who built the great republic of the West —Jefferson and the other Enlightenment fathers who defined themselves as opposites of the savages indicted by their Declaration and founded a middle-class European-derived state, a "fragment culture" in Louis Hartz's terms, under which Native Americans were natives and not Americans. Proper Americans were not necessarily of English stock—Turner went out of his way to insist that they were "of a composite nationality," cited the Scottish–Irish and the Germans as illustrations, and warned against "misinterpreting the fact that there is a common English speech in America into a belief that the stock is also English." Nevertheless, as his examples and melting-pot metaphor made unmistakable, the *sine qua non,* the irreducible prerequisite of being an American, was to be of *European* stock.

Each frontier, said Turner, "was won by a series of Indian wars." Beyond each of those fundamental boundaries of American life was what had to be conquered: "hostile Indians and the stubborn wilderness." Native Americans were strictly equivalent to Jefferson's Stony mountains that had to be overcome. In Turner's pages they subsisted as natural objects, extensions of the nature that had to be subdued, impediments that had to be removed from the irresistible march to the West. Their "primitive Indian life" had to be destroyed—their "savagery" made to disintegrate—so each territory along the way could progress onto the next "higher stage." The most they could do was to be formidable obstacles, thereby making the white settlers' line of advance the sort of anticipation of the John F. Kennedy Special Warfare School that Turner proudly pronounced it: "a military training school, keeping alive the power of resistance to aggression [*sic*], and developing the stalwart and rugged qualities of the frontiersman."

"The great backwoodsman" Daniel Boone was a case in point. Boone was pure empire-builder and hardly some sort of white-Indian in Turner's hands, as the historian traced his route from Pennsylvania to the upper Yadkin, to the Red River country of Kentucky, and beyond to Missouri; then followed his son, who became a mountain man in a still farther West; and ended with "his grandson, Col. A. J. Boone, of Colorado, [who] was a power among the Indians of the Rocky Mountains, and was appointed an agent by the Government. Kit Carson's mother was a Boone. Thus this family epitomizes the backwoodsman's advance across the continent." Impatient of restraints, scornful of "older society," and antisocial in tendency, Turner's stalwart Boones were inland Crusoes who cut their way through Indians to ever more remote wilderness islands and repeated on their individual level the collective flight from Europe, from the multilayered historical process thought of as "the bondage of the past," and from kith and kin. Turner's epic of migrations was the history of social escape.

Without the avid rancor of genial James Hall, Turner added his monumental chapter to the national metaphysics of Indian-hating. At bottom, that doctrinal hate rested on the collective refusal to conceive of Native Americans as persons, a refusal Turner shared in full measure. No less than the judge did he glorify Indian-killers as pathfinders of "civilization," glorify their mastery over every dusky tribe, and throw sheaves of patriotic rhetoric over the real human bodies left behind.

In truth, Turner's "meeting point between savagery and civilization" put Hall's inept formulations into a more sophisticated or "civilized" framework. His magic margin was an imperishably vivid expression of the color line that has whipped so tragically through American life. On one side were the Children of Light, the light of the Gospel, of Enlightenment institutions, law and order, progress, philanthropy, freedom, Americanization, modernization, forced urbanization, the lot. On the other side were the Children of Darkness, "savages" who stood in the way of the redemption and the rationalization of the world—from the Puritan's fiends to be exorcised, to Turner's prime exhibits of "savagery," and on to Lansdale's "Special Targets"; from those living in "the darkness of heathenism" or "the gloom of ignorance," down to those carrying on "the old nightmare of savagery." No doubt all peoples dehumanize their enemies to varying extents, but the Children of Light had a head start with those who seemed to them from the outset ferocious animals ("ravening wolves"), the color of evil, dark reminders of the wilderness they had set out to conquer in themselves and in the world; in short, Native Americans were to the Puritans and their descendants unwelcome mementos of their own mortality. Turner's color line was the supreme expression by a historian of all the other expressions before and since by novelists, poets, playwrights, pulp writers, painters, sculptors, and film directors. It separated the cowboys from the Indians by making the latter easily recognizable dark targets, especially if they had war paint on to boot. It unmistakably shaped national patterns of violence by establishing *whom* one could kill under propitious circumstances and thereby represented a prime source of the American way of inflicting death.

Herein resided the deeper significance of the frontier. In each and every West, place itself was infinitely less important—especially to those so alienated from the land—than what the white settlers brought in their heads and hearts to that particular place. At each magic margin, their metaphysics of Indian-hating underwent a seemingly confirmatory "perennial rebirth." Rooted in fears and prejudices buried deep in the Western psyche, their metaphysics became a time-tested doctrine, an ideology, and an integral component of U.S. nationalism. I do not offer this reading of the significance of the frontier to take the place of Turner's as the master key to American life. To my mind, the intricate interrelationships among a society's subterranean emotions, its channeling of these into myths that seemingly give coherence to the past and bearing for the future, and its institutional means of making that destiny manifest—all these complex and reciprocal relationships foredoom any such attempt. I do suggest these metaphysics helped shape (and were shaped by) political, social, and economic structures; provided substance for fundamental declarations of doctrine, law, and policy; established core themes in literature, sculpture, painting, and film; and helped determine how individuals perceived their world and acted in it. It gave them their astonishing assurance, for instance, that they had a right to be in every West they could "win"—from Boone and Crockett in Kentucky and Tennessee, Custer and the 7th Cavalry in the Black Hills, Worcester and later Lansdale in the Philippines, and on down to the secret agent and Lodge and Calley in Indochina. (Said the ambassador: "Some day we may have to decide how much it is worth to deny Viet-Nam to Hanoi and Peking—regardless of what the Vietnamese may think." Said the lieutenant: *"I'm an American officer and I belong in South Vietnam."* Why South Vietnam? Well, why South Dakota?)

All along, the obverse of Indian-hating had been the metaphysics of empire-building—the backwoods "captain in the vanguard of conquering civilization" merely became the overseas outrider of the same empire. Far out on the boundless watery prairies of the Pacific, the twin metaphysics became nation-building and native-hating. But was there no terminus, no ultimately remote West beyond which the metaphysics lost their power to uproot and destroy? Not if one accepted Turner's definition in "The Problem of the West": "The West, at bottom, is a form of society, rather than an area." It was that final dimension of the problem that future legions of COIN experts and nation-builders seized upon in their attempt to make the Far East into the Farthest West, as Herman Melville, with his uncanny foresight, had known they would. And in his 1893 essay, Turner had approvingly quoted F. J. Grund, who had predicted in 1836 "that the universal disposition of Americans to emigrate to the western wilderness, in order to enlarge their dominion over inanimate nature" was "destined to go on until a physical barrier must finally obstruct its progress." So it did go on, for some twelve decades, until it finally ran into a people barrier, with the horrifying momentum and consequences we have just witnessed.

The sober truth was that the white man's burden of Winning the West was crushing global folly. The West was quite literally nowhere—or everywhere, which was to say the same thing. For Homer's Greeks and North American tribal peoples alike, the West was the land beyond, Spiritland, the land of mystery, of death and of life eternal. It was not a Dark and Bloody Ground to be "won." But for Anglo-Americans it was exactly that, the latest conquest. Yet how could they conclusively "win" it? If the West was at bottom a form of society, as Turner contended, then on our round earth, Winning the West amounted to no less than winning the world. It could be finally and decisively "won" only by rationalizing (Americanizing, Westernizing, modernizing) the world, and that meant conquering the land beyond, banishing mystery, and negating or extirpating other peoples, so the whole would be subject to the regimented reason of one settlement culture with its professedly self-evident middle-class values.

He would indeed be a rash prophet who should assert that the metaphysics of Indian-hating will be lightly cast aside. The Vietnam veterans came home from the war to the Pentagon Papers, Watergate, the CIA scandals, and all the rest—all of which provided compelling evidence that the methods and instrumentalities designed originally for dealing with natives had beat them home—or rather had never left. "Plausible deniability" had become a way of life or, more precisely, the standard operating procedure of successive administrations for telling lies to the people as though they were natives in a distant colony. As for those "locals" traditionally regarded as natives, W. Mark Felt, J. Edgar Hoover's close associate in the FBI, advocated in 1970 that the agency's activities against insurgent campus groups be stepped up: "These violence-oriented black and white savages are at war with the Government and the American people." Meanwhile the FBI harassed and disrupted native groups (Native American, Afro-American, Chicano, Puerto Rican), promoted violence among them, sought to discredit or destroy black nationalists and civil rights leaders—as in the case of Martin Luther King, Jr.—provided the CIA with intelligence on Indian rights activities, and at one point issued a nationwide alert against the "terrorists" in the American Indian Movement (AIM). Under its Operation Chaos (1967–74), the CIA itself undertook domestic disruption and had already targeted AIM for infiltration and perhaps other activities. And during Wounded Knee 1973, the seventy-one-day occupation by the Oglala Sioux and by AIM of the hamlet in South Dakota where so many of their people had been slaughtered in 1890, the Pentagon made available COIN personnel, techniques, and weapons to help put down this latest Indian uprising. It had always been impossible to keep the empire off in distant colonies.

On the night of April 26, 1973, the following radio transmission of a federal marshal was monitored and was later published in *Voices from Wounded Knee* (1974):

> If we see one of them long-haired hippie dudes jump out of that bunker in front of us with long black hair with pigtails, we got a Search 4 [search party] out there, and if we capture him, first off we're gonna cut off his hair. [p. 206]

As in John Endicott's day of longhairs and shorthairs, jollity and gloom were contending for an empire, and now as then the firepower was all on gloom's side.

Happily, however, there has always been a countertradition to brighten this gloom of the Children of Light, a subterranean tradition fittingly rooted in the emotions driven underground and surfacing periodically as invasions of Merry Mounters, familists, millenarians, anarchists, Flower Children, and others from the remote recesses said to be the lurking places of nymphs and fauns. From the Puritan's "howling wilderness" to and beyond Turner's "stubborn wilderness," this countertradition has perennially been reborn.

Critical support for this countercultural underground came from such Ishmaels as Thomas Morton, Henry David Thoreau, George Catlin, Herman Melville, and Mary Austin. In varying measure, all were able to recognize the "Injun" within, all challenged the metaphysics and thereby found strength to give of themselves to the land and its original indwellers. Even those dancing Indian-clad hippies and yippies in Central Park on May Day 1968, alluded to in the opening chapter of this book, were in their wayward way paying tribute to the first Americans and seeking to come to grips with the land. And for all his angularities and ambivalences, the poet Robert Lowell, alive when this work began and, sadly, dead now as it ends, reexamined the past and brought himself to recognize that his heritage included ancestors willing to fence their gardens with the red man's bones and to make three thousand miles of continent "a marsh of blood!" Without comparable courage in facing the self and transcending the past, Lowell's heirs, all of us, will be condemned to go on Indian-hating and empire-building throughout the third and final period of American history.

But Native Americans have lightened the gloom most of all.

And they have not yet vanished, as those who stood up at Wounded Knee in 1973 reminded the country. Enduring the projections of evil heaped upon them by whites, who in so doing exorcised their own demons, they were still willing to share with us their wisdom and rootedness in the earth, which for them was never a howling wilderness. Before the white man came, said Sioux Chief Standing Bear, the "earth was bountiful and we were surrounded with the blessings of the Great Mystery." "The lands wait," said Vine Deloria, Jr., four decades later, "for those who can discern their rhythms." They still wait. So let us dance a new earth into being, proposed Lame Deer, the Sioux holy man: "now not only the Indians but everybody has become an 'endangered species.' So let the Indians help you bring on a new earth without pollution or war." By word and example, Native Americans have been reminding Anglo-Americans of their lack of respect for all living things, of their lost communal sanity, and lost wholeness —of how not to see with the eyes and mind only but, as Lame Deer put it, also with "the eyes of the heart." With their help, Americans of all colors might just conceivably dance into being a really new period in their history.

Notes and Bibliographical Essay

Chapter I
The Maypole of Merry Mount

Following the lead of John Endecott's descendants, the family name was changed to Endicott in 1724, as it appears in Nathaniel Hawthorne's short story and in these pages. For *Endecott and the Red Cross*, see the trilogy *The Old Glory* (New York: Farrar, Straus, & Giroux, 1965). Just before opening night Lowell explained to Richard Gilman that the play had to do with basic American experience, more particularly with "the tensions and antinomies rising from the hidden painfulness of our origins, the contradictions of our freedom and self-definition, the losses that all aggressive gains entail"—*New York Times*, May 5, 1968. Michael Kammen quoted Lowell and celebrated him for this "movingly bifarious view of the North American experience" and for setting Morton and Endicott "as archetypal American rivals, libertine and Puritan, Dionysiac and Apollonian, the establishmentarian and the nonconformist"—*People of Paradox* (New York: Vintage, 1973), p. 111. Largely absent from this enthusiasm for antinomies and paradoxes was the dimension of power: who coerced, deported, or exterminated whom under the colonial theocracy and later nation state? Or more directly, "our" origins, "our" freedom? *Whose?* The painfulness of "our" origins was not in fact hidden from Morton or the Pequots.

"The Maypole of Merry Mount" appeared in the *Token* for 1836 before it was included in the first series of *Twice-Told Tales* in 1837. Of all Hawthorne's many critics, Frederick C. Crews has seen most clearly the symbolic significance of his whipping-post metaphor—see Crews's *Sins of the Fathers* (New York: Oxford

University Press, 1966). On the historical dimension, Hawthorne set his forest drama of jollity versus gloom within two incompatible views of nature and wildness, with one going back through these children of Pan to the Greek classics and the other through the Saints to the Old Testament Hebrews—for a remarkable essay that traces these benign and malign traditions into our period, see Hayden White, "The Forms of Wildness: Archaeology of an Idea," in *The Wild Man Within: An Image in Western Thought from the Renaissance to Romanticism*, ed. Edward Dudley and Maximillian E. Novak (Pittsburgh: University of Pittsburgh Press, 1972), pp. 3–38.

Chapter II
Thomas Morton

Thomas Morton's *New English Canaan or New Canaan* was originally printed in Amsterdam by Jacob Frederick Stam in 1637. The 1883 edition of Charles Francis Adams, Jr., that appeared in *Publications of the Prince Society*, XIV, is invaluable for both biographical and textual data, on which I have drawn freely, and for the inadvertent evidence the introduction and notes provide of the continuity of Indian-hating: Editor Adams's view of the "savages" was of a piece with his great-grandfather's and with Bradford's. The defensive rumor, circulated by Governors Thomas Dudley and Bradford, among others, that Morton was under "foul suspicion of murther" in England was tracked down and put to rest by Charles E. Banks, "Thomas Morton of Merry Mount," *Proceedings of the Massachusetts Historical Society*, LVIII (December 1924), 147–93. Samuel Maverick, like Morton an "old planter" who knew the land well and was under no ideological compulsion to regard it as a desolate waste, held that the *New English Canaan* "indeed was the truest discription of New England as then it was that euer I saw"—"A Briefe Description of New England and the Severall Towns Therein" (1660), *New England Historical and Genealogical Record*, XXXIX (1885), 40. On the "crime" of selling arms to the Indians, see Neal Salisbury's "Red Puritans: The 'Praying Indians' of Massachusetts Bay and John Eliot," *William and Mary Quarterly*, XXXI (January 1974), 41, and also his "Conquest of the 'Savage': Puritans, Puritan Missionaries, and Indians, 1620–1680," Ph.D. diss., University of California, Los Angeles, 1972. As early as 1650 Eliot advocated arming the Christian Indians so they could defend themselves against "savages"; in the 1660s the commissioners of the United Colonies actually allocated missionary funds for that purpose. So long as they enlisted on the side of "civilization" in its struggle against "savagery," Indians might have arms after all.

Chapter III
John Endicott

"Filial" in the strictest sense was Lawrence Shaw Mayo's account of "our Puritan ancestors" in *John Endecott: A Biography* (Cambridge: Harvard University Press, 1936); equally so was Stephen Salisbury's *Antiquarian Papers: Memorial of Gov. John Endecott* (Worcester, Mass.: Chas. Hamilton, 1879).

Illuminating for Endicott's attitude toward women is Charles Francis Adams, Jr., ed., *Antinomianism in the Colony of Massachusetts Bay, 1636–1638* (Boston: The Prince Society, 1894), especially Governor Hutchinson's appendix, "The Examination of Mrs. Anne Hutchinson at the court at Newtown," pp. 235–84, her "trial" in the Boston church, pp. 285–336, and of course John Winthrop's "Short Story of the Rise, Reign, and Ruine of the *Antinomians, Familists, and Libertines,*" pp. 67–233.

William Bradford's "Wickedness Breaks Forth" was dated 1642—*Of Plymouth Plantation*, ed. Samuel Eliot Morison (New York: Alfred A. Knopf, 1952), pp. 316–23. For Freud's discussion of the barriers without which sexual instinct "would break all bounds and the laboriously erected structure of civilization would be swept away," see *A General Introduction to Psychoanalysis* (New York: Permabooks, 1955), p. 321. His dam metaphor appeared in *Three Essays on the Theory of Sexuality* (1905): "One gets an impression from civilized children that the construction of these dams is a product of education" (*Standard Edition*, ed. James Strachey [London: Hogarth Press, 1953–74], VII, 177–78). For some of the consequences of all these repressions, see *Civilization and Its Discontents* (New York: W. W. Norton, 1962); also Herbert Marcuse, *Eros and Civilization* (Boston: Beacon Press, 1955).

Winthrop's remarkable "Relation of His Religious Experience," dated January 1637, is in *Winthrop Papers* (Boston: Massachusetts Historical Society, 1943), III, 338–44. Other of his writings that I have quoted or cited are easily located by date in the *Papers* or in *Winthrop's Journal*, ed. James Kendall Hosmer (New York: Charles Scribner's Sons, 1908), 2 vols. For the motives that led him to emigrate, see Edmund S. Morgan, *The Puritan Dilemma: The Story of John Winthrop* (Boston: Little, Brown, 1958). For his English counterpart, see Thomas Carlyle, ed., *Oliver Cromwell's Letters and Speeches* (New York: Wiley & Putnam, 1845), 2 vols.

Conversations with Mark Neuman, colleague and English historian, and the arguments of Neal Salisbury in an unpublished conference paper, "Inside Out: Perception and Projection in the Puritans' Encounter with Indians," may have helped me avoid what the latter laments as the general tendency of American scholars "to play down the settlers' backgrounds, emphasizing [instead] that new psychological responses were generated by the perception of unfamiliar people in unfamiliar settings." The Saints did indeed arrive in New England with a fully articulated ideology and with particular origins in the English class structure. Still the best single statement of what that meant is Max Weber's astonishingly durable *Protestant Ethic and the Spirit of Capitalism*, trans. Talcott Parsons (New York: Charles Scribner's Sons, 1958—orig. 1904–5). Weber saw that Calvinism's "tendency to form a community worked itself out very largely in the world outside the Church organizations ordained by God. Here the belief that the Christian proved . . . his state of grace by action *in majorem Dei gloriam* was decisive, and the sharp condemnation of idolatry of the flesh and of all dependence on personal relations to other men was bound unperceived to direct this energy into the field of objective (impersonal) activity" (p. 224n). Here the backed-up or surplus energy, libidinal energy in Freudian terms, took the route Bradford thoroughly approved of instead of the one that so horrified him. And thus redirected, it helped the Saints gain their reputation as coldly self-righteous colonizers, though it broke out of this new channel on occasion in flash floods of killings and hurtings. But within the channel of impersonal activity, this energy flowed toward modern capitalism, toward our mechanized, bureaucratized violence, and toward the concurrent exaltation of the natural sciences—Robert K. Merton explored the origins of this third branch in "Puritanism, Pietism and Science," *Social Theory and Social Structure* (Glencoe, Ill.: Free Press, 1949), pp. 329–46. See also Michael Walzer's *Revolution of the Saints* (New York: Atheneum, 1968). For Christopher Hill's work, see *The Century of Revolution* (Edinburgh: Thomas Nelson, 1961); *Society and Puritanism in Pre-Revolutionary England* (New York: Schocken Books, 1964); and the breathtaking *World Turned Upside Down: Radical Ideas during the English Revolution* (New York: Viking Press, 1972), on which I have drawn freely. My understanding of what it meant for the colonies to be established by a middle-class fragment from English society has also benefited from Louis

Hartz's *Founding of New Societies* (New York: Harcourt, Brace & World, 1964).

Chapter IV
The Pequot War

Crucial for any understanding of the war are the contemporary narratives of John Mason, John Underhill, Philip Vincent, and Lion Gardiner, published in the *Collections* of the Massachusetts Historical Society and conveniently assembled by Charles Orr, ed., *History of the Pequot War* (Cleveland: Hellman-Taylor Company, 1897). Captain Edward Johnson's account was virtually contemporary, since it was originally published in 1653; see J. Franklin Jameson, ed., *Johnson's Wonder-Working Providence, 1628–1651* (New York: Charles Scribner's Sons, 1910). Published originally in 1677 but useful for the continuity of Indian-hating is William Hubbard's *The History of the Indian Wars in New England*, II, ed. Samuel G. Drake (Roxbury, Mass.: W. Elliot Woodward, 1865).

Elias Canetti's fine discussion of "first victims" is in *Crowds and Power*, trans. Carol Stewart (New York: Viking Press, 1963), p. 138; "survival as a passion" has obvious relevance to the Puritan sense of elation at seeing all the dead bodies at the Pequot fort (pp. 227, 392). Canetti also dealt with the way rulers manipulate fear of "hostile packs" (p. 457), and as if in illustration of his analysis, Herbert Milton Sylvester in *Indian Wars of New England* (Boston: W. B. Clarke Company, 1910), I, 294, quoted Hubbard on "those savage wolves" and himself referred to the Pequots as a "pack of hungry wolves."

Useful for placing the tribe in its proper context is T. J. C. Brasser, "The Coastal Algonkians: People of the First Frontiers," in *North American Indians in Historical Perspective*, ed. Eleanor Burke Leacock and Nancy Oestreich Lurie (New York: Random House, 1971). The traditional account of Pequot origins I outline in the text has been questioned recently by scholars who believe *Pequot* and *Mohegan* to have been interchangeable terms and doubt they were a branch of the Hudson River Mahicans—see Bert Salwen, "A Tentative 'In Situ' Solution to the Mohegan-Pequot Problem," in *An Introduction to the Archaeology and History of the Connecticut Valley Indian*, ed. William R. Young (Springfield [Mass.] Museum of Science, 1969), I, 81–88. The late Perry Miller contended, I think wrongly, that Roger Williams was the only colonist in New England who could and did treat Indian culture with respect—see *Roger Williams: His Contribution to the American Tradition* (New York: Atheneum, 1962). That contention was elaborated upon by Jack L. Davis, "Roger Williams among the Narragansett Indians," *New England Quarterly*, XLIII (1970), 593–604.

Chapter V
The Legacy of the Pequot War

For John Underhill's subsequent career as a professional Indian-killer, see John W. DeForest, *History of the Indians of Connecticut* (Hartford: Wm. Jas. Hamershey, 1853). In *Fathers and Children: Andrew Jackson and the Subjugation of the American Indian* (New York: Alfred A. Knopf, 1975), Michael Paul Rogin described the "primitive accumulation" of Creek and Cherokee lands, a process completed much earlier of course in the Pequot country. In addition, my indebtedness to Roy Harvey Pearce will be clear to anyone who has read *Savages of America: A Study of the Indian and the Idea of Civilization* (Baltimore: Johns Hopkins Press, 1953).

In 1537 Pope Paul III forbade Christians to think of Indians as less than men. A century later Winthrop recorded that "we received a letter at the general court from the magistrates of Connecticut and New Haven and of Aquiday, wherein they declared their dislike of such as would have the Indians rooted out, as being of the cursed race of Ham" (*Journal*, October 1640). Though Winthrop went on to note that the Massachusetts authorities had agreed with the other magistrates, he had himself shortly before, as we have seen, taken as a given that all Indians were "a common enimie." But the point here is that both items more than suggest that Christians had in fact commenced to think of Indians as a separate species or "cursed race"—as Freud said somewhere, there was never a need to prohibit something no one desired to do.

Lewis Hanke discussed Paul's bull in "Pope Paul III and the American Indians," *Harvard Theological Review*, XXX (1937), 65–102. Hanke dealt at length with Las Casas and the debate at Valladolid in *Aristotle and the American Indians* (London: Hollis & Carter, 1959). This important mid-sixteenth century controversy over whether Indians were truly men was largely neglected by Margaret T. Hodgen in *Early Anthropology in the Sixteenth and Seventeenth Centuries* (Philadelphia: University of Pennsylvania

Press, 1964). Her survey, ably researched and most useful, was from my point of view badly misconceived, especially in its acceptance of the conventional wisdom that "racialism in its familiar nineteenth- and twentieth-century sense of the term was all but nonexistent." It is likely that precisely this sense of the term is badly in need of revision. Though published too late for me to draw on, Ronald Sanders's *Lost Tribes and Promised Lands: The Origins of American Racism* (Boston: Little, Brown, 1978) is a welcome step toward that necessary revision. By tracing the roots of American racism back to the fifteenth and sixteenth centuries, Sanders provides evidence from Spain and Portugal that supports arguments I advance here. Carried abroad by the first colonizing powers, racism as a reality predated by centuries racism as a word.

The problem is terribly important and terribly tangled. Part of the tangle stems from the fact that, as Winthrop Jordan has remarked, *racism* "is terribly hard to define"—for the best study of its origins, though the term does not appear once, see Jordan's *White over Black* (Chapel Hill: University of North Carolina Press, 1968). In *Race Relations* (London: Tavistock, 1967), Michael Banton attempted to clear away terminological troubles by distinguishing between *racism*, as the theory, and *racialism*, as the practice. This distinction is enticing, but as I have argued, both were present in the first Anglo-American settlements, though to be sure theory was not there in its fully developed form.

On the relative unimportance of theory, see Otare Mannoni's fascinating *Prospero and Caliban* (New York: Frederick A. Praeger, 1964). (For one of the most untheoretical racists of all time, see George L. Mosse's discussion of Adolf Hitler in *Toward the Final Solution: A History of European Racism* [New York: Howard Fertig, 1978], pp. 202–9.) Philip Mason's *Prospero's Magic: Some Thoughts on Class and Race* (London: Oxford University Press, 1962) probes for origins in the unconscious. The concurrent dispossession and subjugation of the "black" Irish furnish instructive parallels and contrasts: As early as 1700 one writer said the Irish in "Our Western Plantations" were "distinguished by the Ignominious Epithet of White Negroes"—quoted by Edward D. Snyder, "The Wild Irish: A Study of Some English Satires against the Irish, Scots, and Welsh," *Modern Philology*, XVII (April 1920), 147–85. The English assumption of racial superiority over the Irish did not lead to full-blown racism because the Irish were never really "black," never more than "white Negroes." Their ethnocentric assumption owed part of its appeal to the contrast drawn by adventurers during the Tudor and Stuart periods, L. P. Curtis, Jr., maintained in his penetrating *Anglo-Saxons and Celts: A Study of Anti-Irish Prej-*

udice in Victorian England (Conference on British Studies at the University of Bridgeport, Conn., 1968). And though the English were centuries away from colonizing Australia, when they ultimately went down under they acted not unlike the Puritans of Massachusetts Bay, parallels across time and space suggesting not so much the importance of chronology and theory as the permanency of English forms of treating, feeling, and viewing nonwhite peoples—see F. S. Stevens, ed., *Racism: The Australian Experience* (Sydney: Australia and New Zealand Book Company, 1971), 2 vols. In fine, I think the evidence overwhelming that racism long antedated its nineteenth- and twentieth-century formulations but trust the reader will make charitable allowances for my provisional definition. (Relevant to this argument are two essays that appeared after it was written: G. E. Thomas, "Puritans, Indians, and the Concept of Race," *New England Quarterly*, XLVIII [March 1975], 3–27; and Philip L. Berg, "Racism and the Puritan Mind," *Phylon*, XXXVI [March 1975], 1–7.)

In *Puritanism in America: New Culture in a New World* (New York: Viking Press, 1973) Larzer Ziff helpfully related the Puritans' economic and political interests to their needs to have sinners under the lash and to find an outlet for righteousness baffled during the controversy over Anne Hutchinson and the Antinomian heresy; his discussion of their "cold violence" is less persuasive. Edgar Wind's *Pagan Mysteries in the Renaissance* (London: Faber and Faber, 1958) does much to unearth the "hidden misteries" in Morton's *New English Canaan*.

As the Abiezer Coppe quotation suggests, there was virtually a point-to-point correlation between Morton's royalist libertarianism and the radical views of the group (1649–51) contemporaries called Ranters. Both were in fundamental opposition to the body-denying Protestant ethic, of course, and threw Dionysiac orgies with plenty of sack or beer, lascivious songs, and in the words of a Ranter critic, "downright bawdry and dancing." Ranters drew directly on Familism, thus indirectly on German Anabaptism, and as Christopher Hill observed, may or may not have taken some ideas second-hand from Italian Neoplatonism. Morton, I have ventured, probably drew directly on the latter; he evinced values associated with the aristocracy but possibly was influenced as well by the underground, lower-class radicalism from which the Ranters emerged. Though he was dead and there is no evidence Ranters knew of his *New English Canaan*, Hill's discussion of the political links between Ranters and Royalists in the 1650s underlines the affinities (*The World Turned Upside Down* [New York: Viking Press, 1972], pp. 161, 273–75, 294, 332).

Chapter VI
Timothy Dwight of Greenfield Hill

Timothy Dwight's *Greenfield Hill: A Poem* was originally published in New York by Childs and Swaine in 1794. *The Major Poems of Timothy Dwight* (Gainesville, Florida: Scholars' Facsimiles & Reprints, 1969) makes it easily accessible, along with *The Conquest of Canäan* and "America." Vernon Louis Parrington's *Connecticut Wits* (New York: Harcourt, Brace, 1926) is still worth consulting, as is his section on Dwight in *Main Currents in American Thought* (New York: Harcourt, Brace, 1927, 1930), I, 360–63. See also Kenneth Silverman, *Timothy Dwight* (New York: Twayne Publishers, 1969).

Chapter VII
John Adams

The uncompleted draft of Adams's letter to Jefferson may be found by date (February 2, 1813), as may their completed correspondence, in Lester J. Cappon's ably edited *The Adams–Jefferson Letters* (Chapel Hill: University of North Carolina Press, 1959), II. For Adams's letters to Benjamin Rush (also easily located by date), see the convenient collection *The Spur of Fame*, ed. John A. Schutz and Douglass Adair (San Marino, Calif.: Huntington Library, 1966). The letter to Judge Tudor (September 23, 1818) is in *The Works of John Adams*, ed. Charles Francis Adams (Boston: Little, Brown, 1856), X, 359–62. In his justification of the colonists' land policy, Adams echoed an important address of his son John Quincy, delivered in 1802; the younger Adams's formulations were discussed by Glenn Tucker in *Tecumseh: Vision of Glory* (Indianapolis: Bobbs-Merrill, 1956), p. 141. For evidence that the Adams family was not abnormal in their dedication to slashing down forests, see Hans Huth, *Nature and the American* (Berkeley: University of California Press, 1957). In *The Writer and the Shaman: A Morphology of the American Indian*, trans. Raymond Rosenthal (New York: Harcourt Brace Jovanovich, 1973), Elémire Zolla noted that Adams came up with "a striking intuition" in seeing similarities between the Indians and the Platonists. (An Italian scholar, Zolla himself had striking intuitions that more than retrieve his work from its occasional crotchetiness.)

I must add a qualifying word: Comparisons between the thought of native peoples and that coming out of Greek philosophy are tricky—see Stanley Diamond's "Plato and the Definition of the Primitive," in *In Search of the Primitive: A Critique of Civilization* (New Brunswick, N.J.: Transaction Books, 1974), pp. 176–202. Adams's quarrel was not with the statism of the *Republic* and the *Laws* but with Plato's theory of Ideas or Forms as the metareality. This mystical theory, as Adams saw it, had some affinity with Indian religion, for after all, contrary to current conventional wisdom, native peoples did think in abstractions. But they firmly grounded their abstractions in their sensuous experience and thereby avoided any sharp Platonic split between perceptions and conceptions. This not-insignificant difference suggests that Adams exaggerated the similarity of outlooks he so detested.

Chapter VIII
Thomas Jefferson

The parenthetical references after Jefferson's papers are to *The Writings of Thomas Jefferson*, Memorial Edition, ed. Andrew A. Lipscomb and Albert Ellery Bergh (Washington: Thomas Jefferson Memorial Association, 1903–4), 20 vols. Despite its known weaknesses, this Memorial Edition remains the most inclusive completed collection—unfortunately the great Jefferson industry at Princeton only recently reached March 31, 1791, with its nineteenth volume (Julian P. Boyd et al., eds., *The Papers of Thomas Jefferson* [Princeton University Press, 1950–]). Thus the textual citations (usually date, volume, and page) are to the Memorial Edition, save for the Adams–Jefferson letters (cited by date only) and to the latter's *Notes on the State of Virginia*, for which see William Peden's definitive edition (Chapel Hill: University of North Carolina Press, 1955).

Vernon Louis Parrington built his monumental *Main Currents of American Thought* (New York: Harcourt, Brace, 1927, 1930) on the cornerstone of his devotion to Jeffersonian "liberalism." Merrill D. Peterson traced reverberations of the tradition in Parrington and others in his more recent rhapsody, *The Jeffersonian Image in the American Mind* (New York: Oxford University Press, 1960). Leonard Levy's irreverence in *Jefferson and Civil Liberties: The Darker Side* (Cambridge: Harvard University Press, 1963) was anticipated by Henry Adams in *History of the United States during the*

Administrations of Jefferson and Madison (New York: Charles Scribner's Sons, 1889–90), 9 vols.—the irony verges on satire in Adams's discussion of Jefferson and the Indians. Not hypocrisy but benevolence is the theme of two recent assessments of Jefferson's Indian policy: Francis Paul Prucha, *American Indian Policy in the Formative Years* (Cambridge: Harvard University Press, 1962); and Bernard W. Sheehan, *Seeds of Extinction: Jeffersonian Philanthropy and the American Indian* (Chapel Hill: University of North Carolina Press, 1973). Two older studies remain, I think, more forthright and penetrating: Annie Heloise Abel, "The History of Events Resulting in Indian Consolidation West of the Mississippi," *Annual Report of the American Historical Association, 1906* (Washington: Government Printing Office, 1908), I, 233–450; and George Dewey Harmon, *Sixty Years of Indian Affairs . . . 1789–1850* (Chapel Hill: University of North Carolina Press, 1941). Winthrop Jordan's *White over Black* was first published by the same press in 1968—I have been quoting the Pelican edition of 1969. David Brion Davis's *Problem of Slavery in the Age of Revolution, 1770–1823* (Ithaca: Cornell University Press, 1975), pp. 168–84, 194–95, presents Jefferson as a trimmer, at best, on antislavery; it deals less satisfactorily with his racism, perhaps in part because Indians are left out of the discussion.

For Virginia laws on miscegenation, see William W. Hening, *Statutes at Large* (Richmond, 1810–23). Jefferson said nothing, of course, about "amalgamation" with Indian slaves or slaves with an Indian parent—for what is still the best study of this obscure subject, see Almon Wheeler Lauber, *Indian Slavery in Colonial Times within the Present Limits of the United States* (New York: Columbia University, 1913), esp. pp. 252–54. James Hugo Johnston's *Race Relations in Virginia & Miscegenation in the South, 1776–1860* (Amherst: University of Massachusetts Press, 1970), originally a 1937 University of Chicago Ph.D. diss., contains a wealth of data on colonial and early national racial attitudes: see "Indian Relations," pp. 269–92, wherein he argued that "legal marriages with the Indians became as unthinkable for the average white man as were legal marriages with the Negro," since red and black "both shared the antipathies of the white man." Winthrop Jordan, who contributed the foreword to Johnston's study, still maintained the contrary, finding no pervasive tension in the sexual union of red and white but instead colonists "willing to allow, even advocate, intermarriage with the Indians—an unheard of proposition concerning Negroes. Patrick Henry pushed a bill through two readings in the Virginia House which offered bounties (if that is the proper term) for children of Indian–white marriages" (*White over Black*, p. 163). True, in 1784 Henry supported his extraordinary bill during a time of "Indian troubles," but the act died in the Assembly. John Marshall had supported it strongly as "advantageous to the country" and was bitterly disappointed by its failure. In December 1784 he spoke directly to the question of whether tensions were aroused by unions of red and white: "Our prejudices however," he wrote James Monroe, "oppose themselves to our interests, and operate too powerfully for them" (Albert J. Beveridge, *Life of John Marshall* [Boston: Houghton Mifflin, 1916], I, 240–41; also Robert Douthat Meade, *Patrick Henry: Practical Revolutionary* [Philadelphia: J. B. Lippincott, 1969], pp. 264–65). For the marriage act (Bill no. 86) reported out by Jefferson's Committee of Revisors, see Julian P. Boyd, ed., *The Papers of Thomas Jefferson* (Princeton: Princeton University Press, 1950), II, 556–58: It nullified marriages between free whites and slaves, among whom were Indians, and mulattoes, among whom were also many Indians or persons of Indian ancestry. Manifestly Jefferson did not seize the opportunity to promote amalgamation and at no point cared or dared to take a step like Henry's. Byrd's advocacy of intermarriage appeared in his *History of the Dividing Line:* see Louis B. Wright, ed., *The Prose Works of William Byrd of Westover* (Cambridge: Harvard University Press, 1966), pp. 160–61.

Chapter IX
Jefferson, II: Benevolence Betrayed

For biographical data on Tenskwatawa I have leaned heavily on Glenn Tucker's "Voice in the Wilderness," in *Tecumseh: Vision of Glory* (Indianapolis: Bobbs-Merrill, 1956), pp. 89–104 et passim. Indian biographer John M. Oskison provided other evidence in *Tecumseh and His Times* (New York: G. P. Putnam's Sons, 1938). See also Benjamin Drake, *Life of Tecumseh and His Brother the Prophet* (New York: Arno Press, 1969—orig. 1841); Moses Dawson, *A Historical Narrative of the Civil and Military Services of Major General William H. Harrison* (Cincinnati: M. Dawson, 1824); and Freeman Cleaves, *Old Tippecanoe* (New York: Charles Scribner's Sons, 1939). For Smohalla and the Columbia River dancers, see Click Relander, *Drummers and Dreamers* (Caldwell, Idaho: Caxton Printers, 1956). For the influence of the Shakers still farther west, see H. G. Barnett, *Indian Shakers* (Carbondale: Southern Illinois University Press, 1957).

Anthony F. C. Wallace, "Revitalization Movements: Some Theoretical Considerations for Their Comparative Study," *American Anthropologist*, LVIII (April 1956), 264–81, is the primary source for the revitalization concept; but see also his introduction to James Mooney's *Ghost-Dance Religion and the Sioux Outbreak of 1890* (Chicago: Phoenix Books, 1965), and of course his *Death and Rebirth of the Seneca* (New York: Vintage Books, 1972). For the global context of the reaction of aborigines to white racism and exploitation, see the Italian scholar Vittorio Lanternari's *Religions of the Oppressed: A Study of Modern Messianic Cults*, trans. Lisa Sergio (New York: Alfred A. Knopf, 1963).

My understanding of Jefferson's paternalism and Prospero-like relationship with the Indians owes much to K. M. Abenheimer, "Shakespeare's 'Tempest': A Psychological Analysis," *Psychoanalytic Review*, XXXIII (October 1946), 399–415. On this topic two other works cited earlier are also illuminating: Otare Mannoni, *Prospero and Caliban* (New York: Frederick A. Praeger, 1964), and Philip Mason, *Prospero's Magic* (London: Oxford University Press, 1962). Michael Paul Rogin in "Liberal Society and the Indian Question," *Politics and Society*, II (May 1971), 269–312, used psychoanalysis to probe for the meaning of such paternalism in Jacksonian America.

Carl L. Becker's *Declaration of Independence* (New York: Vintage Books, 1958) remains the best overall study. Julian P. Boyd's *Declaration of Independence* (Princeton: Princeton University Press, 1945) is a careful study of the evolution of the text. Boyd's notes in *The Papers of Thomas Jefferson* (Princeton: Princeton University Press, 1950), I, 299–308, 413–17, incorporate discovery of additional information on Jefferson's drafting of the document.

Chapter X
Driving Indians into Jefferson's Stony Mountains

In *Society and Puritanism in Pre-Revolutionary England* (New York: Schocken Books, 1964), p. 85, Christopher Hill found it "one of history's little ironies that the liberty poles of the American Revolution descended from the Maypole."

Reactions within the Monroe cabinet to Jackson's Florida caper are best followed (by date) in the *Memoirs of John Quincy Adams* (Philadelphia: J. B. Lippincott, 1875), IV, 87–212 passim—Adams's devout hope he would prove equal to the task of justifying Jackson was entered on November 8, 1818 (IV, 168). The dispatch to Minister Erving of November 28, 1818, is in the *Writings of John Quincy Adams* (New York: Macmillan, 1916), VI, 474–502.

Letters and orders to and from Jackson may be found (by date) in the *Correspondence of Andrew Jackson*, ed. John Spencer Bassett (Washington, D.C.: Carnegie Institution, 1927)—most of the Florida papers are in volume II. Bassett's *Life of Andrew Jackson* (New York: Macmillan, 1931—orig. 2 vols., 1911) grants that the Indians might have made peace "if Americans had not coveted Florida" (p. 251) but is useful primarily for tracing into the twentieth century attitudes toward the mercilessness of "sullen savages" (p. 237). By contrast relatively free of cant is James Parton's *Life of Andrew Jackson* (New York: Mason Brothers, 1860), 3 vols.—General David B. Mitchell, the Creek agent, made his deposition on white and red aggressions, prior to the attack on Fowltown, before a Senate committee in February 1819 (II, 430n). For Old Hickory as the embodiment of the image and spirit of his countrymen, see John William Ward, *Andrew Jackson: Symbol for an Age* (New York: Oxford University Press, 1955). More satisfactory on Jackson and the Indians, and unequaled for understanding his inner demons, is Michael Paul Rogin's *Fathers and Children* (New York: Alfred A. Knopf, 1975).

Jackson's view of himself as like one of "the Iseralites of old in the wilderness" is another datum on the massive continuity with the colonial past. Self-anointed heirs of ancient Israel, John Endicott and the Saints sought to establish a redeemer nation in the West; two centuries later Andrew Jackson was a secularized version of their mission. Given his upbringing in the South, he was not as neat an illustration of the legacy as John Quincy Adams, and probably the pervasiveness of "the myth of America" was more uneven than Sacvan Bercovitch's study of rhetoric would suggest: see *The Puritan Origins of the American Self* (New Haven: Yale University Press, 1975). Still, had Bercovitch not relied exclusively on hermeneutics and looked further for ties between the Puritanism of the seventeenth century and the nationalism of the nineteenth, he would have found evidence in Jackson to buttress his discussion of Emerson and Lincoln.

Alexander Arbuthnot made the mistake of liking Indians: "They have been ill treated by the English," read one of his diary entries, "and robbed by the Americans"—quoted in Marquis James, *Andrew Jackson: The Border Captain* (New York: Literary Guild, 1933), p. 310. He was sympathetically discussed in Parton's *Life of Andrew Jackson* and in Edwin C. McReynolds, *The Seminoles* (Norman: University of Oklahoma Press, 1957), pp. 78–86. For Hillis Hadjo and his execution, see Carolyn Thomas Foreman, *Indians*

Abroad, 1493–1938 (Norman: University of Oklahoma Press, 1943), pp. 114–18. For his relationship with Tecumseh, see John Dunn Hunter, *Memoirs of a Captivity among the Indians of North America*, ed. Richard Drinnon (New York: Schocken Books, 1973), pp. 27–31, 108, 242n.

Among historians who see the national government's humane Indian policy as having been scuttled by greedy frontier rogues, Francis Paul Prucha is best known: see *American Indian Policy in the Formative Years* (Cambridge: Harvard University Press, 1962). The contrary is argued in Richard Drinnon's *White Savage: The Case of John Dunn Hunter* (New York: Schocken Books, 1972). The working partnership of Adams and Jackson in 1818 was like the earlier one of Jefferson and Harrison, and like that of a number of later figures—for a revealing glimpse into the symbiotic relationship of capital and countryside on Indian dispossession, see Secretary of War Lewis Cass's "Private and Confidential" letter to His Excellency Wilson Lumpkin, Governor of Georgia, December 24, 1832, Record Group 75, Microfilm 7, Roll 1, National Archives.

For a discussion of the White Paper as Adams's "great gun [that] cleared the air everywhere," see Samuel Flagg Bemis, *John Quincy Adams and the Foundations of American Foreign Policy* (New York: Alfred A. Knopf, 1949), p. 328. For Monroe's statement of administration objectives, see his letters to Madison (February 7, 1819) and to Richard Rush (March 7, 1819), in *Writings of James Monroe*, ed. Stanislaus Murray Hamilton (New York: G. P. Putnam's Sons, 1902), VI, 87–92. See also Harry Ammon, *James Monroe* (New York: McGraw-Hill, 1971), pp. 409–25. Jefferson's letter to Monroe of January 18, 1819 (about Adams's White Paper), is not in the Memorial Edition but in Paul Leicester Ford, ed., *Writings of Thomas Jefferson* (New York: G. P. Putnam's Sons, 1892–99), X, 122. On Adams and Native Americans after he left the White House, see Lynn Hudson Parsons, " 'A Perpetual Harrow upon My Feelings': John Quincy Adams and the American Indian," *New England Quarterly*, XLVI (September 1973), 339–46.

Madison's undated letter to Lafayette is in *Letters and Other Writings of James Madison* (New York: R. Worthington, 1884), III, 239–40—this four-volume Congressional Edition also contains his other letters to Lafayette and his letters to Monroe, Frances Wright, et al., easily located by date. *The Federalist No. 10* may be conveniently read in Jacob E. Cooke, ed., *The Federalist* (Middletown, Conn.: Wesleyan University Press, 1961), pp. 56–65. For a celebration of Madison's profundity as a political economist, see Charles A. Beard, *The Economic Basis of Politics* (New York: Alfred A. Knopf, 1945), pp. 16–19 et passim. Beard was a fine historian and flexible enough to see that military power might disturb or overturn property-based politics; he never seems to have realized, however, that the model he took from Madison came out of the European experience and thus fitted European class conflicts but meshed not at all with fundamental cultural and racial divisions in the Americas.

The Monroe Doctrine's underside was explored long ago by Richard Rollin Stenberg's "American Imperialism in the Southwest, 1800–1837," Ph.D. diss., University of Texas, 1932, pp. 146–69. Richard W. Van Alstyne, in his more recent *Rising American Empire* (Chicago: Quadrangle Books, 1965), depicted the Monroe Doctrine as the cornerstone of United States assumption of hegemony in the New World; for the sources and patterns of this national ideology, see also his *Genesis of American Nationalism* (Waltham, Mass.: Blaisdell, 1970). For a discussion of the Monroe Doctrine and its potent thrust into Texas and Mexico, see Richard Drinnon, "The Metaphysics of Empire-Building: American Imperialism in the Age of Jefferson and Monroe," *Massachusetts Review*, XVI (Autumn 1975), 666–88.

Chapter XI
Westward Ho! with James Kirke Paulding

Horatio Greenough's explicit intention in the *Rescue Group*, as recounted in a letter by a friend of his family and quoted by Roy Harvey Pearce, addressed itself directly to the "merciless savage" theme we have been pursuing—see *Savages of America* (Baltimore: Johns Hopkins Press, 1953).

Abraham Lincoln's Indian-hating Uncle Mordecai was sympathetically discussed by John G. Nicolay and John Hay, *Abraham Lincoln* (New York: Century, 1890), I, 21; the "instinct" that impelled the pioneers to "people" Kentucky was later identified as "that Anglo-Saxon lust of land which seems inseparable from the race" (I, 15–16). John Filson's *The Discovery, Settlement and Present State of Kentucke,* originally published in 1784 and many times reprinted, was reissued in 1966 by University Microfilms at Ann Arbor. For the "fearless buffaloes" and "bloody footsteps" references, see pp. 51, 80.

Paulding's letter to George Bancroft and those to other correspondents discussed in this chapter may be easily located by date in Ralph M. Aderman, ed., *Letters of James Kirke Paulding* (Madison: University of Wisconsin Press, 1962). For biographical details and bibliography, see Amos L. Herold, *James Kirke Paulding: Versatile American* (New York: Columbia University Press, 1926). For his works discussed here, see *The Backwoodsman* (Philadelphia: M. Thomas, 1818), *The Dutchman's Fireside* (New York: J. & J. Harper, 1831), and *Westward Ho!* (New York: J. & J. Harper, 1832). In the last, Paulding acknowledged his particular obligations to Timothy Flint's *Recollections of the Last Ten Years, Passed in Occasional Residences and Journeyings in the Valley of the Mississippi* (Boston, 1826). Arlin Turner, "James Kirke Paulding and Timothy Flint," *Mississippi Valley Historical Review*, XXXIV (June 1947), 105–11, maintained Paulding must have written *Westward Ho!* "with the *Recollections* open before him." He may as well have drawn on Flint's "Indian Fighter" that appeared in the *Token* for 1830. G. Harrison Orians, "The Indian Hater in Early American Fiction," *Journal of American History*, XXVII (1933), 34–44, discussed Flint, Paulding, James Hall, N. M. Hentz, and others and concluded surprisingly that, like many other persons whose energies flow in a single channel, the Indian-hater was "not an unmitigated force for good." No, not "unmitigated," but as Sir William Johnson said, he had his uses.

Joel R. Poinsett, Van Buren's secretary of war, was a South Carolinian slaveholder who saw himself as the apostle of liberty to Spanish America—for an analysis of his attitude toward Indians and Mexicans and his role in United States expansionism see Richard Drinnon, *White Savage* (New York: Schocken Books, 1972), pp. 187–93. Lieutenant Wilkes's instructions and the "Dred Scott" ms. are in Paulding's *Letters*, pp. 223–32, 392–93. For Wilkes's account, see his *Narrative of the United States Exploring Expedition*, 5 vols. (Philadelphia: Lea and Blanchard, 1845). William H. Goetzmann summarized the undertaking in *Exploration and Empire* (New York: Alfred A. Knopf, 1966), pp. 233–40. Best for placing it in its appropriate context is Richard W. Van Alstyne's *Rising American Empire* (Chicago: Quadrangle, 1965), pp. 106–7, 126–27, 173–74.

Chapter XII
An American Romance in Color: William Gilmore Simms

William Gilmore Simms's "Daniel Boon: The First Hunter of Kentucky" was originally published in the periodical he edited, the *Southern and Western Magazine*, I (April 1845), 225–42, and included with "Literature and Art among the American Aborigines," "The Writings of James Fenimore Cooper," and other pieces in *Views and Reviews in American Literature, History and Fiction*, ed. C. Hugh Holman (Cambridge: Harvard University Press, 1962). His perception of the nature of white prejudices, paralleling Montesquieu's, appeared in his essay on Indian literature and art. "The First Hunter" contained equally brilliant lines, most notably his realization that Boone "was not merely a hunter. He was on a mission. The spiritual sense was strong in him. He felt the union between his inner [nature] and the nature of the visible world, and yearned for their intimate communion" (pp. 156–57). Unhappily Simms did not develop this to the further realization that Boone was yearning for what the Indians had. In his extraordinary "The Discovery of Kentucky," William Carlos Williams pursued the point, maintaining that "if the land were to be possessed it must be as the Indian possessed it. Boone saw the truth of the Red Man, not an aberrant type, treacherous and anti-white to be feared and exterminated, but as a natural expression of the place, the Indian himself as 'right,' the flower of his world." More relevant here than whether Boone actually saw the Indian as the natural flower of his world is the fact that Simms sensed man's natural yearning to be at one with the land. Eighty years after Simms, at all events, William Carlos Williams followed in his footsteps first by using Filson, then by denouncing him as an "asinine chronicler," and finally by demonstrating how deeply he had been moved by the archetypal narrative—see *In the American Grain* (New York: New Directions, 1956), pp. 130–39.

Otare Mannoni discussed Robinson Crusoe and "the lure of a world without men" in *Prospero and Caliban* (New York: Frederick A. Praeger, 1964), pp. 97–105. Simms was quite capable of thinking of Boone as a sort of land Crusoe—in his essay on Cooper he used seagoing imagery in his consideration of "Hawkeye, the land sailor of Mr. Cooper" (*Views*, p. 272). For Simms's *City Gazette* denigration of the Indian "as he is," see Mary Oliphant, Alfred Odell, and T. C. Eaves, eds., *The Letters of William Gilmore Simms* (Columbia: University of South Carolina Press, 1952), I, 29–30.

For a modern edition of *The Yemassee*, see that edited by Alexander Cowie (New York: American Book, 1937). J. V. Ridgely's *William Gilmore Simms* (New York: Twayne Publishers, 1962) is a useful recent survey, though it stresses his later emphatic role as a "Southron" at the expense of his earlier nationalism and insistence on "Americanism in Literature" (*Views*, pp. 7–29).

Chapter XIII
Nicks in the Woods: Robert Montgomery Bird

For the doubleness of Boone's legend, see Henry Nash Smith's important chapter "Daniel Boone: Empire Builder or Philosopher of Primitivism?" in *Virgin Land* (Cambridge: Harvard University Press, 1950). The best biography is John Bakeless's *Daniel Boone: Master of the Wilderness* (New York: William Morrow, 1939). Daniel Bryan's epic *Mountain Muse: Comprising the Adventures of Daniel Boone . . .* (Harrisonburg, Va.: Davidson and Bourne, 1813) cast his relative by marriage as the advance agent of progress, bringing "civilization" to wilderness Kentucky "When nought but Beasts and bloody Indians dwelt/Throughout the mighty waste. . . ." Good on the relationship of Boone and other pathfinders to Colonel Richard Henderson and other land speculators is Arthur K. Moore's *Frontier Mind: A Cultural Analysis of the Kentucky Frontiersman* (Lexington: University of Kentucky Press, 1957). W. Eugene Hollon's *Frontier Violence: Another Look* (New York: Oxford University Press, 1974) merits mention for the thesis that the violent society in the "settlements" produced violence on the frontier, and not vice versa.

D. H. Lawrence might have asked of model Boone what he asked and hauntingly answered about Cooper's Natty Bumppo: "What sort of a white man is he? Why, he is a man with a gun. He is a killer, a slayer. Patient and gentle as he is, he is a slayer. Self-effacing, self-forgetting, still he is a killer"—*Studies in Classic American Literature* (New York: Doubleday Anchor, 1953), p. 69. Natty was a reluctant killer in the tradition of Boone, as were so many strong silent men from the Wild West since, including Owen Wister's *Virginian*—see Leslie Fiedler's *Love and Death in the American Novel* (Cleveland: World, 1962), pp. 255–56. Fiedler pointed out that the theme was plagiarized by the movie *High Noon*, "in which Gary Cooper, in whom myth after myth has become flesh, incarnates the Virginian, too." In the film *Sergeant York*, Cooper also fittingly incarnated in celluloid the World War I hero whose great-great-grandfather hunted 'coon with Davy Crockett in the Valley of Three Forks of the Wolf—see *Sergeant York: His Own Life Story and War Diary*, ed. Tom Skeyhill (Garden City, N.Y.: Doubleday, Doran, 1928). I am indebted to Fiedler for his insights into the legend of gentle killers. Colonel Frank Triplett's *Conquering the Wilderness* (New York: N. D. Thompson, 1883) has long been a prime source for such eagerly unreluctant killers as Adam Poe.

Biographical data on Robert Montgomery Bird may be found in Cecil B. Williams, ed., *Nick of the Woods or The Jibbenainosay: A Tale of Kentucky* (New York: American Book, 1939). Williams competently traced the evolution of the novel in the Bird mss. and placed it in the context of Bird's life and other writings, though he failed to pin down the novelist's indebtedness to John Filson's *Kentucke*. For identification of the novelist as a Pennsylvanian, see, for example, R. W. B. Lewis, *The American Adam* (Chicago: University of Chicago Press, 1955), p. 105, where he appears as "Robert Montgomery Bird of Philadelphia . . . a physician of restless temperament."

Constance Rourke's remark in *American Humor* (New York: Doubleday Anchor, 1953) that Bird "tapped a deep fund of western comic talk and character" inadvertently emphasized the callousness and sadism of that humor. Elsewhere in her study she developed the insight that the frontiersman's "prevailing hysteria was shown in his insensate habit of killing more game then [*sic*] he needed, or of shooting the hundreds of pigeons that blackened the sky from a blind wish to exhibit power or a blinder purpose to obliterate the wilderness" (pp. 42, 160). That Indians, blacks, Mexicans, and Chinese were so often the butts of this exuberant blindness led B. A. Botkin to comment on the "essential viciousness of many of our folk heroes"—see his *Treasury of American Folklore* (New York: Crown, 1944) for the Mike Fink stories discussed in the text. See also Walter Blair, *Native American Humor* (New York: American Book, 1937), for a discussion of Thomas Bangs Thorpe, John S. Robb, et al., and especially for an analysis of the "box-like structure" of genteel realism that framed the frontier tales—I have suggested the same device provided a "civilized" frame for classic Indian-haters. Mark Twain was of course master of the technique—in his art the culmination of Southwestern humor—and fittingly enough "an absolute Indian hater" himself—see Leslie Fiedler, *The Return of the Vanishing American* (New York: Stein and Day, 1968), pp. 122–24. The euphemism of "emptying" the Eastern states of Indians is Francis Paul Prucha's in *American Indian Policy in the Formative Years* (Cambridge: Harvard University Press, 1962), chap. 10.

Some years back, one of the major influences on my generation's understanding of the United States was Vernon Louis Parrington's *Main Currents in American Thought* (New York: Harcourt, Brace, 1927, 1930). It still deservedly ranks as a monument to the tradition of Jeffersonian and Jacksonian liberalism; only in the present study have I become aware of the extent to which it also stands as a monument to that tradition's blindness to racism and its openness to the implicit assumptions and attitudes of Indian-hating. Parrington's staggeringly insensitive comments on Bird's "blood-letting" were of a piece with his failure to invite Miantonomo, Metacom, Pontiac, Logan, Tecumseh, Sequoya, Sitting Bull, Chief Joseph, or any other Indians into his 1,300 pages of main currents.

Decades ago Thomas Raynesford Lounsbury observed in *James Fenimore Cooper* (Boston: Houghton Mifflin, 1884), p. 55, that most of the novelist's Indians were "depicted as crafty, bloodthirsty, and merciless." But critics continued to ignore this truth, so that in April 1906 W. C. Brownell exclaimed in *Scribner's* how extraordinary it was that Cooper's "assumed idealization of the Indian" had become a firmly established convention—quoted in Gregory Lansing Paine, "The Indians of the Leather-Stocking Tales," *Studies in Philology*, XXIII (January 1926), 37. Since Cooper's "typical" Indian was bad like Magua and not good like Uncas, it was and is extraordinary, and it has to do with conventions about "primitivism" and "Noble Savages" that produce what amounts to literary Indian-hating: The attack is always on Cooper's "idealized" Delawares and Pawnees and not on his maligned Iroquois and Sioux. For a first-rate analysis of the novelist's dependence on John Heckewelder, see Paul A. W. Wallace, "Cooper's Indians," in *James Fenimore Cooper: A Reappraisal*, ed. Mary E. Cunningham (Cooperstown, N.Y.: New York State Historical Association, 1954), pp. 423–46. My understanding of Cooper's ambivalences has also benefited from Smith's chapter on Leatherstocking in *Virgin Land* and from Roy Harvey Pearce's earlier "The Leatherstocking Tales Re-Examined," *South Atlantic Quarterly*, XLVI (October 1947), 524–36.

As governor of the Michigan Territory and negotiator of treaties relieving the Indians of millions of acres of land, General Lewis Cass enjoyed an unearned reputation as an "Indian expert." The lines of his onslaught against Cooper shaped those since—see his "Indians of North America," *North American Review*, XXII (January 1826), 67; "Review of William Rawle's *A Vindication of the Rev. Mr. Heckewelder's History of the Indian Nations*," ibid., XXVI (April 1828), 357–403; and "Removal of the Indians," ibid., XXX (January 1830), 62–121. For Francis Parkman's attack, see ibid., LXXIV (January 1852), 150–52. Since Parkman himself thought of the Indians as forest beasts, as "man, wolf, and devil, all in one," he spoke sympathetically of Bird's "spirited" *Nick* for its thematic development "of that unquenchable, indiscriminate hate which Indian outrages can awaken in those who have suffered them"—*Conspiracy of Pontiac* (Boston: Little, Brown, 1929), II, 127. The best discussion of Parkman's Indians is David Levin's *History as Romantic Art* (Stanford, Calif.: Stanford University Press, 1959), pp. 132–41; see also Robert Shulman, "Parkman's Indians and American Violence," *Massachusetts Review*, XII (1971), 220–39. See Richard Drinnon's *White Savage* (New York: Schocken Books, 1972), pp. 233–46, for a discussion in greater detail of Cooper's indebtedness to Heckewelder and also to John Dunn Hunter, and for the hostile reactions of Parkman and in particular of Cass.

Chapter XIV
Friend of the Indian: Colonel McKenney

Abbreviated references in this and the next chapter are to the following sources:

(T) Thomas L. McKenney, *Sketches of a Tour to the Lakes, of the Character and Customs of the Chippeway Indians, and of Incidents Connected with the Treaty of Fond du Lac* (Baltimore: Fielding Lucas, Jr., 1827).

(M) ———, *Memoirs, Official and Personal; with Sketches of Travels among the Northern and Southern Indians; Embracing a War Excursion, and Descriptions of Scenes along the Western Borders* (New York: Paine and Burgess, 1846—2nd ed., 2 vols. in one).

(ITNA) ——— and James Hall, *The Indian Tribes of North America with Biographical Sketches and Anecdotes of the Principal Chiefs*, ed. Frederick Webb Hodge et al., 3 vols. (Edinburgh: John Grant, 1933—orig. 1836–44).

(H) James D. Horan, *The McKenney–Hall Portrait Gallery of American Indians* (New York: Crown, 1972)—some useful information.

(V) Herman J. Viola, *Thomas L. McKenney, Architect of America's Early Indian Policy: 1816–1830* (Chicago: Swallow Press, 1974)—uncritical but contains valuable biographical data and quotations from Indian Office correspondence.

(Date) Charles Francis Adams, ed., *Memoirs of John Quincy Adams* (Philadelphia: J. B. Lippincott, 1875).

(Date) *Letters and Other Writings of James Madison*, Congressional Edition (New York: R. Worthington, 1884), III, 487–88, 515–16. In the microfilm of the James Madison Papers, First Series, Vol. LXXIV, folio 85, in the Library of Congress, is a McKenney letter of April 26, 1825, that establishes him as the author of the (Washington) *Daily National Journal* article of that date quoted in the text—an unsigned contribution that is a good summary of what McKenney meant by "civilization." (In *White Savage* [New York: Schocken Books, 1972], p. 65, I erroneously thought him the probable author of a *National Intelligencer* article of the same date.)

(NA, RG75, M21, R3, 328) National Archives, Record Group 75, Microcopy 21, Roll 3, page 328—or of course other roll and page numbers. The Microcopy number of the Bureau of Indian Affairs, Letters Sent, is 21; of Letters Received it is 234.

McKenney's "civilization" program must be placed in perspective. The bill of 1819 appropriated $10,000 annually to that end; Congress appropriated $27,000 or thereabouts for the single treaty Cass and McKenney negotiated at Fond du Lac (V, 139). His *Tour* appeared under the amiable guise of letters to a "friend," who was in fact the secretary of war; the dedication underlined his sycophancy by extolling Barbour "as a Citizen, and Patriot, and as a public functionary." McKenney was only nominally the senior author of *Indian Tribes of North America*. Hall did most of the writing but worked up materials supplied him by McKenney and paraphrased or quoted the latter on individuals and events, including the treaty council with the Creeks, the self-reliance test of Lee Compere, the Yuchi, and the like. The *Tour* was reviewed in the *North American Review*, XXV (October 1827), 334–52; editor Jared Sparks of that journal reviewed the first numbers of *Indian Tribes*, ibid., XLVII (July 1838), 134–48, and the *Memoirs*, ibid., LXIII (October 1846), 481–96.

For a biography of the famous castaway and evidence of Daniel Defoe's use of him as a model for *Life and Adventures of Robinson Crusoe* (1719), see John Howell, *Life and Adventures of Alexander Selkirk* (Edinburgh: Oliver & Boyd, 1829). The commissioners' patriotic grimness, rage for order, and drive for military ascendancy contrasted sharply with the Ojibwas' pleasure in games and unrestrained movements of their bodies. But that is not to say their festival dances were frivolous or lacking in their own kind of seriousness. "The concept of play merges quite naturally with that of holiness," Johan Huizinga observed. "Any Prelude of Bach, any line of tragedy proves it. . . . Primitive, or let us say, archaic ritual is thus sacred play, indispensable for the well-being of the community, fecund of cosmic insight and social development . . . "—*Homo Ludens: A Study of the Play-Element in Culture* (Boston: Beacon Press, 1955), p. 25. For Huizinga's discussion of the dance as the purest form of sacred play, see pp. 164–65. But Cass and McKenney had deliberately buried the child within themselves, or so they thought, and thus had alienated themselves from the sensual world in which "the child and the poet are at home with the savage" (p. 26).

In *American Indian Policy in the Formative Years: The Indian Trade and Intercourse Acts, 1790–1834* (Cambridge: Harvard University Press, 1962), Francis Paul Prucha seemingly assigned a later date for McKenney's "dramatic shift" from *incorporation* to *removal:* "The tour of the Indian country that he made in 1827 opened his eyes to the degradation of the eastern tribes, and when asked to report in 1830 on the previous eight years of operation of the program of civilizing the Indians, he no longer considered salvation possible in the present location of the tribes." This deepened the chronological error and wrenched McKenney from his institutional context: As head of the Bureau of Indian Affairs in the War Department bureaucracy, he implemented and sometimes helped shape, as in 1819, but never made administration policy toward native peoples. Bernard Sheehan more cautiously dated McKenney's shift in "the mid-1820s": "He solidified his change of opinion while touring the tribal areas in 1826 and 1827, after which he lobbied publicly in favor of removal"—*Seeds of Extinction: Jeffersonian Philanthropy and the American Indian* (New York: W. W. Norton, 1974), p. 250. McKenney's change of opinion was seemingly firm enough before 1826 and was publicly expressed prior to his tour of 1827. Otherwise, Sheehan followed the general lines of Prucha's interpretation.

McKenney's estimate of Cass as "the best informed man in the United States on Indian Affairs" was quoted by William T. Hagan, *The Sac and Fox Indians* (Norman: University of Oklahoma Press, 1958), p. 135. The best biography remains Andrew C. McLaughlin's laudatory *Lewis Cass* (Boston: Houghton, Mifflin, 1891). Frank B. Woodford's *Lewis Cass: The Last Jeffersonian* (New

Brunswick, N.J.: Rutgers University Press, 1950) is bathetic but still contains some valid data—for Cass's reference to auxiliaries as "pet Indians," see pp. 93–94. The best sources for Cass's ideas about and attitudes toward Native Americans are his three *North American Review* articles cited on page 489.

In *Notes on the State of Virginia*, ed. William Peden (Chapel Hill: University of North Carolina Press, 1955), Thomas Jefferson had pronounced the position of Native American women unenviable: "The women are submitted to unjust drudgery. This I believe is the case with every barbarous people. With such, force is law. The stronger sex imposes on the weaker. It is civilization alone which replaces the women in the enjoyment of their natural equality" (p. 60). McKenney made this a main prop of his "civilization" program and thereby gave a "feminist" dimension to his attack on Indian cultures. It was also one of the "Indian customs" taken up in the mission schools. Robert F. Berkhofer, Jr., quoted the English composition of one native girl: "The Indian man never works. His wife always does all the work. . . . The Indian woman has hard times when she has to do all this work." From all this you would think white women in Jefferson's native Virginia and in McKenney's Maryland reveled "in the enjoyment of their natural equality." This revealing misplaced concern also recurred like a refrain in the McKenney–Hall *Indian Tribes* (e.g., ITNA, I, 46, 249). A modern discussion of sex roles among the Iroquois appears in Anthony F. C. Wallace's *Death and Rebirth of the Seneca* (New York: Vintage Books, 1972), pp. 21–48. For an amusing and important illustration of how whites unwittingly revealed their own patriarchal attitudes in their multiple misunderstandings of native women, see Paul A. W. Wallace, "Cooper's Indians," in *James Fenimore Cooper: A Reappraisal*, ed. Mary E. Cunningham (Cooperstown, N.Y.: New York State Historical Association, 1954), pp. 423–46. For other sources, see Beatrice Medicine, "The Role of Women in Native American Societies: A Bibliography," *Indian Historian*, VIII (August 1975), 50–54.

No less than McKenney did Cass take a stand against forcible removal and express confidence that "the great body of the Indians within our states . . . are anxious to remove" (Woodford, p. 182). But he moved into the administration as secretary of war shortly after McKenney was forced out and to him fell responsibility for implementing the first phase of forcible removal. By 1835 he had to know the majority were unwilling to go, for Major W. M. Davis, appointed to enroll the Cherokees, wrote him a strong letter pointing out that nine-tenths of them would reject the treaty of New Echota if it were referred to them—James Mooney, "Myths of the Cherokees," *Nineteenth Annual Report of the Bureau of American Ethnology* (Washington: Government Printing Office, 1900), p. 126.

Auburn Prison was the deadly flower of Puritan and Enlightenment efforts to rationalize the world. McKenney was heir of this tradition not only through Madison and Jefferson but also through David Brainerd and John Eliot, back to the Pilgrims of Plymouth: "Who can doubt that Providence guided them to that spot?" asked the colonel (M, II, 70). McKenney linked that past with a present that shared his gushing enthusiasm for institutions of total control: "Philanthropy has become for them," observed Gustave de Beaumont and Alexis de Tocqueville, "a kind of profession, and they have caught the *monomanie* of the penitentiary system, which seems to them the remedy for all the evils of society"—see *On the Penitentiary System in the United States* (Philadelphia: Carey, Lea & Blanchard, 1833). For the larger context of this attempt to banish impulse, disorder, irregularity, and unreason from the world, see Michel Foucault, *Madness and Civilization: A History of Insanity in the Age of Reason*, trans. Richard Howard (New York: Random, 1965). Still more directly relevant is Foucault's *Discipline and Punish: The Birth of the Prison*, trans. Alan Sheridan (New York: Pantheon Books, 1977). For the American context, see David J. Rothman, *Discovery of the Asylum: Social Order and Disorder in the New Republic* (Boston: Little, Brown, 1971), esp. "The Invention of the Penitentiary," pp. 79–108. Regimented reason in Auburn had the same bright, shining, Anglo-Saxon face it showed in McKenney's mission schools, where the identical Puritan virtues were compulsory—see Robert F. Berkhofer, Jr., *Salvation and the Savage* (Lexington: University of Kentucky Press, 1965), pp. 16–43. For insight into parallels between the correctional treatment model for prisoners and for Indians, see American Friends Service Committee, *Struggle for Justice* (New York: Hill & Wang, 1971), pp. 43–44. The parallels extended to the public schools, of course, for the values indoctrinated therein demonstrated the same massive continuity with the Puritan past—see Ruth Miller Elson, *Guardians of Tradition: American Schoolbooks in the Nineteenth Century* (Lincoln: University of Nebraska Press, 1964). A study remains to be written of the inner identities of "the Great Confinement"—as Foucault called it—of the mad and the errant and the young, and of the efforts of McKenney and his compatriots to expatriate blacks and remove reds. In each instance white society moved to sequester, expel, or remove the many faces of unreason, and in each instance the end was to rid the world of its human idiosyncrasies. The underlying model was the army.

McKenney showed shrewdness but little originality in his hostility to "the Indian tongue." In the late sixteenth century the bishop of Avila had put the point bluntly to Queen Isabella: "Your Majesty, language is the perfect instrument of empire"—quoted in Stephen J. Greenblatt's extraordinary "Learning to Curse: Aspects of Linguistic Colonialism in the Sixteenth Century," in *First Images of America*, ed. Fredi Chiappelli et al. (Berkeley: University of California Press, 1976), II, 561–80. McKenney was thus in the mainstream of colonialism and within that larger context might be considered a contributor to "the unwritten manual on old colonial behavior," as Stanley Diamond called it, from which a British director of a Nigerian tin company read him a lesson in 1959: "Africans, he said, were children"—*In Search of the Primitive: A Critique of Civilization* (New Brunswick, N.J.: Transaction Books, 1974), p. 85. Like McKenney, colonial officials in Africa waged war on the real family by pulling native children out of it and subjecting them to the pseudo-paternal authority of civil structure. Diamond designated this ongoing worldwide phenomenon "kin-civil conflict" (pp. 180, 363). In the attack on the family, that "psychic transmission belt between the generations," and then on human nature itself, "we recognize civilization" (pp. 121, 180, 209). My understanding of American imperialism owes much to Diamond's discussion of the colonial mentality in Africa and elsewhere.

Notwithstanding obvious differences, the parallels between McKenney's paternalistic attitudes toward the Indians and toward Ben Hanson and other blacks should be underlined. He treated both reds and blacks as childlike and thereby denied them full humanity. For an analysis of such white paternalism, see Herbert G. Gutman's impressive *Black Family in Slavery and Freedom, 1750–1925* (New York: Pantheon, 1976), pp. 291–303. As Gutman noted, "neither the Civil War nor the Thirteenth Amendment emancipated northern whites from ideological currents that assigned inferior status to nineteenth-century blacks, women, and working-class men." But the same also held for reds, and their putative infantilism predated that of the others in North America. In this context even a historian as able as Gutman wrote as though the Indians really had vanished, though he himself later quoted the main resolution from the prestigious first Lake Mohonk Conference on the Negro Question (1890): "The Christian home is the civilizer. Ultimately, in the homes of the colored people the problem of the colored race will be settled. . . . The one-room cabin is the social curse of the Negro race, as is the reservation tepee that of the Indian, and the overcrowded tenement-room [that] of our city slums" (p. 539). Gutman missed an opportunity here to relate this assertion to others no less egregious that came out of preceding Mohonk Conferences on the Indian "problem"—Lake Mohonk House was run by the Quaker Smiley brothers, one of whom (Albert K.) was on the Board of Indian Commissioners. Of critical importance in gaining acceptance of *forcible* allotment of Indian lands under the Dawes Act (1887) was the conference of 1885, attended by well-known philanthropists who were in a very real sense McKenney's heirs. (An adequate survey appears in Larry E. Burgess, "The Lake Mohonk Conferences on the Indian, 1883–1916," Ph.D. diss., Claremont Graduate School, 1972.) Historians of the Afro-Americans and of the Native Americans need to come together more frequently to compare notes on the white racism that has victimized both blacks and reds.

Color defined the problem for McKenney: "Whence comes this decay, and final disappearing of the red before the white man? It comes not of the color, nor of physical or moral malformation; nor of destiny . . ." (M, I, 332). Engagingly unselfconscious, the colonel affirmed the importance of color even while denying its relevance: "You are a full blooded Choctaw," read his letter to a student at a mission school, "and yet, altho' some men say Indians, to learn, must have white blood in them, yet your letter proves that the Indian's blood is no blacker or thicker than the white man's" (V, 189). *Indian Tribes* repeatedly posited a correlation between darkness of skin and depth of "savagery," with William Gilmore Simms's Yemassees, or rather the survivors, being darkest of all, and a band of Seminoles, "who are descended from them, betray their origin by the dark colour of their skins" (ITNA, II, 336). Not surprisingly the especially dark Seminoles were especially fierce and may have owed their very outcast origins to their "impatience of the restraints even of savage life" (ITNA, II, 201). Such dark plumage and utter unrestraint made them blend in especially well with the other fauna of their swamps and deep woods and helped in what Hayden White called "the hidden, or repressed, *identification* of the natives of the New World with natural objects (that is to say, their *de*humanization) to be used (consumed, transformed, or destroyed) as their conquerors (or owners) desired"—"The Noble Savage Theme as a Fetish," *First Images of America*, I, 124. Such natural objects were McKenney's "fledglings of the forest," and they were in every sense "*our* Indians." Pet Indians.

Chapter XV
Professional Westerner: Judge Hall

An explanation of the abbreviated references appears on page 490. Stanley Diamond's reference to museums as imperial warehouses is in his *In Search of the Primitive* (New Brunswick, N.J.: Transaction Books, 1974), p. 216. For Johnson Jones Hooper's crafty backwoods confidence man, see *Adventures of Captain Simon Suggs* (Louisville, Ky.: Lost Cause Press, 1958—orig. Philadelphia, 1843).

The McKenney–Hall *Indian Tribes* may be instructively contrasted with George Catlin's two-volume *Letters and Notes on the Manners, Customs, and Condition of the North American Indians* (New York: Wiley and Putnam, 1841). The four hundred illustrations and painstaking field notes therein provided truly authentic materials on the trans-Mississippi tribes and revealed why Hall was so anxious to have Catlin join him and McKenney in their makeshift venture. For Catlin's experiences exhibiting his work abroad, see his *Notes of Eight Years' Travels and Residences in Europe* (New York: George Catlin, 1848). The best biography remains Lloyd Haberly's *Pursuit of the Horizon: A Life of George Catlin, Painter & Recorder of the American Indian* (New York: Macmillan, 1948), the primary source of my biographical data. Unlike the colonel and the judge, Catlin approached Indian cultures with respect, and that made him a very rare white man indeed—modern anthropologists regard his *O-Kee-pa: A Religious Ceremony* (Philadelphia: J. B. Lippincott, 1867) as a classic in the ethnology of North America.

In 1855, collecting data for his *Cyclopaedia of American Literature*, Evert A. Duyckinck wrote the most prominent writers in the United States for biographical information. Judge Hall replied at length, with much more about himself than Duyckinck could use. David Donald found the document in the New York Public Library and edited it as "The Autobiography of James Hall, Western Literary Pioneer," *Ohio State Archaeological and Historical Quarterly*, LVI (1947), 295–304. Other biographical data appear in John T. Flanagan, *James Hall: Literary Pioneer of the Ohio Valley* (Minneapolis: University of Minnesota Press, 1941).

Mulford Q. Sibley had an illuminating discussion of Thomas Aquinas and the notion of a "just war" in *Political Ideas and Ideologies* (New York: Harper & Row, 1970), pp. 229–52. On the pervasiveness of Indian-hating in Judge Hall's adopted state, see

Cecil Eby's searing *"That Disgraceful Affair," the Black Hawk War* (New York: W.W. Norton, 1973).

Lincoln's biographers John G. Nicolay and John Hay provided valuable evidence on Indian-hating in Illinois, as they had on the earlier years in Kentucky, and on the continuity of the phenomenon down to the 1890s—see *Abraham Lincoln: A History* (New York: Century, 1890), in the ten volumes of which Indians mattered so little they had no index entry of their own, while there appeared of course "Lincoln, Mordecai . . . hatred of the Indians, I, 21." Unlike his uncle, Lincoln was not an active hater, though he grimly ordered Governor Alexander Ramsey of Minnesota to "attend to the Indians" when the starving Sioux, defrauded by Agent Lucius C. Walker, went on the warpath in the summer of 1862 (Roy P. Basler, ed., *Collected Works of Abraham Lincoln* [New Brunswick, N.J.: Rutgers University Press, 1953], V, 396–97). Following this uprising some three hundred of the Sioux were condemned to death. Lincoln showed mercy or rather relative mercy by commuting the sentences of all but thirty-eight, whom he ordered hanged. Their mass execution at Mankato in December 1862 was an overture to the Plains Indians Wars that were to continue down to Wounded Knee in the last days of 1890. For Lincoln's Indian policy generally, see *Collected Works,* V, 397n, 493n, 525–26, 542–43; VI, 6–7; see especially his "Message to the Senate, December 11, 1862," on "the late Indian barbarities in the State of Minnesota," V, 550–51. (For a useful survey of this hitherto neglected topic, see David A. Nichols's *Lincoln and the Indian: Civil War Policy and Politics* [Columbia: University of Missouri Press, 1978].)

In the Department of Interior, to which the Bureau of Indian Affairs had been moved, Lincoln appropriately placed the Reverend James Mitchell of Indiana, another philanthropist, as his agent for emigration of the blacks. Though Lincoln's colonization projects went further than the schemes of Jefferson and Monroe, they were equally vacuous and equally revealing—for Lincoln's views of black inferiority and of the necessity of making the United States a white man's country, see *Collected Works,* II, 255–56; III, 145–46, 178–86; V, 370–75. Lincoln's racism was discussed under the interesting heading "White Nationalism" by George M. Fredrickson in *The Black Image in the White Mind* (New York: Harper & Row, 1971), pp. 149–51.

That during the Seminole War the United States became "a trafficker in human flesh" was the sad conclusion of Andrew C. McLaughlin in *Lewis Cass* (Boston: Houghton, Mifflin, 1891), p. 159. Fairfax Downey, a modern military historian, contended the Seminole War "presented an example of guerrilla warfare as fought today"—that is, in Indochina and elsewhere—and drew the lesson that "only after the Army adopted counter-guerrilla tactics, did it prevail"—*Indian Wars of the United States Army, 1776–1865* (New York: Doubleday, 1963), pp. 116–17.

On Major Stephen Harriman Long's famous map the vast area between the Rockies and the Missouri was labeled the GREAT DESERT and bore the legend: *"The Great Desert is frequented by roving bands of Indians who have no fixed place of residence but roam from place to place in quest of game."* But even there cartographical contradictions appeared, in the "Pawnee vil." and "Oto & Missouri vil." located on the Platte. In Edwin James's narrative of the expedition, Major Long more cautiously declared the region west of Council Bluffs "almost wholly unfit for cultivation, and of course uninhabitable by a people depending upon agriculture for their subsistence"—James, *Account of an Expedition from Pittsburgh to the Rocky Mountains Performed in the Years 1819, 20 . . . ,* in Reuben Gold Thwaites, ed., *Early Western Travels, 1748–1846,* vols. XIV–XVII (Cleveland: Arthur H. Clark, 1905), XVII, 147. But Long's contemporaries quickly translated this to mean that if it was "wholly unfit for cultivation," it was, as Hall put it, "uninhabitable by human beings" and therefore just the place for the roving bands of Indians.

Of Yankee Sargeant Prentiss as the ideal Mississippian, W. J. Cash in *The Mind of the South* (New York: Vintage, 1960), p. 75, wondered: "Do I need to add that the politician universally succeeds in the measure in which he is able to embody, in deeds or in words, the essence, not of what his clients are strictly, but of their dreams of themselves?" With these careful qualifications, Cash might have been discussing Easterner Hall in Illinois embodying in himself the dreams of clients in the Old Northwest.

The title of John T. Flanagan's *James Hall: Literary Pioneer of the Ohio Valley* ironically echoed the many titles of the judge's quintessential book on the West—for a listing of those titles that leaves the bibliographical problem undisturbed, see pp. 207–8. Hall's *Letters from the West* (London: Henry Colburn, 1828) is available in a modern reproduction (Gainesville, Fla.: Scholars' Facsimiles & Reprints, 1967). For "The Indian Hater" who made his first appearance in 1828, see *Wilderness and the Warpath* (New York: Wiley and Putnam, 1846). *Wilderness and the Warpath* was recently reprinted (New York: Garrett Press, 1969). Melville's chief source on Indian-hating and on Moredock was Hall's *Sketches of History, Life, and Manners in the West,* 2 vols. (Philadelphia, 1835); it was reprinted as *The Romance of Western History* (Cincinnati:

R. Clarke, 1885), chap. XXVI, and is perhaps more accessible under that title.

In *Worcester* v. *Georgia* (6 Peters 515–96, 1832) Chief Justice John Marshall said for the majority of the Supreme Court that "the Indian nations had always been considered as distinct, independent, political communities, retaining their original natural rights, as the undisputed possessors of the soil, from time immemorial. . . . The very term 'nation,' so generally applied to them, means 'a people distinct from others.' " Hall refused absolutely to view them as a distinct *people*, as *nations* with their own governments, though he deviously maintained he was not disposed to dissent from the "high authority" of Marshall's opinion (ITNA, III, 224–28). In reality his might-makes-right position was identical with Jackson's and for that matter McKenney's, when as superintendent of the Indian service the colonel had acted to keep the Cherokees from framing their own constitution. Furthermore, notwithstanding Marshall's opinion, Hall's denial of Indian humanity had the ample precedents discussed earlier. Relevant on this point was Joseph Story's enormously influential *Commentaries on the Constitution* (Boston: Hilliard & Gray, 1833): In the beginning, maintained Marshall's colleague, European nations had granted Indians "a mere right of occupancy. As infidels, heathen, and savages, they were not allowed to possess the prerogatives belonging to absolute, sovereign, and independent nations. The territory over which they wandered, and which they used for their temporary and fugitive purposes, was, in respect to Christians, deemed as if it were inhabited only by brute animals" (Section 152). That was Hall's position exactly and, despite the Marshall Court, Jackson's and the country's.

From before William Bradford to beyond James Hall, the illusion of "unpeopled wilderness" had the term *territorium nullius* as its direct counterpart in international law. Unpeopled lands, uninhabited or inhabited only by isolated individuals incapable of united political action, were *territoria nullius* and thus might be seized without legal (or moral) qualms by members of the "civilized" family of nations. The centuries-long debate over when territory was "vacant" always revealed more about the debaters' assumptions and attitudes than it did about native peoples living in vacuo. In the increasingly frenetic scramble of the imperial powers for Africa and Asia, hard-line racist views such as Hall's became increasingly respectable. For example, John Westlake, the distinguished English authority on international law, defined *nations* as only those with governments recognized by the international family of nations and that meant "all the states of European blood, that is all the European and American states except Turkey, and of Japan." Apart from these awkward exceptions to European blood, "civilization" was thus visibly white and the nation state what Rudyard Kipling would call "the White Man's Burden." For Westlake, see *International Law* (Cambridge: Cambridge University Press, 1910), part I, chap. 5; *Collected Papers of John Westlake on Public International Law*, ed. L. Oppenheim (Cambridge: Cambridge University Press, 1914), chap. 9. For a useful survey of this most revealing terrain, see M. F. Lindley's *Acquisition and Government of Backward Territory in International Law* (London: Longmans, Green, 1926).

My parenthetical page references to Melville's novel are to *The Confidence-Man: His Masquerade*, ed. Elizabeth S. Foster (New York: Hendricks House, 1954). Foster's notes are very helpful, but her misreading of Hall led her astray in dealing with the critically important "Metaphysics of Indian-Hating." I have benefited most from Edwin Fussell's reading of the novel in *Frontier: American Literature and the American West* (Princeton: Princeton University Press, 1965), pp. 303–26; Fussell mentioned the Western steamboat *Fidelity* listed in Hall's *Statistics of the West* (1837). In addition to the discussion of Melville in Roy Harvey Pearce's *Savages of America: A Study of the Indian and the Idea of Civilization* (Baltimore: Johns Hopkins Press, 1953), see also his "Metaphysics of Indian-Hating," *Ethnohistory*, IV (Winter 1957), 27–40.

Chapter XVI
The American Rhythm: Mary Austin

Stevenson's chapter "Despised Races" was first published by *Longman's Magazine* in 1883 and then in *Across the Plains* (London: Chatto and Windus, 1892). But see James D. Hart, ed., *From Scotland to Silverado* (Cambridge: Harvard University Press, 1966), pp. 100–47, where it appears restored by materials deleted from the first edition. The passage on Crusoe and the expurgated sentence are on p. 108.

Philosopher Josiah Royce assigned responsibility for the viciousness of the early days in California to two factors: "a general sense of irresponsibility, and . . . a diseased local exaggeration of the common national feeling towards foreigners"—*California from the Conquest in 1846. . . .* (Boston: Houghton, Mifflin, 1886), p. 276. What Stevenson found typical of "all Western America" was indeed a regional exaggeration of the common national animosity we have thus far traced to the Midwest, but white treatment of dark-skinned peoples went beyond xenophobia to a racism that became more inflamed as it bumped across the continent and down the Pacific slope. The massacres of Native Americans there were extensions and accelerations of patterns formed some three thousand miles back. Of the most conservatively estimated 100,000 Indians in California at the time of the Gold Rush —in 1974 Sherburne Cook estimated the aboriginal population as 375,000—some 70,000 were killed or exterminated by disease and starvation ten years later, a slaughter so wholesale Hubert H. Bancroft called it "one of the last human hunts of civilization, and the basest and most brutal of them all"—for an analysis of race prejudice against Indians, Mexicans, and Chinese, see Bancroft, *Works* (San Francisco: History, 1882–90), XXV, 561 et passim. For a more recent analysis of the human hunt, see Sherburne Cook, *The Conflict between California Indians and White Civilization* (Berkeley: University of California Press, 1976), especially part III, "The American Invasion," pp. 255–361.

Manifestly westward the course of epithets took their way. The "ravening wolves" of Endicott's Massachusetts became the "consumptive wolves" of Boone's Kentucky and the "wild varments" of Glass's upper Missouri. As Indian-hating went westward in space the metaphor became more disdainful. I have already quoted John House on his murder of an Indian baby at the mouth of the Bad Axe in 1832: "Kill the nits, and you'll have no lice" (see text, p. 199n). To the West those words were echoed in 1864 by Colonel J. M. Chivington, who commenced butchery of the Cheyennes at Sand Creek by ordering his Colorado militiamen: "Kill and scalp all, big and little; nits make lice" (quoted by William Meyer, *Native Americans* [New York: International Publishers, 1971], p. 32). Still farther West, in the foothills of Mount Lassen that same year, Yanas were exterminated to the same refrain: "We must kill them big and little," said a pest-controller; "nits will be lice" (quoted by Theodora Kroeber, *Ishi in Two Worlds* [Berkeley: University of California Press, 1964], p. 76). Hence the deflected epithet "hideous vermin" hit the despised Chinese with transcontinental momentum.

Frank M. Pixley's testimony of 1876 before the congressional committee was quoted by Jacobus tenBroek, Edward N. Barnhart, and Floyd W. Matson, *Prejudice, War, and the Constitution* (Berkeley: University of California Press, 1954), p. 21. For a good analysis of the racist underpinnings of the anti-Japanese heritage in California, see ibid., pp. 11–67. Still more critically important are Roger Daniels's *The Politics of Prejudice* (New York: Atheneum,

1968), and *Concentration Camps USA* (New York: Holt, Rinehart and Winston, 1971).

Of Mary Austin's thirty-some books (and over two hundred articles), her first was perhaps the best: *The Land of Little Rain* (Boston: Houghton, Mifflin, 1903). Her autobiography, *Earth Horizon* (Boston: Houghton, Mifflin, 1932), revealed her curious blend of acute sensitivity and no less notable obtuseness. Her stress on the pioneer women in her background reflected her feminist rebellion against the male-dominated 1880s and 1890s, and her later participation in the suffrage and birth-control activities of Mabel Dodge Luhan and her circle in Manhattan—from 1912 Mary Austin divided her time between Carmel and New York until she moved permanently to Santa Fe in 1924. Her awesome capacity to misunderstand people and what they stood for was illustrated by her "Marxian Socialist" label for anarchist Emma Goldman, whom she said she liked: "for Emma, modern history began with Marx," a contention roughly equivalent to saying that for an atheist, modern history began with Christ. On the other hand, she probably quite rightly remembered that William James evidenced quick interest in Paiute prayer techniques when she told the philosopher about them in an Oakland interview and "that he did confirm my own experience that prayer is not merely an emotional reaction but a creative motion" (pp. 282–83).

Perhaps she was right as well that as a woman her "times" helped lead her to *The American Rhythm: Studies and Reëxpressions of Amerindian Songs* (New York: Cooper Square Publishers, 1970—orig. Houghton, Mifflin, 1923, 1930). For a summary of the views therein, see her "American Indian Dance Drama," *Yale Review*, XIX (June 1930), 732–45. For a medicine man's corroboration that Indian dancing was Indian praying, see John Fire/Lame Deer and Richard Erdoes, *Lame Deer: Seeker of Visions* (New York: Simon and Schuster, 1972), pp. 243–44 et passim; see also Sioux Chief Standing Bear's *Land of the Spotted Eagle* (Boston: Houghton, Mifflin, 1933).

For a critical but sympathetic assessment of her ideas, see Dudley Wynn, "Mary Austin, Woman Alone," *Virginia Quarterly Review*, XIII (Spring 1937), 243–56. Elémire Zolla has a thoughtful discussion of her in *The Writer and the Shaman* (New York: Harcourt Brace Jovanovich, 1973), pp. 187–92. The best source of biographical data is her autobiography. T. M. Pearce's *Mary Hunter Austin* (New York: Twayne, 1965) contains information rendered less useful by elementary errors of chronology.

An important new study has recently appeared whose subtitle underlines Mary Austin's theme: Reginald and Gladys Laubin,

Indian Dances of North America: Their Importance to Indian Life (Norman: University of Oklahoma Press, 1976). Though the Laubins appear unaware of it, they demonstrably follow the path she blazed, as their epigraph shows: "Observation and study of Indian dances in earlier days might have revealed the very soul of the people, for they were at one and the same time the focal point of all their material culture and the highest expression of their mystical yearnings" (p. ix).

Chapter XVII
The Manifest Destiny of John Fiske

By now the reader would expect John Fiske to have said a word or two about Crusoe's island as he passed through that moonscape in Wyoming. He did not but had earlier on a swing through Maine that turned up a number of family friends: "Can't find much Robinson Crusoe in these days," he wrote his wife (L, p. 418); "all the desolate islands are colonized with your next-door neighbor's second cousins!"—this and parenthetical references in the text are to *The Letters of John Fiske*, ed. Ethel F. Fisk (New York: Macmillan, 1940). Deficiencies in this selection of letters may be partly overcome by use of John Spencer Clark's *Life and Letters of John Fiske*, 2 vols. (Boston: Houghton Mifflin, 1917). "Manifest Destiny" was published in *Harper's New Monthly Magazine*, LXX (March 1885), 578–90, and in *American Political Ideas Viewed from the Standpoint of Universal History* (API) (Boston: Houghton Mifflin, 1885)—the edition of 1913, ed. John Spencer Clark, included "The Story of a New England Town," pp. 147–84, published first in the *Atlantic Monthly*, LXXXVI (December 1900), 722–35. Fiske's "Story" recapitulated ideas set forth in *The Beginnings of New England* (Boston: Houghton Mifflin, 1889) and earlier historical writings.

In "Some Analogies between Calvinism and Darwinism," *Bibliotheca Sacra*, XXXVII (January 1880), 48–76, George Frederick Wright, the Calvinist editor and noted geologist, perceptively focused on the parallels of these supposedly antithetical doctrines and rightly observed that long before Darwin the Calvinist had "stood in the breach, and defended the doctrine that order is an essential attribute of the divine mind." Wright was quoted by Milton Berman in his outstanding *John Fiske: The Evolution of a Popularizer* (Cambridge: Harvard University Press, 1961), p. 183. Berman was especially good in placing Fiske within the changing configurations of Protestant denominations and sects; he was not effective in dealing with his subject's nativism and racism; the passages from Fiske's letters on Native Americans figured not at all in his considerations.

Fiske's sole contribution to Darwinism, "The Meaning of Infancy," appeared in *Excursions of an Evolutionist* (Boston: Houghton Mifflin, 1883), pp. 279–91, a collection of essays that also included "Our Aryan Forefathers," pp. 68–96. Still useful for the period is Richard Hofstadter's *Social Darwinism in American Thought* (Boston: Beacon Press, 1955), though it exaggerates the impact of evolution on ways of thinking and feeling in the United States. Clearly Darwinism and Christianity were not the implacable enemies they appeared to alarmed souls at the time. With a few relatively minor adjustments, Protestants and Catholics accommodated themselves to the new doctrine, with the help of such reconcilers as Fiske, in large measure because it was, like Locke's *tabula rasa*, a blank sheet on which anyone could write virtually at will. That they did, theists and atheists alike, nationalists and antistatists alike, such laissez-faire champions as Fiske and William Graham Sumner, and such socialists as Karl Marx, who quickly found in the *Origin* "a basis in natural science for the class struggle in history"—quoted by Gertrude Himmelfarb in *Darwin and the Darwinian Revolution* (Garden City, N.Y.: Doubleday, 1959), p. 398. She demonstrated that the so-called crisis of faith antedated Darwinism, making it a very conservative revolution indeed for such supporters as Thomas Henry Huxley, Spencer, and Fiske, who really experienced "not the shock of discovery but rather the shock of recognition. They were so easily converted because there was little to be converted to" (p. 423). That applied with vengeance to the United States in the Age of the Spoilsmen. Darwinism merely provided Andrew Carnegie and other robber barons with a rationale in natural science for what they were already doing so well. Henry Ward Beecher's preference for such "design by wholesale" was quoted by Himmelfarb, p. 373.

In "The Dark-Skinned Savage: The Image of Primitive Man in Evolutionary Anthropology," *Race, Culture, and Evolution* (New York: Free Press, 1968), pp. 110–32, George W. Stocking, Jr., sensitively traced Darwin's dependence on the stage-of-society theorists of the eighteenth century, related his biological evolutionism to current polygenist race notions, and concluded: "In turn-of-the-century [1900] evolutionary thinking, savagery, dark skin, and a small brain and incoherent mind were, for many, all part of the single evolutionary picture of 'primitive' man, who even yet walked the earth." If so, then evolutionary thinking had not changed racism all that much. A wealth of evidence in Part One of the present study demonstrates that "savagery" and dark skin were parts of a single image for the Puritans, and in Parts Two and Three that "savagery," dark skin, and incoherent (childish) mind were parts of a single image for Anglo-Americans from Timothy Dwight to McKenney and Hall. As for the 1900 stress on a small brain, that was part of the message John Augustine Smith of Virginia had brought back from Europe in 1808 along with his newly earned medical degree. He assured students in New York, Winthrop Jordan wrote, that "the Negro's brain was firmer and smaller (by one-thirteenth) than the European's"—*White over Black* (Baltimore: Penguin, 1969), pp. 505–6. Even the relatively recent "at least implicitly organic" stress on *races* instead of *nations* or *peoples* had all sorts of precedents in the ante-Darwin past, as when Henry Clay told John Quincy Adams and the rest of his cabinet that the Indians were not, "as a race, worth preserving. He considered them as essentially inferior to the Anglo-Saxon race, which were now taking their place on this continent. They were not an improvable breed" (text, p. 179). Clay's assertion of Anglo-Saxon superiority and imputation of Indian inferiority were more than implicitly organic.

Attracted to William Paley's arguments from design—that is, as a watch implied a watchmaker, the creation implied a designer, the Deity—because of their logic and Euclidean clarity, Darwin had a view of "civilization" and of "savagery"—such as that of the scarcely human Feugians—that never changed in its essentials. On his visit to Sydney in January 1836, for instance, he was pleased by the progress Europeans had brought to backward Australia, especially liked the regular streets and roads built on the Macadam principle, and wrote admiringly to his sister: "This is really a wonderful Colony; ancient Rome, in her imperial grandeur, would not have been ashamed of such an offspring"—quoted by Bernard Smith in *European Vision and the South Pacific, 1768–1850* (London: Oxford University Press, 1960), p. 216, a remarkable work that showed how the European rage for order went hand in hand with hatred for the dark-skinned aborigines. The consequence was an antipodean Manifest Destiny with a racist core: Racism, declared Humphrey McQueen, "is the most important single component of Australian nationalism"—*A New Britannia* (Melbourne: Penguin, 1970), p. 42. For what progress did to native peoples on that continent, see C. C. Rowley, *The Destruction of Aboriginal Society* (Canberra: Australian National University Press, 1970); and F. S. Stevens, ed., *Racism: The Australian Experience*, 2 vols. (Sydney: Australia and New Zealand Book, 1971).

As many writers have pointed out, Darwin had in mind pigeons when he formulated his *Origin* subtitle: *The Preservation of Favoured Races in the Struggle for Life* (1859). But as previously noted, he himself applied the principle to human races in *The Descent of Man* (1871). In the latter work he shifted from *natural selection* to *sexual selection* to account for racial differences. Sexual selection was based on standards of beauty that over generations established characteristics distinguishing one race from another. *Color* was basic to the standards—Himmelfarb quoted Darwin's ready sympathy for readers who might think it "a monstrous supposition that the jet blackness of the negro has been gained through sexual selection," but so it was (*Darwin*, p. 344). In similar fashion and no less scientifically, William Gilmore Simms had invoked esthetic criteria in 1835 and highlighted color, "perceptible to our most ready sentinel, the sight," to explain the formation of inferior castes (text, p. 145). Despite his aversion to slavery, Darwin found the Negro's jet blackness no less monstrous than had Simms. Like his biology, Darwin's racism evolved out of his background as Simms's had out of his. For an account that placed Darwinian racism *within* its Victorian context, see L. P. Curtis, Jr., *Anglo-Saxons and Celts* (Bridgeport, Conn.: University of Bridgeport, 1968). (My quotation of Edward Augustus Freeman's proposal for getting rid of Irishmen and Negroes comes from Curtis's valuable chapter "Anglo-Saxonist Historiography," p. 81.) Curtis accepted the proposition that "race was perhaps the strongest supporting mechanism of the British Empire" (p. 149), a view that dovetails with evidence set forth in the present study on the rise of the Anglo-American empire. See also Bradford Perkins, *The Great Rapprochement: England and the United States, 1895–1914* (New York: Atheneum, 1968), pp. 51–52, 74–88—Fiske was one of the key figures in bringing this rapprochement about.

For the Manifest Destiny of the Canadian branch of the English-speaking race, as the evolutionist would have it, see Palmer Patterson, "The Colonial Parallel: A View of Indian History," *Ethnohistory*, XVIII (Winter 1971), 1–17, and his *Canadian Indian: A History since 1500* (Don Mills, Ontario: Collier–Macmillan, 1972), both of which emphasize the continuity of an imperial conquest that left behind internal colonies of native peoples from coast to coast. For imperialism and Afro-Americans in this period, see Rayford W. Logan, *The Betrayal of the Negro: From Rutherford B. Hayes to Woodrow Wilson* (New York: Collier, 1968), esp. pp. 242–75. And for the ongoing assault against First Americans, see the convenient collection of documents edited by Wilcomb E. Washburn, *The Assault on Indian Tribalism: The General Allotment Law*

(Dawes Act) of 1887 (Philadelphia: J. B. Lippincott, 1975). The attack was spurred on by the Reverend Doctor Lyman Abbott, who "frequently quoted and praised Fiske" (Berman, *John Fiske*, p. 180); Abbott was editor of the influential *Christian Union* (later *The Outlook*) and forcefully insisted "The Reservation Must Go" in the issue of July 16, 1885: "Barbarism has no rights which civilization is bound to respect." At the end of the century he was still hammering away at the need for complete abolition of reservations and exposure of Indians to the marketplace economy: "Treat them as we have treated the negro. As a race the African is less competent than the Indian, but we do not shut the negroes up in reservations and put them in charge of politically appointed parents called agents"—"Our Indian Problem," *North American Review*, CLXVII (December 1898), 719–28. Concurrently Abbott enthusiastically approved annexation of the Philippines and hailed McKinley's policy as "the new Monroe Doctrine, the new imperialism, the imperialism of liberty"—*Outlook*, August 27, 1898. For an overview of Abbott and other Lake Mohonk liberals, see Wilbert A. Ahearn, "Assimilationist Racism: The Case of the 'Friends of the Indian,' " *Journal of Ethnic Studies*, IV (Summer 1976), 23–32. For U.S. Indian policy after the turn of the century, see Robert F. Berkhofer, Jr., *The White Man's Indian* (New York: Alfred A. Knopf, 1978), pp. 176–97.

In fine, Darwinism did for racism in the United States precisely what it did for the possessive individualism of the marketplace: it legitimated ongoing prejudices, practices, and policies. I do not deny, of course, that it led to reformulations of racist thought in more explicitly biological terms, nor that its presumably neutral base in the natural sciences encouraged the spread of racist concepts in the social sciences and humanities. I do assert that the racism—along with insatiable acquisitiveness, spread-eagle patriotism, and religious bigotry—that took the United States out to those rich and fertile islands in the Pacific was of a piece with the racism that went back far beyond Darwin to the beginnings of the imperial conquest of North America—see my discussion of racism in the 1600s, pp. 48–53. Only in the superficial sense that its implementation took soldiers, missionaries, and businessmen across salt water was it a "new imperialism."

Rich irony lies in the fact that John Fiske had a keener sense than Richard Hofstadter, Walter LaFeber, George M. Fredrickson, and other more recent historians of the unbroken continuity of fundamental attitudes bequeathed by "Our Aryan Forefathers." For example, in the standard *Expansionists of 1898: The Acquisition of Hawaii and the Spanish Islands* (Baltimore: Johns Hopkins Press, 1936), Julius W. Pratt had a chapter titled "The New Manifest Destiny" that might more accurately have been titled "The *Extension* of Manifest Destiny."

In *Manifest Destiny and Mission in American History* (New York: Vintage, 1966), Frederick Merk, Frederick Jackson Turner's successor at Harvard, stated baldly: "The appearance of a demand for overseas colonies coincided in time with a revival of racism in the United States" (p. 237). Like the old-time religion, racism revived on suitable occasions, apparently—a curious view of a continuous phenomenon. Merk even questioned, somewhat contradictorily, whether racism did in fact flourish during the 1890s and submitted a reading of Fiske that can only be termed an exercise in mystification. Merk contended that for Fiske's benign concept of world federalism his "Manifest Destiny" title was merely a foil: "Actually it was derisive of Manifest Destiny in the sense in which the term had been used in the 1840's. It devoted some paragraphs to poking fun at the idea of a United States extending from pole to pole" (p. 239). Actually Fiske solemnly and reverentially incorporated the expansionism of the 1840s on his way to a millennium when the English race, so called, would rule the world. His enthusiastic audiences from coast to coast heard very differently than the professor read.

For Fiske's nativism, see his extended defense of his presidency of the Immigration Restriction League (L, pp. 666–71). He believed the "new" immigrants were "beaten men from beaten races" but never went as far as James H. Patten of the league, who later called them "kindred or brownish" races and thus threats to white America—quoted by Barbara Miller Solomon, *Ancestors and Immigrants* (Cambridge: Harvard University Press, 1956), p. 126, the best source on the league and Fiske's connection with it. See also John Higham, *Strangers in the Land* (New Brunswick, N.J.: Rutgers University Press, 1955); Thomas F. Gossett, *Race: The History of an Idea in America* (New York: Schocken Books, 1965), pp. 287–338.

Fiske's view of the Capitol at Hartford from "sacred ground" and his nationalization of Calvin may perhaps be better understood in the light of Robert N. Bellah's "Civil Religion in America," *Daedalus*, XCVI (Winter 1967), 1–21. For a symposium on this provocative theme, see Russell E. Richey and Donald G. Jones, eds., *American Civil Religion* (New York: Harper & Row, 1974). See also "The Problem of National Destiny," *Newberry Library Bulletin*, IV (August 1957), 165–91.

Chapter XVIII
Outcast of the Islands: Henry Adams

Abbreviated references in this chapter are to the following sources:

(Cater) Harold Dean Cater, ed., *Henry Adams and His Friends: A Collection of His Unpublished Letters* (Boston: Houghton Mifflin, 1947).

(E) *The Education of Henry Adams* (Boston: Houghton Mifflin, 1961—orig. 1918).

(L) Worthington Chauncey Ford, ed., *Letters of Henry Adams*, 2 vols. (Boston: Houghton Mifflin, 1930, 1938).

(Rem) John La Farge, *Reminiscences of the South Seas* (Garden City, N.Y.: Doubleday, Page, 1912)—passages from La Farge's travel diary were published earlier in *Scribner's Magazine*, XXIX (January–June 1901), 670–84 and XXX (July–December 1901), 69–83.

(Samuels) Ernest Samuels, *Henry Adams: The Major Phase* (Cambridge: Harvard University Press, 1964)—the best biography; but see also Elizabeth Stevenson, *Henry Adams* (New York: Collier, 1961); and, for Adams's working ideas, William H. Jordy, *Henry Adams: Scientific Historian* (New Haven: Yale University Press, 1952).

(TD) Tyler Dennett, *John Hay: From Poetry to Politics* (New York: Dodd, Mead, 1934).

(WRT) William Roscoe Thayer, *The Life and Letters of John Hay*, 2 vols. (Boston: Houghton Mifflin, 1915).

Adams's nutshell history lesson stuck, for La Farge later declared emphatically, in his own voice, "the Pacific should be ours, and it must be" (Rem, p. 278). Meanwhile, his instructor's view of the importance of the Western Sea had changed. In terms no less sweeping than his previous dicta, Adams wrote Henry Cabot Lodge from Sydney:

> On the whole, I am satisfied that America has no future in the Pacific. She can turn south, indeed, but after all, the west coast of South America offers very little field. Her best chance is Siberia.

> Russia will probably go to pieces; she is rotten and decrepit to the core, and must pass through a bankruptcy, political and moral. If it can be delayed another twenty-five years, we could Americanise Siberia, and this is the only possible work that I can see still open on a scale equal to American means. [August 4, 1891; L, I, 511]

Siberia turned out to be beyond American means, but destiny offered up instead those Pacific islands as stepping-stones to the fabled China market and to the hundreds of millions waiting to be Christianized and Americanized. Nevertheless, Tyler Dennett concluded that John Hay—who listed as one of his great disappointments not having joined Adams and La Farge in touring a part of the world that had always fascinated him—may well have found in the former's expert opinion, just quoted, the basis for his Far Eastern policy (TD, p. 289).

The very title of one of Gauguin's paintings of native women, *Non dépourvuez de sentiment* ("Not Devoid of Feeling") rebuked Adams's contention that they had "no longings and very brief passions." Gauguin had the wisdom to know that man "trails his double after him." Unlike Adams, who never stopped fleeing from that fact, he faced it directly, acknowledging in himself two natures, "the Indian and the sensitive man"; he was a "savage-in-spite of myself," a formulation that calls to mind Henry David Thoreau's still more felicitous phrasing: "Inside the civilized man stands the savage still in the place of honor." Gauguin's writings were studded with insights into missionaries, government officials, and the Crusoe theme we have pursued—see *Paul Gauguin's Intimate Journals*, trans. Van Wyck Brooks (New York: Crown, 1936). Useful for biographical data is Lawrence and Elisabeth Hanson's *Noble Savage: The Life of Paul Gauguin* (New York: Random, 1954). See also Wayne Andersen's *Gauguin's Paradise* (New York: Viking Press, 1971).

Chapter XIX
The Open Door of John Hay

Abbreviated references in this chapter are to the following sources:

(Add) *Addresses of John Hay* (New York: Century, 1907).

(CPW) *The Complete Poetical Works of John Hay* (Boston: Houghton Mifflin, 1917)—therein appeared Hay's *Pike County*

Ballads and Other Pieces, originally published in Boston by James R. Osgood in 1871, with "Banty Tim" appearing in *The Poetical Works* on pp. 10–13.

(E) *The Education of Henry Adams* (Boston: Houghton Mifflin, 1961).

(KJC) Kenton J. Clymer, *John Hay: The Gentleman as Diplomat* (Ann Arbor: University of Michigan Press, 1975)—a recent and more useful supplement to the other biographies than Howard I. Kushner and Anne Hummel Sherrill, *John Milton Hay: The Union of Poetry and Politics* (Boston: Twayne, 1977).

(L) Worthington Chauncey Ford, ed., *Letters of Henry Adams* (Boston: Houghton Mifflin, 1930, 1938).

(TD) Tyler Dennett, *John Hay: From Poetry to Politics* (New York: Dodd, Mead, 1934)—the subtitle misleads, but this is still the standard biography written by a former State Department adviser sympathetic to his subject and to the official point of view.

(WRT) William Roscoe Thayer, *The Life and Letters of John Hay*, 2 vols. (Boston: Houghton Mifflin, 1915)—invaluable for Hay's letters, since the three-volume collection edited by his widow (1908) is untrustworthy, and for the biographer's expression of ethnocentric and racist views similar to his subject's and on occasion even more frankly put, as in his dismissal of the altruism of the white man's burden: "In reality, the White Man was not a philanthropist: he would treat the Black, Yellow, or Brown Man humanely if it was convenient, but if the dark-skinned resisted, the White Man would destroy him" (II, 250).

Like his biography, Kenton J. Clymer's "Anti-Semitism in the Late Nineteenth Century: The Case of John Hay," *American Jewish Quarterly*, LX (June 1971), 344–54, related to but did not deal with Hay's diary passage on the Viennese ghetto. Clymer pointed out that Hay did not have the vitriolic anti-Semitism of his friend Adams, that as secretary of state he brought pressure on Rumania on behalf of Jews, and that the Jewish community in the United States regarded him as a champion, with B'nai B'rith proposing after his death a monument in his memory. But as Clymer also observed, Hay hoped Jews would support the Republican party and in every instance involving them acted under pressure, reluctantly, and without the sympathy they attributed to him, as a 1902 letter to a subordinate showed: Roosevelt was greatly pleased over the protest to Rumania, Hay wrote, "and the Hebrews—poor dears! all over the country think we are bully boys" (KJC, p. 77).

In the division of labor with Nicolay, Hay was responsible for the first draft of the biographical chapters of *Lincoln* (New York: Century, 1890), and that meant he was the responsible or most responsible author for the saga of the Lincoln family in Kentucky and Illinois, and for the lines on Uncle Mordecai (cf. TD, pp. 137–38). Hay's Crusoe letter to Adams pledging to "search till death for the garden of Eden" is quoted in Thurman Wilkins's *Clarence King* (New York: Macmillan, 1958), pp. 315–16.

On his trip to Florida in 1864 Hay referred to blacks as "niggers," "darkeys," and "contrabands," as well as "negroes" (WRT, I, 161, 166, 157, 162). "Contrabands" were blacks who had become contrabands of war by escaping to or being brought within the Union lines. Thus in 1871 Hay wrote Nicolay that " 'Banty Tim' . . . touches the contraband" (WRT, I, 359). In addition to Hay's later references to "dagoes" and to "Chinks," mentioned in the text, in his diary Japanese were of course "Japs" (WRT, II, 374).

In Hay's novel *unwholesome* was everything not white and middle-class and sexually repressed. But no matter how hard he tried to *de-nigrare*, to blacken his Celt villain, he could not effectively put him beyond the pale of white society. And although nineteenth-century class and sexual beliefs reinforced racial prejudices, as Herbert G. Gutman maintained in *The Black Family in Slavery and Freedom, 1750–1925* (New York: Pantheon, 1976), pp. 295–96, *race* was not subsumed by *class*, as Gutman's approving quotation from sociologist Lewis Copeland would indicate. Beliefs that Afro-Americans lack "restraint" and are without the benefits of "high culture," Copeland has observed, "are simply another form of those fictions almost universally applied to the lower classes of society. Wherever one social class looks down upon another as inferior, members of the latter are regarded as brutish in nature and vulgar." True enough, but racial prejudice is *not* "simply another form" of class bias and the mess Hay made of *The Bread-winners*, I venture, illustrated that truth when he tried to merge the two, overloaded his circuit, and blacked out his Light Show.

If Hay's "Partnership in Beneficence" of the United States and Great Britain was inevitable, so "in the very nature of things" that "no man and no group of men can prevent it," then one wonders why he had to work so hard to strengthen the "bonds of union among the two great branches of our race." (For the classic account of such fantastic geopolitical determinism, see Albert K. Weinberg, *Manifest Destiny* [Baltimore: Johns Hopkins Press, 1935], esp. "The White Man's Burden," pp. 283–323.) Predestined empire conveniently meant dispossession and destruction without individual or collective responsibility. If responsibility had to be placed somewhere, then Providence was the guilty party, as in McKinley's declaration: "Congress can declare war but a higher Power decrees its bounds and fixes its relations and responsibilities"—quoted by David Healy in his useful *US Expansionism: The Imperialist Urge in the 1890s* (Madison: University of Wisconsin Press, 1970), p. 65.

For McKinley's War Message (April 11, 1898), see *Papers Relating to the Foreign Relations of the United States, 1898* (Washington: Government Printing Office, 1901), pp. 750–60. For Cleveland's message before leaving office, see the same source for 1896, published in 1897, pp. xxix–xxxvi.

For sources by and about John Quincy Adams, see p. 483. Samuel Flagg Bemis was the biographer, you will recall, with sufficient insight to see Adams's Monroe Doctrine as "a voice of Manifest Destiny." Bemis was also the authority who labeled the 1890s "the great aberration"—see his *Diplomatic History of the United States* (New York: Rinehart and Winston, 1955), pp. 463–75. For the fundamental continuity of U.S. expansionism on through that decade, see Richard W. Van Alstyne, *The Rising American Empire* (Chicago: Quadrangle, 1965). As the text indicates, I accept the argument of Van Alstyne, Tyler Dennett, and others that the Open Door policy had been adopted long before Hay's publicity stunt at the end of the century.

For Seward, see George E. Baker, ed., *The Works of William H. Seward*, 5 vols. (Boston: Houghton Mifflin, 1853–84), esp. I, 247–49, and IV, 124, 319. For Seward's veneration of Adams, see his *Life and Public Services of John Quincy Adams* (Auburn, N.Y.: Derby, Miller, 1849); see also his *Autobiography with a Memoir . . . by His Son, Frederick W. Seward*, 3 vols. (New York: Derby and Miller, 1891). My understanding of Seward leans heavily on the analysis of Van Alstyne, pp. 176–77; of Walter LaFeber, *The New Empire* (Ithaca: Cornell University Press, 1963), pp. 24–32; and of Charles Vevier, "American Continentalism: An Idea of Expansion, 1845–1910," *American Historical Review*, LXV (January 1960), 323–35—Vevier saw the reciprocal relationships of the internal continental empire and the overseas empire of markets and strategic bases, and helpfully related Seward's expansionism to that of other midcentury champions of the westward-moving empire, such as William Gilpin, Asa Whitney, Perry McDonough Collins, et al.

For Thomas Hart Benton's remarkable 1846 speech, see *Congressional Documents*, 29th Cong., 1st Sess., pp. 851–919. For samples of Benton's views of the inferior merciless savages, see his *Thirty Years in the U.S. Senate* (New York: D. Appleton, 1854, 1856), I, 58–64, 163–66, 624–26, 690–94, and II, 72–82.

Since women were denied formal political participation, they rarely figure in discussions of the rising empire. Nevertheless, those such as Charlotte Perkins Gilman and the others who joined in extolling the advance of "superior" over "inferior" peoples surely were not without influence—for Gilman's racism, see her "Suggestion on the Negro Problem," *American Journal of Sociology*, XIV (July 1908), 78–85. In a quieter way Clara Stone Hay regularly attended meetings of the Home Missionary Society (TD, pp. 156–57) and no doubt shared and reinforced her husband's views of native peoples. Ida Saxton McKinley was perhaps still more religious and, despite her invalidism, actively concerned about the pagan Igorots, a mountain tribe of northern Luzon we shall have occasion to discuss in another context; the president's wife was eager to convert these heathens and may well have hardened her husband's determination to hold on to the Philippine Islands—see Julius W. Pratt, *Expansionists of 1898* (Baltimore: Johns Hopkins Press, 1936), pp. 316, 334.

Chapter XX
Insular Expert: Professor Worcester

Abbreviated references in this chapter are to the following sources:

(KJC) Kenton J. Clymer, *John Hay* (Ann Arbor: University of Michigan Press, 1975).

(Knot) Professor Dean C. Worcester, "Knotty Problems of the Philippines," *Century Magazine*, LVI (October 1898), 873–79.

(Phil) ———, *The Philippine Islands and Their People* (New York: Macmillan, 1898, 1899).

(PP&P) ———, *The Philippines, Past and Present* (New York: Macmillan, 1914, 1921). To the posthumous 1930 edition Joseph Ralston Hayden, James Orin Murfin Professor of Political Science at the University of Michigan, added four new chapters and on pp. 3–79 a "Biographical

Sketch" of Worcester. The latter contains indispensable data—Arthur Stanwood Pier's chapter on Worcester in *American Apostles to the Philippines* (Boston: Beacon Press, 1950), pp. 69–82, is derivative and of interest merely as an exhibit of how the "days of empire" came down into the mid-twentieth century as an inspiration for young Anglo-Americans. The biography by Hayden (1887–1945) is important in its revelation of the continuity of influential attitudes toward Filipinos. An understudy of Worcester, Hayden himself went into the insular government as vice-governor (1933–35) and with his *Philippines: A Study in National Development* (New York: Macmillan, 1942, 1945, 1955) became an insular expert in his own right. Albeit with certain modifications of his discussions of "the backward Malay nation," Hayden remained a Worcesterine retentionist into the 1940s, as the concluding sentence of his *Philippines* showed: The United States "dare not leave its task in the Philippines half done" (p. 805).

(Rep) *Report of the Philippine Commission to the President* (or the Schurman report), I, January 31, 1900 (Washington: Government Printing Office, 1900).

For McKinley's confession to the Methodist ministers, see Charles S. Olcott, *Life of William McKinley* (Boston: Houghton Mifflin, 1916), II, 109–11. Walter Millis discussed McKinley's intercourse with Providence and his off-year election tour in *The Martial Spirit* (New York: Literary Guild, 1931), pp. 382–84. H. Wayne Morgan argued in *William McKinley and His America* (Syracuse, N.Y.: Syracuse University Press, 1963), pp. 388–412, that the president was inclined from the first to keep the islands.

President Taft's assessment of Worcester was passed on to James H. Blount, a former aide of General Arthur MacArthur and former district judge in the Philippines—see his chapter " 'Non-Christian' Worcester" in *The American Occupation of the Philippines, 1898–1912* (New York: Oriole Editions, 1973—orig. 1912), pp. 571–86. Blount was not fond of the professor, calling him with some exaggeration "the direst calamity that has befallen the Filipinos since the American occupation" and also "the P. T. Barnum of the 'non-Christian tribe' industry." Despite its occasional inaccuracies, Blount's work remains valuable for its insights, his own belated anti-annexationism, and his views of nonwhites.

Worcester's doubt that many generations of European or Anglo-American children could be reared in the Philippines suggested he had been reading Benjamin Kidd's *Control of the Tropics* (1898), except the Englishman's book was published concurrently with his own; Kidd's influential essay, "The United States and the Control of the Tropics," appeared in *Atlantic Monthly*, LXXXII (December 1898), 721–27. Kidd advanced a variant of John Fiske's old Manifest Destiny and of Hay's current "partnership in beneficence": The United States and Great Britain had to take control of nonwhite tropical peoples, who were childishly incapable of governing themselves. Their burden was complicated by the fact that whites could not successfully transplant themselves to the physically degenerating torrid zone—hence the need for white males to return home periodically to their own kind and clime or to develop another alternative, such as the insular government's summer capital at Baguio. Since I have discovered no evidence Worcester read Kidd, the chances are he developed his views of the tropics independently. Like Anglo-Saxonism and the white man's burden, such racist notions were in the air. For an important contemporary analysis of their relation to imperialism, see the economist Henry Parker Willis's *Our Philippine Problem: A Study of American Colonial Policy* (New York: Henry Holt, 1905), pp. 1–28.

The quotation of Jean-Paul Sartre is the first sentence of his preface to Frantz Fanon's *The Wretched of the Earth*, trans. Constance Farrington (New York: Grove Press, 1968). Noteworthy is the fact that Fanon's lines on the "colonial vocabulary" might have been written with Worcester in mind. When the latter described Filipinos as indolent children, he saw them as settlers or colonizers have always seen natives, as "those children who seem to belong to nobody, that laziness stretched out in the sun, that vegetative rhythm of life"—*Wretched*, pp. 42–43.

For a discussion of Merritt, Otis, and the other career officers who had been Indian-fighters, see Leon Wolff, *Little Brown Brother: America's Forgotten Bid for Empire Which Cost 250,000 Lives* (London: Longmans, 1961). The U.S. Consul Wildman quotation on treating Filipinos as if they were Indians is from Wolff, p. 112. The parallels extended to tactics, as General Emilio Aguinaldo observed in his account of the famous incident in which General Gregorio del Pilar lost his life defending a mountain pass in northern Luzon. The boy general was killed by a soldier from an American company of sharpshooters who had, "Indian-fighting fashion, clambered up the cliff"—Aguinaldo and Vincente Albano Pacis, *A Second Look at America* (New York: Robert Speller & Sons, 1957), p. 110. With a final ironic twist, the Macabebes, descendants of Mexican Indians brought to the Philippines by the Spanish, also served the United States as mercenaries who made excellent scouts—see Usha Mahajani, *Philippine Nationalism: External Challenge and Filipino Response, 1565–1946* (St. Lucia, Queensland: University of Queensland Press, 1971), p. 198; Aguinaldo and Pacis, *Second Look*, p. 124.

For Colonel Charles Denby, Worcester's ally on the Schurman Commission, see Richard W. Van Alstyne, *The Rising American Empire* (Chicago: Quadrangle, 1965), pp. 181–83. For thirteen years following his appointment in 1885, Denby had served business interests in China from his diplomatic post in Peking. "Fancy what would happen to the cotton trade if every Chinese wore a shirt!" he once exclaimed. "Well, the missionaries are teaching them to wear shirts!" In the February 1899 issue of the *Forum* he was blunt about the white man's burden: "The cold practical question remains: will the possession of these islands benefit us as a nation? If they will not, set them free tomorrow and let their peoples, if they please, cut each other's throats"—quoted by Moorfield Storey and Marcial P. Lichauco, *The Conquest of the Philippines by the United States* (New York: G. P. Putnam's Sons, 1926), p. 43. For Schurman's views, see his *Philippine Affairs: A Retrospect and Outlook* (New York: Charles Scribner's Sons, 1902). As Mahajani pointed out, Schurman later deplored use of the term *tribes* for Filipinos, preferring instead "the Philippine nation"—*Philippine Nationalism*, p. 229n. But that was after he brought himself to sign the report that commonly bore his name and gave the term wide and lasting circulation—Mahajani's excellent analysis of the report itself appears on pp. 225–29.

Aguinaldo's careful misreading of the U.S. Constitution was understandable. It was, after all, identical with that advanced on December 6, 1898, by Senator George G. Vest of Missouri in his joint resolution affirming that the United States had no power under the Constitution to acquire territories to be held and governed permanently as colonies. Vest's resolution went down to defeat after a debate in which he was answered by Senators Orville H. Platt of Connecticut and Henry Cabot Lodge of Massachusetts with the same arguments as those advanced in the Schurman report—see the *Congressional Record*, XXXII, 20, 93–96, 287–88. This view of the virtually absolute power of Congress to legislate for territories was then ratified by the Supreme Court in the Insular Cases, a series of decisions (1901–5) that in effect allowed the United States to become a colonial power so long as it called its colonies "territories." The court also ruled the Constitution did not follow the flag into territory that was not "incorporated"; in *Dorr* v. *United States* (195 U.S. 138; 1904) it ruled the Philippines not incorporated and therefore without the constitutional protection of trial by jury—see Frederic R. Coudert, "The Evolution of the Doctrine of Territorial Incorporation," *American Law Review*, LX (November–December 1926), 801–64. In 1905 Filipinos had about the same legal position as that of the Sioux on the Pine Ridge Reservation in South Dakota. They too were not citizens but subjects of a colonizing power that could extend or withhold basic human rights as it saw fit.

For McKinley's October 14, 1899, declaration that the United States was opposed by only "a portion of one *tribe*," see Mahajani, *Philippine Nationalism*, pp. 166–67. That month Root said the same thing in Chicago in a concoction that was pure Worcestershire sauce—see Blount, *American Occupation*, p. 528. In 1901 Governor-General Taft finally admitted that "the word 'tribe' gives an erroneous impression. There is no tribal relation among the Filipinos. . . . I cannot tell the difference between an Ilocano and a Tagalog or a Visayan. . . . To me all Filipinos are alike"—quoted by Storey and Lichauco, *Conquest of the Philippines*, p. 173. Yet the following year in the hearings before Lodge's Committee on the Philippines he fell back into Worcester's ethnology.

For Theodore Roosevelt's speech in Grand Rapids, Michigan, and his "Issues of 1900," the widely circulated letter accepting the Republican vice-presidential nomination, see Hermann Hagedorn, ed., *Works of Theodore Roosevelt* (New York: Charles Scribner's Sons, 1926), XIV, 537–60; see also his "Strenuous Life" speech in Chicago on April 10, 1899, XV, 267–81. The Rough Rider had his own Fiske- and Hay-like version of the Light Show: "Peace has come through the last century to large sections of the earth," he declared, "because the civilized races have spread over the earth's dark places"—quoted by Howard K. Beale, *Theodore Roosevelt and the Rise of America to World Power* (Baltimore: Johns Hopkins Press, 1956), p. 72.

As David Healy pointed out in *US Expansionism* (Madison: University of Wisconsin Press, 1970), in 1902 William T. Stead, English editor of the *Review of Reviews*, "entitled a special supplement of his review *The Americanization of the World*" (p. 30). This theme, already advanced by Anglo-American writers discussed earlier, was picked up in one guise or another by U.S. court historians and of course by Worcester. His biographer Joseph Ralston Hayden built it into the subtitle of his authoritative work on the Philippines: *A Study in National Development*. Most recently Peter W. Stanley put it in his title in slightly varied form: *A Nation in the Making: The Philippines and the United States, 1899–1921* (Cambridge: Harvard University Press, 1974). Stanley preferred "modernization" to "Americanization" but evidenced the same bland assumption that what was good for the United States was good for the world.

In 1902 the name of Worcester's Bureau of Non-Christian Tribes became the Ethnological Survey for the Philippine Islands; it eventually wound up in the Bureau of Science, where it remained under his direction. Joseph Ralston Hayden incorrectly contended that the Bureau of Non-Christian Tribes established under the Jones Act of 1916 (Sect. 22) had "no connection with the first bureau of that name" (PP&P, p. 425n). The name was the conceptual connection that demonstrated the longevity of Worcester's tribal misanalysis.

For Worcester's articles that helped perpetuate this misanalysis, see the following: "The Malay Pirates of the Philippines, with Observations from Personal Experience," *Century Magazine*, LVI (September 1898), 690–702; "The Non-Christian Tribes of Northern Luzon," *Philippine Journal of Science*, I (1906); "Field Sports among the Wild Men of Luzon," *National Geographic Magazine*, XXII (March 1911); most picturesque of all, "Head-Hunters of Northern Luzon," ibid., XXIII (September 1912), 833–930; and "The Non-Christian Peoples of the Philippine Islands with an Account of What Has Been Done for Them under American Rule," ibid., XXIV (November 1913), 1157–1256. Needless to say, the last three articles were profusely illustrated with photographs of spear-wielding, dog-eating warriors, of pygmies, and of nubile native maidens bare to the waist.

R. F. Barton's reference to the "colonial attitude" of Filipinos toward tribespeople appeared in his *Autobiographies of Three Pagans* (New York: University Books, 1963—orig. 1938), p. xlii. Barton went out to the Philippines in 1906 and shortly thereafter commenced his life work with the mountain tribes of northern Luzon. Though his first years there overlapped Worcester's, he brought to his study of tribes very different attitudes and ideas. The fascinating *Autobiographies* revealed him capable of treating natives with respect as persons; they in turn shared their life worlds with him. The three Ifugaos, for instance, obviously divided the life of their tribe into before- and after-Worcester periods or eras before and after *order*, which was what the professor's "non-Christian" policy was all about. "Formerly a priest was seldom or never called to officiate for non-related persons," explained Ngídulu, an Ifugao priest, "but now that the Americans have established 'order' it is becoming more frequent to invite priests from outside the kinship group" (pp. 112–113; cf. 38–39, 93). For the Kalingas, see Barton's posthumous *The Kalingas: Their Institutions and Custom Law* (Chicago: University of Chicago Press, 1949); a more recent study by a Native American is Edward P. Dozier's *Mountain Arbiters: The Changing Life of a Philippine Hill People* (Tuscon: University of Arizona Press, 1966). For a survey of ethnological evidence on these and the other tribes, see Marcelo Tangco, "Racial and Cultural History of the Filipinos," *Philippine Social Science Review*, X (May 1938), 110–27. For more recent data, see Nena Vreeland et al., *Area Handbook for the Philippines* (Washington: Government Printing Office, 1976). Worcester's project of "discovering" new tribes came on down into the 1970s, when outsiders chanced upon the gentle Tasadays in the caves of Mindanao.

Chapter XXI
The Strenuous Life Abroad

Abbreviated references in this and the following chapter are to the following sources:

(DIL) John Basset Moore, *A Digest of International Law* (Washington: Government Printing Office, 1906), VII.

(MS) Moorfield Storey and Julian Codman, *"Marked Severities" in Philippine Warfare* (Boston: Geo. H. Ellis, 1902). Once an assistant of Senator Charles Sumner and a former president of the American Bar Association, Storey was a distinguished and committed opponent of the Philippine–American War. This contemporary analysis of Root's attempts to minimize, conceal, and palliate U.S. atrocities remains most helpful, especially for those without the time or the stomach for plowing through the *Hearings* listed below. See also Storey and Marcial P. Lichauco, *The Conquest of the Philippines by the United States, 1898–1925* (New York: G. P. Putnam's Sons, 1926).

(SD 205) Senate Document 205, *Charges of Cruelty, etc. . . . Letter from the Secretary of War Relative to the Reports and Charges in the Public Press of Cruelty and Oppression Exercised by Our Soldiers Toward Natives of the Philippines*, 57th Cong., 1st Sess., 1902. In the face of mounting charges against the army, Lodge requested and received this "letter" from Root. It was in two parts, the larger (II) consisting of trials of Filipinos for cruelty against Filipinos. But see especially Exhibit F in Part I, 42–44, for a table showing the token courts-martial of a total of ten officers. One was convicted of "firing into town, and looting" and sentenced to a "reprimand." Lieutenant Bissell Thomas was found guilty of "assaulting prisoners and cruelty"; the court remarked that his cruelty had been "very severe and amounted almost to acute torture"; his sentence was a fine of $300 "and reprimand." More appropriate was the disposition of the case of First Lieutenant Preston Brown, who was found guilty of "killing a prisoner of war" and sentenced to dismissal from the service and confinement "at hard labor for 5 years"—Root contemptuously did not inform the senators or the public that Brown's sentence had already been commuted (January 27, 1902) to a loss of thirty-five places in the army list and forfeiture of half his pay for nine months (Storey and Lichauco, *Conquest*, pp. 136–37). Murder of natives came cheap —if the army was on trial for the "marked severities" that even General Nelson A. Miles charged, it and the secretary of war were not exactly contrite defendants.

(SD 331) Senate Document 331, *Hearings before the Senate Committee on the Philippine Islands*, 57th Cong., 1st Sess., 1902. This invaluable document is in three parts with consecutive pagination. For ease in locating testimony and other materials, the abbreviated references list the parts as though they were volumes—thus (III, 2984).

(UM) Usha Mahajani, *Philippine Nationalism: External Challenge and Filipino Response, 1565–1946* (St. Lucia, Queensland: University of Queensland Press, 1971).

(WF) *The World's Fair: Comprising the Official Photographic Views of the Universal Exposition Held in Saint Louis, 1904. Commemorating the Acquisition of the Louisiana Territory.* With introduction and descriptions by Walter B. Stevens, Secretary Louisiana Purchase Exposition (Saint Louis: N. D. Thompson, n.d. [1904]).

For the Rough Rider's "Strenuous Life" sermon, see Hermann Hagedorn, ed., *Works of Theodore Roosevelt* (New York: Charles Scribner's Sons, 1925), XV, 281. Albert Jeremiah Beveridge's "Star of Empire" is in his *Meaning of the Times* (Indianapolis: Bobbs-Merrill, 1908), p. 128. For K. M. Abenheimer's insight into omniscient and omnipotent fathers who wind up playing the devil, see his "Shakespeare's 'Tempest': A Psychological Analysis," *Psychoanalytic Review*, XXXIII (October 1946), 405.

Still the point of departure for serious consideration of the opposition to overseas empire is Fred Harvey Harrington's "The Anti-Imperialist Movement in the United States, 1898–1900," *Mississippi Valley Historical Review*, XXII (September 1935), 211–30. Useful for the New England Anti-Imperialist League is Daniel Boone Schirmer, *Republic or Empire: American Resistance to the Philippine War* (Cambridge, Mass.: Schenkman, 1972). See also Robert L. Beisner's *Twelve against Empire: The Anti-Imperialists, 1898–1900* (New York: McGraw-Hill, 1971). For the expansionist leanings of the anti-annexationists, see David Healy, *US Expansionism: The Imperialist Urge in the 1890s* (Madison: University of Wisconsin Press, 1970), pp. 55–56. Of importance for their racism is Christopher Lasch's "Anti-Imperialists, the Philippines, and the Inequality of Man," *Journal of Southern History*, XXIV (August 1958), 319–31. See also Rayford W. Logan, *The Betrayal of the Negro* (New York: Collier, 1965), pp. 272–73; Rubin Francis Weston, *Racism in U.S. Imperialism: The Influence of Racial Assumptions on American Foreign Policy* (Columbia: University of South Carolina Press, 1972), pp. 17, 78, 113, et passim. For an outstanding account of blacks and the issue of imperialism, see Willard B. Gatewood, Jr., *Black Americans and the White Man's Burden* (Urbana: University of Illinois Press, 1975), esp. pp. 180–260.

Charles Francis Adams's *"Imperialism" and "The Tracks of Our Forefathers"* (Boston: Dana Estes, 1899) revealed a historical imagination capable of summarizing in a few pages the killings and hurtings we have pursued across the continent and beyond. Without Morton to chastise, as I have remarked, Adams unflinchingly faced and embraced the "racial antipathy" that led the forefathers and their sons to exterminate nonwhites. Erving Winslow's denial of any analogy between acquisition of the Philippines and acquisition of mainland territories was quoted by John W. Rollins, "The Anti-Imperialists and Twentieth Century American Foreign Policy," *Studies on the Left*, III (1962), 9–24. Schirmer, *Republic or Empire*, pp. 8–11, discussed the Pilgrim and Puritan background of leaders of the New England Anti-Imperialist League.

Senator Hoar's correspondent who found he had been misled about the Filipinos being "savages no better than our Indians" went on to observe that half his regiment had gone with General Lawton to Santa Cruz: "The boys came back with different ideas, denied they were savages, and confessed that they did not want to fight them anymore. . . . After this trip neither the 'nigger' nor 'Indian' talk made them enthusiastic soldiers" (*Congressional Record*, XXXIII, 714).

For Whitelaw Reid's contention that the U.S. title to the Philippines was as "indisputable" as its title to California, see Thomas F. Gossett, *Race* (New York: Schocken Books, 1965), pp. 314–15. On November 3, 1898, Reid recorded that William Rufus Day, head of their delegation of peace commissioners, had argued against the United States laying claim to any territory by right of conquest: "I told him," wrote Reid, "it would not do to argue against the validity of such a title, since we had no other to the United States . . ."—H. Wayne Morgan, ed., *Making Peace with Spain: The Diary of Whitelaw Reid, September–December, 1898* (Austin: University of Texas Press, 1965), p. 134.

For Beveridge's "March of the Flag," see his *Meaning of the Times*, pp. 47–57. Beveridge of course saw continuity in that march, devoutly hoping the flag that was to float over the Philippines "be the banner that Taylor unfurled in Texas and Fremont carried to the coast." Hoar's contrary view of the Monroe Doctrine was quoted by Robert L. Beisner, *Twelve against Empire*, p. 162; and the opposing views of Cleveland and Reed, by Harrington in "The Anti-Imperialist Movement." For Roosevelt's rebuttal, see Hagedorn, ed., *Works*, XVI, 551, 556, 557.

On the black soldiers quoted in the text, see Gatewood, *Black Americans and the White Man's Burden*, pp. 230–31, 261–92.

The reference to "Weylerian cruelty" was to the policy of General Valeriano Weyler y Nicolau, who in February 1896 issued a proclamation that all Cubans were to "reconcentrate themselves" immediately in towns garrisoned by his troops. Eight days after the proclamation, natives found outside those towns were to be considered rebels and tried as such—U.S. generals went well beyond this Weylerian cruelty.

The Roosevelt quotation on General Chaffee is from Howard K. Beale, *Theodore Roosevelt and the Rise of America to World Power* (Baltimore: Johns Hopkins Press, 1956), p. 67. For Lodge's handling of the hearings and suppression of evidence, see Karl Schriftgiesser, *The Gentleman from Massachusetts: Henry Cabot Lodge* (Boston: Little, Brown, 1944), pp. 202–3; and Claude G. Bowers, *Beveridge and the Progressive Era* (Boston: Houghton Mifflin, 1932), pp. 179–82 and esp. 202.

In *Henry Cabot Lodge* (New York: Alfred A. Knopf, 1953), pp. 209–10, John A. Garraty invoked the merciless savage theme to palliate U.S. atrocities: "When war degenerates into guerrilla fighting in tropical jungles against a cruel and savage foe, 'civilized' soldiers sometimes adopt the methods of their opponents." A variation of this palliation is the statement of Peter W. Stanley that the Philippine–American conflict "dragged on through successive stages, punctuated by almost countless atrocities on both sides"—*A Nation in the Making* (Cambridge: Harvard University Press, 1974), p. 51. Uncountable and unanalyzable atrocities on both sides thus seemingly canceled themselves out, one way to avoid having to take them into account. Albert Camus mentioned somewhere this phenomenon of atrocity trading: to his indictment of Stalin's slave labor camps, a French Communist would respond by invoking the lynchings of black Americans. I should make clear my own distaste for this bartering of atrocities, by whomever committed. The water cure described in the text was despicable torture when utilized *by* U.S. forces in the Philippine–American War and when utilized *against* U.S. forces on the same archipelago in World War II. But the fact that Filipino atrocities occurred in their own land while they resisted subjugation by alien invaders has to be taken into account and the record scrutinized to see who was doing what to whom. That record bears out General Aguinaldo as both more balanced and more direct than Garraty and Stanley in his charge that U.S. troops had given his people "the water-cure and killed them after capture. They reconcentrated the population and killed able-bodied men. I cannot deny that some of my men might have committed comparable offenses against the Americans also. But the fact is that in general we meticulously followed civilized practices"—Emilio Aguinaldo and Vincente Albano Pacis, *A Second Look at America* (New York: Robert Speller & Sons, 1957), p. 79. And no trade of atrocities lies behind what may be stated flatly: no one at the time or since has produced "kill-and-burn" orders from generals of the Filipino guerrillas.

Best for understanding George A. Dewey is his own testimony in the *Hearings* (SD 331, III, 2226–84)—with this one hardly need consult Aguinaldo's "Dewey, Pratt and I" in *Second Look*, pp. 29–39. See also the admiral's *Autobiography* (New York: Charles Scribner's Sons, 1913).

For the careers of Generals Merritt, Otis, MacArthur, Chaffee, Bell, Smith, and all the other former professional Indian-fighters, see F. B. Heitman, *Historical Register and Dictionary of the United States Army* (Washington: Government Printing Office, 1903). Before all the killings and hurtings started in earnest, Goldwin F. Smith had predicted in "The Moral of the Cuban War," *Forum*, XXVI (November 1898), 284–90, that treatment of Indians would provide the pattern for subjugating Filipinos: If there were tribes similar to the Apaches, said this prophet, "Uncle Sam will handle them in his accustomed style."

For the official "report of the action at Balangiga," see Senate Document 331, II, 1594–98; General Robert P. Hughes forwarded the report and list of the dead with a communication in which he mentioned some good news, a detachment that had "killed more barefoots than any full company on the island" (II, 1599). For a recent shoddy attempt to discredit the testimony of William J. Gibbs, see Joseph L. Schott, *The Ordeal of Samar* (Indianapolis: Bobbs-Merrill, 1964), pp. 160–62. For "the fear of God" in Samar, see James H. Blount, *The American Occupation of the Philippines, 1898–1912* (New York: Oriole Editions, 1973), pp. 378, 391. Like other military men in the islands, Blount regarded General Jacob H. Smith as a scapegoat for the sins of General Chaffee et al. (pp. 379–80; cf. MS, p. 117). Smith's "kill-and-burn" orders went beyond anything that could be proved against General Tomoyuki Yamashita, the commander of the Japanese forces in the Philippines, yet at the end of World War II the military commission of another General MacArthur tried him as a "war criminal" for "brutal" atrocities and sentenced him to be not admonished but executed—see A. Frank Reel, *The Case of General Yamashita* (Chicago: University of Chicago Press, 1949), pp. 32, 106, 109.

For an account of Major Waller's disastrous march and trial, see Schott's *Ordeal of Samar*. The "ordeal" in the title was not that of the natives nor even of the island but of hero Waller and his marines. Used cautiously, for it contains elementary errors of fact, the book is valuable for the marines' role in the "burn-and-kill" campaign and as an exhibit of how Indian-hating came on down into the 1960s. Of Brigadier General William H. Bisbee, who presided over Waller's court-martial, for instance, Schott observed admiringly that "the old bald-headed ranker . . . had shot his share of Indians in his career" (p. 174). Heavy and misplaced reliance on Schott for data mars the otherwise suggestive essay of Stuart C. Miller, "Our Mylai of 1900: Americans in the Philippine Insurrection," *Transaction*, VII (September 1970), 17–28.

For Sherman and the Fort Phil Kearny "massacre," see Robert G. Athearn, *William Tecumseh Sherman and the Settlement of the West* (Norman: University of Oklahoma Press, 1956), pp. 98–99. An indulgent biographer, Athearn maintained Sherman did not mean what he wrote to the commander of the army, a curious view of a general who is perhaps best remembered for his comment to the mayor of Atlanta: "War is cruelty, and you cannot refine it." For a penetrating essay on Sherman, see Edmund Wilson's *Patriotic Gore* (New York: Oxford University Press, 1966), pp. 174–218. For William J. Fetterman as a hot-headed would-be Indian-killer, see S. L. A. Marshall, *Crimsoned Prairie* (New York: Charles Scribner's Sons, 1972), pp. 57–73. For Washington's "Instructions to Major General John Sullivan," see John C. Fitzpatrick, ed., *Writings of George Washington* (Washington: Government Printing Office, 1936), XV, 189–93. The best account of Sullivan's destruction in the lands of the Six Nations is William L. Stone, *Life of Joseph Brant—Thayendanegea* (New York: George Dearborn, 1838), II, 1–40. For a modern account of how the Iroquois lands were made into "rural slums," see Anthony F. C. Wallace, *The Death and Rebirth of the Seneca* (New York: Random House, 1969), pp. 141–45.

Chapter XXII
The Strenuous Life at Home

McKinley's comment on "our priceless principles" was quoted by Moorfield Storey and Marcial P. Lichauco, *The Conquest of the Philippines by the United States* (New York: G. P. Putnam's Sons, 1926), p. 7. His view of expositions as the "timekeepers of progress" appeared as the epigraph of *World's Fair* (WF)—for an explanation of this and other abbreviated references in this chapter, see p. 521.

The quotations from Tocqueville on "the singular fixity of certain principles" may be found in his *Oeuvres, papiers et correspondances*, ed. J. P. Mayer (Paris: Gallimard, 1951–70), I, part 2, 264, 262. My translation generally follows but differs at points from that of Henry Reeve et al.—see Phillips Bradley, ed., *Democracy in America* (New York: Alfred A. Knopf, 1945), II, 257, 255.

The fair statuary carried a message no open-eyed visitor could misinterpret: what Helen Hunt Jackson had called "a century of dishonor" (in her book of that title [1881]) was really a century of glory, a national epic of subjugation and dispossession of natives, mainland and insular, that yet moved warm-hearted citizens of the republic to feel "something of the plaintive in it." For murals glorifying this progress, see the descriptions in Walter Williams, ed., *The State of Missouri: An Autobiography* (Columbia: E. W. Stephens, 1904), pp. 582–84. For Governor Pennypacker's noteworthy speech, see James H. Lambert, *The Story of Pennsylvania at the World's Fair* (Philadelphia: Pennsylvania Commission, 1905), I, 261–86.

For the bids to rent Aguinaldo from the War Department, see

Philip C. Jessup, *Elihu Root* (New York: Dodd, Mead, 1938), I, 335. I am much obliged to Frederick W. Turner III for pointing out that Kate Chopin, author of the hauntingly suggestive novel about "mixed" marriages and sex, *The Awakening* (1899), attended the exposition regularly until she went home one day to die; Scott Joplin also presumably attended a gathering there of ragtime pianists. The sounds of the fair as well as its sights must have been fascinating—along with the beat of Indian and African drums, the polyrhythms of ragtime provided a fitting backdrop commentary on "the noise of progress."

For the historic meeting of Geronimo and the Igorots, see *Geronimo: His Own Story*, edited by S. M. Barrett and newly edited by Frederick W. Turner III (New York: Ballantine Books, 1971), p. 176. Angie Debo had a good account of the Apache at the fair in *Geronimo: The Man, His Time, His Place* (Norman: University of Oklahoma Press, 1976), pp. 408–17. While a prisoner of war at Fort Sill, Debo noted, Geronimo became "a commercial property" that was billed by showmen as "the Apache terror" or "the tiger of the human race" (p. 423). No doubt he was most grotesquely exploited in 1905 during the course of the annual meeting of the National Editorial Association. The editors took excursion trains out from Oklahoma City to the Miller Brothers' 101 Ranch, where Geronimo was featured in "the last buffalo hunt." Mounted on an automobile owned by a Chicago visitor, Geronimo and other last hunters killed animals hauled in from Colonel Charles Goodnight's ranch in the Texas panhandle. And on that note sounded the last hurrah of the Wild West.

No doubt the most historically significant exploitation of Geronimo was his command presence at Theodore Roosevelt's inaugural parade a year after the fair: "I wanted to give the people a good show," explained the president-elect and had no need to specify who "the people" were. The pageantry of his inaugural underlined the thesis of his *Winning of the West:* Geronimo, Quanah Parker the Comanche, and four other traditional or "wild" Indians led off and were followed by a unit of "civilization" in the form of tamed, disciplined cadets from the Carlisle Indian School. Afterward Geronimo had an audience at the White House and there pleaded to be allowed to return to Arizona: "I pray you to cut the ropes and make me free. Let me die in my own country, an old man who has been punished enough and is free." But given the Indian-hating of "the people" in Arizona, the president feared his return would lead to bloodshed and had to refuse. "It is best for you to stay where you are," ruled Roosevelt and assured Geronimo he did so with "no feeling against you" (Debo, pp. 417–21).

In addition to the Igorots, Negritos, and other tribespeople, some forty "distinguished" Filipinos were in St. Louis as an honorary board of commissioners. Of the prominent Federalista members, Peter W. Stanley observed that they "discovered what Taft had always known, that Americans didn't want colored and allegedly backward Filipinos as compatriots"—*A Nation in the Making* (Cambridge: Harvard University Press, 1974), p. 116. That was no doubt true and Taft had already said as much. Usha Mahajani remarked that "even Taft, under whose close supervision the statehood plank had been adopted, opposed the plan at the Senate hearings in 1902 on the grounds that it had no value in the Philippines" (UM, pp. 330–31). Philanthropists at the 1911 Lake Mohonk Conference of Friends of the Indian and Other Dependent Peoples were enthusiastically invited by Edward B. Bruce, the Manila businessman, to invest their money in the islands, since prosperous conditions prevailed there and Americanization had extended to the "infrastructure," to use Stanley's term—see pp. 142–43, 295.

For Twain's biting "Comments on the Killing of 600 Moros," see Maxwell Geismar, ed., *Mark Twain and the Three R's: Race, Religion, Revolution* (Indianapolis: Bobbs-Merrill, 1973), pp. 20–28; see also Geismar's *Mark Twain: An American Prophet* (Boston: Houghton Mifflin, 1970), pp. 220–22.

For Stimson's and Hoover's determination to carry the white man's burden on past 1932, see Richard Nelson Current, *Secretary Stimson: A Study in Statecraft* (New Brunswick, N.J.: Rutgers University Press, 1954), p. 120.

Chapter XXIII
The Occident Express: From the Bay Colony to Indochina

Abbreviated references in this chapter are to the following sources:

(GPP) Gravel Edition, *Pentagon Papers* (Boston: Beacon Press, 1971), 4 vols.

(NEF) Alden T. Vaughan, *New England Frontier: Puritans and Indians, 1620–1675* (Boston: Little, Brown, 1965).

(NYT) *New York Times* Edition, *Pentagon Papers* (New York: Bantam, 1971).

(PP) Alden T. Vaughan, "Pequots and Puritans: The Causes of the War of 1637," *William and Mary Quarterly*, XXI (April 1964), 256–69.

The Lyndon B. Johnson epigraph is from his Johns Hopkins University speech, *Department of State Bulletin*, LII (April 26, 1965), 606–10.

N. Scott Momaday's essay "The Morality of Indian Hating" appeared in *Ramparts*, III (Summer 1964), 29–40.

William Wood's *New Englands Prospect*, published originally in 1634, was reprinted in the Prince Society Publications, I (1865), and by B. Franklin in New York (1967). Vaughan listed Wood in the bibliography of *New England Frontier* and quoted him in the text but never acknowledged his rare good words about the Pequots.

Francis Jennings pointed out that the killer of John Oldham was identified by Roger Williams as the Narragansett sachem Audsah and that the suspect probably took refuge with the Narragansetts' tributary Eastern Niantics—*The Invasion of America: Indians, Colonialism, and the Cant of Conquest* (Chapel Hill: University of North Carolina Press, 1975), p. 209; for decisive criticism of Vaughan's handling of his "copious" sources, see pp. 187n, 194n, and esp. 267n. On the racially segregated Christian Indians in the Bay Colony, see Neal Salisbury's important "Red Puritans: The 'Praying Indians' of Massachusetts Bay and John Eliot," *William and Mary Quarterly*, XXXI (January 1974), 27–54; for a first-rate analysis of Eliot, see Salisbury's "Conquest of the 'Savage': Puritans, Puritan Missionaries, and Indians, 1620–1680," Ph.D. diss., University of California, Los Angeles, 1972, pp. 152–67.

Edmund S. Morgan's review had to be especially gratifying to Vaughan, since Vaughan had taken a stand in his text beside those who had labored so hard to dispel misconceptions about the Puritans: "Samuel Eliot Morison, Perry Miller, Edmund Morgan and others" (p. 323). Certainly Vaughan was in the same tradition as Morgan and for that matter Morison, who introduced Douglas Edward Leach's *Flintlock and Tomahawk: New England in King Philip's War* (New York: W. W. Norton, 1966—orig. 1958) by stressing its timeliness: "In view of our recent experiences of warfare," wrote the admiral, "and of the many instances today of backward peoples getting enlarged notions of nationalism and turning ferociously on Europeans who have attempted to civilize them, this early conflict of the same nature cannot help but be of interest. . . . Behind King Philip's War was the clash of a relatively advanced race with savages, an occurrence not uncommon in history." All too evidently "savages" are like the poor—but Morison's observation was prescient, in a way, for in 1958 Vietnam was hardly more than a cloud on the most distant horizon.

For Thoreau on Indian-hating historians, see Bradford Torrey and Francis H. Allen, eds., *Journal of Henry D. Thoreau* (Boston: Houghton Mifflin, 1949), XI, 437–38. Of course Melville's observation comes from his Indian-hating chapter in *The Confidence-Man* (see p. 501).

In *Pragmatic Illusions: The Presidential Politics of John F. Kennedy* (New York: David McKay, 1976), p. 144, Bruce Miroff wrote that "to protect the process of modernization from those who would seek to wreck it, a new military strategy—counterinsurgency—was developed." Wrong about its brand-newness, Miroff was quite right about the confluence of COIN and modernization. Walt Whitman Rostow was a case in point, of course; another was Eugene Staley, builder of strategic hamlets and author of *The Future of Underdeveloped Countries* (New York: Praeger, 1961), "one of the standard texts on economic and social development"—*Pragmatic Illusions*, p. 313n. Miroff's discussion of "Global Liberalism" (pp. 110–66) is first-rate. Also helpful is Richard J. Walton's *Cold War and Counterrevolution: The Foreign Policy of John F. Kennedy* (New York: Viking Press, 1972). For the test-of-will temper of the New Frontier, see Kennedy's *Profiles in Courage* (New York: Harper & Row, 1964). For the view from his White House, see Arthur M. Schlesinger, Jr., *A Thousand Days* (Boston: Houghton Mifflin, 1965) and Theodore Sorenson, *Kennedy* (New York: Bantam, 1966).

On Richard Giles Stilwell as "a former CIA man," see David Halberstam's *Best and the Brightest* (Greenwich, Conn.: Fawcett, 1973), pp. 308–9, 342–44. L. Fletcher Prouty reproduced the Stilwell document in *The Secret Team: The CIA and Its Allies in Control of the United States and the World* (Englewood Cliffs, N.J.: Prentice-Hall, 1973), pp. 442–79.

In *The Acquisition and Government of Backward Territory in International Law* (London: Longmans, Green, 1926) M. F. Lindley pointed out that historically "trusteeship" and "wardship" had been construed in "a broad ethical sense to mean the duties which the advanced peoples collectively owe to backward races in general—the 'white man's burden,' or, as the Covenant of the League of Nations puts it in the 'Mandates' Article, 'a sacred trust of civilization' " (p. 329). While understandably not stressing the ethics involved, Stilwell strenuously urged assumption of that burden by advanced Westerners and in effect treated "less developed" peoples as wards. The behavioral science jargon of Stilwell and of Vaughan made their language different from that of the drafters of the Covenant of the League of Nations; otherwise their assumptions were fundamentally similar. And other evidence aside, insofar as the historian and the general identified "civilization" and the West—"all the states of European blood," in the words of John Westlake (see p. 500)—their views were *ab initio* racist.

The stockadelike camp of the Army Special Forces at Nhatrang was described by Horace Sutton in "The Ghostly War of the Green Berets," *Saturday Review*, October 18, 1969.

In the style of a Norman Rockwell cover or a Frederic Remington drawing, editor Harold H. Martin of the *Saturday Evening Post* invoked the past with the strategic-hamlet stockades, with references to "Stone Age" hill tribesmen who "look more like American Indians than Orientals, and they are the most skillful hunters and trackers since the Apaches," and with an assertion about the guerrilla war that was true even if it did appear in the *Post*: "Contrary to popular belief," wrote Martin, "it is not a type of war that we are by temperament and training ill-prepared to fight. Throughout history Americans have fought as guerrillas and against guerrillas" (CCXXXV [November 24, 1962], 14–17). The alleged standing orders of Rogers's Rangers posted outside Admiral Felt's Honolulu headquarters were reproduced by John Mecklin in *Mission in Torment: An Intimate Account of the U.S. Role in Vietnam* (New York: Doubleday, 1965), pp. 27–28. For the original standing orders and discussion of their context, see John Ellis, *A Short History of Guerrilla Warfare* (New York: St. Martin's Press, 1976), p. 50; and F. Sully, *Age of the Guerrilla* (New York: Avon, 1968), pp. 198–99.

Maxwell Taylor's testimony before the Fulbright committee in 1966 was reproduced by Senator Wayne Morse in *The Truth about Vietnam: Report on the U.S. Senate Hearings* (San Diego, Calif.: Greenleaf Classics, 1966), pp. 266–67. On Taylor generally the *Pentagon Papers* are best, but see his *Swords and Plowshares* (New York: W. W. Norton, 1972)—it is misleading factually but inadvertently revealing about the imperial mood of the New Frontier.

Francis FitzGerald's characterization of the CIA Phoenix program is from *Fire in the Lake: The Vietnamese and the Americans in Vietnam* (New York: Vintage Books, 1973), p. 550; see also Morton H. Halperin et al., *The Lawless State: The Crimes of the U.S. Intelligence Agencies* (New York: Penguin Books, 1976), p. 46.

Halberstam quoted the exchange of General Curtis E. LeMay and McGeorge Bundy in *The Best and the Brightest*, p. 560. Also unintentionally revealing is LeMay's *Mission with LeMay* (New York: Doubleday, 1965).

In addition to Samuel P. Huntington's "The Bases of Accommodation," *Foreign Affairs*, XLVI (July 1968), 642–56, see his introduction ("Social Science and Vietnam") to a series of articles by behaviorists in *Asian Survey*, VII (August 1967), 503–6. An authority on civil–military relations, Huntington has unfavorably contrasted sprawling civilian communities with the virtually total order of West Point—see his *Soldier and the State* (Cambridge: Harvard University Press, 1957). It is safe to say that Huntington liked disorder no more than Vaughan. Defense analysts, as they thought of themselves, prided themselves on their tough-minded realism. Herman ("Thinking-the-unthinkable") Kahn, for instance, obviously did not subscribe to Huntington's forced-draft urbanization but revealed himself no less impervious to Vietnamese suffering. After a briefing on the economic situation there, Kahn said to his informant: "I see what you mean. We have corrupted the cities. Now, perhaps we can corrupt the countryside as well"—quoted by FitzGerald, *Fire in the Lake*, p. 470. For trenchant criticism of such figures, see Noam Chomsky's *American Power and the New Mandarins* (New York: Vintage Books, 1969).

Chapter XXIV
The Ugly American

The Ugly American by William J. Lederer and Eugene Burdick was published originally by W. W. Norton in 1958; in 1960 a paperback edition by Crest Books made it available to many more readers. For the "slashing" review, see *Time*, LXXII (October 6, 1958), 92. Joseph Buttinger, the former Austrian socialist politician, contributed a long, rambling critique of the book to *Dissent* VI (Summer 1959), 317–67.

Lederer and Burdick's "dedicated Americans" were forerunners of those who enlisted a few years later in Kennedy's Peace Corps. Earnest volunteers in that New Frontier venture acted out the recommendation of the authors in their "factual epilogue": Volunteers "must be willing to risk their comforts and—in some lands—their health. They must go equipped to apply a positive policy promulgated by a clear-thinking government. They must speak the language of the land of their assignment, and *they must be more expert in its problems than are the natives.*" My italics underscore what I mean by covert racism.

Chapter XXV
The Secret Agent: Edward Geary Lansdale

Abbreviated references in this and the following chapters are to these sources:

(AAP) Senate Select [Church] Committee to Study Governmental Operations with Respect to Intelligence Activities, *Alleged Assassination Plots Involving Foreign Leaders: An Interim Report*, 94th Cong., 1st Sess. (Washington: Government Printing Office, 1975).

(CIA) Victor Marchetti and John D. Marks, *The CIA and the Cult of Intelligence* (New York: Alfred A. Knopf, 1974).

(CW) Joseph Burkholder Smith, *Portrait of a Cold Warrior* (New York: G. P. Putnam's Sons, 1976). These "second thoughts of a top CIA agent" are uneven but invaluable for understanding Lansdale's role in the Far East. They are also useful for documenting the blind cold war ideology driving such operatives and their racist attraction to what Professor Worcester used to describe as the "strangely firm grip" of the tropics, or as these successors put it, the "operationally ripe" Third World. In Singapore Smith and his wife mixed class with caste and location: "By the time the great events in Vietnam began absorbing my working hours almost completely, we had begun to enjoy the particular charm of being upper-class white residents of Asia" (p. 181).

(DOD) U. S. Department of Defense [DOD], *United States–Vietnam Relations, 1945–1967* (Washington: Government Printing Office, 1971), 12 vols. The Gravel Edition (GPP) listed below is indispensable for materials the DOD censor excised from this offset edition of the Pentagon Papers. On the other hand these volumes contain three times as many documents as the Gravel Edition, and that means the collections must be collated to fill in gaps. For important Lansdale reports and memoranda herein, see X, 1307–10;, 1329–31; and XI, 1–12, 22–56, 157–58, 175–76, 427.

(FF) Frances FitzGerald, *Fire in the Lake: The Vietnamese and the Americans in Vietnam* (New York: Vintage Books, 1973)

—the best nonfictional account of the United States in Indochina.

(GPP) Gravel Edition, *Pentagon Papers* (Boston: Beacon Press, 1971), 4 vols. Collation of this edition with the DOD publication listed above reveals what the government censors tried to keep secret. They excised two of Lansdale's key documents, his diary-form "Report on Covert Saigon Mission in 1954 and 1955" and his memorandum for General Maxwell D. Taylor "Resources for Unconventional Warfare, S. E. Asia," undated but apparently from July 1961; both appear herein: I, 573–83; and II, 643–49.

(HUK) Benedict J. Kerkvliet, *The Huk Rebellion: A Study of Peasant Revolt in the Philippines* (Berkeley: University of California Press, 1977). This exemplary study makes most of the other works on the rebellion seem exercises in cold war demonology. Perhaps the most egregious was that of Alvin H. Scaff, as his title forewarned: *The Philippine Answer to Communism* (Stanford, Calif.: Stanford University Press, 1955)—Scaff was in the islands in 1953–54 as a Fulbright professor, but if he knew of Lansdale's critical role, he did not share that knowledge with his readers. For a leading Huk's account, see Luis Taruc's *Born of the People* (New York: International Publishers, 1953) and *He Who Rides the Tiger* (London: Geoffrey Chapman, 1967).

(IMW) Edward Geary Lansdale, *In the Midst of Wars: An American's Mission to Southeast Asia* (New York: Harper & Row, 1972). See also his "Viet Nam: Do We Understand Revolution?" *Foreign Affairs*, XLIII (October 1964), 75–86; "Letter to the Editor," in Robert Manning and Michael Janeway, eds., *Who We Are: An Atlantic Chronicle of the United States and Vietnam* (Boston: Little, Brown, 1965); "Two Steps to Get Us Out of Vietnam," *Look*, March 4, 1969; and for his participation in a postmortem conducted by the war managers in 1973–74, W. Scott Thompson and Donaldson D. Frizzell, eds., *The Lessons of Vietnam* (New York: Crane, Russak, 1977). His most recent attempt to justify the Vietnam War and draw the proper COIN lessons from it is his "Thoughts about a Past War," the foreword to *A Short History of the Vietnam War*, ed. Allan R. Millett (Bloomington: University of Indiana Press, 1978).

(ST) L. Fletcher Prouty, *The Secret Team: The CIA and Its Allies in Control of the United States and the World* (Englewood Cliffs, N.J.: Prentice-Hall, 1973). As the DOD focal-point officer for covert operations, Prouty worked with both the CIA and the Joint Chiefs and was in a position to observe Lansdale and other Company (i.e., CIA) "moles" or "plants" in the Pentagon and other federal agencies. Despite his unfortunate (premature?) subtitle and tendency to overstate his case, Prouty made a valuable contribution to our understanding of the vexed question of in-house agents.

As noted, "The Report the President Wanted Published," *Saturday Evening Post,* CCXXXIV (May 20, 1961), 31, 69–70, was merely the annex of the report by Lansdale that was presented to the president by Walt Whitman Rostow, who had become the secret agent's sponsor within the administration. For Kennedy's reaction to the full report, see Arthur M. Schlesinger, Jr., *A Thousand Days* (Boston: Houghton Mifflin, 1965), pp. 320, 341–42. Tongue not visibly in cheek, Schlesinger characterized Lansdale as "an imaginative officer"; an OSS veteran himself, the historian naturally did not blow the operative's cover by revealing his CIA status. Lansdale's account of guerrilla successes in "conducting savage [*sic*] and elusive warfare against pro-western regimes," (Schlesinger's words) seemingly initiated Kennedy's reading of Mao and Che and his personal supervision of the Special Forces at Fort Bragg.

The Village That Refused to Die and the CARE "Settler Kits" the TV movie prompted were discussed by Robert Scheer in *How the United States Got Involved in Vietnam* (Santa Barbara, Calif.: Center for the Study of Democratic Institutions, 1965), pp. 68–70.

Melville would have been amused that in the 1950s, a century after the publication of *The Confidence-Man,* the CIA novices in Paul Linebarger's Washington COIN seminar were assigned David W. Maurer's *The Big Con* (New York: Pocket Books, 1949) as required reading and told that, while their job and the confidence man's were almost identical, there was this difference: "He wants to fleece his mark out of his money. You want to convince a Chinese, Filipino, an Indonesian, a Malay, a Burmese, a Thai, that what you want him to believe or do for the good of the U.S. government is what he thinks he himself really believes and wants to do" (CW, p. 94).

For General Douglas MacArthur's refusal to have Wild Bill Donovan's agents in the South Pacific, see R. Harris Smith, *OSS: The*

Secret History of America's First Intelligence Agency (Berkeley: University of California Press, 1972), pp. 246–51. The best thing about Anthony Cave Brown's breathless study of MI-6 is the title: *Bodyguard of Lies* (New York: Harper & Row, 1975)—as General Dwight D. Eisenhower said of the British agency's zest for deceptions, it bred "a habit that was later difficult to discard" (p. 74). For OSS indebtedness to MI-6 and for the anglophilia of its top officers, see Smith's *OSS*, pp. 1–35, 163–64. For the origins of the U.S. domestic intelligence agency, see Max Lowenthal's still useful *Federal Bureau of Investigation* (London: Turnstile Press, 1951); for the recent past, see Sanford J. Ungar's *FBI* (Boston: Little, Brown, 1975).

For what Hannah Arendt once called "the foundation legend" of the secret agent, "the legend of the Great Game as told by Rudyard Kipling," see *Kim* (Garden City, N.Y.: Doubleday, Page, 1920—orig. 1901). I discovered belatedly that my reading of the novel finds welcome support in Edmund Wilson's "The Kipling That Nobody Reads," *The Wound and the Bow* (New York: Oxford University Press, 1947), pp. 105–81. For a reading that stresses the novelist's mystical, Buddhist side, a dimension necessarily neglected here, see Vasant A. Shahane, *Rudyard Kipling* (Carbondale: Southern Illinois University Press, 1973). George Orwell ironically observed that Anglo-Indians felt that Kipling had mixed with the "wrong people" in India, "and because of his dark complexion he was wrongly suspected of having a streak of Asiatic blood"—"Rudyard Kipling," *The Orwell Reader* (New York: Harcourt Brace, 1956), pp. 271–83. Angus Wilson's *The Strange Ride of Rudyard Kipling* (New York: Viking Press, 1978) is disappointingly superficial on Kipling's imperialism and on *Kim*.

Lansdale resembled Gary Cooper and Jimmy Stewart somewhat, and that is to say he looked a little like the prototypal clean-cut Anglo-American, with a loose-jointed amiability that was ideal surface for the underlying imperial agent. In the islands, Filipinos sometimes called him "Eagle," he wrote, "from my initials 'EGL' " (IMW, p. 125n). This master of the occult must have chuckled inwardly at the mysterious appropriateness of the nickname. As a visitor to the vast headquarters at Langley, Virginia, might have noted, his employers had placed a mammoth seal in the lobby with the words CENTRAL INTELLIGENCE AGENCY set in the marble floor and with a baleful eagle's head in the center. By initials and by nickname Lansdale had thus left clues to his masquerade, for the face of the colonizer is the face "of a bird of prey seeking with cruel intentness for distant quarry," Carl Jung wrote in his autobiography after visiting the Taos Indians: "All the eagles and other predatory creatures that adorn our coats of arms seem to me apt psychological representatives of our true nature"—quoted by Tom Hayden, *The Love of Possessions Is a Disease with Them* (New York: Holt, Rinehart and Winston, 1972), p. 113.

Chapter XXVI
Covert Savior of the Philippines

For Magsaysay's 1948 trip to Washington on behalf of Filipino veterans, see Carlos P. Romulo and Marvin M. Gray, *The Magsaysay Story* (New York: John Day, 1956), pp. 95–96; Carlos Quirino, *Magsaysay of the Philippines* (Manila: Ramon Magsaysay Memorial Society, 1958), p. 42—see pp. 44–45 for U.S. pressure on President Quirino to make Magsaysay his defense secretary. Needless to note, these studies of the Philippine leader leave much to be desired; their authors would have found it awkward to reveal the extent to which their "man of the people" was actually the man of the CIA.

The temper and outlook of Lansdale's "team" was revealed by Napoleon D. Valeriano and Charles T. R. Bohannan in *Counter-Guerrilla Operations: The Philippine Experience* (New York: Frederick A. Praeger, 1962)—theirs may very well have been one of the "fifteen or sixteen books" the Praeger firm admitted (1967) publishing at the CIA's request—CIA, p. 164. (For an explanation of the abbreviated references, see p. 534.) Amusingly, Valeriano and Bohannan reached back to Professor Worcester and his *Philippines, Past and Present* (1914) for their authority on the Philippine "insurrection" and on the good intentions of that earlier generation of American soldiers, who had somehow managed to persuade "the Philippine people that the United States' presence was in their best interest" (pp. 231–34, 275). Without question these two members of Lansdale's team would have fitted in nicely on Worcester's crew of "absolutely fearless" colonizers.

Richard Critchfield of the *Washington Star* reported Lansdale's characterization of land reform as a "gimmick" in *The Long Charade* (New York: Harcourt, Brace & World, 1968), p. 323. With his own background in advertising, Lansdale knew the importance of cultivating close relationships with journalists, managed their coverage of his operations with considerable finesse, and effectively used some as conduits for CIA handouts (see, for example, GPP, I, 581). Among his oldest friends was Robert Shaplen of the *New Yorker*, to whom he paid tribute for his "detailed" coverage of the campaign against the Huks (IMW, pp. 99, vii). In turn Shaplen wrote of Lansdale as "a remarkable and controversial man" in *The Lost Revolution* (New York: Harper & Row, 1965), pp. 101–4; gave him a puff as a leading COIN expert in *The Road from War* (New York: Harper & Row, 1970), pp. 30–31, 34; and in both instances considerately refrained from mentioning his friend's CIA status.

In his campaign to make Magsaysay known around the world, Lansdale achieved a spectacular publicity triumph when he pried out of the International Lions Clubs an invitation for his candidate to deliver the keynote address at their 1952 convention in Mexico City—Romulo and Gray, *Magsaysay*, pp. 163–69; cf. IMW, pp. 98–99. The ensuing confrontation of Lansdale and E. Howard Hunt, the two current operational heroes of the CIA, in the Mexican capital was little short of historic—see Tad Szulc, *Compulsive Spy: The Strange Career of E. Howard Hunt* (New York: Viking Press, 1974), pp. 66–67. In *Undercover: Memoirs of an American Secret Agent* (New York: Berkeley Publishing, 1974), Hunt ungenerously referred to his rival Lansdale as "a brilliant but erratic CIA station chief in Saigon" (p. 151). And it was probably no coincidence that in *End of a Stripper* (New York: Dell, 1960), one of the novels Hunt wrote under the pseudonym Robert Dietrich, there was a smooth-operating "Representative Lansdale" who bought an election.

Gary Wills dissected the clandestine mentality in "The CIA from Beginning to End," *New York Review of Books*, XXII (January 22, 1976), 23–33. This mentality consigned nonwhites to the generic category of *natives*. The fundamental distinction was between white *man* and colored *native*, with multiple meanings interacting in both words to make them "compacted doctrines," in Raymond Williams's useful phrase: *native* was taken to imply "all subjected peoples are biologically inferior"—*Keywords: A Vocabulary of Culture and Society* (New York: Oxford University Press, 1976), pp. 155–56, 180–81, 184–89. In this negative sense, *native* meant "non-European" or, in U.S. usage, of "non-European" extraction. Along with the gun and the Bible, the term became a weapon of conquest wherever Western imperialists went. It implied a special closeness or identification of the nonwhite and nature.

General Philip H. Sheridan had originally declared to Comanche Chief Tosawi that "the only good Indians I ever saw were dead" but in time his words were filed down to the Anglo-American aphorism usually attributed to him: *The only good Indian is a dead Indian*. The earnest efforts of the central Luzon mayors and of General Castañeda (quoted by Quirino, *Magsaysay of the Phil-*

ippines, p. 51) to apply those words to the Huks demonstrated that even new-caught natives could be brought around to share a core tradition of their overseas mentors. From Sheridan on across the Pacific the aphorism implied its complement: *The only good nature is a conquered nature.* That Marxist William Pomeroy so fully shared that conquistador view of nature and its indwellers underlined the identity of his and Lansdale's basic attitudes and assumptions and their shared faith in perpetual "progress," a faith that "is rooted in, and is indefensible apart from, Judeo-Christian teleology," as Lynn White, Jr., observed in his memorable essay: "The fact that Communists share it merely helps to show what can be demonstrated on many other grounds: that Marxism, like Islam, is a Judeo-Christian heresy. We continue to live, as we have lived for about 1700 years, very largely in a context of Christian axioms"—see "The Historical Roots of Our Ecologic Crisis," *Science,* CLV (March 10, 1967), 1203–7.

After a half century of direct U.S. rule the Philippines had a model colonial economy. After 1946 "independence" transformed the colony to a client state and made U.S. rule indirect (and covert) or what I regard as simply a more sophisticated form of colonialism. But to William J. Pomeroy, accelerated economic penetration, political domination, and military intervention made the Philippines a prime example of "neo-colonialism," as Marxists call that "condition in which a country, supposedly independent, is not the master of its own affairs but remains dominated and exploited by the former colonial powers"—*An American Made Tragedy* (New York: International Publishers, 1974), p. 2; for an able discussion of what happened when U.S. agencies and corporations launched programs to reshape Philippine society (along lines laid out in General Richard Giles Stilwell's seminal document discussed on p. 365), see Pomeroy's "The Road to Dictatorship," pp. 91–128. See also his *American Neo-Colonialism* (New York: International Publishers, 1970) and *The Forest* (New York: International Publishers, 1963). In the mid-1970s estimates of the book value of all U.S. investments in the islands ranged from one to two billion dollars—for a useful survey of the role of the multinationals, see *The Philippines: American Corporations, Martial Law, and Underdevelopment,* A Report Prepared by the Corporate Information Center of the National Council of Churches (1973).

Meanwhile, in 1957 the CIA had scrambled to find a replacement for Magsaysay when their man died in an airplane crash—for this post-Lansdale chapter in nation-building, see Joseph B. Smith's *Cold Warrior,* pp. 249–321. To "get the Philippines back on the track," top officers in the Company came out to review operations in the area at the Far Eastern chiefs of station meetings: "In those days," Smith noted, "these meetings took place annually in the John Hay Air Force Base, the U.S. Armed Forces Rest and Recreation Center for the Far East, in the lovely mountain town of Baguio" (CW, p. 319). Surely Secretary Hay and Professor Worcester would have been touched by this evidence that their works lived on.

In 1972, President Ferdinand E. Marcos clamped martial law down on the Philippines and, to borrow a phrase, ended the long charade. Martial-law dictatorship was necessary, Marcos explained, primarily to combat Communist and Moslem "insurgencies" (*New York Times*, January 21, 1973). By the mid-1970s some 16,000 of the amazingly durable Moros were under arms in Mindanao and the Sulu Archipelago, while a New People's Army of some 2,000 had expanded operations from Luzon to the Visayas, with the island of Samar, that old stamping ground of General Jacob H. ("Hell-Roaring") Smith, a principal area of expansion. After a half century of direct rule and another quarter of indirect, the U.S. "showcase of democracy in Asia" displayed its wares behind steel shutters. And so had the inglorious experiment in nation-building ground to a halt amid the wreckage of peoples and institutions.

Chapter XXVII
Closing the Circle of Empire: Indochina

So much depends on how many layers of the onion we remove. Just one layer below the uppermost, Joseph B. Smith could defensibly argue that the U.S. entanglement in Vietnam began in Manila in the early 1950s with Lansdale's Magsaysay triumph and not in Saigon in the early 1960s (CW, p. 101)—for an explanation of the abbreviated references, see p. 534. Several layers deeper Harrison E. Salisbury of the *New York Times* could also argue defensibly that "in a sense, the Philippines was where it all began [in 1898]—the flawed, distorted, and confused American adventure on the far side of the Pacific"—see Salisbury's foreword to Raul S. Manglapus's *Philippines: The Silenced Democracy* (Maryknoll, N.Y.: Orbis Books, 1976). Down toward the core, as I like to think our pilgrimage down through the centuries has revealed, one sees that "it all began" at least as far back as the origins of the Anglo-American trek westward.

To be sure, U.S. fighting men had already been on the Asian

continent as part of the international expedition against the Boxers in 1900 and more recently as about four-fifths of the United Nations' "police action" against North Korea (1950–53). But it was only after Lansdale's arrival in June 1954 that the American empire established a beachhead of its own in Southeast Asia.

My reference to colonialism as "an association of the philanthropic, the pious and the profitable" comes from Ronald Segal, *The Race War* (New York: Viking Press, 1966), p. 44.

In their preoccupation with the "unprecedented victory of Asian over European" at Dien Bien Phu, the Pentagon analysts had forgotten the Russo-Japanese War and the Japanese victory over the Russian fleet at the Battle of Tsushima Strait in May 1905. But the Viet Minh victory was virtually unprecedented and sent out shock waves of fear lest those proverbial yellow "hordes" appear on the horizon.

Biographical data on Lucien Conein appear in David Halberstam's *Best and the Brightest* (Greenwich, Conn.: Fawcett, 1973), pp. 350–51, my source for the distaste with which Conein was regarded by the U.S. command in Saigon for having "gone too native." (Later Conein rose to a top post in Richard Nixon's Drug Enforcement Administration, where he reportedly put his talents to good use in planning the final solution of the narcotics problem, "a full-fledged assassination program directed against the traffickers"—Jonathan Marshall, *Inquiry*, I [November 27, 1978], 29.)

For a discussion of the Lansdale team's violation of the international-law rights of North Vietnam, see Richard A. Falk, "CIA Covert Operations and International Law" in *The CIA File*, ed. Robert L. Borosage and John Marks (New York: Grossman, 1976), pp. 144–46. For the CIA use of the tribal peoples of Indochina as cannon fodder, see Fred Branfman's "The President's Army," ibid., pp. 46–78; see also Ralph L. Stavins, Richard J. Barnet, and Marcus Raskin, *Washington Plans an Aggressive War* (New York: Random House, 1971), pp. 61–63.

Chapter XXVIII
The Quiet American

The veterans' legions Lansdale helped organize in the Philippines, Vietnam, and Laos tied in with the schemes of Cord Meyer, head of the CIA's International Organizations Division, to use such quasi-military groups as anti-Communist fronts (CW, p. 252) —for an explanation of the abbreviated references, see p. 534.

In "A Harlot High and Low: Reconnoitering through the Secret Government," *New York*, IX (August 16, 1976), 23–46, Norman Mailer insightfully considered the "relative identity" of such CIA agents as Lansdale: "To live with a role is to live as an actor—so soon as the role is more satisfying than the life, all clear boundaries of identity are lost." What was Lansdale? Where was the inner life that might have stood in a tensional relationship with his secret role? That he had *become* the latter helps account for what was after all bizarre behavior. Presumably the point of deception is to hide the truth, yet Lansdale brought out his memoir *after* the publication of the *Pentagon Papers* had made some of his more flagrant falsities relatively easy to expose. He literally lived his cover even when it became counterproductive to do so. "The more successful a liar is," Hannah Arendt once observed, "the more people he has convinced, the more likely it is that he will end by believing his own lies"—"Lying in Politics: Reflections on the Pentagon Papers," *New York Review of Books*, XVII (November 18, 1971), 36. In that sense Lansdale was only one of the more spectacular casualties of that national habit of self-deception so decisively revealed by the secret history of the Vietnam War. In this context one can begin to understand the firm conviction of Lansdale, his team, and a long line of compatriots going back past Woodrow Wilson to Jefferson and beyond, the conviction that Americans could never become colonialists like the French and the British. By substituting abstractions for realities, Anglo-Americans disguised *national* interventions as exercises in *international* spiritual leadership. Out of this lethal tradition came the secret agent and his erstwhile superior Robert Kennedy, who claimed for his countrymen on March 17, 1968, "our right to the spiritual direction of the planet"—for this quotation and that in the text on the fundamental ambiguities of a nationalism that parades as selfless internationalism, see Claude Julien, *America's Empire* (New York: Pantheon Books, 1971), pp. 7, 31, et passim.

In addition to Wesley R. Fishel's *New Leader* piece cited in the text, see his "Free Vietnam since Geneva," *Yale Review*, LXIX (Autumn 1959), 68–79: "It is the one Asian area where Communism has been rolled back, and rolled back without war." On Fishel and other American Friends of Vietnam, see Robert Scheer, *How the United States Got Involved in Vietnam* (Santa Barbara, Calif.: Center for the Study of Democratic Institutions, 1965); and Bernard B. Fall, *The Two Viet-Nams* (New York: Frederick A. Praeger, 1967), pp. 242–45.

In *End of a War: Indochina, 1954* (New York: Frederick A. Praeger, 1969), Philippe Devillers and Jean Lacouture cited "direct and well-informed sources" who told them that Bao Dai had put Diem at the head of his government in part because "Diem was the candidate of the Department of State (and even of the Dulles brothers)" (p. 224n). They also commented on the fact that shortly after Lansdale arrived in Saigon, General William J. Donovan, his old boss from OSS days, passed through to become ambassador to Thailand (p. 211), all of which strongly suggested that Lansdale had been dispatched to give Diem his backing while Donovan built up an anti-Communist bastion in nearby Bangkok. Indeed, Lansdale himself may have played a significant part in the choice of Diem, for on his 1953 visit he had met Ngo Dinh Nhu, after which Nhu had founded the Movement of National Union for Independence and Peace and had campaigned to make his brother Diem head of a new regime (p. 211). Very likely Lansdale arrived already committed to the Ngo family and with the intention of having Diem succeed Bao Dai as head of state.

For General Collins's angry accusation that Lansdale had incited "a mutiny" to save Diem, see Robert Shaplen, *The Lost Revolution* (New York: Harper & Row, 1965), p. 124. Lansdale gave his friend Shaplen a slightly different account of Diem's reaction to the news that their "true friend" Trinh Minh Thé had been shot in the head and killed. In this earlier version, the secret proconsul witnessed tears come into Diem's eyes as he begged forgiveness for what he had just said: "It was the only time I ever saw him give in to his emotions. Impulsively, I put my hand on his shoulder" (pp. 125–26). Cradling the sobbing statesman in his arms, the memoir version, made a better story. For a discussion of Thé's mysterious death and the allegations he was murdered, see Joseph Buttinger, *Vietnam: A Dragon Embattled* (New York: Frederick A. Praeger, 1967), II, 882–84, 1064–65, 1112.

Chapter XXIX
The New Frontier

The epigraph is from Kennedy's nomination acceptance—Arthur M. Schlesinger, Jr., *A Thousand Days* (Boston: Houghton Mifflin, 1965), pp. 60–61.

Well into the 1970s Lansdale still felt it had been a mistake to let Prince Norodom Sihanouk stay in power as long as he did. In a postmortem of the war managers, he complained that "the enemy" had made "continuing and constant use of sanctuaries across the border in Cambodia, a gift from our self-imposed political decision to limit the geographical boundaries of the war, which allowed them to ready units for combat and rest them up afterward"—*The Lessons of Vietnam*, ed. W. Scott Thompson and Donaldson D. Frizzell (New York: Crane, Russak, 1977), p. 127. For the Cambodian leader's side of their conflict, see Norodom Sihanouk, *My War with the CIA* (New York: Pantheon Books, 1973).

For the David Halberstam quotations, see *The Best and the Brightest* (Greenwich, Conn.: Fawcett, 1973), pp. 158, 159. In *Washington Plans an Aggressive War* (New York: Random House, 1971), pp. 77–78, Ralph Stavins noted that Lansdale's appointment as assistant to the secretary of defense for Special Operations was perceived as a threat by the Joint Chiefs, who feared losing control over the secret war in Vietnam. They "set up their own counter-insurgency agency by creating a Special Assistant for Counter-Insurgency and Special Activities (SACSA). Victor Krulak, the first 'SACSA,' a former Marine Corps general and an astute politician who was referred to as 'the brute,' undercut Lansdale at every turn until Lansdale was called a 'paper tiger.' " The Pentagon analysts also suggested that by then Kennedy may have "had second thoughts on Lansdale, aside from State's objections on bureaucratic grounds" (DOD, II, 44)—for an explanation of the abbreviated references, see p. 534.

What was Lansdale? Robert S. McNamara's uncertainty about whom the secret agent was working for, even though he was officially his special COIN assistant, came out amusingly in his testimony before the Church committee: "On May 30 [1975] in connection with my inquiries to determine exactly who General Lansdale was working for at the time of August 1962," McNamara said, "I called . . . Ros Gilpatric . . . and during my conversation with Mr. Gilpatric I asked him specifically what Lansdale was working for in August '62 and Mr. Gilpatric stated that he was not working for either himself, that is Gilpatric, or me in August '62, but rather for the committee that was dealing with the MONGOOSE operation" (AAP, p. 159n). Seemingly McNamara never wondered if Lansdale were also working for the Company and responsive to the wishes (if not under the immediate direction) of John McCone, Dulles's successor as DCI. Nor apparently did Gilpatric, though he had recommended Lansdale for promotion to brigadier general and had worked closely with him on the projected task force.

Almost certainly McNamara proposed the assassination of Castro at the August 10, 1962, meeting of the SGA. Like Lansdale, William Harvey testified that the secretary of defense had asked, "shouldn't we consider the elimination or assassination" of Castro (AAP, p. 164)? In "The Kennedy Vendetta: How the CIA Waged a Secret War against Castro," *Harper's*, CCLI (August 1975), 61, Taylor Branch and George Crile III quoted Richard Goodwin on McNamara's advocacy of killing: "I was surprised and appalled to hear McNamara propose this. It was at the close of a Cuba task force session, and he said that Castro's assassination was the only productive way of dealing with Cuba." Goodwin next told the staff of the Church committee that this shocking proposal was "etched on his memory," yet six weeks later he testified that he was unable to say who proposed the assassination and maintained he had been misquoted by Branch and Crile (AAP, pp. 321–22).

Gary Wills observed that the Church committee, like the CIA, considered ordinary native lives as cheap—see "The CIA from Beginning to End," *New York Review of Books*, XXII (January 22, 1976), 28.

On the Lansdale team turning around on Diem and leading the assault against him, see Halberstam, *Best and Brightest*, p. 342.

In *Twenty Years and Twenty Days* (New York: Stein and Day, 1976), pp. 74–75, Nguyen Cao Ky quoted his comment to kingmaker Lansdale. By then Ky was living in Alexandria, Virginia, as a neighbor of the secret agent. For Ky's young Hitler image and Lansdale's refurbishing of it, see George McTurnan Kahin and John W. Lewis, *The United States in Vietnam* (New York: Delta, 1967), pp. 242–43, 263; and Richard Critchfield, *The Long Charade: Political Subversion in the Vietnam War* (New York: Harcourt, Brace & World, 1968), pp. 164–70, 214–17—see p. 390 for Nguyen Van Thieu's refusal to deal with Lansdale.

Daniel Ellsberg discussed his former team leader in *Papers on the War* (New York: Simon and Schuster, 1972), pp. 15, 27–28, 138–42, 195–96; see also his interview in *Look*, October 5, 1971, pp. 31–42. For Lansdale's disappointed comment on Ellsberg, see *Newsweek*, June 28, 1971, p. 16.

In *The War Managers* (Hanover, N.H.: University Press of New England, 1977), Douglas Kinnard reproduced Lansdale's astrology memorandum for Bunker on pp. 89–90.

On his return from Vietnam in 1968—now perhaps really "retired"—Lansdale proposed to open a Freedom Studies Center, or cold war college, that would send psywar and paramilitary teams to foreign countries at the request of their governments or "acceptable third parties." These covert philanthropists would help Third World states "resolve problems of concern to freedom." According to Alexander Kendrick, "the Lansdale idea was rejected by Congress as 'too competitive' with its own House Un-American Activities Committee, but defense contractors and other interested parties promptly subscribed the necessary funds. The Center had on its advisory board in 1971 the Vice-President, three cabinet members, nine senators, twenty-eight congressmen, and six governors, and the Army chief of staff, who had been the commander in Vietnam, spoke at one of its dinners"—*The Wound Within: America in the Vietnam Years, 1945–1974* (Boston: Little, Brown, 1974), p. 229. Manifestly Lansdale was terminally addicted to patriotic "fun and games."

Chapter XXX
The Problem of the West

CIA analyst Frank Snepp's *Decent Interval: An Insider's Account of Saigon's Indecent End* (New York: Random House, 1977) is invaluable for its portraits of frustrated empire-builders. A legatee of Lansdale and all the other can-do enthusiasts who believed the country might not have been "lost" had American know-how been applied properly, Snepp was moved by his anger and self-pity to pen revealing profiles of such officials and their families. "An American wife," for instance, ran over a Vietnamese picket outside the commissary gates in her impatience to get inside and in her hatred for "the little beasties veering all over the highway as if it were their country. . . . Something snapped, as at My Lai" (p. 47). The swollen racism came to a head at 3:45 A.M. on April 30, 1975, when the ambassador decided all the Vietnamese waiting in the Embassy for evacuation "were to be herded out into the courtyard. The helo-lift off the roof was now to be reserved for Americans only" (p. 558). A white face was the precious passport to the roof and flight back to the Free World.

Historian James C. Thomson, Jr., served in the White House and at State from 1961 to 1966. Close to the processes of power, he had insight into those who wielded it, especially into the covert racism of their "traditional Western sense that there are so many Asians, after all; that Asians have a fatalism about life and a disregard for its loss; and that they are cruel and barbaric to their own people; and they are very different from us (and all look alike?)." An East Asia specialist himself, he suggested that such subliminal views raised the subliminal question of whether Asians, particularly peasants and most particularly Communists, "are really people—like you and me. To put the matter another way: would we have pursued quite such policies—and quite such military tactics—if the Vietnamese were white?"—"How Could Vietnam Happen?" *Atlantic*, CCXXI (April 1968), 51. Such a well-put question calls for a direct response: *No*. See also Thomson's "Indochina: The Collapse, the Horror and How It Happened," *International Herald Tribune*, April 21, 1975.

The color of the corpses: This definition of "Vietnamization" by Ellsworth Bunker was quoted by John W. Dower, "Asia and the New Nixon Doctrine: The New Face of Empire," in *Open Secret: The Kissinger–Nixon Doctrine in Asia*, eds. Virginia Brodine and Mark Selden (New York: Harper & Row, 1972), p. 132. Ambassador Bunker's remark appeared originally in *Le Figaro*, January 15, 1970 (weekly edition).

Arthur M. Schlesinger, Jr., paraphrased Kennedy's warning against "a white man's war" in *A Thousand Days* (Boston: Houghton Mifflin, 1965), p. 547. Earlier, Senator Lyndon B. Johnson had taken the same stand against "sending American GIs into the mud and muck of Indochina in a bloodletting spree to perpetuate colonialism and white man's exploitation in Asia"—quoted by Thomas Lippman, *Washington Post*, March 4, 1975.

When he returned to the Pentagon as army chief of staff, Westmoreland found haircut policy "one of my toughest decisions." The general proved himself a modern soldier, less inflexible on this issue than the iron-willed John Endicott: "The anguish I underwent on the decision was almost farcical. In the end I settled for an increase in hair length that probably satisfied neither side completely but was acceptable enough to both to defuse the issue" —*A Soldier Reports* (Garden City, N.Y.: Doubleday, 1976), p. 371. The model general was too sanguine: the issue had a long-burning fuse in American history.

The proposal to solve the Montagnard problem like "we" solved the Indian problem was quoted by Tom Hayden in *The Love of Possession Is a Disease with Them* (New York: Holt, Rinehart and Winston, 1972), p. 110—the italics are probably Hayden's. Hugh Manke, head of the International Voluntary Service, testified that Vietnamese colonialists, such as Premier Ky's wife, were buying up tribal lands in a fashion "painfully reminiscent of the activities of American pioneers with regard to the Indian tribes" (ibid.). Frank Snepp reported that the CIA simply left its Montagnard "friendlies" behind at the end: "of the estimated 5,000 montagnards who had escaped from the highlands in mid- and late-March, only around eighty had been logged in at the Guamian reception center as of 8 July 1975"—*Decent Interval*, p. 566n.

For the U.S. Mission's response to press critics, see David Halberstam, *The Best and the Brightest* (Greenwich, Conn.: Fawcett,

1973), pp. 253–56 et passim. AP correspondent Malcolm Browne's Little Big Horn allusion was quoted by John Mecklin, *Mission in Torment* (Garden City, N.Y.: Doubleday, 1965), pp. 131–32.

For the uncovering of My Lai, see Seymour M. Hersh, *My Lai 4* (New York: Random House, 1970); and Richard Hammer, *One Morning in the War* (New York: Coward, McCann & Geoghegan, 1970). Evidence that the antiwar movement helped make My Lai known and spurred the press to relax its self-censorship was discussed by Noam Chomsky, *At War with Asia* (London: Fontana/Collins, 1971), pp. 81–82. For the cover-up in Congress, see the report of the Armed Services Investigating Subcommittee of the Committee on Armed Services, House of Representatives, July 15, 1970, *Investigation of the My Lai Incident* (Washington: Government Printing Office, 1970)—the Hébert subcommittee's conclusion quoted in the text appeared on p. 53. Indispensable for the Peers panel's report is of course Seymour M. Hersh's *Cover-Up* (New York: Random House, 1972).

When Robert Jay Lifton discussed the MGR in "The 'Gook Syndrome' and 'Numbed Warfare,' " in *Saturday Review*, December 1972, pp. 66–72, a reader pointed out that the term "gook" had been used long before by American forces to abuse natives and in particular directed his attention to an article in the *Nation* of July 10, 1920, wherein the writer quoted leathernecks in Haiti using the nickname to express their contempt: "I have heard officers wearing the United States uniform in the interior of Haiti talk of 'bumping off' (i.e., killing) 'Gooks' as if it were a variety of sport like duck hunting"—see Lifton's "Gooks and Men" in *Home from the War* (New York: Simon and Schuster, 1973), pp. 189–216. The evidence leads back to the unbroken continuity of such terms of dehumanization, the most lethal of which have always been intertwined with color and race. For a first-rate analysis of the implications of the army's phrase "Oriental human beings," see Richard Hammer, *The Court-Martial of Lt. Calley* (New York: Coward, McCann & Geoghegan, 1971), pp. 46–47, 349.

In *Nuremberg and Vietnam: An American Tragedy* (Chicago: Quadrangle Books, 1970), Telford Taylor discussed the MGR and the general issue of racism, and still expressed the view that all Asians tended to look alike to "our" soldiers. But the general did put the issue squarely: if "to come home alive" required "the slaughter of all the Vietnamese who might be sympathetic to the Vietcong, then all our talk of 'pacification' . . . is the sheerest hypocrisy, and we had better acknowledge at once that we are prepared to do what we hanged and imprisoned Japanese and German generals for doing" (p. 169). General Westmoreland became a little uneasy when Taylor appeared on a television talk show and held that if the same standards had been applied to the My Lai trial as had been applied to the trial and execution (in the Philippines) of WWII General Tomoyuki Yamashita, "there would be a very strong possibility that they [myself and civilian government officials] would come to the same end he did"—*A Soldier Reports*, pp. 378–79. Taylor was guilty of "an emotional outburst," concluded Westmoreland, convinced to the end that his actions in Vietnam would stand every test "before both the bar of justice and the court of history."

Every general is entitled to his day in Clio's court, a day that came to Yamashita too late for it to do him much good—for compelling evidence that Taylor was right and that the Japanese general was executed (1946) for actions less demonstrably grave than those that earned Generals Jacob H. Smith (1902) and Samuel W. Koster (1971) reprimands and General Westmoreland promotion to army chief of staff (1968), see A. Frank Reel's *Case of General Yamashita* (Chicago: University of Chicago Press, 1949). On the issue of "command responsibility" and the responsibility of generals and others for the slaughter in Vietnam, see Taylor's *Nuremberg and Vietnam;* Richard A. Falk, Gabriel Kolko, and Robert Jay Lifton, eds., *Crimes of War* (New York: Random House, 1971); John Duffet, ed., *Against the Crime of Silence* (New York: Clarion, 1970); and Clergy and Laymen Concerned about Vietnam, *In the Name of America* (Annandale, Va.: Turnpike, 1968).

For the Calley quotations in the text—the italics are his—see John Sack, *Lieutenant Calley: His Own Story* (New York: Viking Press, 1971), pp. 3, 31, 104–5. Best for placing the case in its historical context is Richard Hammer's *Court-Martial of Lt. Calley*—see especially pp. 391–92. See also the *New York Times*, March 31, 1971, for the lieutenant's contention he was not killing people, "just a philosophy in a man's mind. That was my enemy out there," words that must have made Graham Greene wince.

The only good —— is a dead ——: The quotations below the colloquy from Joseph Strick's film *Interviews with My Lai Veterans* (1971) come seriatim from Robert Jay Lifton, *Home from the War*, p. 47; and *San Francisco Chronicle*, December 1, 1969—see also the "There Are No Rules" letter to the editor in the *Chronicle* of December 31, 1969; and Charles Levy, *Spoils of War* (Boston: Houghton Mifflin, 1974), p. 26.

The testimony of the hundred honorably discharged veterans reproduced in the *Congressional Record*, CXVII (April 6–7, 1971), 2825–900, 2903–36, also appeared as Vietnam Veterans against the

War, *The Winter Soldier Investigation: An Inquiry into American War Crimes* (Boston: Beacon Press, 1972).

In "Ecocide in The American Global Frontiers," *Akwesasne Notes*, VI (Early Summer 1974), 26–27, Navaho Ben Muneta reported that the U.S. Air Force was trying to sell Agent Orange, a military herbicide so lethal it was banned from Vietnam, to South American countries, especially Brazil, where it "would further devastate the Amazon Indians." For a survey of the ecocide in Indochina, see Stockholm International Peace Research Institute, *Ecological Consequences of the Second Indochina War* (Stockholm: Almqvist & Wiksell, 1976). Most illuminating on the origins of the attitudes and assumptions that led to all this devastation is Lynn White, Jr., "The Historical Roots of Our Ecologic Crisis," *Science*, CLV (March 10, 1967), 1203–7.

Originally Frederick Jackson Turner's "The Problem of the West" was published in *Atlantic*, LXXVIII (September 1896), 289–97. It and his celebrated 1893 paper were reproduced in *The Frontier in American History* (New York: Henry Holt, 1920), pp. 1–38, 205–21. By linking democracy to "free" land, the problem of the West became acute for Turner when the westward advance of his frontier "caused the free land to disappear." As Henry Nash Smith has observed: "What then was to become of democracy? The difficulty was the greater because in associating democracy with free land he had inevitably linked it also with the idea of nature as a source of spiritual values"—see Smith's germinative discussion of Turner in *Virgin Land: The American West as Symbol and Myth* (Cambridge: Harvard University Press, 1970—orig. 1950), pp. 250–60. This deeper linkage of democracy and nature heightened Turner's quandary and indeed anticipated formulations of the problem that would be posed as the ecologic crisis became more desperate: *With the ongoing rape of nature, what then was to become a source of spiritual values?*

Turner saw the world solely through Anglo-American eyes, and that meant he could not really perceive the Native American as a person. His were like all the other "interior perspectives of the American fragment which have prevailed in historical study," in Louis Hartz's words. "But, of course, once we get outside the national fragment the very fact that the Indian was thus eliminated, could thus be neglected by the European logic of the culture, becomes a matter of very great importance." By expanding abroad, furthermore, outriders of the national fragment bumped up against the revolutionary process and ideologies they had fled Europe in the first place to escape, Crusoe-like, with "natives," no less, reminding them of their long flight. "Must we not see the continuous significance of the dwindling Indian challenge?" asked Hartz. "Life outside the United States fragment is bound to convert the North American Indian, and his fate, into matters of the deepest historical interest." My indebtedness to Hartz is considerablé—see his *Founding of New Societies* (New York: Harcourt, Brace & World, 1964), pp. 1–99, and "A Comparative Study of Fragment Cultures" in *Violence in America: Historical and Comparative Perspectives*, ed. Hugh Davis Graham and Ted Robert Gurr (Washington: Government Printing Office, 1969), I, 87–100.

In the 1960s J. Edgar Hoover and his agents set out to destroy Martin Luther King, Jr., and looked around for a suitable replacement, the "right kind of a Negro . . . to advance to positions of national leadership," as though they were McKenney and Cass picking out a suitable chief for the Menominees in the nineteenth century or Lansdale and other agents seeking a proper leader for the Cambodians—see David Wise, "FBI: The Lawbreakers" in *The American Police State* (New York: Random House, 1976), pp. 274–321. For the CIA's domestic Operation Chaos, see Jerry J. Berman and Morton H. Halperin, eds., *The Abuses of the Intelligence Agencies* (Washington, D.C.: Center for National Security Studies, 1975), pp. 124–28—the Rockefeller report on the CIA revealed that the American Indian Movement (AIM) was one of its domestic targets. For a summary of the FBI's memorandum for the CIA on AIM, see Morton H. Halperin et al., *The Lawless State: The Crimes of the U.S. Intelligence Agencies* (New York: Penguin Books, 1976), pp. 240–41.

Voices from Wounded Knee, 1973 was published by *Akwesasne Notes* in 1974. The quotations of Native Americans come from the following: Chief Standing Bear, *Land of the Spotted Eagle* (Boston: Houghton Mifflin, 1933), p. xix; Vine Deloria, Jr., *God Is Red* (New York: Grosset & Dunlap, 1973), p. 301; and John Fire/Lame Deer and Richard Erdoes, *Lame Deer: Seeker of Visions* (New York: Simon and Schuster, 1972), pp. 39, 69, 235. Lame Deer defined a *wićaśa wakan*, or holy man, as one "who feels the grief of others. A death anywhere makes me feel poorer."

Index

Grateful acknowledgment is made to the following for permission to reprint previously published material:

The Folger Shakespeare Library: Illustration "Bishop Fish" from *Icones animalium* (1560) by Konrad Gesner. Reprinted by permission.

Harcourt Brace Jovanovich, Inc., and Faber and Faber Limited: Excerpt from "Children of Light" from *Lord Weary's Castle, Poems 1938–1949* by Robert Lowell. Copyright 1946 and renewed 1974 by Robert Lowell. World rights excluding the United States and Canada administered by Faber and Faber Limited. Reprinted by permission of Harcourt Brace Jovanovich, Inc., and Faber and Faber Limited.

Life Picture Service: Photograph "My Lai Victims" by Ron Haeberle from *Life Magazine*, December 5, 1969, issue. Copyright © 1969 by Time Inc. Reprinted by permission.

AP/Wide World Photos, Inc.: Photograph of Edward G. Lansdale. Reprinted by permission.